OXFORD MEDICAL PUBLICATIONS

Neuropharmacology of Serotonin

Neuropharmacology of Serotonin

EDITED BY

A. RICHARD GREEN
MRC Clinical Pharmacology Unit
Radcliffe Infirmary
Oxford

OXFORD   NEW YORK   TORONTO   MELBOURNE
OXFORD UNIVERSITY PRESS
1985

Oxford University Press, Walton Street, Oxford OX2 6DP

London New York Toronto
Delhi Bombay Calcutta Madras Karachi
Kuala Lumpur Singapore Hong Kong Tokyo
Nairobi Dar es Salaam Cape Town
Melbourne Auckland

and associated companies in
Beirut Berlin Ibadan Mexico City Nicosia

Oxford is a trade mark of Oxford University Press

Published in the United States
by Oxford University Press, New York

British Library Cataloguing in Publication Data
Neuropharmacology of serotonin.—(Oxford
medical publications)
1. Neuropharmacology   2. Serotonin
I. Green, A. Richard
591.1'88   RM315
ISBN 0-19-261471-1

Library of Congress Cataloging in Publication Data
Main entry under title:
Neuropharmacology of serotonin.
(Oxford medical publications)
Includes bibliographies and index.
1. Serotonin—Physiological effect. 2. Neuropharmacology.
I. Green, A. Richard. II. Series.
[DNLM: 1. Nervous System—drug effects. 2. Serotonin—
pharmacodynamics. QV 126 N493]
QP801.S4N48 1985   615'.78   84-25534
ISBN 0-19-261471-1

Set by Cotswold Typesetting Ltd, Cheltenham
Printed in Great Britain by
J. W. Arrowsmith Ltd., Bristol

For Susan, Judith, and Kirsten

Preface

Just over a year ago I was approached by Oxford University Press with a suggestion that I organize and edit a monograph on serotonin neuro-pharmacology. Being a monograph meant that my brief was not that of organizing a comprehensive review but rather a book on those areas which seemed worthy of particular attention. After discussing the idea with colleagues, I drew up a proposal and then wrote to various neuropharma-cologists working actively on serotonin. Their response was enthusiastic, all feeling that a marked revival of interest in this monoamine is now occurring and that a monograph would therefore be of value both to those currently engaged in serotonin research and those merely wishing to get up to date.

The authors were not required to produce reviews but rather to discuss critically those areas of research they felt to be of the greatest interest or controversy. For this reason, the authors in this book are all scientists currently working in the areas they are reviewing.

Not all readers are going to agree with my selection of topics. However, I hope they will nevertheless agree that all authors have put considerable effort into providing interesting and stimulating chapters which will stand for some time as major commentaries on their chosen areas and for this I would like to thank them.

Finally, I would like also to express my thanks to the staff of Oxford University Press for their enthusiastic inception of this project and for helping it through to completion.

Oxford   A.R.G.
April 1984

Contents

x   Contents

4   Characterization of serotonin receptor binding sites
     JOSÉE E. LEYSEN

5   Inositol phospholipid breakdown as an index of serotonin
     receptor function
     MICHAEL C. W. MINCHIN

6   Effects of antidepressant drugs on cerebral serotonin receptor
     mechanisms
     SVEN-OVE ÖGREN and KJELL FUXE

List of contributors

PIERRE BLIER

Neuroscience Research Center and Psychiatric Research Center, University of Montréal, P.O. Box 6128—Station A, Montréal, Quebéc, Canada H3C 3J7

MIKE BRILEY

Department of Biochemical Pharmacology, Centre de Recherches Pierre Fabre, 17 avenue Jean Moulin, 81106 Castres, France

CLAUDE de MONTIGNY

Neuroscience Research Center and Psychiatric Research Center, University of Montréal, P.O. Box 6128—Station A, Montréal, Québec, Canada H3L 3J7

RAY W. FULLER

Lilly Research Laboratories, Eli Lilly and Company, Indianapolis IA 46285, USA

KJELL FUXE

Department of Histology, Karolinska Institutet, Stockholm, Sweden

COLIN R. GARDNER

Roussel Laboratories, Kingfisher Drive, Covingham, Swindon SN3 5BZ, England

RICHARD A. GLENNON

Department of Pharmaceutical Chemistry, School of Pharmacy, Medical College of Virginia, Virginia Commonwealth University, Richmond VA 23298, USA

A. RICHARD GREEN

MRC Clinical Pharmacology Unit, Radcliffe Infirmary, Oxford OX2 6HE, England

DAVID J. HEAL

MRC Clinical Pharmacology Unit, Radcliffe Infirmary, Oxford OX2 6HE, England

BARRY L. JACOBS

Department of Psychology, Princeton University, Princeton NJ 08544, USA

JOSÉE E. LEYSEN

Department of Biochemical Pharmacology, Janssen Pharmaceutica, B-2340, Beerse, Belgium

CHARLES A. MARSDEN

Department of Physiology and Pharmacology, Medical School, Queen's Medical Centre, Clifton Boulevard, Nottingham NG7 2UH, England

MICHAEL C. W. MINCHIN

MRC Clinical Pharmacology Unit, Radcliffe Infirmary, Oxford OX2 6HE, England

CHANTEL MORET

Department of Biochemical Pharmacology, Centre de Recherches Pierre Fabre, 17 avenue Jean Moulin, 81106 Castres, France

SVEN-OVE ÖGREN

Research and Development Laboratories, Astra Läkemedal AB, Södertälje, Sweden

ROBERT J. WALKER

Department of Neurophysiology, School of Biochemical and Physiological Sciences, Medical and Biological Sciences Building, Bassett Crescent East, Southampton SO9 3TU, England

List of abbreviations

The abbreviations given below are used in most chapters, except Chapter 10 where slightly different, but defined, abbreviations are used. These have not been changed or listed below because of the large number involved. Also, note below the abbreviations used for binding sites and electro-physiologically defined receptors.

B_{max}	Maximum number of binding sites
BOL	2-Bromolysergic acid diethylamide (2-BromoLSD)
cAMP	cyclic 3'5'-adenosine monophosphate (cyclic AMP)
CSF	Cerebrospinal fluid
DA	Dopamine
5,6-DHT	5,6-Dihydroxytryptamine
5,7-DHT	5,7-Dihydroxytryptamine
DMT	Dimethyltryptamine
DOM	1-(2,5-Dimethoxy-4-methylphenyl)-2-aminopropane
DOPAC	3,4-Dihydroxyphenylacetic acid
DRN	Dorsal raphe nucleus
ECS	Electroconvulsive shock
ECT	Electroconvulsive therapy
GABA	γ-Aminobutyric acid
5-HIAA	5-Hydroxyindoleacetic acid
5-HT_1	} Receptors defined by binding studies
5-HT_2	} (see Section 14.2)
5-HTP	5-Hydroxytryptophan
HVA	Homovanillic acid
Hz	Hertz
IP	Inositol-1-phosphate
IP_2	Inositol-1,4-biphosphate
IP_3	Inositol-1,4,5-triphosphate
K_D	Dissociation constant

K_i	Equilibrium dissociation constant
LC EC	High pressure liquid chromatography (electrochemical detection)
LSD	Lysergic acid diethylamide
MAO	Monoamine oxidase
MAOI	Monoamine oxidase inhibitor
5-MeODMT	5-Methoxy-N,N-dimethyltryptamine
5-MeOT	5-Methoxytryptamine
α-MeT	α-Methyltryptamine
MK-212	6-Chloro-2-[1-piperazinyl] pyrazine
MRN	Median raphe nucleus
NA	Noradrenaline
NCS	Nucleus centralis superior
NRP	Nucleus raphe pallidus
8-OH-DPAT	8-Hydroxy-N,N-dipropyl-2-aminotetralin
PAG	Periaqueductal grey
PCA	p-Chloroamphetamine
PCPA	p-Chlorophenylalanine
PI	Phosphatidylinositol
PIP	Phosphatidylinositol-4-phosphate
PIP_2	Phosphatidylinositol-4,5-biphosphate
REM	Rapid eye movement
RU 24969	5-Methoxy-3(1,2,3,6-tetrahydropyridin-4-yl)1H indole
S_1	
S_2	Receptors defined by electrophysiological studies (See sections 7.1.2 and 14.2)
S_3	
TCA	Tricyclic antidepressant
TRH	Thyrotropin releasing hormone

1

Drugs altering serotonin synthesis and metabolism

RAY W. FULLER

1.1 Introduction

Drugs that directly inhibit or accelerate serotonin synthesis or degradation are useful for altering serotonergic function, and drugs that modify serotonergic function through actions on receptors may secondarily alter the rate of serotonin synthesis. Several methods for demonstrating changes in the rate of serotonin synthesis or its metabolism are available and have been used in studying effects of enzyme inhibitors, precursors, direct and indirect agonists, and antagonists. As a result, some mechanisms for physiological regulation of serotonin turnover and several means of pharmacological intervention have been identified.

1.2 Pathway of serotonin synthesis and metabolism

Serotonin is formed from the amino acid tryptophan, for which the major source is dietary protein, as shown in Fig. 1.1. The first enzyme in this metabolic pathway, tryptophan 5-hydroxylase (EC 1.14.16.4, L-tryptophan, tetrahydropteridine: oxygen oxidoreductase [5-hydroxylating]) is a specific enzyme believed to be present only in serotonin-synthesizing cells. The intermediate product, 5-hydroxy-L-tryptophan (5-HTP), is normally decarboxylated almost immediately it is formed and exists in tissues at concentrations so low as to be almost impossible to measure under ordinary circumstances. The enzyme that decarboxylates 5-HTP, aromatic L-amino acid decarboxylase (EC 4.1.1.28), is a non-specific enzyme rather ubiquitously distributed. Decarboxylation removes the asymmetry on the amine-bearing carbon so serotonin is not optically active, i.e. does not exist in stereoisomeric forms.

Serotonin is held in storage granules within neurones, before being released by nerve impulses into the synaptic cleft. Membrane uptake

1

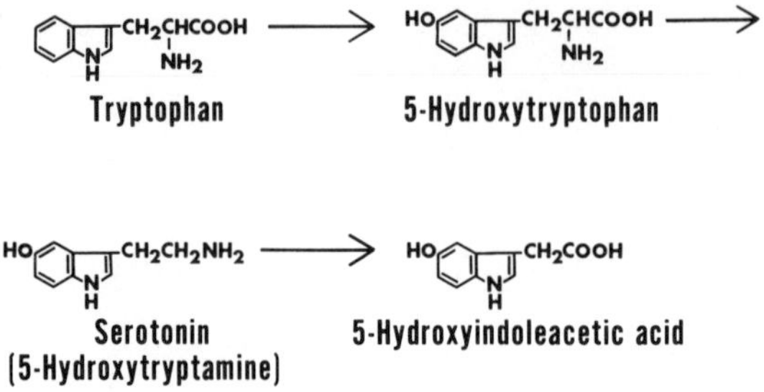

2   Ray W. Fuller

Fig. 1.1   Serotonin formation and metabolism.

carriers present on the serotonin nerve terminal then remove serotonin from the synaptic cleft by transporting it back into the serotonin-forming neurone. The physiological purpose of this transport is one of inactivation, that is removal of serotonin from contact with synaptic receptors, rather than one of preservation, since serotonin synthesis is apparently able to supply all the serotonin needed under normal circumstances.

Serotonin may become exposed to monoamine oxidase (MAO) (EC 1.4.3.4, amine: oxygen oxidoreductase [deaminating] [flavin-containing]) either before release, or after neuronal re-uptake. Probably most of the oxidation of neuronal serotonin occurs within those neurons that form it. MAO is recognized to exist in more than one form. Type A MAO is defined as the MAO form most susceptible to inhibition by drugs like clorgyline, LY51641 and harmaline (Neff and Yang 1974), and type A MAO prefers substrates like serotonin and noradrenaline. Type B MAO, on the other hand, is preferentially inhibited by drugs like deprenyl (selegiline) and destroys amines like phenylethylamine and benzylamine most rapidly. Recently Levitt, Pintar and Breakefield (1982) have used antisera both to serotonin and type B MAO in double immunofluorescence studies and shown that serotonin-containing neurones also contain type B MAO. Although similar identification of type A MAO within serotonin-containing neurones has not been made, it seems highly probable that these neurones do contain type A MAO. Serotonin is a much better substrate for type A MAO than for type B MAO *in vitro* (Fowler and Tipton 1982), and most pharmacological studies have indicated that the enzymic degradation of serotonin *in vivo* occurs almost exclusively by type A MAO (Ortmann, Waldmeier, Radeke, Felner and Delini-Stula 1980; Schoepp and Azzaro 1981). Nevertheless type B MAO is capable of acting on serotonin *in vitro* (Fowler and Tipton 1982), albeit at a much slower rate, and there is evidence that type B MAO does contribute to serotonin degradation *in vivo* when type A MAO has been inhibited (Green and Youdim 1975; Ashkenazi, Finberg and Youdim 1983).

The immediate product formed when serotonin is oxidized by MAO is 5-hydroxyindoleacetaldehyde. This aldehyde is mainly oxidized to 5-hydroxyindoleacetic acid (5-HIAA), shown in Fig. 1.1, which is the major product of serotonin metabolism, but can also be reduced to the alcohol, 5-hydroxytryptophol (Cheifetz and Warsh 1980; Diggory, Ceaser, Hazelby and Taylor 1979a; Diggory, Ceaser and Morgan 1979b).

1.3   Methods of measuring serotonin synthesis and metabolism

The most widely used methods of measuring serotonin turnover are listed in Table 1.1. Introduction of radioactive serotonin by injecting it intra-ventricularly and letting it be taken up by neurones permits labelling of intraneuronal serotonin stores, but this method is not widely used due primarily to the ability of serotonin to enter cells other than serotonin neurones and the difficulty of labelling all serotonin neurones uniformly. Conversion of radioactive tryptophan to radioactive serotonin is one of the best methods of measuring turnover but requires measurement of specific activities of both precursor and product. Incorporation of isotopic oxygen into serotonin is a method that has found limited use.

Table 1.1   Methods of measuring serotonin synthesis and turnover

Isotopic methods	Non-isotopic methods
Disappearance of tritiated serotonin (Schildkraut *et al.* 1969)	5-HIAA levels
	5-HIAA accumulation after probenecid (Neff and Tozer 1968)
Conversion of radioactive tryptophan to serotonin (Neff *et al.* 1971)	
	5-HIAA decline after MAO inhibition (Neff and Tozer 1968)
Conversion of $^{18}O_2$ to serotonin (Galli *et al.* 1978)	
	serotonin decline after synthesis inhibition (Fuller *et al.* 1974; Hjorth *et al.* 1982)
	5-HTP accumulation after decarboxylase inhibition (Carlsson *et al.* 1972)

A simple index of serotonin turnover is the increase or decrease in 5-HIAA concentration when serotonin concentration itself is not changed. While this measurement can indicate an increase or decrease in serotonin turnover and has proved convenient for studying duration of effects and for single-point measurements in human brain tissue or cerebrospinal fluid, it cannot be used to calculate actual rates of turnover. The rate of accumulation of 5-HIAA when its efflux from the brain is blocked with probenecid or the rate of disappearance of 5-HIAA when its formation is blocked with an MAO inhibitor provide two ways of calculating serotonin turnover rates. Similarly, the rate of accumulation of serotonin after MAO inhibition or the rate of decline in serotonin following inhibition of tryptophan hydroxylase with *p*-chlorophenylalanine or other inhibitors permits calculation of serotonin turnover. Although steady state levels of 5-HTP are so low as to be unmeasured by most techniques, the rapid accumulation of 5-HTP after decarboxylase inhibition provides a good method for measuring the *in vivo* rates of tryptophan hydroxylation, which in steady state conditions equals the rate of serotonin synthesis. Most of these methods have been used in studying drug effects on serotonin turnover.

1.4   Drugs that inhibit serotonin synthesis

The rate-limiting enzyme and the enzyme most specifically involved in serotonin biosynthesis is tryptophan hydroxylase, so inhibitors of this enzyme represent the most useful means of decreasing serotonin formation. The most widely used inhibitor of serotonin biosynthesis has been *p*-chlorophenylalanine (PCPA) (Koe and Weissman 1966). PCPA is only a weak, competitive inhibitor of tryptophan hydroxylase *in vitro,* but *in vivo* it leads to irreversible and long-lasting inhibition of the enzyme (Jequier, Lovenberg and Sjoerdsma 1967). The mechanism for this *in vivo* effect of PCPA remains unknown. An earlier suggestion that PCPA was incorporated into the enzyme molecule near the active site to produce a catalytically inactive enzyme has been discounted (Gal and Whitacre 1982).

PCPA has been used in a large number of functional studies to deplete serotonin concentration since it has the advantage that a single dose of PCPA can deplete serotonin concentration for several days. Against this there is the disadvantage that high doses have to be used. PCPA is slowly metabolized, since the primary route of phenylalanine metabolism in the rat, *para*-hydroxylation, is prevented by the chloro substituent. Koe and Weissman (1966) showed that PCPA disappeared from rat plasma with a half-life of about 3 days, and we have found a half-life of PCPA in dog plasma of 28 hours. Twenty-four hours after PCPA treatment in rats, we found decreased plasma concentrations of all other amino acids except valine, which was not significantly changed, and phenylalanine (which increased due to block of its *para*-hydroxylation to tyrosine). Persistence of high levels of PCPA influences amino acid transport (A. Tagliamonte, P. Tagliamonte, Corsini, Mereu and Gessa 1973*b*), and this is one reason that PCPA affects synthesis of brain catecholamines as well as serotonin.

Other inhibitors of tryptophan hydroxylase include tryptophan analogues, especially those containing a 6-halogen substituent, and various catechol-containing compounds (McGeer and Peters 1969). Catechols inhibit hydroxylating enzymes requiring a tetrahydropteridine cofactor by competition with the tetrahydropteridine. Among catechol-containing compounds, α-propyldopacetamide is one that inhibits tryptophan hydroxylase *in vivo* as well as *in vitro*. This compound has been used in turnover studies in which the rate of decline in brain serotonin following its injection to inhibit tryptophan hydroxylation is measured (Hjorth, Carlsson, Lindberg, Sanchez, Wikstrom, Arvidsson, Hacksell and Nilsson 1982), although it has limited use in functional studies as it also inhibits tyrosine hydroxylation, and therefore catecholamine biosynthesis (Johnson, Kim, Platz and Mickelson 1968; Fuxe, Holmstedt and Jonsson 1972).

6-Fluorotryptophan produces a rapid decrease in brain serotonin concentration which persists for several hours but not longer than about 24 hours (E. G. McGeer, Peters and P. L. McGeer 1968). This compound is therefore useful for producing short duration decreases in serotonin concentration. It has the advantage over catechol-containing inhibitors of tryptophan hydroxylase in that it does not cause any substantial depletion of brain catecholamines (McGeer *et al.* 1968).

Brain tryptophan hydroxylase is also inhibited *in vivo* by halogenated analogues of amphetamine including *p*-chloroamphetamine (Sanders-Bush and Sulser 1970) and fenfluramine (Fuller, Snoddy and Hemrick 1978*a*). The mechanism of this inhibition is not understood, since these compounds are poor inhibitors of the enzyme *in vitro* (Koe and Corkey 1976). Tryptophan hydroxylase inhibition is not the only mechanism by which these compounds decrease brain serotonin concentration *in vivo*, since they also are potent serotonin releasers (Marsden, Conti, Strope, Curzon and Adams 1979).

1.5 Drugs that reduce the rate of serotonin synthesis secondarily to receptor stimulation

A large number of compounds have the ability to mimic the action of serotonin on tissue receptors. These direct agonists include indoles closely related in structure to serotonin; among these, 5-MeODMT is one that has been used frequently due to its ability to penetrate the blood–brain barrier and its relative metabolic stability compared to serotonin itself (Fuxe *et al.* 1972). Some other indole-containing serotonin agonists synthesized by medicinal chemists include indorenate (TR3369) (Safdy, Kurchacova, Schut, Vidrio and Hong 1982) and RU 24969 (Fig. 10.5) (Euvrard and Boissier 1980; Green, Guy and Gardner 1984). Arylpiperazines make up one group of non-indoles that have serotonin agonist activity. Quipazine,

2-(1-piperazinyl)quinoline (Fig. 10.5) was the first such compound to be described (Hong, Sancilio, Vargas and Pardo 1969; Hamon, Bourgoin, Enjalbert, Bockaert, Hery, Ternaux and Glowinski 1976). Others in this group include MK-212 (6-chloro-2-[1-piperazinyl]pyrazine; Fig. 10.5) (Clineschmidt 1979), m-trifluoromethylphenylpiperazine (Fig. 10.5) (Fuller, Snoddy, Mason and Molloy 1978b) and m-chlorophenylpiperazine (Fuller, Snoddy, Mason and Owen 1981). A recently described serotonin agonist that contains neither an indole nor a piperazine moiety is 8-hydroxy-N,N-dipropyl-2-aminotetralin (Hjorth et al. 1982; Fig. 10.5). All of these direct-acting serotonin agonists decrease the rate of serotonin synthesis and turnover in the brain. This decreased turnover has been demonstrated (see above references for each compound) using the techniques listed in Table 1.1 and in some cases a parallel reduction in the rate of firing of serotonin neurones has been demonstrated electrophysiologically (de Montigny and Aghajanian 1977; Blier and de Montigny 1983).

In addition to direct agonists, serotonin uptake inhibitors which act indirectly as agonists by increasing serotonin concentrations within the synaptic cleft also decrease serotonin turnover in vivo. Initially such effects were demonstrated with uptake inhibitors like imipramine, amitriptyline and chlorimipramine, which are relatively non-selective in that they block catecholamine uptake as well as serotonin uptake in vivo (Corrodi and Fuxe 1969; Meek and Werdinius 1970; Svensson 1978). In recent years, numerous compounds have been described that selectively inhibit serotonin uptake and not catecholamine uptake both in vitro and in vivo. These agents, including fluoxetine (Fuller, Perry and Molloy 1974a), zimelidine (Ross, Hall, Renyi and Westerlund 1981), citalopram (Hyttel, 1977), fluvoxamine (Claassen, Davies, Hertting and Placheta 1977) and femoxetine (Buus Lassen, Squires, Christensen and Molander 1975), all decrease serotonin turnover in the brain (see Carlsson and Lindqvist, 1978, and the above references). Serotonin uptake inhibitors also reduce the rate of firing of serotonin neurones in the raphe (Aghajanian and Wang 1978; Clemens, Sawyer and Cerimele 1977).

Reinhard and Wurtman (1977) have called attention to the fact that re-uptake of serotonin from brain synapses precedes its metabolism to 5-HIAA, so the ability of uptake inhibitors to prevent that re-uptake might account for the reduction in 5-HIAA concentration and in the accumulation of 5-HIAA after probenecid injection. However, the other methods of measuring serotonin formation and turnover have also been used to demonstrate that serotonin synthesis is reduced concomitant with the reduction in serotonin neurone firing after serotonin uptake inhibition (see Fuller and Wong 1977). The only exception has been when the effect of uptake inhibitors on brain serotonin accumulation following MAO inhibition has been studied. Independent investigations (Meek and Fuxe 1971; Fuller and Steinberg 1976; Hyttel 1977) have revealed that the rate of serotonin accumulation after MAO inhibition is not reduced by uptake inhibitors that do decrease serotonin formation under all other conditions. These findings led Marco and Meek (1979) to suggest that measurement of changes in serotonin accumulation after pargyline is a poor index of changes in serotonin turnover or the firing rate of serotonin neurones.

Apart from uptake inhibitors, another class of indirect serotonin agonists consists of serotonin-releasing drugs. p-Chloroamphetamine and fenfluramine are examples of releasers that are relatively specific for serotonin as opposed to catecholamines. p-Chloroamphetamine and fenfluramine produce several effects acutely that indicate they stimulate synaptic receptors by releasing serotonin from intraneuronal granular stores (Trulson and Jacobs 1976). Recently we presented evidence that these two drugs decrease the rate of serotonin synthesis as measured by the rate of 5-HTP accumulation after decarboxylase inhibition (Fuller and Perry 1983). Earlier literature had reported decreased serotonin turnover after p-chloroamphetamine but not fenfluramine.

The reduction in serotonin turnover by direct and indirect agonists may result from stimulation of presynaptic autoreceptors on serotonin neurones or from stimulation of postsynaptic receptors which leads to trans-neuronal feedback influences on the serotonin neurones. Recently it has become possible to study autoreceptors that modulate brain serotonin release *in vitro*, and agents like 5-MeODMT, m-chlorophenylpiperazine and m-trifluoromethylphenylpiperazine have been shown to mimic the action of serotonin at these receptors (Baumann and Waldmeier 1981; Martin and Sanders-Bush 1982). Paradoxically, quipazine behaves as an antagonist at these receptors, whereas quipazine decreases serotonin turnover *in vivo* exactly as do the other drugs that were agonists. For further details of these experiments see Chapter 2.

1.6   Drugs that accelerate serotonin synthesis

Since neither of the two enzymes in serotonin biosynthesis is saturated with its amino acid substrate under normal physiological conditions, administration of either L-tryptophan or 5-HTP will increase the rate of serotonin formation (Moir and Eccleston 1968). Administration of each of these amino acids has been useful in functional studies as a means of increasing serotonin formation and release into the synaptic cleft. Each of the two precursors has a set of advantages and disadvantages associated with its use in studies of this type.

The major advantages of tryptophan are that it is generally thought to be converted to serotonin only in cells that normally form serotonin (Aghajanian and Asher 1971), i.e. that contain tryptophan-5-hydroxylase, and that it is a normal dietary constituent. Disadvantages include the facts that reasonably high doses are required for modest increases in brain serotonin concentration (Carlsson and Lindqvist 1978) or for functional effects attributed to serotonin (Sved, Van Itallie and Fernstrom 1982), that a small percentage of tryptophan is metabolized to serotonin, that other biologically active substances such as tryptamine can be formed from tryptophan (Warsh, Coscina, Godse and Chan 1979) and that high doses of tryptophan can cause other biological effects such as aggregation of ribosomes and enhancement of protein synthesis (Sidransky, Bongiorno, Sarma and Verney 1967).

5-HTP has the advantage of being a more efficient precursor to serotonin; compared to tryptophan, smaller doses of 5-HTP produce larger increases in serotonin. A serious disadvantage of 5-HTP is that it can be converted to serotonin by any cell containing aromatic amino acid decarboxylase; thus serotonin can be formed from 5-HTP in cells that normally do not form serotonin. Consequently 5-HTP can affect other neurones besides serotonin neurones, such as by releasing catecholamines through formation of serotonin within catecholamine neurones (Fuxe, Butcher and Engel 1971; Ng, Chase, Colburn and Kopin 1972). There is some evidence that this problem is not avoided altogether even using L-tryptophan, as serotonin has been found in catecholamine nerve terminals of rats treated with tryptophan in combination with an MAO inhibitor (Arbuthnott, Eccleston, Laszlo and Nicolaou 1981).

For functional studies, the disadvantage of 5-HTP can be overcome more readily than can the disadvantages of tryptophan. For instance, a selective inhibitor of serotonin uptake such as fluoxetine potentiates those effects of 5-HTP that are mediated by serotonin synapses (Fuller 1982) but not other effects of 5-HTP (Fuller and Perry 1981). Thus, when given with a serotonin uptake inhibitor, 5-HTP can be used in much smaller doses below the range at which serotonin is formed non-specifically in substantial amounts (Fuxe et al. 1971). The combination of 5-HTP at a low dose, together with an uptake inhibitor seems to be a suitable treatment regime for increasing serotonin formation and concentration in the synaptic cleft in studies intended to increase serotonergic transmission.

Certain drugs may accelerate brain serotonin synthesis by increasing brain concentrations of tryptophan. The entry of tryptophan into brain is influenced by the ratio of tryptophan concentration in plasma to the concentration of other large neutral amino acids that compete with tryptophan for uptake into the brain (Fernstrom 1981). This ratio may be influenced by drugs directly if they alter either tryptophan concentration or the concentration of the other amino acids, or by a variety of indirect means such as alteration of carbohydrate metabolism leading to insulin secretion. Insulin increases the ratio of tryptophan to competing large neutral amino acids in plasma, thereby increasing brain tryptophan concentration and serotonin biosynthesis. Drugs like caffeine, for instance, may in this way accelerate brain serotonin synthesis secondarily to insulin release (Schlosberg, Fernstrom, Kopczynski, Cusack and Gillis (1981).

Tryptophan is unique among amino acids in that it exists in plasma not in free form but largely bound to albumin (McMenamy and Oncley 1958). Drugs may alter the ratio of free/bound tryptophan in plasma, such as by increasing circulating non-esterified fatty acids which compete with tryptophan for binding sites on albumin (Curzon, Friedel and Knott 1973). An increase in free tryptophan concentration has been suggested to be a factor that results in increased tryptophan entry into brain (Tagliamonte, Biggio and Gessa 1971), and some drugs that increase brain tryptophan conversion to serotonin have been suggested to act via this mechanism (Tagliamonte, Biggio, Vargiu and Gessa 1973a; Bourgoin, Hery, Ternaux and Hamon 1975; Badawy 1982). However, others have argued that free tryptophan concentration in plasma is not an important determinant of tryptophan uptake into brain and that only the ratio of total tryptophan concentration to the concentration of the competing amino acids is important in this regard (Fernstrom, Hirsch and Faller 1978). This controversy has been reviewed in detail elsewhere (Green 1978).

Brain monoamine turnover generally is increased by antagonists and decreased by agonists, effects that may be mediated partly by actions on presynaptic autoreceptors and partly by trans-neuronal feedback influences on transmitter synthesis and release and on neuronal firing. For example, dopamine turnover is increased by antagonists and decreased by agonists, and the same is true for noradrenaline turnover. Although serotonin turnover is decreased by agonists, as discussed previously in this chapter, serotonin turnover generally is not increased by antagonists that block certain functional responses to serotonin or to serotonin agonists. Methiothepin at high doses does increase serotonin turnover, but this increase apparently is not entirely attributable to receptor blockade (Jacoby, Shabshelowitz, Fernstrom and Wurtman 1975). Most other serotonin antagonists do not influence serotonin turnover at doses that are usually used. Metergoline has been reported to increase serotonin turnover by some workers (Fuxe, Agnati and Everitt 1975; Invernizzi and Samanin 1981), whereas others (Bourgoin, Artaud, Bockaert, Hery, Glowinski and Hamon 1978) have reported the opposite effect, a decrease in serotonin turnover. It is interesting that methiothepin, the only serotonin antagonist consistently reported to increase serotonin turnover in brain, is the most potent antagonist at autoreceptors that modulate serotonin release *in vitro* (Baumann and Waldmeier 1981; Martin and Sanders-Bush 1982, and see

Table 2.4). One paradox raised by these *in vitro* studies is the finding that quipazine resembes methiothepin in behaving as an antagonist (though weaker in potency than methiothepin) at these receptors that modulate serotonin release *in vitro,* whereas *in vivo* quipazine decreases serotonin turnover, an effect opposite to that of methiothepin. This paradox is not explained at present, but one possibility might be that the *in vivo* effects of quipazine relate more to its ability to inhibit serotonin uptake (Hamon *et al.* 1976) than to direct receptor effects (see also Section 2.5). There is ample evidence, however, that quipazine is capable of stimulating central serotonin receptors directly *in vivo* (see, for example, Neuman and White 1982).

Most of the studies with serotonin antagonists that have shown any effect on brain serotonin turnover have been at extraordinarily high doses of the antagonists. Table 1.2 lists ED_{50} values for five antagonists in blocking those central serotonin receptors that mediate quipazine-induced increases in serum corticosterone concentration in rats (Fuller, Snoddy and Mason 1983). Agents like metergoline, methiothepin and mianserin are very potent, having ED_{50} values well below 1 mg/kg in this system. In

Table 1.2   Doses of serotonin antagonists that block quipazine-induced elevation of serum corticosterone in rats

Antagonist	ED_{50}, mg/kg (i.p.)
Metergoline	0.03
Pirenperone	0.035
Methiothepin	0.21
Mianserin	0.35
LY53857	0.55
Spiperone	1.5

contrast, changes in brain serotonin turnover have been reported only at doses in the range of 2–20 mg/kg for metergoline and methiothepin (Jacoby *et al.* 1975; Fuxe *et al.* 1975; Bourgoin *et al.* 1978; Invernizzi and Samanin 1981), and no effect of mianserin was found even at 10 mg/kg (Green, Hall and Rees 1981). Pirenperone at 0.1 mg/kg has been reported not to alter the rate of serotonin turnover in rat brain (Green, O'Shaughnessy, Hammond, Schachter and Grahame-Smith 1983), although this dose is sufficient to block central serotonin receptors (Table 1.2) (Green *et al.* 1983). Thus in general it appears that blockade of serotonin receptors in brain does not necessarily result in an increase in serotonin turnover.

1.7   Drugs that inhibit serotonin metabolism

The metabolic degradation of serotonin in mammalian brain occurs almost exclusively by the action of MAO, and MAO inhibitors directly inhibit this metabolism. Serotonin is an excellent substrate for type A MAO but a much poorer substrate for type B MAO *in vitro* (Fowler and Tipton 1982), and the oxidation of serotonin in the brain normally occurs by type A MAO (Ashkenazi, Finberg and Youdim 1983). Selective inhibitors of type A MAO, such as clorgyline, harmaline and LY51641, inhibit serotonin oxidation and increase brain concentrations of serotonin, as do non-selective inhibitors at doses capable of inhibiting type A MAO.

Other monoamine transmitters like dopamine, noradrenaline and adrenaline also are oxidized preferentially by type A MAO (Fuller and Hemrick-Luecke 1981), so the use of these inhibitors does not permit selectively influencing serotonin and not other amine transmitters. The possibility of finding MAO inhibitors that selectively localize in serotonin neurones has been suggested (Fuller 1978). Recently Ask, Fagervall and Ross (1983) have described a compound, amiflamine (FLA 336($^+$)), that appears to have this characteristic. Amiflamine is a competitive, reversible inhibitor with selectivity for type A MAO. These workers presented evidence that amiflamine was more potent in inhibiting MAO inside serotonergic neurones than elsewhere. There experiments were based on the long-known ability of reversible inhibitors of MAO to protect against the inactivation of the enzyme by irreversible inhibitors (enzyme-activated inhibitors). Ask and others (1983) have used phenelzine as the irreversible inhibitor and determined the ability of amiflamine to protect against the inactivation of MAO in rat brain by phenelzine, both drugs injected systematically. Twenty-four hours after phenelzine injection, MAO was inactivated if amiflamine had not been given to protect the enzyme, but the direct inhibitory effects of amiflamine no longer persisted. MAO within particular types of neurones was evaluated by preparing synaptosomes from the brains of treated rats and incubating these synaptosomes with radioactive serotonin, noradrenaline or dopamine. These monoamines were actively transported into synaptosomes selectively and then deaminated, e.g. serotonin was transported via the membrane carrier into serotonergic synaptosomes, then deaminated by MAO present in those synaptosomes. Selective inhibitors of uptake were used to distinguish this selective deamination within neurones that accumulated each particular monoamine from non-specific deamination by other synaptosomes. The studies revealed that amiflamine preferentially antagonized the inactivation of MAO within serotonin neurones by phenelzine, meaning that amiflamine itself preferentially inhibited MAO within serotonin neurones. This is an important demonstration of the principle that in addition to the

selectivity of inhibitors for one form of MAO, selectivity of effects can also
be achieved by the preferential localization of the inhibitor within a particular type of neurone in the brain.

We have been able to show a degree of selectivity in an analogous manner
with the longer homologue of *p*-chloroamphetamine, LY87079. Figure 1.2
shows the chemical structures of *p*-chloroamphetamine, LY87079, and
amiflamine. These *para*-substituted amphetamine analogues all appear to
have affinity for the serotonin uptake carrier and presumably are concentrated within serotonin neurones *in vivo*. *p*-Chloroamphetamine leads to
neurotoxic effects on serotonin neurones, effects that are prevented if its
accumulation there is blocked by means of uptake inhibitors like fluoxetine
(Fuller, Perry and Molloy 1975). We had chosen LY87079 because it
retained the affinity of *p*-chloroamphetamine for the serotonin uptake site
and the ability of *p*-chloroamphetamine to inhibit type A MAO but did not
share the neurotoxic activity of *p*-chloroamphetamine (Fuller, Perry, Wong
and Molloy 1974*b*). Thus we anticipated that LY87079 might be localized
in serotonin neurones and preferentially inhibit MAO there. Our approach
was slightly different from that of Ask *et al.* (1983). We determined the
ability of LY87079 to antagonize the pargyline-induced inhibition of serotonin and dopamine metabolism *in vivo*. These effects of pargyline were
assessed by its ability to diminish the increase in brain levels of the serotonin metabolite, 5-HIAA, and of the dopamine metabolites,
3,4-dihydroxyphenylacetic acid (DOPAC) and homovanillic acid (HVA)
when the amines were released by Ro 4-1284, a rapidly acting monoamine
releaser. LY87079 antagonized the effect of pargyline on 5-HIAA more
potently than it antagonized the effects on DOPAC or HVA. Since the
effect on 5-HIAA reflects MAO inhibition within serotonin neurones and
the effect on DOPAC and HVA reflects MAO inhibition within dopamine
neurones, the results suggest that LY87079 preferentially inhibited MAO

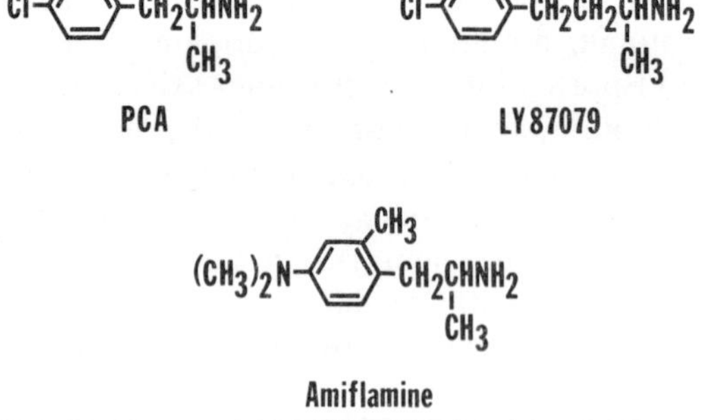

Fig. 1.2 Chemical structures of *p*-chloroamphetamine (PCA), LY87079 and
amiflamine.

within serotonin neurones more than that in dopamine neurones. Probably the mechanism was similar to that suggested for amiflamine, namely selective localization of the reversible inhibitor in serotonin neurones due to its affinity for the membrane uptake carrier.

Neither amiflamine nor LY87079 can be used to inhibit MAO exclusively within serotonin neurones, but they do offer encouragement that such localization may be attainable. Possibly it would be easier to achieve with an irreversible inhibitor, which could be given at low doses to minimize non-specific effects so that effective concentrations would be reached only within serotonin-forming neurones in which the inhibitor would accumulate, and its inactivation of MAO there would be cumulative. At present, however, one must be satisfied with inhibitors that block type A MAO selectively and thus inhibit serotonin metabolism along with the metabolism of other monoamines that are substrates for this enzyme.

1.8 Summary

Serotonin synthesis appears to be coupled to release, i.e. serotonin neuronal firing, under most circumstances. Serotonin synthesis may be inhibited directly by blocking one of the two enzymes involved or may be reduced secondarily to stimulation of synaptic receptors by direct or indirect agonists. Serotonin synthesis may be accelerated by administration of one of the amino acid precursors, tryptophan or 5-HTP, or by treatments that increase brain uptake of tryptophan. MAO inhibitors block serotonin metabolism and increase intraneuronal stores of serotonin. Drugs that decrease or increase serotonin stores are often used to suppress or enhance serotonergic function, respectively. Serotonin turnover and serotonergic function do not always change in the same direction, for instance enhancement of serotonergic function by uptake inhibitors, releasers, or direct agonists results in a decrease in serotonin turnover. With these drugs, measurement of serotonin turnover may be one useful indicator of their actions.

References

Aghajanian, G. K. and Asher, I. M. (1971). Histochemical fluorescence of raphe neurons; selective enhancement by tryptophan. *Science* **172**, 1159.
—— and Wang, R. Y. (1978). Physiology and pharmacology of central serotonergic neurons. In *Psychopharmacology: a generation of progress* (eds. M. A. Lipton, JA. DiMascio and K. F. Killam) p. 171. Raven Press, New York.
Arbuthnott, G. W., Eccleston, D., Laszlo, I. and Nicolaou, N. (1981). Uptake of 5-hydroxytryptamine in the catecholamine containing areas of

the hypothalamus of the rat after treatment with phenelzine and trypto-phan. *Br. J. Pharmac.* **73**, 143.

Ashkenazi, R., Finberg, J. P. M. and Youdim, M. B. H. (1983). Behavioural hyperactivity in rats treated with selective monoamine oxidase inhibitors and LM 5008, a selective 5-hydroxytryptamine uptake blocker. *Br. J. Pharmac.* **79**, 765.

Ask, A-L., Fagervall, I. and Ross, S. B. (1983). Evidence for a selective inhibition by FLA 336($^+$) of the monoamine oxidase in serotonergic neurones in the rat brain. *Acta pharmac. toxic.* **51**, 395.

Badawy, A. A-B. (1982). Mechanisms of elevation of rat brain tryptophan concentration by various doses of salicylate. *Br. J. Pharmac.* **76**, 211.

Baumann, P. A. and Waldmeier, P. C. (1981). Further evidence for negative feedback control of serotonin release in the central nervous system. *Naunyn-Schmiedeberg's Arch. Pharmac.* **317**, 36.

Blier, P. and de Montigny, C. (1983). Effects of quipazine on pre- and postsynaptic serotonin receptors: single cell studies in the rat CNS. *Neuropharmacology* **22**, 495.

Bourgoin, S., Hery, F., Ternaux, J. P. and Hamon, M. (1975). Effects of benzodiazepines on the binding of tryptophan in serum. Consequences of 5-hydroxyindoles concentrations in the rat brain. *Psychopharmac. Commun.* **1**, 209.

—— Artaud, F., Bockaert, J., Hery, F., Glowinski, J. and Hamon, M. (1978). Paradoxical decrease of brain 5-HT turnover by metergoline, a central 5-HT receptor blocker. *Naunyn-Schmiedeberg's Arch. Pharmac.* **302**, 313.

Buus Lassen, J., Squires, R. F., Christensen, J. A. and Molander, L. (1975). Neurochemical and pharmacological studies on a new 5HT-uptake inhibitor, FG4963, with potential antidepressant properties. *Psycho-pharmacologia* **42**, 21.

Carlsson, A. and Lindqvist, M. (1978). Effects of antidepressant agents on the synthesis of brain monoamines. *J. Neural Transm.* **43**, 73.

—— Davis, J. N., Kehr, W., Lindqvist, M. and Atack, C. V. (1972). Simul-taneous measurement of tyrosine and tryptophan hydroxylase activities in brain *in vivo* using an inhibitor of the aromatic amino acid decar-boxylase. *Naunyn-Schmiedeberg's Arch. Pharmac.* **275**, 153.

Cheifetz, S. and Warsh, J. J. (1980). Occurrence and distribution of 5-hydroxytryptophol in the rat. *J. Neurochem.* **34**, 1093.

Claassen, V., Davies, J. E., Hertting, G. and Placheta, P. (1977). Fluvoxa-mine, a specific 5-hydroxytryptamine uptake inhibitor. *Br. J. Pharmac.* **60**, 505.

Clemens, J. A., Sawyer, B. D. and Cerimele, B. (1977). Further evidence that serotonin is a neurotransmitter involved in the control of prolactin secretion. *Endocrinology* **100**, 692.

Clineschmidt, B. V. (1979). MK-212: A serotonin-like agonist in the CNS. *Gen. Pharmac.* **10**, 287.

Corrodi, H. and Fuxe, K. (1969). Decreased turnover in central 5-HT nerve terminals induced by antidepressant drugs of the imipramine type. *Eur. J. Pharmac.* **7**, 56.

Curzon, G., Friedel, J. and Knott, P. J. (1973). The effect of fatty acids on the binding of tryptophan to plasma protein. *Nature* **242**, 198.

de Montigny, C. and Aghajanian, G. K. (1977). Preferential action of 5-methoxytryptamine and 5-methoxydimethyltryptamine on presynaptic serotonin receptors: a comparative iontophoretic study with LSD and serotonin. *Neuropharmacology* **16**, 811.

Diggory, G. L., Ceasar, P. M. and Morgan, R. M. (1979*b*). The regional metabolism of 5-hydroxytryptamine in mouse brain *in vitro. Life Sci.* **24**, 1939.

―― ―― Hazelby, D. and Taylor, K. T. (1979*a*). Endogenous 5-hydroxytryptophol in mouse brain. *J. Neurochem.* **32**, 1323.

Euvrard, C. and Boissier, J. R. (1980). Biochemical assessment of the central 5-HT agonist activity of RU 24969 (a piperidinyl indole). *Eur. J. Pharmac.* **63**, 65.

Fernstrom, J. D. (1981). Dietary precursors and brain neurotransmitter formation. *Ann. Rev. Med.* **32**, 413.

―― Hirsch, M. J. and Faller, D. V. (1976). Tryptophan concentrations in rat brain. Failure to correlate with free serum tryptophan or its ratio to the sum of other serum neutral amino acids. *Biochem. J.* **160**, 589.

Fowler, C. J. and Tipton, K. F. (1982). Deamination of 5-hydroxytryptamine by both forms of monoamine oxidase in the rat brain. *J. Neurochem.* **38**, 733.

Fuller, R. W. (1978). Selectivity among monoamine oxidase inhibitors and its possible importance for development of antidepressant drugs. *Prog. Neuro-Psychopharmac.* **2**, 303.

―― (1982). Functional consequences of inhibiting serotonin uptake with fluoxetine in rats. In *Serotonin in biological psychiatry* (eds. B. T. Ho, J. C. Schoolar and E. Usdin) p. 219. Raven Press, New York.

―― and Hemrick-Luecke, S. K. (1981). Elevation of epinephrine concentration in rat brain by LY51641, a selective inhibitor of type A monoamine oxidase. *Res. Commun. Chem. Path. Pharmac.* **32**, 207.

―― and Perry, K. W. (1983). Decreased accumulation of brain 5-hydroxytryptophan after decarboxylase inhibition in rats treated with fenfluramine, norfenfluramine or *p*-chloroamphetamine. *J. Pharm. Pharmac.* **35**, 597.

―― ―― (1981). Elevation of 3,4-dihydroxyphenylacetic acid concentration by L-5-hydroxytryptophan in control and fluoxetine-pretreated rats. *J. Pharm. Pharmac.* **33**, 406.

—— and Steinberg, M. (1976). Regulation of enzymes that synthesize neurotransmitter monoamines. *Adv. Enz. Regul.* **14,** 347.

—— and Wong, D. T. (1977). Inhibition of serotonin reuptake. *Fed. Proc.* **36,** 2154.

—— Perry, K. W. and Molloy, B. B. (1975). Reversible and irreversible phases of serotonin depletion by 4-chloroamphetamine. *Eur. J. Pharmac.* **33,** 119.

—— —— —— (1974*a*). Effect of an uptake inhibitor on serotonin metabolism in rat brain: studies with 3-(*p*-trifluoromethylphenoxy)-*N*-methyl-3-phenylpropylamine (Lilly 110140). *Life Sci.* **15,** 1161.

—— Snoddy, H. D. and Hemrick, S. K. (1978*a*). Effects of fenfluramine and norfenfluramine on brain serotonin metabolism in rats. *Proc. Soc. Exp. Biol. Med.* **157,** 202.

—— —— and Mason, N. R. (1983). Antagonism of the quipazine-induced elevation of serm corticosterone in rats by ketanserin, pirenperone and other antagonists of $5HT_2$ receptors. *Fed. Proc.* **42,** 459.

—— Perry, K. W., Wong, D. T. and Molloy, B. B. (1974*b*). Effects of some homologues of 4-chloroamphetamine on brain serotonin metabolism. *Neuropharmacology* **13,** 609.

—— Snoddy, H. D., Mason, N. R. and Molloy, B. B. (1978*b*). Effect of 1-(*m*-trifluoromethylphenyl)piperazine on ³H-serotonin binding to membranes from rat brain *in vitro* and on serotonin turnover in rat brain *in vivo. Eur. J. Pharmac.* **52,** 11.

—— —— —— and Owen, J. E. (1981). Disposition and pharmacological effects of *m*-chlorophenylpiperazine in rats. *Neuropharmacology* **20,** 155.

Fuxe, K., Agnati, L. and Everitt, B. (1975). Effects of methergoline on central monoamine neurones. Evidence for a selective blockade of central 5-HT receptors. *Neurosci. Lett.* **1,** 283.

—— Butcher, L. L. and Engel, J. (1971). DL-5-Hydroxytryptophan-induced changes in central monoamine neurons after peripheral decarboxylase inhibition. *J. Pharm. Pharmac.* **23,** 420.

—— Holmstedt, B., and Jonsson, G. (1972). Effects of 5-methoxy-*N,N*-dimethyltryptamine on central monoamine neurons. *Eur. J. Pharmac.* **19,** 25.

Gal, E. M. and Whitacre, D. H. (1982). Mechanism of irreversible inactivation of phenylalanine-4- and tryptophan-5-hydroxylase by [4-36Cl,2-14C]*p*-chlorophenylalanine: A revision. *Neurochem. Res.* **7,** 13.

Galli, C., Commisiong, J. W., Costa, E. and Neff, N. H. (1978). Incorporation of $^{18}O_2$ into brain serotonin *in vivo* as a procedure for estimating turnover: A feasibility study in animals. *Life Sci.* **22,** 473.

Green, A. R. (1978). The effects of dietary tryptophan and its peripheral metabolism on brain 5-hydroxytryptamine synthesis and function. In *Essays in neurochemistry and neuropharmacology*(eds. M. B. H. Youdim, W. Lovenberg, D. F. Sharman and J. R. Lagnado) Vol. 3, 104. John Wiley, Chichester.

—— Guy, A. P. and Gardner, C. R. (1984). The behavioural effects of RU 24969, a suggested 5-HT$_1$ receptor agonist, and the effect on the behaviour of treatment with antidepressants. *Neuropharmacology* **23**, 655.

—— Hall, J. E. and Rees, A. R. (1981). A behavioural and biochemical study in rats of 5-hydroxytryptamine receptor agonists and antagonists, with observations on structure-activity requirements for the agonists. *Br. J. Pharmac.* **73**, 703.

—— O'Shaughnessy, K., Hammond, M., Schachter, M. and Grahame-Smith, D. G. (1983). Inhibition of 5-hydroxytryptamine-mediated behaviour by the putative 5-HT$_2$ antagonist pirenperone. *Neuropharmacology* **22**, 573.

Hamon, M., Bourgoin, S., Enjalbert, A., Bockaert, J., Hery, F., Ternaux, J. P. and Glowinski, J. (1976). The effects of quipazine on 5-HT metabolism in the rat brain. *Naunyn-Schmiedeberg's Arch. Pharmac.* **294**, 99.

Hjorth, S., Carlsson, A., Lindberg, P., Sanchez, D., Wikstrom, H., Arvidsson, L.-E., Hacksell, U. and Nilsson, J. L. G. (1982). 8-Hydroxy-2-(di-*n*-propylamino)tetralin, 8-OH-DPAT, a potent and selective simplified ergot congener with central 5-HT-receptor stimulating activity. *J. Neural Transm.* **55**, 169.

Hong, E., Sancilio, L. F., Vargas, R. and Pardo, E. G. (1969). Similarities between the pharmacological actions of quipazine and serotonin. *Eur. J. Pharmac.* **6**, 274.

Hyttel, J. (1977). Effect of a selective 5-HT uptake inhibitor—Lu 10-171— on rat brain 5-HT turnover. *Acta pharmac. toxic.* **40**, 439.

Invernizzi, R. and Samanin, R. (1981). Effects of metergoline on regional serotonin metabolism in the rate brain. *Pharmac. Res. Commun.* **13**, 511.

Jacoby, J. H., Shabshelowitz, H., Fernstrom, J. D. and Wurtman, R. J. (1975). The mechanisms by which methiothepin, a putative serotonin receptor antagonist, increases brain 5-hydroxyindole levels. *J. Pharmac. Exp. Ther.* **195**, 257.

Jequier, E., Lovenberg, W. and Sjoerdsma, A. (1967). Tryptophan hydroxylase inhibition: The mechanism by which *p*-chlorophenylalanine depletes rat brain serotonin. *Mol. Pharmac.* **3**, 274.

Johnson, G. A., Kim, E. G., Platz, P. A. and Mickelson, M. M. (1968). Comparative aspects of tyrosine hydroxylase and tryptophan hydroxylase inhibition: arterenones and dihydroxyphenylacetamide (H 22/54). *Biochem. Pharmac.* **17**, 403.

Koe, B. K. and Corkey, R. F., Jr. (1976). Inhibition of rat brain tryptophan hydroxylation with *p*-chloroamphetamine. *Biochem. Pharmac.* **25**, 31.
—— and Weissman, A. (1966). *p*-Chlorophenylalanine: a specific depletor of brain serotonin. *J. Pharmac. Exp. Ther.* **154**, 499.
Levitt, P., Pintar, J. E. and Breakefield, X. O. (1982). Immunocytochemical demonstration of monoamine oxidase B in brain astrocytes and serotonergic neurons. *Proc. Nat. Acad. Sci.* **79**, 6385.
Marco, E. J. and Meek, J. L. (1979). The effects of antidepressants on serotonin turnover in discrete regions of rat brain. *Naunyn-Schmiedeberg's Arch. Pharmac.* **306**, 75.
Marsden, C. A., Conti, J., Strope, E., Curzon, G. and Adams, R. N. (1979). Monitoring 5-hydroxytryptamine release in the brain of the freely moving unanaesthetized rat using *in vivo* voltammetry. *Brain Res.* **171**, 85.
Martin, L. L. and Sanders-Bush, E. (1982). Comparison of the pharmacological characteristics of $5HT_1$ and $5HT_2$ binding sites with those of serotonin autoreceptors which modulate serotonin release. *Naunyn-Schmiedeberg's Arch. Pharmac.* **321**, 165.
McGeer, E. G. and Peters, D. A. V. (1969). *In vitro* screen of inhibitors of rat brain serotonin synthesis. *Can. J. Biochem.* **47**, 501.
—— Peters, D. A. V. and McGeer, P. L. (1968). Inhibition of rat brain tryptophan hydroxylase by 6-halotryptophans. *Life Sci.* **7**, 605.
McMenamy, R. H. and Oncley, J. L. (1958). The specific binding of L-tryptophan to serum albumin. *J. Biol. Chem.* **233**, 1436.
Meek, J. L. and Fuxe, K. (1971). Serotonin accumulation after monoamine oxidase inhibition. Effects of decreased impulse flow and of some antidepressants and hallucinogens. *Biochem. Pharmac.* **20**, 693.
—— and Werdinius, B. (1970). Hydroxytryptamine turnover decreased by the antidepressant drug chlorimipramine. *J. Pharm. Pharmac.* **22**, 141.
Moir, A. T. B. and Eccleston, D. (1968). The effects of precursor loading in the cerebral metabolism of 5-hydroxyindoles. *J. Neurochem.* **15**, 1093.
Neff, N. H., Spano, P. F., Groppetti, A., Wang, C. T. and Costa, E. (1971). A simple procedure for calculating the synthesis rate of norepinephrine, dopamine and serotonin in rat brain. *J. Pharmac. Exp. Ther.* **176**, 701.
—— and Tozer, T. N. (1968). *In vivo* measurement of brain serotonin turnover. *Adv. Pharmac.* **6A**, 97.
—— and Yang, H.-Y. T. (1974). Another look at the monoamine oxidase inhibitor drugs. *Life Sci.* **14**, 2061.
Neuman, R. S. and White, S. R. (1982). Serotonin-like actions of quipazine and CPP on spinal motoneurones. *Eur. J. Pharmac.* **81**, 49.
Ng, L. K. Y., Chase, T. N., Colburn, R. W. and Kopin, I. J. (1972). Release of [^{3}H]dopamine by L-5-hydroxytryptophan. *Brain Res.* **45**, 499.
Ortmann, R., Waldmeier, P. C., Radeke, E., Felner, A. and Delini-Stula, A.

(1980). The effects of 5-HT uptake- and MAO-inhibitors on L-5-HTP-induced excitation in rats. *Naunyn-Schmiedeberg's Arch. Pharmac.* **311,** 185.

Reinhard, J. F., Jr. and Wurtman, R. J. (1977). Relation between brain 5-HIAA levels and the release of serotonin into brain synapses. *Life Sci.* **21,** 1741.

Ross, S. B., Hall, H., Renyi, A. L. and Westerlund, D. (1981). Effects of zimelidine on serotonergic and noradrenergic neurons after repeated administration in the rat. *Psychopharmacology* **72,** 219.

Safdy, M. E., Kurchacova, E., Schut, R. N., Vidrio, H. and Hong, E. (1982). Tryptophan analogues. 1. Synthesis and antihypertensive activity of positional isomers. *J. Med. Chem.* **25,** 723.

Sanders-Bush, E. and Sulser, F. (1970). *p*-Chloroamphetamine: *In vivo* investigations on the mechanism of action of the selective depletion of cerebral serotonin. *J. Pharmac. Exp. Ther.* **175,** 419.

Schildkraut, J. J., Schanberg, S. M., Breese, G. R. and Kopin, I. J. (1969). Effects of psychoactive drugs on the metabolism of intracisternally administered serotonin in rat brain. *Biochem. Pharmac.* **18,** 1971.

Schlosberg, A. J., Fernstrom, J. D., Kopczynski, M. C., Cusack, B. M. and Gillis, M. A. (1981). Acute effects of caffeine injection of neutral amino acids and brain monoamine levels in rats. *Life Sci.* **29,** 1973.

Schoepp, D. D. and Azzaro, A. J. (1981). Specificity of endogenous substrates for types A and B oxidase in rat striatum- *J. Neurochem.* **36,** 2025.

Sidransky, H., Bongiorno, M., Sarma, D. S. R. and Verney, E. (1967). The influence of tryptophan on hepatic polyribosomes and protein synthesis in fasted mice. *Biochem. Biophys. Res. Commun.* **27,** 242.

Sved, A. F., Van Itallie, C. M., and Fernstrom, J. D. (1982). Studies on the antihypertensive action of L-tryptophan. *J. Pharmac. Exp. Ther.* **221,** 329.

Svensson, T. H. (1978). Attenuated feed-back inhibition of brain serotonin synthesis following chronic administration of imipramine. *Naunyn-Schmiedeberg's Arch. Pharmac.* **302,** 115.

Tagliamonte, A., Biggio, G. and Gessa, G. L. (1971). Possible role of "free" plasma tryptophan in controlling brain tryptophan concentrations. *Riv. Farmac. Terap.* **11,** 251.

―――― Vargiu, L. and Gessa, G. L. (1973*a*). Increase of brain tryptophan and stimulation of serotonin synthesis by salicylate. *J. Neurochem.* **20,** 909.

―― Tagliamonte, P., Corsini, G. U., Mereu, G. P. and Gessa, G. L. (1973*b*). Decreased conversion of tyrosine to catecholamines in the brain of rats treated with *p*-chlorophenylalanine. *J. Pharm. Pharmac.* **25,** 101.

Trulson, M. E. and Jacobs, B. L. (1976). Behavioural evidence for the rapid release of CNS serotonin by PCA and fenfluramine. *Eur. J. Pharmac.* **36,** 149.

Warsh, J. J., Coscina, D. V., Godse, D. D. and Chan, P. W. (1979). Dependence of brain tryptamine formation on tryptophan availability *J. Neurochem.* **32,** 1191.

2

Pharmacology of the serotonin autoreceptor

CHANTAL MORET

2.1 Introduction

The current tendency of multiplying the number of receptor sub-types in the brain does not necessarily facilitate our understanding of their biological roles although it may be the easy way out when one's results do not fit the existing hypothesis. The concept of presynaptic autoreceptors was probably born under such circumstances. The subsequent demonstration of solid evidence supporting the concept has, however, now led to its wide acceptance in pharmacology.

α-Adrenergic autoreceptors were the first presynaptic receptors to be demonstrated and have been most thoroughly studied, both in the peripheral and in the central nervous system (Langer 1981; Starke 1981). These receptors are involved in a negative feedback mechanism through which the neurotransmitter, noradrenaline, can regulate its own release. Thus α_2-adrenoceptor agonists inhibit noradrenaline release during nerve stimulation, while α_2-adrenoceptor antagonists enhance the stimulation-evoked release of the transmitter.

The evidence in support of the presynaptic location of the α_2-adrenoceptors involved in the autoregulation of noradrenergic neurotransmission is based principally on experiments of destruction of postsynaptic membranes (for example Filinger, Langer, Perec and Stefano 1978). Under these conditions the autoregulation is not affected. Furthermore the use of superfused cortical synaptosomes (Mulder, de Langen, de Regt and Hogenboom 1978) has confirmed that the α_2-adrenergic autoreceptor-mediated negative feedback control of noradrenaline release is localized on noradrenergic presynaptic nerve terminals.

Presynaptic autoreceptors have now been described for a number of other neurotransmitters such as acetylcholine (Szerb 1979), dopamine (Hertting, Reimann, Zumstein, Jackisch and Starke 1979), GABA (Kamal,

21

Arbilla and Langer 1979), histamine (Arrang, Garbarg and Schwartz 1983) and serotonin.

In this chapter I will review the evidence for the existence of a serotonin autoreceptor in terms of its characterization, its biological role and the possible exploitation of this receptor as a therapeutic target.

2.2 Methodology

In vivo and *in vitro* methods of studying the serotonergic autoreceptor are available.

2.2.1   In vivo *methods.*

Although it is theoretically possible to study serotonin autoreceptors by animal behavioural methods, the current lack of drugs acting specifically on the presynaptic serotonergic autoreceptor makes such studies difficult to conceive at the present time.

Microiontophoresis and voltammetry have also been used to study the serotonin autoreceptor. Since the chapters in this book by de Montigny (Chapter 7) and by Marsden (Chapter 9) are dedicated to these subjects they will not be described here.

The *in vivo* release of [^{3}H]-serotonin may be measured by using a push-pull cannula (Héry, Simonnet, Bourgoin, Soubrié, Artaud, Hamon and Glowinski 1979; Héry and Ternaux 1981; Bourgoin, Soubrié, Artaud, Reisine and Glowinski 1981). This consists of implanting stereotaxically a push-pull cannula composed of two concentric tubes into the studied part of the brain. An artificial cerebrospinal fluid is delivered continuously using a pump. This method allows the estimation of the release of [^{3}H]-serotonin as it is being endogenously synthesized from [^{3}H]-tryptophan. Superfusate fractions are collected and [^{3}H]-serotonin determined. Test drugs are added to the perfusion fluid and their effects on the level of [^{3}H]-serotonin estimated. Although this method is attractive by the fact that it measures an *in vivo* phenomenon, it is laborious and relatively insensitive. The majority of the studies on serotonin autoreceptors have used *in vitro* release experiments. The present chapter will therefore concentrate on the latter technique.

2.2.2   In vitro *release experiments*

In vitro release studies are generally carried out in brain slices or synaptosomes. Rats are the most commonly used species and tissues rich in serotonin such as cortex, hypothalamus, striatum and hippocampus are usually chosen. After preparation of slices of thickness from 0.3 mm to 0.5

mm with a McIlwain chopper as described by Farnebo and Hamberger (1971), these tissues are incubated under oxygenation at 37°C in a physiological salt solution containing [³H]-serotonin for a period of 6 to 60 min depending on the investigators. The concentration of [³H]-serotonin generally used is 0.1 μM since at this concentration the labelled transmitter is preferentially taken up into serotonin neurones (Shaskan and Snyder 1970). The slices are then transferred to superfusion chambers and superfused at a rate of 0.3–0.5 ml/min for 30–110 min. Similarly rat brain synaptosomes have been used. These are incubated for 10 min with 0.1 μM [³H]-serotonin and then transferred to slightly modified superfusion chambers as described by Raiteri, Angelini and Levi (1974).

The efflux of tritium from the tissue decreases with time following a multiphasic, exponential course which becomes essentially constant after a certain period of time. When this phase is reached, the tissues are, in general, stimulated either electrically with parameters which vary depending on the investigators (1–100 Hz; 12–20 mA; 2–4 ms duration), or by depolarization by a high concentration of potassium (15–50 mM) for a period of 1 to 10 minutes. Usually two stimulations are applied, separated by the time necessary to return to the basal level. The first one is usually considered as the control, the test drug being added to the superfusing medium before the second stimulation. The superfusate is collected in 1–5 min fractions and the radioactivity in the tissues and in the superfusate samples measured by liquid scintillation counting.

Tritium efflux into each superfusate sample is normally calculated as the fraction of the tritium content of the tissue at the onset of the respective collection period (fractional rate of outflow) (Taube, Starke and Borowski 1977). The effect of drugs on the basal outflow of tritium is estimated from the basal values in the sample preceding the stimulation period, in comparison with the basal value of the control. The depolarization-evoked release is calculated as the additional release evoked by stimulation. This is calculated as per cent of the tritium content of the slices at the start of stimulation. In order to quantify drug-induced changes of depolarized tritium overflow, the ratio of the release evoked by the second stimulation and that evoked by the first (S_2/S_1) is determined. This is illustrated in Fig. 2.1. Some authors express the effect of drugs as percentage of control values.

In these release experiments it should be pointed out that total radioactivity is measured, which means that [³H]-serotonin and its deaminated metabolite, 5-HIAA are included. During stimulation, however, the total tritium overflow consists principally of unmetabolized [³H]-serotonin (Chase, Katz and Kopin 1969; Baumann and Waldmeier 1981; Collard, Cassidy, Pye and Taylor 1981; Göthert and Schlicker 1983). Thus changes in stimulated [³H]-overflow reflect changes in [³H]-serotonin release from

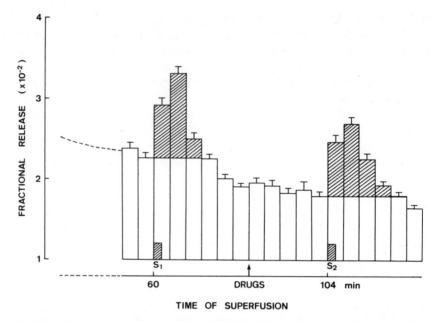

Fig. 2.1   Electrically evoked release of [³H]-serotonin from slices of the rat hypothalamus. After incubation with [³H]-serotonin 0.1 μM during 30 min and a period of
washing of 60 min with Krebs solution, the slices, continuously superfused, were
stimulated twice (S_1 and S_2) for 2 min (3 Hz, 2 ms, 20 mA). The test drugs were
added 20 min before S_2. Ordinate: Amount of tritium released per 4 min sample as a
fraction of the total tissue tritium content at the onset of the respective collection
period. Histograms represent each superfusate sample collected at 4 min intervals.
The hatched parts represent the proportion of [³H]-serotonin release evoked by
stimulation. The dotted line represents the decrease of radioactivity to reach a stable
level before eliciting the first stimulation. Shown are mean values ±S.E.M. of 8
experiments. (C. Moret, unpublished results)

serotonergic neurones. In addition, the proportion of unmetabolized serotonin released by stimulation is increased by uptake inhibition. The rate of
superfusion is however also an important factor in the rate of metabolism
of the neurotransmitter.

2.3   Demonstration of the serotonin autoreceptor

To my knowledge there is, to date, no evidence to suggest the existence of
serotonin autoreceptor in the periphery. In the brain, two types of serotonin
autoreceptors have been described. Those located on the serotonin-
containing neurones in the raphe nuclei, as indicated by the
electrophysiological experiments of Haigler and Aghajanian (1977),
inhibit the spontaneous firing of the raphe cells. The other autoreceptors

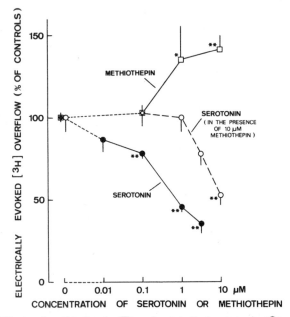

Fig. 2.2   Effects of methiothepin (□) and unlabelled serotonin (●) on tritium over-flow evoked by electrical stimulation (3 Hz for 2 min) from rat brain cortex slices preincubated with [³H]-serotonin, and interaction of serotonin with methiothepin (○). Paroxetine (3.2 μM; an inhibitor of neuronal uptake of serotonin) was present throughout superfusion. Means ±S.E. of 4–15 experiments. *p <0.05, **p <0.005. (Reproduced and modified with permission from Göthert 1982)

are located presynaptically on serotonergic nerve terminals and inhibit the stimulation-induced release of serotonin. It is these latter receptors which will be discussed in this chapter.

In 1974, Farnebo and Hamberger and Hamon, Bourgoin, Jagger and Glowinski, suggested the existence of serotonin receptors on the seroton-ergic nerve fibres when experiments on brain slices revealed that serotonin receptor agonists such as LSD, ergocornine and 2-brom-α-ergocryptine decreased the depolarization-evoked release of [³H]-serotonin, whereas the serotonin receptor antagonist methiothepin increased it.

Cerrito and Raiteri (1979, 1980) have shown that serotonergic auto-receptors can be stimulated by exogenous serotonin to mediate a decrease in the potassium-induced [³H]-serotonin release from rat hypothalamic synaptosomes, and that methiothepin counteracts this inhibitory effect of serotonin. Similar results have also been obtained with exogenous sero-tonin inhibiting the electrically evoked release of [³H]-serotonin from rat brain cortex slices (Göthert and Weinheimer 1979; Baumann and

Waldmeier 1981; Mounsey, Brady, Carroll, Fisher and Middlemiss 1982) or from rat hypothalamic slices (Cox and Ennis 1982). Göthert (1980) showed that the concentration-response curve of unlabelled serotonin is shifted to the right by methiothepin (Fig. 2.2). The neurotransmitter serotonin thus appears to regulate its own release via a negative feedback mechanism, mediated through the activation of inhibitory autoreceptors. Therefore, the increase in serotonin release from brain slices which occurs when methiothepin is given alone (Fig. 2.2) is probably due to the disinhibition of the release mechanism from blockade by the autoreceptors.

It should be pointed out that, in order to observe an inhibition of [^{3}H]-serotonin release by exogenous serotonin, it is necessary to inhibit the neuronal uptake of the amine. The presence of an uptake inhibitor is required to abolish displacement of [^{3}H]-serotonin from its intraneuronal storage sites by the exogenous serotonin taken up. Paroxetine, as used by Göthert (1980), or citalopram, as used by Baumann and Waldmeier (1981), are very suitable for blocking the effect of unlabelled on basal efflux. When LSD is used as the receptor agonist there is no need to inhibit neuronal uptake of the serotonergic neurotransmitter. LSD has been shown to reduce the stimulation-induced overflow of [^{3}H]-serotonin from rat brain slices (Katz and Kopin 1969; Chase et al. 1969; Farnebo and Hamberger 1971), but these authors used rather high concentrations of LSD (1 μM and higher), and the frequency of stimulation used to elicit [^{3}H]-serotonin release was also fairly high (10 and 100 Hz). Similar results were obtained for potassium-evoked release of [^{3}H]-serotonin from brain slices (Hamon et al. 1974; Bourgoin, Artaud, Enjalbert, Héry, Glowinski and Hamon 1977). These authors suggested that LSD reduces serotonergic neurotransmission through the stimulation of inhibitory receptors.

More pronounced inhibitory effects of LSD were found on potassium evoked [^{3}H]-serotonin release from slices of rat raphe nuclei (Ennis and Cox 1982) and on stimulation-induced [^{3}H]-serotonin overflow from rat cortex slices (Baumann and Waldmeier 1981). A similar potency for LSD was obtained by Langer and Moret (1982) in rat hypothalamic slices when [^{3}H]-serotonin release was elicited electrically (Fig. 2.3). As already described for exogenous serotonin (Fig. 2.2), methiothepin shifts to the right the concentration response curve for LSD on [^{3}H]-serotonin release evoked by electrical stimulation in a concentration-dependent manner (Langer and Moret 1982). The parallelism of the curves (Fig. 2.3) suggests that the antagonism of the LSD-induced inhibition by methiothepin at the level of the serotonin autoreceptor might involve a competitive interaction at the same receptor site, as suggested by Göthert (1980), for the inhibition by exogenous serotonin (Fig. 2.2). It is curious to note that the minimal concentration of methiothepin capable of counteracting the inhibitory effect of LSD is 0.1 μM (Fig. 2.3) but 1 μM in the case of serotonin (Göthert 1980)

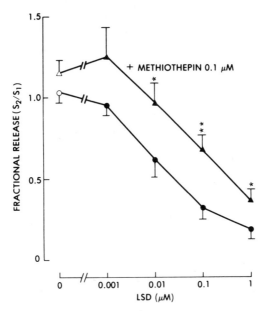

Fig. 2.3 Antagonism by methiothepin (▲) of the inhibition by LSD (●) of the release of [³H]-serotonin elicited by electrical stimulation (3 Hz for 2 min, 20 mA, 2 ms) from slices of the rat hypothalamus. Open symbols are respective controls. Each point represents mean values ±S.E.M. of at least six experiments per group. *p <0.05, **p <0.01 when compared to the corresponding value obtained in the absence of methiothepin. (Reproduced with permission from Langer and Moret 1982. © 1982, Am. Soc. of Parenteral and Enteral Nutrition)

(each gives an approximately 10-fold shift). In a classical competitive situation the action of an antagonist is independent of the agonist used. This paradox may be explained by the fact that two different tissues are used, cortex and hypothalamus, but personally I think that there may be a more interesting explanation.

It is well known that LSD can stimulate α-adrenoceptors (Gillespie and McGrath 1975), as well as dopamine receptors (Von Hungen, Roberts and Hill 1974; Da Prada, Saner, Burkard, Bartholini and Pletscher 1975). It was therefore necessary to eliminate the possibility that the inhibitory effect of LSD on serotonin release was due to the stimulation of these two types of receptors. The presence of inhibitory α-adrenoceptors on the serotonergic nerve terminals has been demonstrated in the rat cortex (Göthert and Huth 1980), hippocampus (Frankhuyzen and Mulder 1980) and hypothalamus (Galzin, Moret and Langer 1984). Langer and Moret (1982) have shown however that phentolamine, an α-adrenoceptor antagonist, did not modify the inhibitory effect of LSD on [³H]-serotonin release elicited by electrical stimulation, indicating that LSD did not inhibit serotonergic neurotransmission by acting on the α-adrenoceptors present on the serotonin nerve terminals of the rat hypothalamus. Previously Göthert and Huth (1980) had shown that the inhibitory effect of exogenous serotonin on stimulation-evoked release of [³H]-serotonin was not antagonized by phentolamine. Similarly, sulpiride which blocks presynaptic dopamine receptors (Langer, Arbilla and Kamal 1980a; Arbilla and Langer 1981), did not modify the inhibition by LSD of electrically induced overflow of [³H]-serotonin (Langer and Moret 1982). LSD therefore does not appear to act on dopamine receptors to reduce the electrically evoked release of [³H]-serotonin. From these studies, it would seem that LSD acts in this system as a potent serotonergic agonist in spite of its α-adrenoceptor and dopaminergic receptor activities.

As in all biological and physiological reactions where the role of calcium is primordial, the modulation of serotonin release by serotonin auto-receptors is not possible if this cation is absent from the medium. In 1969, Chase *et al.* reported that [³H]-serotonin release from rat brain slices was not calcium-dependent. It may be that this was because of the very high frequency of electrical stimulation used (100 Hz). More recently however, it has been shown that the potassium or electrical stimulation-induced overflow of [³H]-serotonin from rat brain synaptosomes reduced in the absence of calcium (Mulder, Van Den Berg and Stoof 1975; Lane and Aprison 1977; Collard *et al.* 1981). Similar results have been obtained in rat cerebral cortex slices (Farnebo 1971; Göthert 1980; Mounsey *et al.* 1982) and rat hypothalamic slices (Langer and Moret 1982). Göthert (1980) showed that the release of serotonin evoked by electrical stimulation depended on the calcium concentration in the superfusion fluid, indicating that the availability of calcium for stimulus-release coupling appears to be increased by raising the extracellular calcium concentration and that serotonin release is controlled by changes in the availability of calcium ions. He also showed that as the calcium concentration in the superfusion fluid was lowered the inhibitory effect of exogenous serotonin on stimulated-evoked [³H]-overflow became more pronounced (Göthert 1980). The role of calcium in the release modulation by serotonin auto-receptors was further demonstrated by experiments in which [³H]-serotonin release was stimulated by introduction of calcium ions into a calcium free, potassium-rich superfusion fluid. This calcium-evoked overflow was a function of the calcium concentration applied. Under these conditions exogenous serotonin inhibited and methiothepin increased calcium-evoked [³H]-overflow to the same extent as observed in the experiments using electrical stimulation. The author concluded that the actions of serotonin and methiothepin were mediated by changing the availability of calcium ions for stimulus-release coupling, probably by decreasing the affinity of the voltage-sensitive permeability channel of the cell membrane for calcium ions.

Further support for the proposal of Göthert (1980) on the role of calcium ions in the autoregulation of serotonin release was provided by Langer and Moret (1982). They compared the release of [^{3}H]-serotonin from slices of the rat hypothalamus induced by electrical stimulation with that produced by the addition of fenfluramine. Fenfluramine causes a depletion of serotonin from brain stores (Neckers, Bertilsson and Costa 1976) and its effects on tryptaminergic terminals can be compared with those of tyramine on catecholaminergic nerve endings which displaces catecholamines from vesicular storage sites through a calcium-independent mechanism (Lindmar and Muscholl 1965). In contrast to the results obtained with electrical stimulation, the release elicited by fenfluramine was not reduced by the absence of calcium ions in the external medium (Table 2.1). Furthermore, LSD and methiothepin were completely ineffective at modifying the [^{3}H]-serotonin release elicited by fenfluramine (Fig. 2.4) (Langer and Moret 1982). These studies demonstrated that the modulation of serotonin release via autoreceptors does not operate in a system which is independent of calcium. Similarly, Homan and Ziance (1981), using chopped rat cerebral cortex, found that the release of [^{3}H]-serotonin induced by d-amphetamine was not dependent on calcium, whereas the potassium-evoked release was calcium-dependent.

Table 2.1 Calcium dependency of [^{3}H]-serotonin release elicited by electrical stimulation but not by fenfluramine from slices of the rat hypothalamus

Experimental group	n	Fractional release $\times 10^{-2}$	
		With Ca^{2+}	Without Ca^{2+}
Electrical stimulation	3	1.91±0.27	0.04±0.04*
Fenfluramine	6	2.03±0.61	2.80±1.01

Fractional release during the first period of electrical stimulation or exposure to fenfluramine in the presence of calcium (1.3 mM) is compared with the fractional release during the second period of electrical stimulation or exposure to fenfluramine in the absence of calcium. Shown are mean values ±S.E.M. n, number of experiments per group; *, p <0.005 when compared to the corresponding stimulation in the presence of calcium (Student's t test). (Modified and calculated from Langer and Moret 1982)

The frequency of stimulation is an important parameter for the sensitivity of autoreceptors. The overflow per pulse has been shown to decrease with increasing frequency of stimulation (Göthert 1980; Baumann and Waldmeier 1981). Furthermore, the inhibitory effect of exogenous serotonin on electrically evoked [^{3}H]-serotonin release was more pronounced

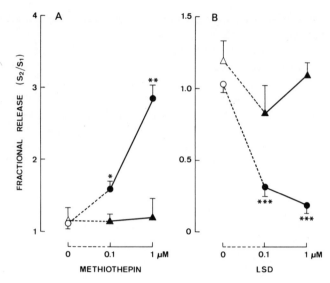

Fig. 2.4 Effect of methiothepin (A) and LSD (B) on [³H]-serotonin release from slices of the rat hypothalamus. Ordinates: fraction of the total tissue radioactivity released by a 2 min period of electrical stimulation (3 Hz, 20 mA, 2 ms) (●) or by exposure to fenfluramine (10 μM) (▲); expressed as the ratio (S_2/S_1) obtained between the second period carried out in the presence of the drug (S_2) and the first period (S_1). Open symbols: respective controls. Abscissae: concentration of methiothepin and LSD, which were added to the medium 20 min before S_2. Each point represents mean values ±S.E.M. of at least 4 experiments per group. *p <0.05, **p <0.005, ***p <0.001 when compared to the corresponding control value. (Drawn from data in Langer and Moret 1982)

at lower frequencies of stimulation as well as at lower calcium concentrations in the superfusion medium (Göthert 1980). The intraneuronal accumulation of calcium is increased by both high extracellular calcium concentrations and by high frequencies of stimulation. The relationship between intraneuronal calcium and release follows saturation kinetics (Starke 1977); thus if the intraneuronal concentration of calcium reaches the saturation level, either as the result of a high extracellular calcium concentration or a high frequency of electrical impulses or both, the release becomes maximal. Under these conditions, the serotonin concentration in the synaptic cleft rises because the rate of release exceeds the rate of the two clearing processes, re-uptake and diffusion. This leads to an increase in the number of autoreceptors occupied by the released serotonin, and a consequent reduction in the effect of exogenous amine. A similar inverse relationship between potency of autoreceptor modulators and calcium concentration in the extracellular medium or frequency of stimulation has previously been observed in central and peripheral noradrenergic neurones (Langer 1977; Starke 1977; Westfall 1977).

2.4   Localization of the serotonin autoreceptor

Throughout the preceding section, no mention has been made as to where the serotonergic autoreceptor site is located. The word 'presynaptic' has been avoided, on purpose, since it was more logical to first describe the serotonin autoreceptor before being preoccupied by its localization. Electrophysiological experiments have shown (Haigler and Aghajanian 1977) that some serotonergic autoreceptors are located on the serotonin-containing neurones in the dorsal raphe nucleus. Release experiments have also been carried out with slices prepared from the raphe nuclei of the rat and pre-loaded with [^{3}H]-serotonin (Ennis and Cox 1982). The cell bodies of the serotonergic neurones located in the raphe nuclei project into various brain regions such as the cortex, hippocampus, hypothalamus and corpus striatum. Therefore in these regions which contain only serotonergic fibres and nerve endings, autoreceptors are found at the synapse and more precisely presynaptically. Another piece of evidence in favour of the pre-synaptic location of the serotonergic autoreceptor is given by the results of experiments using hypothalamic synaptosomes by Cerrito and Raiteri (1979, 1980) who reasoned that if autoreceptors are specific presynaptic sites, it should be possible to demonstrate the modulation of serotonin release from synaptosomes. The problem of the localization of autorecep-tors in synaptosomes is simplified, with respect to other more complex, but more intact, nervous tissue preparations such as brain slices. Synaptosomes like brain slices are nevertheless a heterogeneous popula-tion containing several neurotransmitters. In superfused synaptosomes, however, any transmitter being released from a different population of nerve endings is removed very effectively by the flow of the superfusion fluid. Therefore endogenous transmitters do not appear to reach concen-trations high enough to activate synaptosomal receptors. The serotonergic autoreceptors therefore remain totally available to agonists added in con-trolled amounts to the superfusion fluid. The fact that autoreceptor antagonist, methiothepin, by itself, does not facilitate the depolarization-evoked release in superfused synaptosomes is consistent with the lack of inhibition of the release system by endogenous serotonin (Cerrito and Raiteri 1979, 1980).

Göthert and Weinheimer (1979) provided additional evidence for the localization of serotonergic autoreceptors on presynaptic side of the synapse. [^{3}H]-serotonin release from cortical slices was evoked by the introduction of calcium after superfusion with a potassium-rich, calcium-free solution. Under these conditions, the inhibition of calcium-induced overflow of [³H]-serotonin by exogenous serotonin was also observed in the presence of tetrodotoxin which blocks sodium channels, which are responsible for action potentials. This result indicates that the response to serotonin was not dependent on the movement of sodium ions as in the propagation of action potentials and that interneurones do not play a role in this effect. Thus the involvement of a negative feedback loop via inter-neurones in the inhibitory effect of serotonin can be excluded since tetrodotoxin would have interrupted it. Thus it may be concluded that these serotonergic autoreceptors are located on the terminal fibres them-selves.

2.5 Characterization of the presynaptic serotonin autoreceptor

As stated previously there is currently no evidence to suggest the existence of peripheral serotonin autoreceptors. Several types of serotonin presynaptic receptors have, however, been found in peripheral tissues (Haefely 1974; Wallis and Woodward 1975; Fozard and Mwaluko 1976). In the rabbit heart, for example, these receptors have been shown to be located on the sympa-thetic nerve terminals and exert their effects by modulating the release of noradrenaline (Fozard and Mwaluko 1976; Fozard and Mobarok Ali 1978). Studies on rabbit heart with indolethylamines (Fozard and Mobarok Ali 1978; Göthert and Dührsen 1979) compared with those of Göthert and Schlicker (1983) on rat brain cortex have indicated that the peripheral presynaptic receptors and central presynaptic autoreceptors are different from each other.

In the central nervous system, multiple receptors for serotonin have been found (Bradley and Briggs 1974; Haigler and Aghajanian 1977; Hamon, Nelson, Herbet and Glowinski 1980). The development of simple tech-niques based on the binding of appropriate ligands to membranes has improved the knowledge of the characteristics of serotonin receptors in the central nervous system. Binding studies have described two types of sero-tonin receptors in the frontal cortex of the rat which are termed 5-HT$_1$ preferentially labelled by [³H]-serotonin and 5-HT$_2$ preferentially labelled by [³H]-spiperone or [³H]-ketanserin. Both receptors are labelled to a similar extent by [³H]-LSD (Peroutka and Snyder 1979, 1982; Leysen 1981; Leysen, Awouters, Kennis, Laduron, Vandenberk and Janssen 1981). This work is described in detail in Chapter 4.

Electrolytic lesion of the raphe nuclei in rats which results in degenera-tion of serotonin neurones does not decrease the density of either of these two brain serotonin binding sites, indicating that they are not located on presynaptic serotonin neurones (Bennet and Snyder 1976; Blackshear, Steranka and Sanders-Bush 1981). Similarly Nelson, Herbet, Bourgoin, Glowinski and Hamon (1978) and Whitaker and Deakin (1981) have shown that the binding characteristics of [^{3}H]-serotonin in rat frontal cortex were identical in control rats and in rats whose serotonergic neurones were destroyed by administration of the serotonin neurotoxin, 5,7-dihydroxy-tryptamine. In contrast, kainic acid lesions of postsynaptic elements in the corpus striatum have been shown to reduce [^{3}H]-serotonin binding (G. Fillion, Beaudoin, Rousselle, Deniau, M. P. Fillion, Dray and Jacob 1979). This is again consistent with a postsynaptic localization of [^{3}H]-serotonin binding sites. Thus there is no evidence for presynaptic serotonin receptors from binding studies.

It could be that the proportion of the presynaptic serotonin autorecep-tors might be too small to be detectable by radioligand binding studies. A similar suggestion has been made about the discrepancy between results of binding studies and functional studies on brain slices or synaptosomes for the autoreceptors on central noradrenergic nerve fibres (Starke 1981).

The characterization of multiple serotonin receptors in brain is the subject of Chapter 4 and its consideration here will be limited to the question of whether the presynaptic receptor resembles the 5-HT$_1$ or 5-HT$_2$ receptor. This question has not yet been answered, because the different authors are not in agreement. Martin and Sanders-Bush (1982a) have clearly shown a significant correlation between the effects of agonists on potassium-evoked [^{3}H]-serotonin release from rat hypothalamic synaptosomes and the binding of [^{3}H]-serotonin but not [^{3}H]-spiperone. Figure 2.5 illustrates that a similarity could exist between the presynaptic serotonin autoreceptor and the 5-HT$_1$ binding site on hypothalamic mem-branes, but not between the autoreceptor and 5-HT$_2$ sites. A similar correla-

Table 2.2 Comparison of the effects of analogues of serotonin on the serotonin autoreceptor and [^{3}H]-serotonin binding

	(A)		(B)
	S_2/S_1 (% of controls)	Conc. of drug used (μM)	IC_{50} (nM
Serotonin	46	1	10
5-MeOT	59	1	50
5,6-DHT	54	10	600
Tryptamine	58	10	1000
5,7-DHT	62	100	>10 000

(A) Electrically evoked tritium overflow from superfused rat brain cortex slices pre-incubated with [^{3}H]-serotonin. Indolethylamines were added before the second stimulation and tritium overflow is given as per cent of the ratio S_2/S_1 in the controls; (B) Displacement of specifically bound [^{3}H]-serotonin from rat cerebral cortex membranes. [^{3}H]-serotonin 7 nM and non-specific binding occurring in the presence of 10 μM unlabelled serotonin were used. (Compiled from data from (A) Göthert and Schlicker 1983, and (B) Bennet and Snyder 1976)

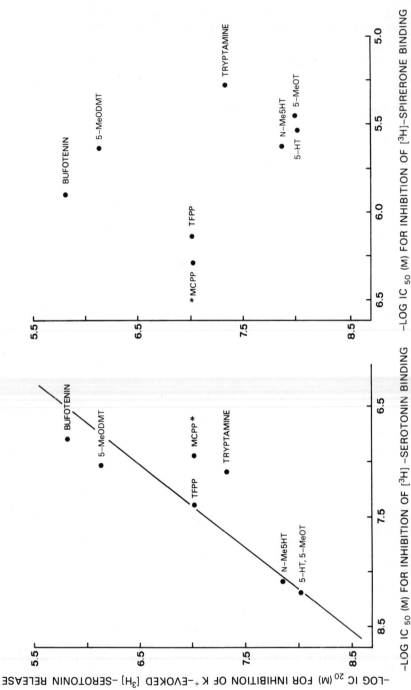

Fig. 2.5 Comparison of the effects of serotonin receptor agonists on the serotonin autoreceptor and [³H]-serotonin binding (on the left) or [³H]-spiperone binding (on the right). Ordinates: –log IC₂₀ (M) for the inhibition by 20% of K⁺-evoked [³H]-serotonin release from hypothalamic synaptosomes. Abscissae: –log IC₅₀ (M) for the inhibition by 50% of specific [³H]-serotonin binding to hypothalamic membranes (on the left) or specific [³H]-spiperone binding to cortical membranes (on the right). *MCPP, point omitted from linear regression analysis. For further details, see Martin and Sanders-Bush (1982a). (Reproduced with permission from Martin and Sanders-Bush 1982a)

tion (Table 2.2) is also clear between the results of Göthert and Schlicker (1983) on the inhibitory potencies of analogues of serotonin on electrically evoked release of [³H]-serotonin and results on 5-HT₁ binding (Bennet and Snyder 1976). Rat cerebral cortex was used in both studies. The results of Mounsey *et al.* (1982) with potassium-evoked [³H]-serotonin release were similar to those of Martin and Sanders-Bush (1982*a*) and Göthert and Schlicker (1983), since they obtained the following relative potencies to inhibit [³H]-serotonin overflow: serotonin = 5-MeOT > 5-MeODMT.

De Montigny and Aghajanian (1977) have demonstrated that 5-MeOT and 5-MeODMT are both more potent than serotonin in depressing the firing rates of serotonergic cells located in the midbrain raphe nuclei through effects on serotonin autoreceptors located on these cell bodies. This is a little in contradiction with the results of the three groups mentioned above since in their studies they found that serotonin is equipotent, or more potent (Göthert and Schlicker 1983), than 5-MeOT and between 10 times (Mounsey *et al.* 1982) to nearly 100 times (Martin and Sanders-Bush, 1982*a*) more potent than 5-MeODMT in suppressing depolarization-evoked [³H]-serotonin release. However, it should be noted that the relative order of potency of hydroxylated tryptamines at the somatodendritic autoreceptors (5-HT > 4-HT > 6-HT) found by Rogawski and Aghajanian (1981) was the same as found by Göthert and Schlicker (1983) for presynaptic serotonin autoreceptors in brain cortex. The differences between the results of de Montigny and Aghajanian (1977) and those of the others could come from the fact that serotonin autoreceptors located on cell bodies of raphe nuclei are not the same as serotonin autoreceptors located on nerve terminals. In an attempt to clarify this, Ennis and Cox compared serotonin autoreceptors present in the rat hypothalamus (Cox and Ennis 1982) with those present in the rat raphe nuclei (Ennis and Cox

Table 2.3 pA_{10} values for the antagonism of 5-methoxytryptamine (*a*), or serotonin (*b*)-induced inhibition of potassium-evoked tritium release from rat slices pre-loaded with [³H]-serotonin

| | pA_{10} values | |
	(A) Raphe nuclei	(B) Hypothalamus
Metergoline	7.14	6.80
Methiothepin	7.26	6.46
Methysergide	5.40	4.34
Cyproheptadine	not active	5.54
Cinanserin	not active	not active

pA_{10}, molar concentration, expressed as a negative logarithm, of antagonist which reduced the IC_{50} of an agonist by 10-fold. (Compiled from data from (B) Cox and Ennis 1982, and (A) Ennis and Cox 1982)

1982), using the same experimental approach. They studied the antagonism of an indolamine-induced inhibition of potassium-evoked [³H]-serotonin release from slices of rat hypothalamus or raphe nuclei preloaded with [³H]-serotonin. It is a pity that the authors used 5-MeOT to inhibit potassium-evoked serotonin release from slices of raphe nuclei (Ennis and Cox 1982) and serotonin in the experiments with slices of rat hypothalamus (Cox and Ennis 1982). Table 2.3 shows that in both tissues metergoline and methiothepin were the most potent antagonists, while methysergide, cyproheptadine and cinanserin were weakly active or completely inactive. Thus the relative order of potency suggests a similarity between these two types of autoreceptor.

Even if the results of de Montigny and Aghajanian (1977) are contradictory with the others, a similarity can nevertheless be deduced between serotonin autoreceptors and 5-HT₁ receptor sites. This conclusion is based on results of relative potencies of agonists. When antagonists are considered, however, major discrepancies appear. Out of the controversy however it is possible to conclude that methiothepin is both an effective antagonist at the autoreceptor and a potent displacer of [³H]-serotonin binding. However, even here the results of Haigler and Aghajanian (1977) are in disagreement. They found that methiothepin fails to block autoreceptors mediating an inhibition of neuronal activity at the level of somatodendritic parts of the serotonin neurones. Again this result would suggest that there may be differences between the autoreceptors in the raphe nuclei and the presynaptic autoreceptors on the nerve terminals.

Most studies have been carried out using slices (Göthert 1980; Schlicker and Göthert 1981; Baumann and Waldmeier 1981; Mounsey et al. 1982) and rat hypothalamic synaptosomes (Martin and Sanders-Bush 1982b). Studies on slices of rat hypothalamus and raphe nuclei have already been commented on (Ennis and Cox 1982; Cox and Ennis 1982).

Table 2.4 shows the relative order of potency of the antagonists on serotonin autoreceptors. Most serotonin receptor antagonists such as methysergide, cyproheptadine, mianserin, cinanserin and metergoline appear to be more or less ineffective in blocking the presynaptic serotonin autoreceptors on nerve terminals and on somato-dendritic parts of the neurone. Quipazine which is generally assumed to be an agonist at postsynaptic receptors (Green et al. 1976; Lansdown, Nash, Preston, Wallis and Williams 1980) also acts as an antagonist at the presynaptic autoreceptors (Baumann and Waldmeier 1981; Schlicker and Göthert 1981; Martin and Sanders-Bush 1982b).

Radioligand binding studies of serotonin receptors have also produced controversial results. In four investigations (Table 2.5) with antagonists, metergoline has a higher affinity for the 5-HT₁ binding sites than methiothepin (Nelson, Herbet, Enjalbert, Bockaert and Hamon 1980; Leysen et

Table 2.4 Relative order of potency of serotonergic antagonists on presynaptic serotonin autoreceptor

(A)	(B)	(C)	(D)	(E)	(F)
Methiothepin	Methiothepin	Methiothepin	Methiothepin	Methiothepin	Metergoline
Metergoline	Quipazine	Quipazine	Cinanserin	Metergoline	Methiothepin
Quipazine	Metergoline	Metergoline*	Mianserin	Mianserin	Cyproheptadine
Cinanserin		Cyproheptadine*	Cyproheptadine*	Methysergide	Methysergide
Methysergide*		Cinanserin*	Metergoline*	Cinanserin*	Cinanserin*
Mianserin*		Mianserin*	Methysergide*	Cyproheptadine*	
Pizotyline*		Methysergide*			
Cyproheptadine*		Pizotyline*			

(A) Drug effects on stimulation-induced [³H]-overflow from rat cortical slices pre-labelled with [³H]-serotonin; (B) Influence of drugs on the concentration-response curve of unlabelled serotonin for its inhibitory effect on the electrically evoked [³H]-serotonin overflow from rat brain cortex slices; (C) Effect of drugs on the inhibition of potassium-induced [³H]-serotonin release by 30 nM serotonin from rat hypothalamic synaptosomes; (D) Effect of drugs on the inhibition of potassium-evoked release of [³H]-serotonin by 1 μM serotonin from rat cortical slices; (E) Effect of drugs on the inhibition of potassium-evoked [³H]-serotonin release by 5-methoxytryptamine from slices of raphe nuclei; (F) Drug effects on the inhibition of potassium-evoked [³H]-serotonin release by serotonin from slices of rat hypothalamus. Compounds are listed in order of potency. The first compound is the most potent.*, inactive. (Data from: (A) Baumann and Waldmeier (1981); (B) Göthert (1980 and Schlicker and Göthert (1981); (C) Martin and Sanders-Bush (1982b); (D) Mounsey et al. 1982); (E) Ennis and Cox (1982); (F) Cox and Ennis (1982).

Table 2.5 Comparison of antagonist potencies at serotonin receptors

	(A)		(B)		(C)		(D)
	K_i (nM)		K_i (nM)		IC_{50} (nM)		IC_{50} (nM)
	5-HT$_1$	5-HT$_2$	5-HT$_1$	5-HT$_2$	5-HT$_1$	5-HT$_2$	5-HT$_1$
Metergoline	20	0.9	9.9	2.1	4.68	3.41	5.8
Methiothepin	62	1.9	300	4.1	22.9	1.03	463
Methysergide	99	12	150	3.1	61.7	8.13	—
Cyproheptadine	700	6.5	1100	2.4	3650	6.68	6000
Cinanserin	3500	41	1800	15	1310	28.5	236
Mianserin	1100	13	—	—	912	12.7	647
Quipazine	—	—	—	—	322	2430	365
Pizotyline	1500	6.5	—	—	1740	4.10	287

5-HT$_1$ receptors were labelled with [^{3}H]-serotonin; 5-HT$_2$ receptors were labelled with [^{3}H]-spiperone. (A) 5-HT$_1$ in rat hippocampal membranes, 5-HT$_2$ in rat frontal cortex membranes; (B) Rat cortical membranes; (C) 5-HT$_1$ in rat hypothalamic membranes, 5-HT$_2$ in rat cortical membranes; (D) 5-HT$_1$ in rat forebrain membranes. Data from: (A) Leysen *et al.* (1981); (B) Peroutka *et al.* (1981); (C) Martin and Sanders-Bush (1982*a*); (D) Nelson *et al.* (1980).

al. 1981; Peroutka, Lebovitz and Snyder 1981; Martin and Sanders-Bush 1982*a*). Cinanserin was by far less potent than methiothepin in some studies (Leysen *et al.* 1981; Peroutka *et al.* 1981; Martin and Sanders-Bush 1982*a*), whereas in another study (Nelson *et al.* 1980) it was more potent. It is also surprising that metergoline which is quite potent at inhibiting [^{3}H]-serotonin binding is so weak at antagonizing serotonin autoreceptors. The same comment can also be made about methysergide. From the results of binding studies it is clear that all antagonists tested are more potent as antagonists on 5-HT$_2$ binding sites than on 5-HT$_1$ binding sites, even methiothepin. Thus it is the data obtained with agonists rather than with antagonists which suggests similarities between the presynaptic serotonin autoreceptors and the postsynaptic 5-HT$_1$ binding sites of the rat brain (see Fozard 1983). The conclusion on the characterization of the serotonin autoreceptor is that it is very difficult to make a conclusion. The great variety of effects of drugs may be due to use of different rat strains, brain regions, tissue preparations (i.e. slices v. synaptosomes). It can be also argued that results from binding studies do not have any pharmacological relevance since with this method no pharmacological response is measured, in contrast to results from release studies where responses are linked with the presence of a receptor and not only with a binding site.

The classification of serotonin receptors into 5-HT$_1$ and 5-HT$_2$ could represent an oversimplification and there might be at least three and probably more serotonin receptor types in the brain (Middlemiss 1982). Deshmukh, Nelson and Yamamura (1982) have determined two serotonin receptor sub-types in rat brain, 5-HT_{1A} and 5-HT_{1B} and recently, Middlemis and Fozard (1983) have shown that 8-hydroxy-2-(di-n-propyl-amino)-tetralin (8-OH-DPAT; Fig. 10.5), which is a centrally active serotonin agonist (Hjorth, Carlsson, Lindberg, Sanchez, Wikström, Arvidsson, Hacksell and Nilsson 1982), selectively displaces [^{3}H]-serotonin from the 5-HT_{1A} binding site in the rat frontal cortex. Gozlan, El Mestikawy, Pichat, Glowinski and Hamon (1983) have shown that in the striatum [^{3}H]-8-OH-DPAT binding sites exhibit a subcellular distribution and pharmacological characteristics usually associated with presynaptic autoreceptors. Furthermore, [^{3}H]-OH-DPAT binding sites in the striatum are lost after the degeneration of serotonergic fibres by 5,7-DHT (Gozlan *et al.* 1983). Further studies should help to clarify the nature of the serotonin autoreceptor. Selective antagonists with high affinity for presynaptic serotonin autoreceptors are also essential tools for further defining the pharmacology of this receptor.

A final point to be borne in mind when considering the characteristics of the serotonin autoreceptor is the potent effect of the putative 5-HT_1 agonist RU 24969 on serotonin synthesis. Most serotonin agonists have only a modest action in decreasing serotonin synthesis in the brain, even at high doses. In contrast, both RU 24969 and 8-OH-DPAT decrease synthesis by about 50 per cent in both rat and mouse brain at doses around 2.5–5 mg/kg (Euvrard and Boissier, 1980; Green, Guy and Gardner 1984; Green and Goodwin, unpublished).

2.6   Roles played by the serotonin autoreceptor

2.6.1   *Interaction with hormones*

Serotonin has been reported to have a stimulatory influence on prolactin release (Clemens, Sawyer and Cerimele 1977; Clemens and Roush 1982). Various manipulations of the serotonergic neuronal systems have been reported to alter prolactin release. For example, lesions of the dorsal raphe reduces prolactin release, while electrical stimulation of the dorsal raphe increases it (Advis, Simpkins, Bennett and Meites 1979). Clemens and Roush (1982) have shown that prolactin release in response to L-5-HTP or morphine, was significantly reduced by pre-treatment of rats with 5-MeODMT. They concluded that 5-MeODMT acts as a presynaptic serotonin autoreceptor agonist and not as a postsynaptic serotonin agonist on the neuronal systems that control prolactin release.

2.6.2   *Effect of* in vivo *drug treatment*

Up to now few studies have been carried out to determine the effect of a chronic drug treatment on the serotonin autoreceptor. Blier and de

Montigny (1980) have shown that tricyclic antidepressant drugs, such as imipramine and desipramine, and atypical antidepressants such as iprindole and femoxetine, administered daily for 14 days, altered neither the responsiveness of dorsal raphe neurones to intravenous injection of LSD nor the effectiveness of microiontophoretic applications of serotonin and LSD. Furthermore, these treatments did not change the mean firing rate of these serotonin neurones. These results suggested that chronic treatment with antidepressant drugs does not alter the sensitivity of the serotonin autoreceptor.

De Montigny and Aghajanian (1978) have reported that the responsiveness of postsynaptic neurones in rat forebrain to serotonin is increased by a long-term administration of tricyclic antidepressant drugs. These two findings further underline the different properties of the pre- and postsynaptic serotonin receptors. Therefore the functioning of the serotonin autoreceptor would not be implicated in the mechanism of action of currently used antidepressants. It has been shown by Hagan and Hughes (1983) that neither acute nor chronic treatment with methiothepin has any effect on serotonin autoreceptors. They studied the electrically evoked [^{3}H]-serotonin release from rat hypothalamic slices. Neither [^{3}H]-serotonin accumulation nor the stimulation-evoked overflow of [^{3}H]-serotonin was modified by acute or chronic methiothepin pre-treatment. Furthermore under these conditions, serotonin autoreceptor sensitivity to exogenous serotonin or exogenous methiothepin was also unmodified.

It would seem that nerve terminal serotonin autoreceptors in rat hypothalamus and somatodendritic serotonin autoreceptors in the dorsal raphe are resistant to pharmacological manipulations of serotonergic transmission *in vivo* in contrast to postsynaptic serotonin sites in the central nervous system which are affected by chronic exposure to serotonin antagonists (Samanin, Mennini, Ferraris, Bendotti and Borsini 1980).

2.6.3   Interaction with the serotonin uptake site

Figure 2.6 shows that the selective inhibitor of serotonin uptake, citalopram, prevents the inhibitory effect of LSD on electrically evoked [^{3}H]-serotonin release *in vitro* from rat hypothalamic slices (Langer and Moret 1982). The fact that in contrast to methiothepin, citalopram did not, by itself, increase the overflow of [^{3}H]-serotonin elicited by electrical stimulation and that citalopram antagonized the effects of LSD in an apparent non-competitive manner (the curves are not parallel), would imply that citalopram is acting at a site other than the serotonin autoreceptor itself. Citalopram is a potent uptake inhibitor with an IC_{50} for the inhibition of [^{3}H]-serotonin uptake in the rat hypothalamus of 40 nM (Langer, Moret, Raisman, Dubocovich and Briley 1980*b*). The increased

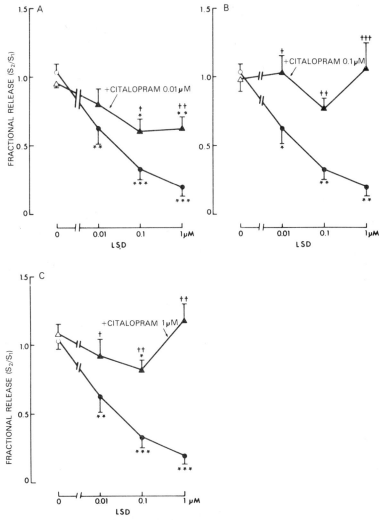

Fig. 2.6   Antagonism by citalopram (▲) of the inhibition by LSD (●) of the release of [³H]-serotonin elicited by electrical stimulation (3 Hz for 2 min, 20 mA, 2 ms) from slices of the rat hypothalamus. Open symbols are respective controls. Each point represents mean values ±S.E.M. of at least four experiments per group. *When compared to the corresponding control value, +when compared to the corresponding value obtained with LSD in the absence of citalopram. A, *p <0.05; **p <0.01; ***p <0.001; +p <0.05; ++p <0.005. B, *p <0.01; **p <0.001; +p <0.05; ++p <0.01; +++p <0.005. C, *p <0.05; **p <0.01; ***p <0.001; +p <0.05; ++p <0.005. (Reproduced with permission from Langer and Moret 1982. © 1982, Am. Soc. of Parenteral and Enteral Nutrition)

amount of amine present in the synaptic cleft produced by this drug might play a role by competing with LSD for the serotonin autoreceptor. If the complete antagonism of the effect of LSD by 0.1 to 1 μM citalopram could be explained by the fact that serotonin uptake was completely inhibited, it is not easy to conclude the same for 0.01 μM citalopram since [³H]-serotonin uptake was only weakly inhibited at this concentration, whereas the effects of LSD were still antagonized. Even if they could not exclude the possibility that there is an interaction between the amount of serotonin, which is increased by citalopram in the synaptic gap and the presynaptic serotonin autoreceptor, the authors were more attracted by the hypothesis of specific interaction between the presynaptic serotonin autoreceptor and the site of neuronal serotonin uptake. This finding may be important from the pharmacological, physiological and therapeutic point of view since many antidepressant drugs are inhibitors of neuronal uptake of serotonin. This is, however, *in vitro,* and several questions remain. What is the effect of a chronic treatment with citalopram? Is the antagonism still there? Even if the chronic treatment with methiothepin was ineffective (Hagan and Hughes 1983) on the serotonin autoreceptor it is worthwhile trying citalopram chronic treatment. As with all interactions, it was tempting to see if a reciprocal interaction existed. Namely does the autoreceptor modulate serotonin uptake? In reply to this question, Briley and Moret (1983) have shown that stimulation or blockade of the autoreceptor does not alter the serotonin uptake process. In fact, neither LSD nor methiothepin modify the potency of citalopram on serotonin uptake. Thus if any interaction exists between the autoreceptor and the serotonin uptake site, the functional control is unidirectional.

2.7   Conclusions

It is clear that serotonin autoreceptors exist in the central nervous system. It is well established that these receptors are located on the presynaptic side of the serotonergic neuronal synapse. The characterization of the serotonin autoreceptor remains, however, unclear. It is only possible to conclude that it resembles the 5-HT$_1$ postsynaptic receptor but it is still not possible to say if they are identical. In order to clarify this question, it is necessary to find new selective drugs, especially presynaptic serotonin autoreceptor antagonists. Many antidepressants are inhibitors of neuronal uptake of serotonin and it has been suggested that they act by increasing the amount of the neurotransmitter in the synaptic gap (however, see Chapters 6 and 12). If this hypothesis is true, a specific presynaptic serotonin antagonist, by increasing serotonin release, might, as Göthert (1982) has suggested, play the same therapeutic role. To test this hypothesis new potent and selective autoreceptor antagonists need to be developed for clinical trials. This is only one example of a possible role for a drug acting at the serotonin auto-receptor in mental illness. Wherever serotonin has been implicated, in other mental illnesses, in neuroendocrine conditions, in pain, sleep, appetite, sexual behaviour, etc. . . . a possible therapeutic role for drugs acting on the serotonin autoreceptor should not be overlooked.

The main questions remaining to be answered concern the physiological role, if any, played by the presynaptic serotonin autoreceptor. *In vitro* and *in vivo* its effects are clear. But physiologically is the serotonin autoreceptor efficient to the same extent in all different regions of the brain? Is the serotonin autoreceptor stimulated under normal levels of firing and release or only in disease states such as depression, for example? In addition, the serotonergic autoreceptor must be placed in its context within the mosaic of neurotransmission control. For example, inhibitory α-adrenoceptors present on the serotonergic nerve terminals (Göthert and Huth 1980; Frankhuyzen and Mulder 1980) undoubtedly also play a role in the control of serotonin release.

The therapeutic potential of autoreceptors in general and of the serotonin autoreceptor in particular, is not yet fully appreciated. In view, however, of the wide involvement of serotonin in central nervous system disorders one can imagine that it is only a question of time before serotonin autoreceptors are therapeutically exploited.

Acknowledgements

The author wishes to thank Mike Briley for his constructive criticism and Martine Dehaye for typing the manuscript.

References

Advis, J. P., Simpkins, J. W., Bennett, J. and Meites, J. (1979). Serotonergic control of prolactin release in male rats. *Life Sci.* **24**, 359.

Arbilla, S. and Langer, S. Z. (1981). Stereoselectivity of presynaptic auto-receptors modulating dopamine release. *Eur. J. Pharmac.* **76**, 345.

Arrang, J. M., Garbarg, M. and Schwartz, J. C. (1983). Auto-inhibition of brain histamine release mediated by a novel class (H_3) of histamine receptor. *Nature* **302**, 832.

Baumann, P. A. and Waldmeier, P. C. (1981). Further evidence for negative feedback control of serotonin release in the central nervous system. *Naunyn-Schmiedeberg's Arch. Pharmac.* **317**, 36.

Bennett, J. P. and Snyder, S. H. (1976). Serotonin and lysergic acid diethyl-amide binding in rat brain membranes: relationship to postsynaptic serotonin receptors. *Molec. Pharmac.* **12**, 373.

Blackshear, M. A., Steranka, L. R. and Sanders-Bush, E. (1981). Multiple serotonin receptors: regional distribution and effect of raphe lesions. *Eur. J. Pharmac.* **76**, 325.

Blier, P. and de Montigny, C. (1980). Effect of chronic tricyclic antidepressant treatment on the serotoninergic autoreceptor. *Naunyn.Schmiedeberg's Arch. Pharmac.* **314**, 123.

Bourgoin, S., Artaud, F., Enjalbert, A., Héry, F., Glowinski, J. and Hamon, M. (1977). Acute changes in central serotonin metabolism induced by the blockade or stimulation of serotoninergic receptors during ontogenesis in the rat. *J. Pharmac. Exp. Ther.* **202**, 519.

—— Soubrié, P., Artaud, F., Reisine, T. D. and Glowinski, J. (1981). Control of 5HT release in the caudate nucleus and the substantia nigra of the cat. *J. Physiol (Paris)* **77**, 303.

Bradley, P. B. and Briggs, I. (1974). Further studies on the mode of action of psychotomimetic drugs: antagonism of the excitatory actions of 5-hydroxytryptamine by methylated derivatives of tryptamine. *Br. J. Pharmac.* **50**, 345.

Briley, M. and Moret, C. (1983). Is 5HT uptake regulated by the 5HT autoreceptor? *Br. J. Pharmac.* **80**, 673P.

Cerrito, F. and Raiteri, M. (1979). Serotonin release is modulated by presynaptic autoreceptors. *Eur. J. Pharmac.* **57**, 427.

—— and Raiteri, M. (1980). Presynaptic autoreceptors control serotonin release from central nerve endings. *Pharmac. Res. Commun.* **12**, 593.

Chase, T. N., Katz, R. I. and Kopin, I. J. (1969). Release of ^{3}H-serotonin from brain slices. *J. Neurochem.* **16**, 607.

Clemens, J. A. and Roush, M. E. (1982). Inhibition of prolactin release by stimulation of presynaptic serotonin autoreceptors. *Life Sci.* **31**, 2641.

—— Sawyer, B. D. and Cerimele, B. (1977). Further evidence that serotonin is a neurotransmitter involved in the control of prolactin secretion. *Endocrinology* **100**, 692.

Collard, K. J., Cassidy, D. M., Pye, M. A. and Taylor, R. M. (1981). The stimulus-induced release of unmetabolized 5-hydroxytryptamine from superfused rat brain synaptosomes. *J. Neurosci. Methods.* **4**, 163.

Cox, B. and Ennis, C. (1982). Characterization of 5-hydroxytryptaminergic autoreceptors in the rat hypothalamus. *J. Pharm. Pharmac.* **34**, 438.

Da Prada, M., Saner, A., Burkard, W. P., Bartholini, G. and Pletscher, A. (1975). Lysergic acid diethylamide: evidence for stimulation of cerebral dopamine receptors. *Brain Res.* **94**, 67.

de Montigny, C. and Aghajanian, G. K. (1977). Preferential action of 5-methoxytryptamine and 5-methoxydimethyltryptamine on presynaptic serotonin receptors: a comparative iontophoretic study with LSD and serotonin. *Neuropharmacology* **16**, 811.

—— and Aghajanian, G. K. (1978). Tricyclic antidepressants: Long-term treatment increases responsivity of rat forebrain neurons to serotonin.

Science **202,** 1303.

Deshmukh, P. P., Nelson, D. L. and Yamamura, H. I. (1982). Localization of 5HT₁ receptor subtypes in rat brain by autoradiography. *Fed. Proc.* **41,** 6238.

Ennis, C. and Cox, B. (1982). Pharmacological evidence for the existence of two distinct serotonin receptors in rat brain *Neuropharmacology* **21,** 41.

Euvrard, L. and Boissier, J. R. (1980). Biochemical assessment of the central agonist activity of RU 24969 (a piperidinyl-indole). *Eur. J. Pharmac.* **63,** 65.

Farnebo, L. O. (1971). Release of monoamines evoked by field stimulation. Studies on some ionic and metabolic requirements. *Acta Physiol. Scand.* **371,** 19.

—— and Hamberger, B. (1971). Drug-induced changes in the release of ³H-monoamines from field stimulated rat brain slices. *Acta Physiol. Scand.* **371,** 35.

—— —— (1974). Regulation of ³H-5-hydroxytryptamine release from rat brain slices. *J. Pharm. Pharmac.* **26,** 642.

Filinger, E. J., Langer, S. Z., Perec, C. J. and Stefano, F. J. E. (1978). Evidence for the presynaptic location of the alpha-adrenoceptors which regulate noradrenaline release in the rat submaxillary gland. *Naunyn-Schmiedeberg's Arch. Pharmac.* **304,** 21.

Fillion, G., Beaudoin, D., Rousselle, J. C., Deniau, J. M., Fillion, M. P., Dray, F. and Jacob, J. (1979). Decrease of ³H-5HT high affinity binding and 5HT adenylate cyclase activation after kainic acid lesion in rat brain striatum. *J. Neurochem.* **33,** 567.

Fozard, J. R. (1983). Functional correlates of 5HT₂ recognition sites. *Trends Pharmac. Sci.* **4,** 288.

—— and Mwaluko, G. M. P. (1976). Mechanism of the indirect sympathomimetic effect of 5-hydroxytryptamine on the isolated heart of the rabbit. *Br. J. Pharmac.* **57,** 115.

—— and Mobarok Ali, T. T. M. (1978). Receptors for 5-hydroxytryptamine on the sympathetic nerves of the rabbit heart. *Naunyn-Schmiedeberg's Arch. Pharmac.* **301,** 223.

Frankhuyzen, A. L. and Mulder, A. H. (1980). Noradrenaline inhibits depolarization-induced ³H-serotonin release from slices of rat hippocampus. *Eur. J. Pharmac.* **63,** 179.

Galzin, A. M., Moret, C. and Langer, S. Z. (1984). Evidence that exogenous but not endogenous norepinephrine activates the presynaptic alpha₂-adrenoceptors on serotonergic nerve endings in the rat hypothalamus. *J. Pharmac. Exp. Ther.* **228,** 725.

Gillespie, J. S. and McGrath, J. C. (1975). The effects of lysergic acid diethylamide on the response to field stimulation of the rat vas deferens and the rat and cat anococcygeus muscles. *Br. J. Pharmac.* **54,** 481.

Göthert, M. (1980). Serotonin-receptor-mediated modulation of Ca^{2+}-dependent 5-hydroxytryptamine release from neurones of the rat brain cortex. *Naunyn-Schmiedeberg's Arch. Pharmac.* **314,** 223.

—— (1982). Modulation of serotonin release in the brain via presynaptic receptors. *Trends Pharmac. Sci.* **3,** 437.

—— and Dührsen, U. (1979). Effects of 5-hydroxytryptamine and related compounds on the sympathetic nerves of the rabbit heart. *Naunyn-Schmiedeberg's Arch. Pharmac.* **308,** 9.

—— and Weinheimer, G. (1979). Extracellular 5-hydroxytryptamine inhibits 5-hydroxytryptamine release from rat brain cortex slices. *Naunyn-Schmiedeberg's Arch. Pharmac.* **310,** 93.

—— and Huth, H. (1980). Alpha-adrenoceptor-mediated modulation of 5-hydroxytryptamine release from rat brain cortex slices. *Naunyn-Schmiedeberg's Arch. Pharmac.* **313,** 21.

—— and Schlicker, E. (1983). Autoreceptor-mediated inhibition of ^{3}H-5-hydroxytryptamine release from rat brain cortex slices by analogues of 5-hydroxytryptamine. *Life Sci.* **32,** 1183.

Gozlan, H., El Mestikawy, S., Pichat, L., Glowinski, J. and Hamon, M. (1983). Identification of presynaptic serotonin autoreceptors using a new ligand: ^{3}H-PAT. *Nature* **305,** 140.

Green, A. R., Guy, A. P. and Gardner, C. R. (1984). The behavioural effects of RU 24969, a suggested 5-HT$_1$ receptor agonist, and the effect on the behaviour of treatment with antidepressants. *Neuropharmacology* **23,** 655.

—— Youdim, M. B. H. and Grahame-Smith, D. G. (1976). Quipazine: its effects on rat brain 5-hydroxytryptamine metabolism, monoamine oxidase activity and behaviour. *Neuropharmacology* **15,** 173.

Haefely, W. (1974). The effects of 5-hydroxytryptamine and some related compounds on the cat superior cervical ganglion in situ. *Naunyn-Schmiedeberg's Arch. Pharmac.* **281,** 145.

Hagan, R. M. and Hughes, I. E. (1983). Lack of effect of chronic methiothepin treatment on 5-hydroxytryptamine autoreceptors. *Br. J. Pharmac.* **80,** 513 p.

Haigler, H. J. and Aghajanian, G. K. (1977). Serotonin receptors in the brain. *Fed. Proc.* **36,** 2159.

Hamon, M., Bourgoin, S., Jagger, J. and Glowinski, J. (1974). Effects of LSD on synthesis and release of 5HT in rat brain slices. *Brain Res.* **69,** 265.

—— Nelson, D. L., Herbet, A. and Glowinski, J. (1980). Multiple receptors for serotonin in the rat brain. In *Receptors for neurotransmitters and peptide hormones.* (eds. G. Pepeu, M. J. Kuhar and S. J. Enna) p. 223. Raven Press, New York.

Hertting, G., Reimann, W., Zumstein, A., Jackisch, R. and Starke, K. (1979). Dopaminergic feedback regulation of dopamine release in slices of the caudate nucleus of the rabbit. In *Advances in the biosciences: pre-synaptic receptors* (eds. S. Z. Langer, K. Starke and M. L. Dubocovich) p. 145. Pergamon Press, Oxford.

Héry, F. and Ternaux, J. P. (1981). Regulation of release processes in central serotoninergic neurons. *J. Physiol. (Paris)* **77**, 287.

—— Simonnet, G., Bourgoin, S., Soubrié, P., Artaud, F., Hamon, M. and Glowinski, J. (1979). Effect of nerve activity on the in vivo release of ^{3}H-serotonin continuously formed from L-^{3}H-tryptophan in the caudate nucleus of the cat. *Brain Res.* **169**, 317.

Hjorth, S., Carlsson, A., Lindberg, P., Sanchez, D., Wikström, H., Arvidsson, L. E., Hacksell, U. and Nilsson, J. L. G. (1982). 8-Hydroxy-2-(di-n-propylamino)-tetralin, 8-OH-DPAT, a potent and selective simplified ergot congener with central 5HT receptor stimulating activity. *J. Neural Trans.* **55**, 169.

Homan, H. D. and Ziance, R. J. (1981). The effects of d-amphetamine and potassium on serotonin release and metabolism in rat cerebral cortex tissue. *Res. Commun. Chem. Path. Pharmac.* **31**, 223.

Kamal, L., Arbilla, S. and Langer, S. Z. (1979). Effects of GABA-receptor agonists on the potassium-evoked release of ^{3}H-GABA from the rat substantia nigra. In *Advances in the biosciences: presynaptic receptors* (eds. S. Z. Langer, K. Starke and M. L. Dubocovich) p. 193. Pergamon Press, Oxford.

Katz, R. I. and Kopin, I. J. (1969). Effect of D-LSD and related compounds on release of norepinephrine-H^3 and serotonin-H^3 evoked from brain slices by electrical stimulation. *Pharmac. Res. Commun.* **1**, 54.

Lane, J. D. and Aprison, M. H. (1977). Calcium-dependent release of endogenous serotonin, dopamine and norepinephrine from nerve end-ings. *Life Sci.* **20**, 665.

Langer, S. Z. (1977). Presynaptic receptors and their role in the regulation of transmitter release. *Br. J. Pharmac.* **60**, 481.

—— (1981). Presynaptic regulation of the release of catecholamines. *Pharmac. Rev.* **32**, 337.

—— and Moret, C. (1982). Citalopram antagonizes the stimulation by lysergic acid diethylamide of presynaptic inhibitory serotonin auto-receptors in the rat hypothalamus. *J. Pharmac. Exp. Ther.* **222**, 220.

—— Arbilla, S. and Kamal, L. (1980a). Autoregulation of noradrenaline and dopamine release through presynaptic receptors. In *Neurotrans-mitters and their receptors* (eds. U. Z. Littauer, Y. Dudai, I. Silman, V. I. Teichberg and Z. Vogel) p. 7. John Wiley, New York.

—— Moret, C., Raisman, R., Dubocovich, M. L. and Briley, M. (1980b). High-affinity ^{3}H-imipramine binding in rat hypothalamus: association with uptake of serotonin but not of norepinephrine. *Science* **210**, 1133.

Lansdown, M. J. R., Nash, H. L., Preston, P. R., Wallis, D. I. and Williams, R. G. (1980). Antagonism of 5-hydroxytryptamine receptors by quipazine. *Br. J. Pharmac.* **68**, 525.

Leysen, J. E. (1981). Serotoninergic receptors in brain tissue: properties and identification of various ^{3}H-ligand binding sites in vitro. *J. Physiol. (Paris)* **77**, 351.

—— Awouters, F., Kennis, L., Laduron, P. M., Vandenberk, J. and Janssen, P. A. J. (1981). Receptor binding profile of R 41 468, a novel antagonist at $5HT_2$ receptors. *Life Sci.* **28**, 1015.

Lindmar, R. and Muscholl, E. (1965). Die Verstärkung der Noradrenaline-wirkung durch Tyramin. *Naunyn-Schmiedeberg's Arch. Exp. Path. Pharmac.* **252**, 122.

Martin, L. L. and Sanders-Bush, E. (1982*a*). Comparison of the pharmacological characteristics of $5HT_1$ and $5HT_2$ binding sites with those of serotonin autoreceptors which modulate serotonin release. *Naunyn-Schmiedeberg's Arch. Pharmac.* **321**, 165.

—— and Sanders-Bush, E. (1982*b*). The serotonin autoreceptor: antagonism by quipazine. *Neuropharmacology* **21**, 445.

Middlemiss, D. N. (1982). Multiple 5-hydroxytryptamine receptors in the central nervous system of the rat. In *Presynaptic receptors: mechanism and function* (ed. J. De Belleroche) p. 46. Ellis Horwood, Chichester.

—— and Fozard, J. R. (1983). 8-hydroxy-2-(di-n-propylamino)-tetralin discriminates between subtypes of the $5HT_1$ recognition site. *Eur. J. Pharmac.* **90**, 151.

Mounsey, I., Brady, K. A., Carroll, J., Fisher, R. and Middlemiss, D. N. (1982). K$^+$-evoked ^{3}H-5HT release from rat frontal cortex slices: the effect of 5HT agonists and antagonists. *Biochem. Pharmac.* **31**, 49.

Mulder, A. H., Van Den Berg, W. B. and Stoof, J. C. (1975). Calcium-dependent release of radiolabelled catecholamines and serotonin from rat brain synaptosomes in a superfusion system. *Brain Res.* **99**, 419.

—— de Langen, C. D. J., de Regt, V. and Hogenboom, F. (1978). Alpha-receptor-mediated modulation of ^{3}H-noradrenaline release from rat brain cortex synaptosomes. *Naunyn-Schmiedeberg's Arch. Pharmac.* **303**, 193.

Neckers, L. M., Bertilsson, L. and Costa, E. (1976). The action of fenfluramine and *p*-chloramphetamine on serotonergic mechanisms: a comparative study in rat brain nuclei. *Neurochem. Res.* **1**, 29.

Nelson, D. L., Herbet, A., Bourgoin, S., Glowinski, J. and Hamon, M. (1978). Characteristics of central 5HT receptors and their adaptive changes following intracerebral 5,7-dihydroxytryptamine administration in the rat. *Molec. Pharmac.* **14**, 983.

—— Herbet, A., Enjalbert, A., Bockaert, J. and Hamon, M. (1980).

Serotonin-sensitive adenylate cyclase and ^{3}H-serotonin binding sites in the CNS of the rat–I. Kinetic parameters and pharmacological properties. *Biochem. Pharmac.* **29**, 2445.

Peroutka, S. J. and Snyder, S. H. (1979). Multiple serotonin receptors: differential binding of ^{3}H-5-hydroxytryptamine, ^{3}H-lysergic acid diethylamide and ^{3}H-spiroperidol. *Molec. Pharmac.* **16**, 687.

—— and Snyder, S. H. (1982). Recognition of multiple serotonin receptor binding sites. *Adv. Biochem. Psychopharmac.* **34**, 155.

—— Lebovitz, R. M. and Snyder, S. H. (1981). Two distinct central serotonin receptors with different physiological functions. *Science* **212**, 827.

Raiteri, M., Angelini, F. and Levi, G. (1974). A simple apparatus for studying the release of neurotransmitters from synaptosomes. *Eur. J. Pharmac.* **25**, 411.

Rogawski, M. A. and Aghajanian, G. K. (1981). Serotonin autoreceptors on dorsal raphe neurons: structure-activity relationships of tryptamine analogs. *J. Neuroscience* **1**, 1148.

Samanin, R., Mennini, T., Ferraris, A., Bendotti, C. and Borsini, F. (1980). Hyper- and hyposensitivity of central serotonin receptors: ^{3}H-serotonin binding and functional studies in the rat. *Brain Res.* **189**, 449.

Schlicker, E. and Göthert, M. (1981). Antagonistic properties of quipazine at presynaptic serotonin receptors and α-adrenoceptors in rat brain cortex slices. *Naunyn-Schmiedeberg's Arch. Pharmac.* **317**, 204.

Shaskan, E. G. and Snyder, S. H. (1970). Kinetics of serotonin accumulation into slices from rat brain: relationship to catecholamine uptake. *J. Pharmac. Exp. Ther.* **175**, 404.

Starke, K. (1977). Regulation of noradrenaline release by presynaptic receptor systems. *Rev. Physiol. Biochem. Pharmac.* **77**, 1.

—— (1981). Presynaptic receptors. *Ann. Rev. Pharmac. Toxicol.* **21**, 7.

Szerb, J. C. (1979). Autoregulation of acetylcholine release. In *Advances in the biosciences: presynaptic receptors* (eds. S. Z. Langer, K. Starke and M. L. Dubocovich) p. 293. Pergamon Press, Oxford.

Taube, H. D., Starke, K. and Borowski, E. (1977). Presynaptic receptor systems on the noradrenergic neurones of rat brain. *Naunyn-Schmiedeberg's Arch. Pharmac.* **299**, 123.

Von Hungen, K., Roberts, S. and Hill, D. F. (1974). LSD as an agonist and antagonist at central dopamine receptors. *Nature* **252**, 588.

Wallis, D. I. and Woodward, B. (1975). Membrane potential changes induced by 5-hydroxytryptamine in the rabbit superior cervical ganglion. *Br. J. Pharmac.* **55**, 199.

Westfall, T. C. (1977). Local regulation of adrenergic neurotransmission. *Physiol. Rev.* **57**, 659.

Whitaker, P. M. and Deakin, J. F. W. (1981). Does ^{3}H-serotonin label presynaptic receptors in rat frontal cortex? *Eur. J. Pharmac.* **73**, 349.

3

Imipramine binding: its relationship with serotonin uptake and depression

MIKE BRILEY

3.1 Introduction

The technique of ligand building has introduced a new term of abuse into pharmacology. I imagine that I am not alone in having been referred to as a 'grinder and binder' by so-called 'real pharmacologists'. Although the term was probably coined largely out of jealousy for the rapid, cheap and precise methodology involved in ligand binding, the implied criticism is not unfounded. The very simplicity and rapidity of ligand binding has led certain researchers to shy away from the more fundamental and technically more demanding problems of the function of the various binding sites they study.

The problem of the binding site without a function is most acute in the case of receptor sub-types defined upon the basis of binding studies and of binding sites for clinically active drugs. The first description, in 1979, of a specific high-affinity binding site for [3H]-imipramine (Raisman, Briley and Langer 1979 *a, b*) immediately posed the problem of what functional role, if any, could be attributed to this new binding site. In an attempt to answer, at least partially, the criticism of being 'grinders and binders', Langer and the author (Langer and Briley 1981) listed the classical criteria for the definition of a specific site of drug action (or receptor) and compared the data available at that time for the [3H]-imipramine binding site (Table 3.1). Even then the list was sufficient to convince us that [3H]-imipramine bound to something more than a non-functional binding site. Evidence accumulated since that time has reinforced the original idea that the [3H]-imipramine binding site plays some functional role and that it is possibly important in the physiopathology of affective disorders.

This chapter traces the development of [3H]-imipramine binding and the attempts to determine its functional significance.

Table 3.1 Criteria for the binding of a radioligand to a specific site of drug action (or a receptor) and the properties of specific ^{3}H-imipramine binding (Langer and Briley 1981).

Criteria	Properties of ^{3}H-imipramine binding
Binding parameters	
High-affinity	$K_D = 4$ nM
Saturable	Yes—Hill coefficient = 0.97
Limited No. of sites (usually (usually 10–100 pmol/g tissue	B_{max} (hypothalamus) = 16 pmol/mg tissue
Rapid kinetics	$T_{1/2} = 5$ min at 0°C and 3.5 nM ^{3}H-imipramine
Affinity from kinetic constants and equilibrium constants should be the same.	K_D (kinetic) = 6.8 nM K_D (equilibrium) = 4.0 nM
Distribution	
Asymmetrical regional distribution	Fivefold difference between the richest and poorest brain regions studied
Asymmetrical tissue distribution	Found only in brain and platelets
Cellular distribution	Not found in glial cells
Selectivity	
Pharmacological selectivity	Only tricyclic antidepressants and 5-HT uptake blockers inhibit the binding with high-affinity.
Stereoselectivity	Z-forms of zimelidine and 10-OH tricyclic antidepressants inhibit with much greater affinity than E-forms
Sensitivity to ions	Sensitive to Na^+ ions
Functional correlations	
Correlation with pharmacological effects	Positive correlation with the inhibition of 5-HT uptake
Correlation with clinicals effects	Positive correlation with mean clinical doses of tricyclic antidepressants
Correlation with pathological conditions	^{3}H-imipramine binding is decreased in the platelets of untreated depressed patients
Existence of agonists and antagonists	Not known
Changed by chronic treatments	Binding decreased by chronic tricyclic antidepressants, chronic electro-convulsive shock and chronic REM sleep deprivation

3.2   Pharmacology of imipramine binding

3.2.1   *Binding characteristics*

Once the importance of carrying out the incubation of brain membranes and [^{3}H]-imipramine at 0°C was realized the high affinity binding of [^{3}H]-imipramine showed itself to have fairly classical binding characteristics. Scatchard plots indicate a single population of high-affinity sites (K_D 4–5 nM) with a maximal binding in the rat cerebral cortex of about 250 fmol/mg protein (Raisman *et al.* 1979*a, b*). Bound [^{3}H]-imipramine can be rapidly displaced by an excess of desipramine, the dissociation constant calculated from the kinetic constants being similar to that found by Scatchard analysis (Raisman, Briley and Langer 1980). Recently a low affinity (micromolar) site has also been described (Reith, Sershen, Allen and Lajtha 1983) but its importance, if any, remains to be established.

3.2.2   *Biochemical characterization*

Insights into the chemical nature of the binding site have come from bio-chemical studies (Kinnier, Chuang, Gwynn and Costa 1981; Wennogle, Beer and Meyerson 1981) which show that [^{3}H]-imipramine binding may be decreased by incubation with proteolytic enzymes such as trypsin and pronase and by phospholipase A_2. Calcium ions (Ca^{2+}) seem to play an important role in the spontaneous decrease in the number of binding sites found after incubation at 37°C. This latter phenomenon probably involves a Ca^{2+}-dependent protease (Kinnier *et al.* 1981). In addition certain authors have described a temperature-sensitive reversible loss of [^{3}H]-imipramine binding sites (Davis, Morris and Tang 1983*a*; Dumbrille-Ross, Morris, Davis and Tang 1983). This was interpreted as possibly involving a conformational change.

The use of specific chemical reagents (Wennogle *et al.* 1981) suggested that the imipramine binding site is a protein with no disulphide bridges or free amino groups at its recognition site. Free sulphydryl groups, however, appear to be important for binding activity.

3.2.3   *Solubilization*

Attempts to solubilize [^{3}H]-imipramine binding sites with triton X-100, strong salt solutions or various detergents have in general not been success-ful (Talvenheimo and Rudnick 1980; Kinnier *et al.* 1981), digitonin being the only detergent found to solubilize the sites (Talvenheimo and Rudnick 1980; Davis, Morris and Tang 1983*b*). The pharmacological character-istics of the solubilized sites appeared to be very similar to those of the membrane-bound form.

3.2.4   *Other antidepressant ligands*

Following the description of specific high affinity sites for [^{3}H]-imipramine a number of other antidepressant ligands have been introduced. Of these only [^{3}H]-nitroimipramine, a slowly reversible ligand (see Section 3.4.1) (Rehavi, Skolnick and Paul 1983*a*), [^{3}H]-norzimelidine (Hall, Ross, Ogren and Gawell 1982) and [^{3}H]-Ro 11-2465 (Burkard 1980) seem to bind to the same site as [^{3}H]-imipramine. Recent reports (Davis, Dumbrille-Ross and Tang 1983*c*; Dumbrille-Ross and Tang 1983) have suggested that [^{3}H]-Ro 11-2465 may, in fact, only bind to a sub-population of [^{3}H]-imipramine binding sites.

[^{3}H]-Desipramine binds to high affinity sites which are probably associated with noradrenaline uptake both in the brain (Hrdina 1981; Hrdina, Elson-Hartman, Robers and Pappas 1981; Lee and Snyder 1981; Lee, Jarich and Snyder 1982, 1983; Rehavi, Skolnick, Hulihan and Paul 1981*a*; Rehavi, Skolnick, Brownstein and Paul 1982*a*) and in the periphery (Langer, Raisman and Briley 1981*b*; Raisman, Sette, Pimoule, Briley and Langer 1982*a*). In many ways the binding of [^{3}H]-imipramine and [^{3}H]-desipramine appear analogous in their relationship with the two monoamine uptake systems. The lack of noradrenaline uptake and thus [^{3}H]-desipramine binding in platelets has not, however, permitted the exploitation of [^{3}H]-desipramine binding in clinical pharmacology as has been the case for [^{3}H]-imipramine binding (see Section 3.6.).

Other antidepressant ligands appear to bind to known neurotransmitter receptors. [^{3}H]-Doxepine binds to histamine-H$_1$ receptors (Taylor and Richelson 1982) while [^{3}H]-mianserin binds mainly to 5-HT$_2$ receptors (Dumbrille-Ross, Tang and Seeman (1980).

3.2.5   *Specificity*

The first indication that the [^{3}H]-imipramine binding site was functionally relevant came from studies of its specificity. [^{3}H]-Imipramine binding can be potently inhibited by most other tricyclic antidepressants (Raisman *et al.* 1980). The stereospecific nature of this inhibition has been demonstrated using a series of hydroxylated derivatives of amitryptyline and nortriptyline (Langer, Raisman and Briley 1980*a*). A significant correlation has been obtained between the mean clinical dose for a number of tricyclic antidepressants and their ability to inhibit [^{3}H]-impramine binding (Briley, Raisman, Sechler, Zarifian and Langer 1980*a*). This correlation has led certain authors to propose antidepressant activities for drugs on the basis of their ability to inhibit imipramine binding (De Monitis, Devoto and Tagliamonte 1982; Fulton, Norman and Burrows 1982). It is, however, now widely accepted that inhibition of [³H]-imipramine binding is neither necessary for, nor indicative of antidepressant activity.

Of the wide range of drugs tested for inhibition of [³H]-imipramine binding (Raisman *et al.* 1980; Langer, Briley, Raisman, Henry and Morselli 1980*b*) no single pharmacological class of drugs other than serotonin uptake inhibitors shows any significant potency. [³H]-imipramine binding sites can thus be clearly differentiated from any of the well characterized neurotransmitter receptors. Furthermore the large number of atypical antidepressants studied has shown that only those inhibiting the uptake of serotonin are potent inhibitors of [³H]-imipramine binding (Raisman *et al.* 1980; Langer *et al.* 1980*b*).

This indication of a possible association between the inhibition of serotonin uptake and the inhibition of [³H]-imipramine binding was rapidly followed up by the formal demonstration that in the rat hypothalamus the inhibition of [³H]-imipramine binding was highly correlated with the inhibition of serotonin uptake (Langer, Moret, Raisman, Dubocovich and Briley 1980*c*). This has subsequently also been shown to be true of mice (cortex: Reith *et al.* 1983) and man (platelets: Paul, Rehavi, Rice, Ittah and Skolnick 1981*a*).

3.3   Localization of imipramine binding

The localization of [³H]-imipramine binding sites corresponds closely to that expected for the serotonin uptake system as the following sections show.

3.3.1   *Tissue and regional distribution*

[³H]-Imipramine binding sites are unevenly distributed in the brain. The highest density of sites in the rat brain have been found in the hypothalamus and cortex and the lowest in the cerebellum (Raisman *et al.* 1979*b*). The distribution of sites in post-mortem human brain appears to be very similar (Rehavi, Paul, Skolnick and Goodwin 1980; Langer, J. Agid, Raisman, Briley and Y. Agid 1981*a*).

Initially it was reported (Langer, Zarifian, Briley, Raisman and Sechter 1982) that the only [³H]-imipramine binding sites to be found in the periphery were located on platelets. Recently, however, Langer and Raisman (1983) reported [³H]-imipramine binding in rat lung, although Kinnier, Chuang and Costa (1981) failed to detect these sites in an earlier study. In addition, Borbe and Zube (1983) found the bovine retina to be a rich source of [³H]-imipramine binding sites. These sites appeared to be the same as those found in the central nervous system.

The fact that the greatest density of [³H]-imipramine binding sites were found in serotonin-rich brain regions suggested a functional relationship with serotonin. Their similar regional distribution in rat brain was confirmed in a detailed study of 23 microdissected brain regions (Palkovits, Raisman, Briley and Langer 1981; see Fig. 3.1). A highly significant correlation was found between the density of [³H]-imipramine binding sites and the endogenous levels of serotonin. Interestingly in two regions, the dorsal raphe and the raphe magnus, where serotonin is localized in cell bodies instead of nerve endings there were proportionally less [³H]-imipramine binding sites (Fig. 3.1).

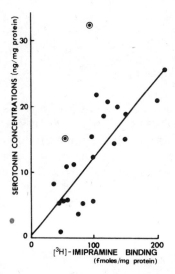

Fig. 3.1   Correlation between [³H]-imipramine binding at 3.5–4.0 nM and serotonin concentrations in 23 microdissected regions of rat brain. ⊙ represents regions (dorsal raphe and raphe magnus) where serotonin is considered to be located in serotonin cell bodies rather than nerve terminals as for other regions. Regression coefficient (excluding dorsal raphe and raphe magnus) r, 0.824; n, 21. (Reproduced with permission from Palkovits *et al.* 1981)

3.3.2   *Autoradiography*

Recent autoradiographic studies using either [³H]-imipramine (Rainbow, Biegon and McEwen 1982; Grabowsky, McCabe and Wamsky 1983) or the slowly dissociating ligand [³H]-nitroimipramine (see Section 3.4.1) (Rainbow and Biegon 1983) have confirmed the parallel distribution of [³H]-imipramine binding sites and serotonin nerve terminals in the rat brain.

3.3.3   *Platelets*

As soon as evidence suggested that [^{3}H]-imipramine binding was associated with serotonin-rich regions the possibility of their existence on blood platelets became an obvious question. Platelets take up, store and metabolize serotonin in a similar manner to synaptosomes or brain slices (Sneddon 1973; Stahl 1977). This similarity has led to their extensive use as a model for central serotonin nerve terminals.

Human platelets possess a high density of high-affinity [^{3}H]-imipramine binding sites with binding characteristics similar to those already described for rat and human brain (Briley, Raisman and Langer 1979; Paul, Rehavi, Skolnick and Goodwin 1980). A detailed comparison of the potency of a wide range of drugs for the inhibition of [^{3}H]-imipramine binding in rat brain, human platelets and human brain (Langer *et al.* 1980*b*; Langer *et al.* 1981*a*; Rehavi *et al.* 1980) has led to the conclusion that the [^{3}H]-imipramine binding sites from all three sources are almost certainly identical.

A study of [^{3}H]-imipramine binding in platelets from 35 healthy volunteers (Langer *et al.* 1980*b*) demonstrated that it was possible to reliably measure [^{3}H]-imipramine binding in platelets obtained from individual donors. Furthermore [^{3}H]-imipramine binding remained relatively constant when measured in samples taken from the same donor over a period of up to five weeks. It has therefore been feasible to study [^{3}H]-imipramine binding in depressed patients throughout the development of their depressive episode (see Section 3.6).   •

3.3.4   *Lesion studies*

The deviation of the two regions of the raphe nucleus from the correlation of serotonin content and [^{3}H]-imipramine binding site density (see Section 3.3.1) suggested that [^{3}H]-imipramine binding sites were associated with nerve terminals rather than the cell bodies. This was subsequently confirmed by studies in which the raphe nucleus was destroyed by sterotaxic electrolytic lesion (Sette, Raisman, Briley and Langer 1981). The density of [^{3}H]-imipramine binding sites in the cortex and hypothalamus was decreased in parallel with the serotonin content. This study was later confirmed (Dumbrille-Ross, Tang and Coscina 1981) and extended. Both Gross, Göthert, Ender and Schüman (1981) and Luine, Frankfurt, Rainbow, Biegon and Azmitia (1983), using the specific serotonin neurotoxin, 5,7-DHT obtained an extensive and parallel reduction in serotonin uptake and [^{3}H]-imipramine binding. Parallel experiments with the specific catecholamine neurotoxin, 6-hydroxydopamine produced only a slight reduction in serotonin uptake and left the number of [^{3}H]-imipramine binding sites virtually untouched. Similarly, Paul *et al.* (1981*a*), using electrolytic lesions showed an excellent correlation between the percent decrease of serotonin uptake and [³H]-imipramine binding when measured in individual hypothalami from lesioned animals.

Recently Hrdina, Pappas, Bialik and Ryan (1982) have shown that whereas neonatal administration of 5,7-DHT decreased serotonin levels in the hippocampus, the concentration in the pons-medulla was increased. In both cases, the density of [³H]-imipramine binding sites paralleled the changes in serotonin. A further study (Brunello, Chuang and Costa 1982*a*) introduced an interesting additional factor. These authors using 5,7-DHT again confirmed the above results in the cortex and hypothalamus but found that serotonin levels were decreased more extensively than the number of [³H]-imipramine binding sites in the hippocampus and striatum (Table 3.2). This difference, if not due to some experimental artefact, may indicate a postsynaptic location for some [³H]-imipramine binding sites in these regions. This interesting possibility seems, however, unlikely since the same authors subsequently published a study using kainic acid lesions of the postsynaptic neurones of the hippocampus (Brunello, Chuang and Costa 1982*b*). This postsynaptic lesion left the [³H]-imipramine binding sites unchanged whereas the postsynaptically located [³H]-mianserin binding sites were extensively reduced.

Table 3.2 Decrease in serotonin levels and [³H]-imipramine binding following 5,7-DHT lesions

	Decrease (%)	
	Serotonin	[³H]-Imipramine binding
Hypothalmus	73.7	80.5
Cortex	53.6	58.9
Hippocampus	56.7	33.2
Striatum	53.4	22.6

Calculated from Brunello *et al.* (1982*a*)

3.3.5   *Cellular distribution*

Examination of the density of [³H]-imipramine binding sites in neuronal and glial fractions prepared from horse striatum (Briley, G. Fillion, Beaudoin, M. P. Fillion and Langer 1980*b*) showed that the sites were concentrated in the neuronal fraction. The small amount of [³H]-imipramine binding found in the glial fraction probably resulted from a minor cross-contamination between the fractions. The absence of [³H]-imipramine binding in glial cells is consistent with the low levels of binding found in the corpus callosum (Palkovits *et al.* 1981), a region which consists primarily of glial cells. Thus although glial cells (Blömstrand and Hamberger 1970) and glial cultures (Suddith, Hutchison and Haber, 1978) have been shown to be capable of taking up serotonin this uptake mechanism does not appear to be associated with [^{3}H]-imipramine binding sites.

The existence of [^{3}H]-imipramine binding in N4TG1 neuroblastoma cells cultures (Kinnier *et al.* 1981) further confirmes the neuronal location of the site.

3.3.6 *Subcellular distribution*

Crude subcellular fractionation of the rat hippocampus showed [^{3}H]-imipramine binding to be associated principally with the synaptosomal fraction with very little binding in the nuclear mitochondrial or microsomal fractions (Kinnier *et al.* 1981). This distribution confirmed the now generally accepted idea that [^{3}H]-imipramine binding sites are located on serotonin nerve terminals. Indeed the section on subcellular distribution would have stopped here if another publication had not added a little confusion for a few months. In 1982 Laduron, Robbyns and Schotte published a study in which they reported that [^{3}H]-imipramine binding was principally located in the nuclear fraction. On the basis of this result and in spite of the accumulated evidence of the studies already described these authors stated 'that ... [^{3}H]-imipramine did not bind to the sites associated with 5-HT uptake; the subcellular fractionation ruled out the possibility that binding is localized, like uptake, on nerve terminals'.

This challenge led several groups, to go back to subcellular fractionation studies. Within a year Rehavi *et al.* (1983*a*) were able to explain this unexpected result. In measuring [^{3}H]-imipramine binding in the fractions at only one ligand concentration and with a very high concentration of unlabelled ligand to define non-specific binding, Laduron *et al.* (1982) had detected a large amount of low affinity non-saturable sodium-independent (see Section 3.4.1) [^{3}H]-imipramine binding in the nuclear fraction. In contrast, high-affinity, saturable, sodium-dependent [^{3}H]-imipramine binding was, principally located in the synaptosomal fraction together with the uptake of serotonin (Rehavi *et al.* 1983*a*).

Similarly Agid, Langer, Raisman, Ruberg, Scatton, Sette and Zivkovic (1983) contribued to the clarification of the problem by confirming that less than 5 per cent of the specific [^{3}H]-imipramine binding was to be found in the nuclear fraction. The majority of the [^{3}H]-imipramine binding was located with the uptake of serotonin in the heavy mitochondrial (synaptosomal) fraction. Furthermore after serotonergic denervation with either 5,7-DHT or a electrolytic lesion of the raphe nucleus the [^{3}H]-imipramine binding in this fraction was substantially decreased.

These authors (Agid *et al.* 1983) expressed surprise at their finding some (37 per cent) [^{3}H]-imipramine binding in the microsomal fraction but not uptake of serotonin and a very low concentration of serotonin. It is likely however that a certain quantity of synaptosomes were ruptured during the preparation leaving [^{3}H]-imipramine binding associated with synapto-somal membrane fragments. These fragments would of course be found in the microsomal fraction and lack any endogenous serotonin or serotonin uptake which would be found only in intact synaptosomes. This explana-tion seems all the more likely since in lesioned animals the amount of [^{3}H]-imipramine binding in the microsomal fraction was greatly reduced (Agid *et al.* 1983).

3.4 The relationship of the imipramine binding site with serotonin uptake

3.4.1 *Is there a functional relationship?*

It is now clear that [^{3}H]-imipramine binding sites are located in serotonin nerve terminals in the brain and on blood platelets in the periphery. The simple fact of the co-existence of [^{3}H]-imipramine binding sites and sero-tonin uptake in the same structures does not, however, necessarily imply any functional association. However in the face of the excellent correlation between the ability of drugs to inhibit [^{3}H]-imipramine binding and sero-tonin uptake (see Section 3.2.6) it is difficult to imagine them as being completely independent. Other recent data have confirmed a probable functional relationship.

A study in developing rats showed a parallel appearance of [^{3}H]-impira-mine binding and high affinity serotonin uptake (Mocchetti, Brunello and Racagni 1982). Before the appearance of [^{3}H]-imipramine binding only a low-affinity uptake of serotonin could be detected. Similarly the hereditary deficiency in platelet serotonin content and storage in Fawn-Hooded rats (Da Prada, L. Pieri, Keller, M. Pieri and Bonetti 1978) has been shown to be accompanied by a reduced number of [^{3}H]-imipramine binding sites (Dumbrille-Ross and Tang 1981; Arora, Tong, Jackman, Stoff and Meltzer 1983). A series of 'irreversible' or, more correctly, very slowly dissociable imipramine derivatives have been recently synthesized (Rehavi, Ittah, Skolnick, Rice and Paul 1982*b*). Both 2,8-dinitroimipramine (Paul *et al.* 1981*a*) and 2-nitroimipramine (Rehavi, Racer, Rice, Skolnick and Paul 1983*b*) have been found to inhibit both [^{3}H]-imipramine binding and sero-tonin uptake with high affinity. Decreases were found in the maximal binding (B_{max}) and maximal uptake (v_{max}) with no changes in the affinity constants (K_D and K_m respectively) as would be expected for an essentially irreversible inhibition (Rehavi, Ittah, Rice, Skolnick, Goodwin and Paul

1981*b*). In addition to their confirmation of the functional relationship between the [³H]-imipramine binding site and serotonin uptake these ligands in their tritiated forms should prove to be highly valuable tools in the further elucidation of this relationship.

Serotonin uptake is known to be a sodium-dependent process (Rudnick 1977). The idea of a functional association of [³H]-imipramine binding with serotonin uptake was again supported by the demonstration that [³H]-imipramine binding was also sodium-dependent (Briley and Langer 1981), sodium ions being required for high-affinity binding. Indeed this sodium-dependency has recently been used by Rehavi *et al.* (1983*a*) to differentiate specific and non-saturable [³H]-imipramine binding in subcellular fractions.

3.4.2   *Do serotonin and imipramine bind to the same site?*

There are currently two schools of thought on the nature of the functional relationship between the [³H]-imipramine binding site and serotonin uptake. The simplest interpretation is that [³H]-imipramine binds to the 'serotonin recognition site' or the 'serotonin-transporter' of the uptake mechanism. This would imply that [³H]-imipramine is simply a convenient, alternative method for measuring serotonin uptake capacity, with the possible major advantage of measuring 'uptake' in frozen tissue such as human post-mortem samples (see Section 3.6.2).

Alternatively the [³H]-imipramine binding site and the serotonin transporter could be different sites but functionally associated with, for example, the imipramine binding site modulating serotonin uptake. The data on the topography and functional association of the two sites discussed above are compatible with both hypotheses. There are, however, a

Table 3.3   Comparison of the affinities of serotonin and imipramine for the [³H]-imipramine binding site and the serotonin uptake site in human platelets and rat brain

	Serotonin uptake	[³H]-imipramine binding
Rat brain		
Imipramine	$K_i = 690$ nM (1)	K_D 5 nM (2)
Serotonin	$K_m = 140$ nM (3)	$K_i = 1430$ nM (4)
Human platelets		
Imipramine	$K_i = 640$ nM (5)	$K_D = 1.4$ nM (5)
Serotonin	$K_m = 110$ nM (6)	$K_i = 2500$ nM (7)

Calculated from: (1) Langer *et al.* 1980*c*, (2) Raisman *et al.* 1980, (3) C. Moret, unpublished results, (4) Sette *et al.* 1983, (5) Paul *et al.* 1980, (6) Ahtee *et al.* 1981, (7) Langer *et al.* 1980*b*.

number of elements that suggest that the second, more complex, hypothesis may be nearer the truth.

First of all, if [³H]-imipramine and serotonin bind to the same site their affinities for this site should be identical and independent of whether they are measured by uptake or binding. As can be seen from Table 3.3, there are some major discrepancies between values obtained by uptake and those obtained by binding. In both rat brain and human platelets imipramine has a more than 100-fold higher affinity for the [³H]-imipramine binding site than for the serotonin uptake site. Serotonin on the other hand has a more than 10-fold greater affinity for the serotonin uptake site than for the [³H]-imipramine binding site. Furthermore in a study of Fawn-Hooded rats, Arora *et al.* (1983) found that although platelet serotonin uptake was unchanged [³H]-imipramine binding was significantly reduced.

In addition two recent clinical studies (see Section 3.6) suggest that in human platelets [³H]-imipramine binding sites and serotonin uptake sites are not identical. In one study, platelets from patients suffering from alcoholic cirrhosis were shown to have severely reduced serotonin uptake. The degree of their platelet [³H]-imipramine binding was however undiminished as compared to healthy volunteers (Ahtee, Briley, Raisman, Lebrec and Langer 1981). In another study, analysis of the individual values of serotonin uptake and [³H]-imipramine binding in platelets from over 40 untreated depressed patients and healthy volunteers showed no correlation between the values of serotonin uptake and [³H]-imipramine binding for each subject (Raisman, Briley, Bouchami, Sechter, Zarifian and Langer 1982*b*).

If, as the above data suggests, there is a modulatory or allosteric relationship between the [³H]-imipramine binding site and the serotonin uptake site, this should be revealed by a careful analysis of the inhibition of [³H]-imipramine binding by serotonin and the inhibition of serotonin uptake by imipramine.

3.4.3   *The nature of the relationship*

The rate of dissociation of a drug-receptor complex is typically a pseudo-first-order reaction. Thus the presence or absence of a competitive inhibitor does not modify this parameter. This was indeed what was found by Sette, Briley and Langer (1983) when they examined the dissociation of [³H]-imipramine bound to rat cortical membranes in the presence or absence of desipramine. In the presence of 100 μM serotonin, however, the dissociation rate was dramatically slowed. Fluoxetine, a non-tricyclic serotonin uptake inhibitor, gave qualitatively similar results.

Wennogle and Meyerson (1983) independently discovered the same phenomenon in human platelets observing that 50 μM serotonin markedly slowed the fluoxetine-induced dissociation of [³H]-imipramine bound to platelet membranes. A variety of other biogenic amines were shown to be inactive. This effect of serotonin in platelets has recently been confirmed by Langer, Raisman and Segonzac (1983) who found that tryptamine, a substrate for serotonin uptake was as effective as serotonin in slowing the dissociation rate of [³H]-imipramine binding.

All three studies were in agreement in interpreting this data as suggesting an allosteric inhibition of [³H]-imipramine binding by serotonin and non-tricyclic compounds acting on the serotonin uptake recognition site as substrates or inhibitors.

Further evidence for an allosteric interaction has come from inhibition studies. Tricyclic antidepressants inhibit the binding of [³H]-imipramine in rat brain in a classical competitive manner (Sette *et al.* 1983) with Hill coefficients close to 1.0 (Table 3.4). Inhibition at various concentrations of [³H]-imipramine showed that the IC_{50} values obtained were proportional to the concentration of [³H]-imipramine used, thus giving a constant K_i value.

The inhibition of [³H]-imipramine binding in rat brain by serotonin was however complex with a Hill coefficient between 0.4 and 0.5 (Table 3.4) (Sette *et al.* 1983). Although one of these authors (Langer *et al.* 1983) subsequently reported that serotonin competitively inhibited [³H]-imipramine binding in human platelets more recent studies (Phillips and Williams 1983) suggested that, as in rat brain, the inhibition of [³H]-imipramine binding by serotonin was also complex in this tissue.

A series of non-tricyclic serotonin uptake inhibitors including fluoxetine were also found to inhibit [³H]-imipramine binding in a complex manner with Hill coefficients less than unity (Table 3.4). This suggests that the site through which serotonin inhibits [³H]-imipramine binding is the serotonin recognition site of the uptake complex. Thus it would appear that tricyclic

Table 3.4 IC_{50} values and Hill coefficients for the inhibition of [³H]-imipramine binding (at 2.5 nM [³H]-imipramine) by a series of tricyclic and non-tricyclic drugs

	IC_{50} (nM)	Hill (n)
Chlorimipramine	23	1.03
Imipramine	33	0.91
Amitriptyline	39	0.93
Nortriptyline	57	0.94
Desipramine	147	1.01
Citalopram	30	0.64
Fluoxetine	50	0.58
Norzimelidine	200	0.55
Paroxetine	250	0.44
Serotonin	2000	0.49

Calculated from Sette *et al.* (1983).

antidepressants bind directly to the [^{3}H]-imipramine site. Serotonin and non-tricyclic serotonin uptake inhibitors, however, seem to bind to the serotonin recognition site of the uptake complex having their inhibitory effect on the binding by an indirect mechanism.

Although the exact nature of the interaction between these two sites is unclear it would seem, as already suggested by displacement studies, (see Section 3.4.3) to involve an allosteric interaction. In the light of this probable allosteric interaction the results of an earlier study (Abbott, Briley, Langer and Sette 1982) are particularly interesting. These authors found that whereas the potencies of a series of tricyclic antidepressants for the inhibition of [^{3}H]-imipramine were only slightly modified (less than 10-fold) in the absence of sodium (Table 3.5), the potencies of serotonin and a series of non-tricyclic serotonin uptake inhibitors were decreased by 40- to 100-fold in the absence of sodium (Table 3.5). It therefore appears that the allosteric interaction between the [^{3}H]-imipramine binding site and the serotonin recognition site of the uptake complex is highly sodium-dependent.

Table 3.5 IC_{50} values for the inhibition of [^{3}H]-imipramine binding (at 2.5 nM [^{3}H]-imipramine) in the presence and absence of 120 nM sodium

	IC_{50} (nM)		$\dfrac{-Na}{+Na}$
	+Na	−Na	
Amitriptyline	35	37	1.1
Chlorimipramine	24	171	7.1
Desipramine	177	153	0.9
Fluoxetine	25	1187	47.5
Citalopram	41	3350	81.7
Paroxetine	66	6075	92.0
Serotonin	3180	139 700	43.9

Calculated from Abbott *et al.* (1982).

The [^{3}H]-imipramine binding site and the serotonin recognition site can be therefore imagined as two different sites associated through a sodium-dependent allosteric mechanism (Fig. 3.2). To date, however, only the interaction of serotonin with the [^{3}H]-imipramine binding site has been demonstrated. The model would, however, also predict that a drug acting at the [^{3}H]-imipramine binding site should modulate serotonin uptake. Thus substrates of serotonin uptake such as tryptamine and its derivatives and the various non-tricyclic inhibitors of serotonin uptake studied so far would be predicted to inhibit serotonin uptake competitively. A tricyclic antidepressant, on the other hand, would inhibit serotonin uptake non-competitively via a sodium-dependent allosteric interaction (Fig. 3.2). At

64   Mike Briley

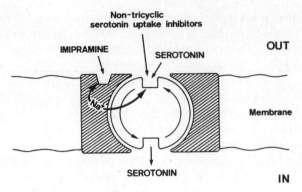

Fig. 3.2 Schematic representation of the proposed interaction between the [³H]-imipramine binding site and the serotonin uptake complex.

the moment there is insufficient rigorous data on the inhibition of serotonin uptake available to permit the testing of this hypothesis. It has, however, been recently suggested that certain tricyclic antidepressants are non-competitive inhibitors of serotonin uptake (Lingjaerde 1979; Wood and Wyllie 1981).

More fundamentally, the proposed model (Fig. 3.2) implies a modulation of the neuronal re-uptake of serotonin by a presynaptic site (the [³H]-imipramine binding site). This is analogous to the modulation of the release of serotonin by the presynaptic autoreceptor (see Chapter 2). If this hypothesis is correct, both release and re-uptake of serotonin would be under the control of independent modulatory systems permitting a very fine control of the synaptic levels of serotonin.

3.5   Modifications of [³H]-imipramine binding

Since the first demonstration that [³H]-imipramine bound with high affinity to a site different from classical neurotransmitter receptors a possible relationship of this site with the biochemical pathogenesis of depression has been hypothesized (Raisman *et al.* 1979*b*). Subsequently a variety of factors thought to be related to the physiopathology of affective disorders have been shown to modify [³H]-imipramine binding.

Chronic administration of desipramine to rats for three weeks has been shown to significantly decrease the density of [³H]-imipramine binding sites without changing the affinity (Raisman *et al.* 1980; Kinnier *et al.* 1980). Although this down-regulation is accompanied, in certain brain regions, by a down-regulation of β-adrenoceptors this is not true for all regions implying that there is no functional relationship between the [³H]-imipramine binding site and the β-adrenoceptor (Kinnier *et al.* 1980).

Down-regulation of [³H]-imipramine binding has also been observed in cat brain and blood platelets following daily administration of imipramine for 4 weeks (Briley, Raisman, Arbilla, Casadamont and Langer 1982). Five weeks of lithium treatment similarly decreased the number of [³H]-imipramine binding sites (Plenge and Mellerup 1982) although these authors were unable to demonstrate any effect of long-term imipramine treatment. Down-regulation was not however produced by chronic administration of atypical antidepressants such as iprindole (Kinnier *et al.* 1980). On the other hand, deprenyl (selegiline) a selective monoamine oxidase type B inhibitor increased [³H]-imipramine binding (Zsilla, Barbaccia, Gandolfi, Knoll and Costa 1983).

Deprivation of paradoxical or rapid eye movement (REM) sleep which has been shown to improve depressive symptoms in man (King 1977) has been shown after 72 hours, to significantly decrease [³H]-imipramine binding in rat brain (Mogilnicka, Arbilla, Depoortere and Langer 1980). Rats conditioned to the 'learned helplessness model' of depression have also been reported to have decreased [³H]-imipramine binding (Petty and Sherman 1982). Furthermore the level of [³H]-imipramine binding increased to normal values within one week of the loss of learned helplessness behaviour. Guisado, Garzon and Del Rio (1981) have reported that rats isolated two weeks after birth for at least 10 months showed increased [³H]-imipramine binding. This was accompanied by a marked hyperlocomotor (manic?) behaviour.

Recent studies have demonstrated a marked circadian rythmn of [³H]-imipramine binding (35 per cent amplitude) in various rat brain nuclei. This rhythm with highest levels at the end of the dark period and lowest levels at the end of the light period is similar to the variation of serotonin turnover (Wirz-Justice, Krauchi, Morimasa, Willener and Feer 1983).

Thus various factors thought to be related to affective disorders are capable of modifying [³H]-imipramine binding. This strongly implies a functional role for the [³H]-imipramine binding site in depression.

3.6   Clinical studies using platelets

3.6.1   *Platelets—really a model for central serotonin neurones?*

From a very early stage, we proposed the working hypothesis that the [³H]-imipramine binding site was, in some way, implicated in the physiopathology of affective disorders.

The discovery of [³H]-imipramine binding sites on human blood platelets (see Section 3.3.3) has opened the way for testing this hypothesis in the best model of all—the depressed patient. However, before considering the results obtained in studies of [³H]-imipramine binding in human platelets it would be judicious to examine the assumption that differences in the [^{3}H]-imipramine binding site seen in platelets from depressed patients reflect changes occurring in the brain. As already discussed (see Section 3.3.3) [^{3}H]-imipramine binding sites in brain and platelets appear to be identical. Evidence that they can be modified in parallel come from studies in cats treated chronically with imipramine for three weeks (Briley *et al.* 1982). Both platelet and brain [^{3}H]-imipramine binding sites were found to be down-regulated to the same degree. Similarly in Fawn-Hooded rats, a strain of rats with defective platelet serotonin storage (see Section 3.4.1) [^{3}H]-imipramine bnding sites were found to be absent or severely reduced in both brain and platelets (Dumbrille-Ross and Tang 1981; Arora *et al.* 1983). Thus the assumption that studies on platelet [^{3}H]-imipramine binding sites can be extrapolated to the brain would therefore not seem to be unreasonable.

The final confirmation that platelet and brain [^{3}H]-imipramine binding sites are under a common, presumably genetic, control has came recently from human post-mortem studies in depressed patients (Stanley, Virgilio and Gershon 1982). This will be further discussed below (see Section 3.7).

3.6.2   Studies on untreated depressed patients

A comparison of the binding of [^{3}H]-imipramine to platelets from 16 untreated severely depressed women with those from 21 female control volunteers showed a significantly lower density of sites in the depressed patients (Briley, Langer, Raisman, Sechler and Zarifian 1980c) This initial report has subsequently been confirmed by the same authors (Raisman *et al.* 1982b) and in three other laborotories (Table 3.6) (Asarch, Shih and

Table 3.6   Summary of studies on the binding of [^{3}H]-imipramine in platelets from depressed patients and control volunteers

Reference	Maximal binding (B_{max}) Mean ± S.E.M. (fmol/mg protein)	
	Depressed	Controls
Briley *et al.* (1982) (includes data from Briley *et al.* (1980c) and Raisman *et al.* (1982b))	318±18 (48)	564±30 (70) $p < 0.0001$
Paul *et al.* (1981b)	318±20 (14)	450±23 (28) $p < 0.001$
Asarch *et al.* (1980)	561±37 (23)	694±93 (16) $p < 0.02$
Suranyi-Cadotte *et al.* (1982)	381±34 (10)	658±33 (17) $p < 0.001$
Mellerup *et al.* (1982)	1190±54 (19)	1010±60 (33) $p < 0.01$

Number of subjects studied shown in brackets.

Kulsar 1980; Paul, Rehavi, Skolnick, Ballenger and Goodwin 1981*b*; Suranyi-Cadotte, Wood, Vasaran, Nair and Schwartz 1982). These results obtained in over 90 depressed patients of both sexes and more than 100 control volunteers have not, however, been replicated by a Danish group (Mellerup, Plenge and Rosenberg 1982) who found a slightly higher density of [^{3}H]-imipramine binding in platelets from depressed patients. The reason of their failure to find the same result is not clear but may result from methodological differences and possibly differences in handling and preparation of the platelets.

Although the patients were drug-free for at least a week before the blood sample was taken, it could be that previous drug treatments could affect the levels of platelet [^{3}H]-imipramine binding. Raisman *et al.* (1982*b*) demonstrated, however, that patients who had recently been taken off tricyclic antidepressants (less than 30 days earlier) had similar [^{3}H]-imipramine binding to those who had been drug-free for a longer period (greater than 30 days) or had never taken tricyclic antidepressants. Furthermore a recent study showed that the [^{3}H]-imipramine values were unaltered whether or not the patient were taking tricyclic antidepressants (Suranyi-Cadotte *et al.* 1982).

The decreased density of [^{3}H]-imipramine binding sites in depressed patients appears to be independent of the type of depression from which the patients were suffering. Platelets from patients diagnosed as reactional or mono- or bipolar endogenous depressives were indistinguishable (Raisman *et al.* 1982*b*). Furthermore the decrease binding in platelets from depressed patients was independent of the severity of the depressive symptoms as expressed by the Hamilton Depression Rating Score (Briley *et al.* 1980*c*). This has also been confirmed by subsequent studies (Asarch *et al.* 1980; Paul *et al.* 1981*b*).

3.6.3   *A state-dependent or a state-independent phenomenon?*

An obvious question leading from the observation of a decreased density of [^{3}H]-imipramine binding in platelets from depressed patient is whether this decrease is related to the depressive state and that [^{3}H]-imipramine binding reverts to normal values when the patient recovers. Alternatively is it a state-independent phenomenon linked genetically to a susceptibility to depression and independent of the presence or absence of a depressive episode?

A longitudinal study of 10 depressed patients treated with imipramine or amitriptyline showed that the level of [^{3}H]-imipramine binding was independent of their clinical improvement and when these patients were discharged from hospital (20 to 130 days after the beginning of their treat-

Table 3.7   Changes in the maximal platelet binding of [^{3}H]-imipramine during clinical improvement of depressed patients

Treatments	B_{max} fmol/mg protein			
	Before	After	Controls	Ref.
Tricyclic antidepressants	283±29 (10)	310±33 (10)	581±37 (39)	(1)
ECT (5 sessions)	294±24 (12)	369±47 (12)	539±40 (23)	(2)
Maprotiline	308±46 (10)	359±51 (10)	539±40 (23)	(2)
Spontaneous remission	306±40 (3)	725±102 (3)	658±33 (17)	(3)
Tricyclic antidepressants	413±39 (7)	612±38 (7)	658±33 (17)	(3)
Euthymic bipolar patients	450±151 (12)	—	440±168 (12)	(4)

Number of subjects studied shown in brackets.
(1) Raisman *et al.* 1082*b*, (2) Gay *et al.* 1983, (3) Suranyi-Cadotte *et al.* 1982, (4) Berrettini *et al.* 1982.

ment) the level of [^{3}H]-imipramine binding was not significantly different from the initial values (Table 3.7) (Raisman *et al.* 1982*b*).

Since tricyclic antidepressants have been shown in animals to down-regulate [^{3}H]-imipramine binding (see Section 3.5) it is possible that continuing treatment with these drugs may prevent an increase in the number of [^{3}H]-imipramine binding sites which would otherwise be linked to the improved depressed state. Similar results were, however, subse-quently obtained in patients treated for 15 days with maprotiline a drug which is a selective inhibitor of noradrenaline uptake and which inhibits [^{3}H]-imipramine binding only at very high concentrations (Table 3.7) (Gay, Langer, Loo, Raisman, Sechler and Zarifian 1983). In addition, 12 patients undergoing at least 6 sessions of electroconvulsive therapy (ECT) and showing a major clinical improvement were also shown to have platelet [^{3}H]-imipramine binding levels similar to those before treatment (Table 3.7) (Gay *et al.* 1983).

The reported finding that there was a marked concordance of platelet [^{3}H]-imipramine binding value within pairs of monozygotic twins also supports a genetic, state-independent regulation of the binding site (S. Paul unpublished results cited in Paul *et al.* 1981*b*).

These data in favour of a state-independent nature of the phenomenon contrast, however, with other recent studies. The remission of 10 depressed patients (3 natural remissions, 7 following tricyclic antidepressant therapy) was found to result three weeks after remission in an increase in platelet [^{3}H]-imipramine binding to values to those measured in control volunteers (Suranyi-Cadotte *et al.* 1982). Similarly a study comparing 12 unmedi-cated euthmic bipolar depressed patients (i.e. depressed patients currently in a normal non-depressed condition) with 12 control volunteers (Berrettini, Nurnberger, Post and Gershon 1982) found no significant difference in their density of platelet [^{3}H]-imipramine binding sites. This latter study would have been more convincing if it had been possible to study the same patients in both the euthymic and depressed states.

Thus, for the moment the state-dependent or -independent nature of the decrease in [^{3}H]-imipramine binding seen in depressed patients is unclear. A possible hypothesis that can reconcile the existing data is that the [^{3}H]-imipramine binding decrease is a state-dependent phenomenon that changes more slowly than clinical signs. Thus, the studies measuring [^{3}H]-imipramine binding immediately after recovery (Raisman *et al.* 1982*b*; Gay *et al.* 1983) would obtain results different from studies measuring the binding some time after remission (Suranyi-Cadotte *et al.* 1982) or in a stabilized euthymic state (Berrettini *et al.* 1982).

3.6.4 *Specificity*

The specificity of the decreased levels of platelet [^{3}H]-imipramine binding in depression is also of fundamental importance in establishing the clinical relevance of this phenomenon. To date, few studies are available. In the one study where the binding of [^{3}H]-imipramine was studied in the platelets of psychiatric patients without affective disorders (Mellerup *et al.* 1982) no decrease was seen in depressed patients (see Section 3.6.2) leaving the other results of this study uninterpretable. A study of non-psychiatric patients suffering from alcoholic cirrhosis (Ahtee *et al.* 1981) showed, however, that although platelet serotonin uptake was decreased [^{3}H]-impramine binding was not significantly different from that measured in control volunteers. It is clear, however, that further studies are required in various psychiatric disorders before any attempt to use decreased platelet [^{3}H]-imipramine binding was a biochemical marker for depression can be taken seriously.

3.7 Clinical studies on post-mortem tissue

The use of platelets in biochemical psychiatry has produced some fascinating data. The interpretation of these data in terms of brain dysfunction is still delicate, however, in spite of apparently convincing evidence that platelets and serotonin neurones share a common genetic control (see Section 3.6.1). It was therefore extremely reassuring when Stanley *et al.* (1982) reported that [^{3}H]-imipramine binding in samples of the frontal cortex of post-mortem brains from 9 suicide victims was markedly reduced as compared to samples from 9 control brains (accident victims). There was no difference in the dissociation constants. A similar finding has been recently reported by E. K. Perry, Marshall, Blessed, Tomlinson and R. H. Perry (1983). Binding of [^{3}H]-imipramine to membranes prepared from the hippocampal and occipital cortex of 11 post-mortem brains from depressed patients was significantly lower than that found in the brains of 14 normals. The level of [^{3}H]-imipramine binding in 13 brains of patients dying from Alzheimers Disease were not significantly different from the normals.

A third study has, however, reported an increased binding of [^{3}H]-imipramine in the brain of suicide victims compared to homicide controls (Meyerson, Wennogle, Abel, Coupet, Lippa, Rauh and Beer 1982). The reason for the difference between this and the other two studies is unclear but different pre-mortem drug histories could be an important factor. Thus, as predicted by the platelet studies described earlier [^{3}H]-imipramine binding appears possibly to be specifically reduced in suicide victims (presumably depressed) and depressed patients.

3.8   Conclusions

At this point with all the known facts assembled I fear that the reader may be expecting a clear concise explanation of the meaning of it all. Unfortunately, there are still too many pieces of the puzzle missing to hope for an overall understanding of the role of [^{3}H]-imipramine binding in depression. Certain conclusions can, however, be drawn and perhaps grouped into three classes . . . the certain (well . . . almost), the probable and the speculative.

3.8.1   *The certain . . .*

It now seems fairly clear that [^{3}H]-imipramine binds to a high-affinity site which, in the brain, is located on serotonin nerve terminals. It is closely related to but not identical with the uptake mechanism for serotonin.

3.8.2   *The probable . . .*

The [^{3}H]-imipramine binding site appears to be a regulatory unit capable of modulating serotonin uptake. Different treatments which may possibly be related to affective disorders have been shown to modify the density of these sites in animals. In man the density of sites is reduced in both platelets and in the brain of depressed patients. In platelets the decreased density of [^{3}II]-imipramine binding sites may be independent of the depressive episode, and related to a genetic susceptibility to depression.

3.8.3 *And the speculation . . .*

The propositions in this section, while being consistent with the majority of the known facts, are purely speculative. If they stimulate thought, debate or new experimentation they will have achieved their goal.

The presence of [^{3}H]-imipramine binding sites associated with the serotonin uptake complex permits a modulation of serotonin uptake either through serotonin itself or an, as yet, undiscovered substance acting competitively at the [^{3}H]-imipramine binding site. This modulation may be considered to be relatively inactive during normal neurotransmission as found in normal subjects and euthymic depressed patients.

It is likely that the firing of serotonin neurones is sometimes decreased either by external factors acting possibly through the neuro-endocrine system or by some internal neurochemical desequilibria. In normal subjects the modulation of serotonin uptake via the imipramine binding system enables the synaptic levels of serotonin to be maintained thereby preventing the appearance of depressive symptoms. Depressed patients have a decreased density of [^{3}H]-imipramine binding sites and thus presumably a decreased modulatory capacity which may be insufficient to maintain normal synaptic serotonin levels thus explaining the appearance of depressive symptoms. This speculation has been supported by some recent results (Barbaccia, Gandolfi, Chuang and Costa 1983) which demonstrate that under conditions which produce a down-regulation of [^{3}H]-imipramine binding sites, the uptake of [^{3}H]-serotonin by hippocampal minces is increased. Furthermore under these conditions the efficiency of imipramine as a serotonin uptake blocker is diminished.

Various biochemical markers reported in depressed patients, such as decreased 5-HIAA levels (Lloyd *et al.* 1974) and up-regulated 5-HT$_2$ receptors in post-mortem brain (Stanley and Mann 1983), are consistent with decreased synaptic levels of serotonin in depression. Tricyclic antidepressants and other inhibitors of serotonin uptake presumably exert their antidepressant effects by reinforcing the natural modulation of the [^{3}H]-imipramine binding complex.

This highly simplified scheme attempts to integrate the available data of serotonin uptake and neurotransmission in depression with the findings from [^{3}H]-imipramine binding studies. As a mere 'grinder and binder' it would be imprudent to attempt to go any further!

Acknowledgements

The author wishes to thank Martine Dehaye for secretarial assistance in preparing the manuscript and Chantal Moret for her stimulating criticisms of it.

72   Mike Briley

References

Abbott, W. M., Briley, M. S., Langer, S. Z. and Sette, M. (1982). Sodium shift of the inhibition of ^{3}H-imipramine binding by serotonin and serotonin uptake blockers but not by tricyclic antidepressant drugs. *Br. J. Pharmac.* **76**, 259.

Agid, Y., Langer, S. Z., Raisman, R., Ruberg, M., Scatton, B., Sette, M. and Zivkovic, B. (1983). Subcellular fractionation of ^{3}H-imipramine binding after chemical or electrolytic lesion of the serotoninergic system. *Br. J. Pharmac.* **79**, 198.

Ahtee, L., Briley, M., Raisman, R., Lebrec, D. and Langer, S. Z. (1981). Reduced uptake of serotonin but unchanged ^{3}H-imipramine binding in the platelets from cirrhotic patients. *Life Sci.* **29**, 2323.

Arora, R. C., Tong, C., Jackman, H. L., Stoff, D. and Meltzer, H. Y. (1983). Serotonin uptake and binding in blood platelets and brain of fawn-hooded and Sprague-Dawley rats. *Life Sci.* **33**, 437.

Asarch, K. B., Shih, J. C. and Kulser, A. (1980). Decreased ^{3}H-imipramine binding in depressed males and females. *Commun. Psychopharmac.* **4**, 425.

Barbaccia, M. L., Gandolfi, O., Chuang, D. M. and Costa, E. (1983). Modulation of neuronal serotonin uptake by a putative endogenous ligand of imipramine recognition sites. *Proc. Nat. Acad. Sci.* **80**, 5134.

Berrettini, W. H., Nurnberger, J. I. Jr., Post, R. M. and Gershon, E. S. (1982). Platelet ^{3}H-imipramine binding in euthymic bipolar patients. *J. Psychiat. Res.* **7**, 215.

Blomstrand, C. and Hamberger, A. (1970). Amino acid incorporation in vitro in proteins of neuronal and glial cell enriched fractions. *J. Neurochem.* **17**, 1187.

Borbe, H. O. and Zube, I. (1983). The detection of specific ^{3}H-imipramine binding in bovine retina. *Brain Research,* **264**, 178.

Briley, M. and Langer, S. Z. (1981). Sodium dependency of ^{3}H-imipramine binding in rat cerebral cortex. *Eur. J. Pharmac.* **72**, 377.

—— Raisman, R. and Langer, S. Z. (1979). Human platelets possess high-affinity binding sites for ^{3}H-imipramine. *Eur. J. Pharmac.* **58**, 347.

—— —— Sechter, D., Zarifian, E. and Langer, S. Z. (1980a). ^{3}H-imipramine binding in human platelets: a new biochemical parameter in depression. *Neuropharmacology* **19**, 1209.

—— —— Arbilla, S., Casadamont, M. and Langer, S. Z. (1982). Concomitant decrease in ^{3}H-imipramine binding in cat brain and platelets after chronic treatment with imipramine. *Eur. J. Pharmac.* **81**, 309.

—— Fillion, G., Beaudoin, D., Fillion, M. P. and Langer, S. Z. (1980b). ^{3}H-imipramine binding in neuronal and glial fractions of horse striatum. *Eur. J. Pharmac.* **64**, 191.

—— Langer, S. Z., Raisman, R., Sechter, D. and Zarifian, E. (1980c). Tritiated imipramine binding sites are decreased in platelets of untreated depressed patients. *Science* **209**, 303.

Brunello, N., Chuang, D. M. and Costa, E. (1982a). Different synaptic location of mianserin and imipramine binding sites. *Science* **215**, 112.

—— —— —— (1982b). Specific binding of ^{3}H-imipramine to structures of rat hippocampus. *Eur. J. Pharmac.* **78**, 383.

Burkard, W. P. (1980). Specific binding in rat brain for a new and potent inhibitor of 5-hydroxytryptamine uptake: Ro 11-2465. *Eur. J. Pharmac.* **61**, 409.

Da Prada, M., Pieri, L., Keller, H. H., Pieri, M. and Bonetti, E. P. (1978). Effects of 5,6-dihydroxytryptamine and 5,7-dihydroxytryptamine on rat central nervous system after intraventricular or intracerebral application and on blood platelets in vitro. *Ann. N. Y. Acad. Sci.* **305**, 595.

Davis, A., Morris, J. M. and Tang, S. W. (1983a). Temperature-sensitive conformational changes in membrane-bound and solubilized ^{3}H-imipramine binding sites. *Eur. J. Pharmac.* **88**, 407.

—— —— —— (1983b). Solubilization and assay of ^{3}H-imipramine binding sites from human platelets. *Eur. J. Pharmac.* **86**, 353.

—— Dumbrille-Ross, A. and Tang, S. W. (1983c). Differentiation between platelet and cortical ^{3}H-imipramine binding sites. *Br. J. Pharmac.* **80**, 664.

De Montis, G. M., Devoto, P. and Tagliamonte, A. (1982). Possible anti-depressant activity of methadone. *Eur. J. Pharmac.* **79**, 145.

Dumbrille-Ross, A. and Tang, S. W. (1981). Absence of high-affinity ^{3}H-imipramine binding in platelets and cerebral cortex of fawn-hooded rats. *Eur. J. Pharmac.* **72**, 137.

—— —— (1983). Binding of ^{3}H-Ro 11-2465. Possible identification of a subclass of ^{3}H-imipramine binding sites. *Molec. Pharmac.* **23**, 607.

—— —— and Seeman, P. (1980). High-affinity binding of ^{3}H-mianserin to rat cerebral cortex. *Eur. J. Pharmac.* **68**, 395.

—— —— and Coscina, D. V. (1981). Differential binding of ^{3}H-imipramine and ^{3}H-mianserin in rat cerebral cortex. *Life Sci.* **29**, 2049.

—— Morris, J., Davis, A. and Tang, S. W. (1983). Temperature-sensitive reversible loss of ^{3}H-imipramine binding sites: evidence suggesting different conformational states. *Eur. J. Pharmac.* **91**, 383.

Fulton, A., Norman, T. and Burrows, G. D. (1982). Ligand binding and platelet uptake studies of loxapine, amoxapine and their 8-hydroxylated derivatives. *J. Affect. Dis.* **4**, 113.

Gay, C., Langer, S. Z., Loo, H., Raisman, R., Sechter, D. and Zarifian, E. (1983). ^{3}H-imipramine binding in platelets: a state-dependent or independent biological marker in depression? *Br. J. Pharmac.* **78**, 57 p.

Grabowsky, K. L., McCabe, R. T. and Wamsley, J. K. (1983). Localization

74   Mike Briley

of ^{3}H-imipramine binding sites in rat brain by light microscopic autoradiography. *Life Sci.* **32**, 2355.

Gross, G., Göthert, M., Ender, H. P. and Schüman, H. J. (1981). ^{3}H-
imipramine binding sites in the rat brain. Selective localization on
serotoninergic neurones. *Naunyn-Schmiedeberg's Arch. Pharmac.* **317**,
310.

Guisado, E., Garzon, J. and Del Rio, J. (1981). Increased ^{3}H-spiroperidol
and ^{3}H-imipramine binding in the brain of rats after long-term isolation.
Eighth International Cong. Pharmac. Tokyo (Abst.) p. 1417.

Hall, H., Ross, S., Ogren, S. O. and Gawell, L. (1982). Binding of a specific
5HT uptake inhibitor, 3H-norzimelidine, to rat brain homogenates. *Eur.
J. Pharmac.* **80**, 281.

Hrdina, P. D. (1981). Pharmacological characterization of ^{3}H-desipramine
binding in rat cerebral cortex. *Prog. Neuro-Psycho-pharmac.* **5**, 553.

—— Elson-Hartman, K., Roberts, D. C. S. and Pappas, B. A. (1981). High-
affinity ^{3}H-desipramine binding in rat cerebral cortex decreases after
selective lesion of noradrenergic neurons with 6-hydroxydopamine. *Eur.
J. Pharmac.* **73**, 375.

—— Pappas, B. A., Bialik, R. J. and Ryan, C. L. (1982). Regulation of ^{3}H-
imipramine binding sites in rat brain regions: effect of neonatal
5,7-dihydroxytryptamine treatment. *Eur. J. Pharmac.* **83**, 343.

King, D. (1977). Pathological and therapeutic consequences of sleep loss: a
review, *Dis. Nerv. Syst.* **38**, 873.

Kinnier, W. J., Chuang, D. M. and Costa, E. (1980). Down regulation of
dihydroalprenolol and imipramine binding sites in brain of rats repeatedly treated with imipramine. *Eur. J. Pharmac.* **67**, 289.

—— Chuang, D. M., Gwynn, G. and Costa, E. (1981). Characteristics and
regulation of high affinity ^{3}H-imipramine binding to rat hippocampal
membranes. *Neuropharmacology* **20**, 411.

Laduron, P. M., Robbyns, M. and Schotte, A. (1982). ^{3}H-desipramine and
^{3}H-imipramine binding are not associated with noradrenaline and serotonin uptake in the brain. *Eur. J. Pharmac.* **78**, 491.

Langer, S. Z. and Briley, M. (1981). High-affinity ^{3}H-imipramine binding: a
new biological tool for studies in depression. *Trends Neurosci.* **4**, 28.

—— and Raisman, R. (1983). Specific high-affinity binding sites for ^{3}H-
imipramine are present in the rat lung. *Br. J. Pharmac.* **80**, 453.

—— —— and Briley, M. S. (1980a). Stereoselective inhibition of ^{3}H-
imipramine binding by antidepressant drugs and their derivatives. *Eur. J.
Pharmac.* **64**, 89.

—— Briley, M. S., Raisman, R., Henry, J. F. and Morselli, P. L. (1980b).
Specific ^{3}H-imipramine binding in human platelets. *Naunyn-
Schmiedeberg's Arch. Pharmac.* **313**, 189.

—— Moret, C., Raisman, R., Dubocovich, M. L. and Briley, M. (1980c).

High-affinity ^{3}H-imipramine binding in rat hypothalamus: Association with uptake of serotonin but not of norepinephrine. *Science* **210,** 1133.

—— Agid, F. J., Raisman, R., Briley, M. and Agid, Y. (1981*a*). Distribution of specific high-affinity binding sites for ^{3}H-imipramine in human brain. *J. Neurochem.* **37,** 267.

—— Raisman, R. and Briley, M. (1981*b*). High-affinity ^{3}H-DMI binding is associated with neuronal noradrenaline uptake in the periphery and the central nervous system. *Eur. J. Pharmac.* **72,** 423.

—— —— and Segonzac, A. (1983). Tryptamine and serotonin inhibit competitively ^{3}H-imipramine binding in human platelets and modify the rate of dissociation. *Br. J. Pharmac.* **80,** 451.

—— Zarifian, E., Briley, M., Raisman, R. and Sechter, D. (1982). High-affinity ^{3}H-imipramine binding: A new biological marker in depression. *Pharmacopsychiat.* **15,** 4.

Lee, C. M. and Snyder, S. H. (1981). Norepinephrine neuronal uptake binding sites in rat brain membranes labelled with ^{3}H-desipramine. *Proc. Natl. Acad. Sci.* **78,** 5250.

—— Javitch, J. A. and Snyder, S. H. (1982). Characterization of ^{3}H-desipramine binding associated with neuronal norepinephrine uptake sites in rat brain membranes. *J. Neuroscience* **2,** 1515.

—— —— —— (1983). Recognition sites for norepinephrine uptake: regulation by neurotransmitter. *Science* **220,** 626.

Lingjaerde, O. (1979). Inhibitory effect of clomipramine and related drugs on serotonin uptake in platelets: more complicated than previously thought. *Psychopharmacology* **61,** 245.

Lloyd, K. G., Farley, I. J., Deck, J. H. N. *et al.* (1974). Serotonin and 5-hydroxyindoleacetic acid in discrete areas of the brainstem of suicide victims and control patients. In *Serotonin: new vistas* (eds. E. Costa, G. L. Gessa and M. Sandler) p. 97. Raven Press, New York.

Luine, V. N., Frankfurt, M., Rainbow, T. C., Biegon, A. and Azmitia, E. (1983). Intrahypothalamic 5,7-dihydroxytryptamine facilitates feminine sexual behaviour and decreases ^{3}H-imipramine binding and 5HT uptake. *Brain Res.* **264,** 344.

Mellerup, E. T., Plenge, P. and Rosenberg, R. (1982). ^{3}H-imipramine binding sites in platelets from psychiatric patients. *Psychiat. Res.* **7,** 221.

Meyerson, L. R., Wennogle, L. P., Abel, M. S., Coupet, J., Lippa, A. S., Rauh, C. E. and Beer, B. (1982). Human brain receptor alterations in suicide victims. *Pharmac. Biochem. Behav.* **17,** 159.

Mocchetti, I., Brunello, N. and Racagni, G. (1982). Ontogenetic study of ^{3}H-imipramine binding sites and serotonin uptake system: indication of possible interdependence. *Eur. J. Pharmac.* **83,** 151.

Mogilnicka, E., Arbilla, S., Depoortere, H. and Langer, S. Z. (1980). Rapid-eye-movement sleep deprivation decreases the density of

^{3}H-dihydroalprenolol and ^{3}H-imipramine binding sites in the rat cerebral cortex. *Eur. J. Pharmac.* **65**, 289.

Palkovits, M., Raisman, R., Briley, M. and Langer, S. Z. (1981). Regional distribution of ^{3}H-imipramine binding in rat brain. *Brain Res.* **210**, 493.

Paul, S. M., Rehavi, M., Skolnick, P. and Goodwin, F. K. (1980). Demonstration of specific "high-affinity" binding sites for ^{3}H-imipramine on human platelets. *Life Sci.* **26**, 953.

—— Rehavi, M., Rice, K. C., Ittah, Y. and Skolnick, P. (1981*a*). Does high affinity ^{3}H-imipramine binding label serotonin reuptake sites in brain and platelet? *Life Sci.* **28**, 2753.

—— —— Skolnick, P., Ballenger, J. C. and Goodwin, F. K. (1981*b*). Depressed patients have decreased binding of ^{3}H-imipramine to the platelet serotonin "transporter". *Arch. Gen. Psychiat.* **38**, 1315.

Perry, E. K., Marshall, E. F., Blessed, G., Tomlinson, B. E. and Perry, R. H. (1983). Decreased imipramine binding in the brains of patients with depressive illness. *Br. J. Psychiat.* **142**, 188.

Petty, F. and Sherman, A. D. (1982). Learned helplessness and imipramine binding (Abstract 64). Presented at the *37th Annual Convention of the Society of Biological Psychiatry,* Toronto, Canada.

Phillips, O. and Williams, D. C. (1983). Allosteric regulation of ^{3}H-imipramine binding to human platelet membranes by 5-hydroxytryptamine. *Br. J. Pharmac.* **80**, 669.

Plenge, P. and Mellerup, E. T. (1982). ^{3}H-imipramine high-affinity binding sites in rat brain. Effects of imipramine and lithium. *Psychopharmacology* **77**, 94.

Rainbow, T. C. and Biegon, A. (1983). Quantitative autoradiography of ^{3}H-nitroimipramine binding sites in rat brain. *Brain Res.* **262**, 319.

—— Biegon, A. and McEwen, B. S. (1982). Autoradiography localization of imipramine binding in rat brain. *Eur. J. Pharmac.* **77**, 363.

Raisman, R., Briley, M. and Langer, S. Z. (1979*a*). High-affinity ^{3}H-imipramine binding in rat cerebral cortex. *Eur. J. Pharmac.* **54**, 307.

—— Briley, M. and Langer, S. Z. (1979*b*). Specific tricyclic antidepressant binding sites in rat brain. *Nature* **281**, 148.

—— Briley, M. S. and Langer, S. Z. (1980). Specific tricyclic antidepressant binding sites in rat brain characterised by high-affinity ^{3}H-imipramine binding. *Eur. J. Pharmac.* **61**, 373.

—— Sette, M., Pimoule, C., Briley, M. and Langer, S. Z. (1982*a*). High-affinity ^{3}H-desipramine binding in the peripheral and central nervous system: a specific site associated with the neuronal uptake of noradrenaline. *Eur. J. Pharmac.* **78**, 345.

—— Briley, M. S., Bouchami, F., Sechter, D., Zarifian, E. and Langer, S. Z. (1982*b*). ^{3}H-imipramine binding and serotonin uptake in platelets from untreated depressed patients and control volunteers. *Psychopharmaco-*

logy **77**, 332.

Rehavi, M., Paul, S. M., Skolnick, P. and Goodwin, F. K. (1980). Demonstration of specific high affinity binding sites for ^{3}H-imipramine in human brain. *Life Sci.* **26**, 2273.

—— Skolnik, P. and Paul, S. M. (1983*a*). Subcellular distribution of high affinity ^{3}H-imipramine binding and ^{3}H-serotonin uptake in rat brain. *Eur. J. Pharmac.* **87**, 335.

—— —— Hulihan, B. and Paul, S. M. (1981*a*). High affinity binding of ^{3}H-desipramie to rat cerebral cortex: relationship to tricyclic antidepressant-induced inhibition of norepinephrine uptake. *Eur. J. Pharmac.* **70**, 597.

—— —— Brownstein, M. J. and Paul, S. M. (1982*a*). High-affinity binding of ^{3}H-desipramine to rat brain: a presynaptic marker for noradrenergic uptake sites. *J. Neurochem.* **38**, 889.

—— Ittah, Y., Rice, K. C., Skolnick, P., Goodwin, F. K. and Paul, S. M. (1981*b*). 2-nitroimipramine: a selective irreversible inhibitor of ^{3}H-serotonin uptake and ^{3}H-desipramine to rat brain: a presynaptic marker for noradrenergic uptake sites. *J. Neurochem.* **38**, 889.

—— —— Skolnick, P., Rice, K. C., and Paul, S. M. (1982*b*). Nitroimipramines—Synthesis and pharmacological effects of potent long-acting inhibitors of ^{3}H-serotonin uptake and ^{3}H-imipramine binding. *Naunyn-Schmiedeberg's Arch. Pharmac.* **320**, 45.

—— Tracer, H., Rice, K., Skolnick, P. and Paul, S. M. (1983*b*). ^{3}H-2-nitroimipramine: a selective "slowly-dissociating" probe of the imipramine binding site ("serotonin transporter") in platelets and brain. *Life Sci.* **32**, 645.

Reith, M. E. A., Sershen, H., Allen, D. and Lajtha, A. (1983). High- and low-affinity binding of ^{3}H-imipramine in mouse cerebral cortex. *J. Neurochem.* **40**, 389.

Rudnick, G. (1977). Active transport of 5-hydroxytryptamine by plasma membrane vesicles isolated from human blood platelets. *J. Biol. Chem.* **252**, 2170.

Sette, M., Raisman, R., Briley, M. and Langer, S. Z. (1981). Localisation of tricyclic antidepressant binding sites on serotonin nerve terminals. *J. Neurochem.* **37**, 40.

—— Briley, M. S. and Langer, S. Z. (1983). Complex inhibition of ^{3}H-imipramine binding by serotonin and nontricyclic serotonin uptake blockers. *J. Neurochem.* **40**, 622.

Sneddon, J. M. (1973). Blood platelets as a model for monoamine-containing neurones. *Prog. in Neurobiol.* **1**, 153.

Stahl, S. M. (1977). The human platelet: a diagnostic and research tool for the study of biogenic amines in psychiatric and neurologic disorders. *Arch. Gen. Psychiat.* **34**, 509.

Stanley, M. and Mann, J. J. (1983). Increased serotonin-2 binding sites in frontal cortex of suicide victims. *Lancet* i, 214.

—— Virgilio, J. and Gershon, S. (1982). Tritiated imipramine binding sites are decreased in the frontal cortex of suicides. *Science* **216**, 1337.

Suddith, R. L., Hutchison, H. T. and Haber, B. (1978). Uptake of biogenic amines by glial cells in culture. A neuronal-like transport system for serotonin. *Life Sci.* **22**, 2179.

Suranyi-Cadotte, B. E., Wood, P. L., Vasavan Nair, N. P. and Schwartz, G. (1982). Normalization of platelet ^{3}H-imipramine binding in depressed patients during remission. *Eur. J. Pharmac.* **85**, 357.

Talvenheimo, J. and Rudnick, G. (1980). Solubilization of the platelet plasma membrane serotonin transporter in an active form. *J. Biol. Chem.* **255**, 8606.

Taylor, J. E. and Richelson, E. (1982). High-affinity binding of ^{3}H-doxepin to histamine H_1-receptors in rat brain: possible identification of a subclass of histamine H_1-receptors. *Eur. J. Pharmac.* **78**, 279.

Wennogle, L. P. and Meyerson, L. R. (1983). Serotonin modulates the dissociation of ^{3}H-imipramine from human platelet recognition sites. *Eur. J. Pharmac.* **86**, 303.

—— Beer, B. and Meyerson, L. R. (1981). Human platelet imipramine recognition sites: Biochemical and pharmacological characterization. *Pharmac. Biochem. Behav.* **15**, 975.

Wirz-Justice, A., Krauchi, K., Morimasa, T., Willener, R. and Freer, H. (1983). Circadian rhythm of ^{3}H-imipramine binding in the rat suprachiasmatic nuclei. *Eur. J. Pharmac.* **87**, 331.

Wood, M. D. and Wyllie, M. G. (1981). The inhibitory action of imipramine on monoamine transport. *Br. J. Pharmac.* **74**, 890.

Zsilla, G., Barbaccia, M. L., Gandolfi, O., Knoll, J. and Costa, E. (1983). (−)-Deprenyl a selective MAO 'B' inhibitor increases ^{3}H-imipramine binding and decreases β-adrenergic receptor function. *Eur. J. Pharmac.* **89**, 111.

4

Characterization of serotonin receptor binding sites

JOSÉE E. LEYSEN

4.1 Historical notes

As early as 1878, Langley defined the receptor concept to explain mutual interaction of drugs on isolated tissues (Langley 1878) and for almost a century, receptor characterization and classification was based on observations of pharmacological or physiological effects elicited by drugs. In the early 1970s, the advent of highly labelled radioactive drugs opened new perspectives for the study of receptors at the molecular level.

Radioligand receptor binding rapidly became a powerful research tool, allowing direct investigation of drug receptor binding properties, receptor localization, receptor coupled mechanisms involved in signal transduction and receptor regulatory mechanisms. However, most important in binding studies is the demonstration of a role for the binding site which has been detected *in vitro* in pharmacological or physiological effects. Unfortunately, this prerequisite, inherent to the receptor concept is often neglected or not fulfilled. Binding data in themselves have served to hypothesize receptors or receptor sub-types, the link with pharmacological receptors being missing.

The first attempts to label serotonin receptors in brain tissue homogenates were made using $[^3H]$-lysergic acid diethylamide (LSD) and applying an equilibrium dialysis method (Farrow and Van Vunakis 1972). The breakthrough, however, came with the use of the rapid filtration technique for separating membrane-bound and free radioactivity. Bennett and Snyder (1975, 1976) studied serotonin receptor binding in mammalian brain tissue using both $[^3H]$-LSD and $[^3H]$-serotonin and detected high affinity (i.e. of nanomolar order) saturable binding sites with both ligands. These early, but extensive studies revealed notable differences between the $[^3H]$-LSD and $[^3H]$-serotonin binding sites with regard to the total number of the binding sites and their regional distribution in the brain.

79

In particular, the higher affinity of indole-like agonists for the [^{3}H]-sero-tonin labelled sites in contrast to the higher affinity of antagonists for [^{3}H]-LSD labelled sites was obvious. At that stage, differences in binding properties between the sites labelled with both ligands were interpreted in terms of a two-state model of the receptor, with [^{3}H]-serotonin labelling the agonist state and [^{3}H]-LSD labelling the antagonist state. One year later it was found that [^{3}H]-spiperone (a neuroleptic used for dopamine receptor binding studies) labelled a population of binding sites in the rat frontal cortex which displayed the features of a serotonin receptor. These sites were shown to be akin to the sites labelled with [^{3}H]-LSD in the cortex. Amongst a variety of neurotransmitters, serotonin showed the highest binding affinity for the sites and the binding affinity of drugs correlated with their potencies to antagonize serotonin-mediated excitatory behaviour in rodents (Leysen and Laduron 1977; Leysen, Niemegeers, Tollenaere and Laduron (1978). This was the first demonstration of a relationship between a binding site and a serotonin-mediated effect. Peroutka and Snyder (1979) challenged the two-state receptor model and supplied evidence that cortical binding sites labelled by [^{3}H]-serotonin and by [^{3}H]-spiperone were distinct molecular entities. The notation 5-HT$_1$-(5-hydroxytryptamine$_1$) or serotonin-S$_1$ sites and 5-HT$_2$- or serotonin-S$_2$ sites was introduced (for discussion of notation see Section 14.2), predominantly based on the distinct binding properties of the sites. [^{3}H]-LSD was found to label both the 5-HT$_1$ and the 5-HT$_2$ sites. A controversial issue in serotonin receptor binding research is, is there a role for the 5-HT$_1$ site? Numerous hypotheses have been put forward, but all appeared to be poorly substantiated and are refutable. On the contrary 5-HT$_2$ sites were shown to mediate distinct functions of serotonin centrally and peripherally.

In this review the following items will be discussed: (*i*) binding properties characterizing 5-HT$_1$ and 5-HT$_2$ sites labelled with various radioligands, (*ii*) the localization of the binding sites in central and peripheral tissues and the sensitivity of the sites to neuronal lesions, (*iii*) the role of the binding sites in *in vivo* and *in vitro* pharmacological effects and in the regulation of neurotransmitter release and neuronal firing, (*iv*) the importance of platelets as a model for studying serotonin receptors, and (*v*) goals and achievements in serotonin receptor site solubilization and purification.

4.2   Current methodology in binding studies

Dissected tissues are homogenized in ice-cold buffered medium and membranes are isolated and washed by centrifugation procedures. The final membrane pellets are suspended in buffer in a high dilution (2–10 mg original wet weight of tissue per ml). Small aliquots of the membrane suspension are incubated with a low concentration (nanomolar order) of a radioactively labelled ligand. The incubation is stopped by rapid filtration of the incubation mixture over glass fibre filters, to separate membrane-bound and free radioactivity. Membrane-bound radioactivity, retained on the glass fibre filters, is counted in a liquid scintillation spectrometer.

Specific binding of the radioactive ligand is defined as the difference in bound radioactivity in assays in the absence and the presence of excess of an unlabelled competitor.

The total number of binding sites (B_{max}) and the binding affinity of the radioactive ligand, given by the equilibrium dissociation constant (K_D) are derived from radioligand binding concentration curves. These are often analysed in Scatchard plots (Scatchard 1949) which are a mathematical transformation of the law of mass action: linear Scatchard plots are indicative of a simple interaction of the labelled ligand with a single binding site, curvilinear Scatchard plots point to complex interactions probably involving multiple binding sites.

Binding affinities of unlabelled drugs are derived from binding inhibition curves, obtained through incubating the radioactive ligand in the presence of increasing concentrations of the unlabelled drug. The IC_{50}-value is the drug concentration causing 50 per cent inhibition of the specific binding of the radioactive ligand. The equilibrium inhibition constant (K_i) is calculated according to Cheng and Prusoff (1973) as $K_i = IC_{50}/[1 + C/K_D]$ where C is the concentration and K_D the equilibrium dissociation constant of the radioactive ligand. Analysis of the shape of the inhibition curves, either graphically, e.g. in Hill plots (see Segal 1975) or by computer dissection (De Lean, Munson and Rodbard 1978, 1979) is applied to obtain information as to whether the labelled and unlabelled drug compete for a single binding site or whether multiple binding sites are possibly involved.

4.3   Binding properties of 5-HT$_1$ sites

4.3.1   [^{3}H]-serotonin binding in mammalian brain tissue

[^{3}H]-Serotonin was found to label high affinity binding sites in mammalian brain membrane preparations. However, the studies of the [^{3}H]-serotonin binding properties have suffered from many technical difficulties. Binding data reported by different investigators are often hard to reconcile because of variations in experimental conditions. Problems which have been encountered have included the following. [^{3}H]-Serotonin is chemically highly unstable and daily purification of the radioligand prior to use in binding assays has been required to improve reproducibility of binding data. [^{3}H]-Serotonin binding is very sensitive to additives in the incubation medium. Addition of 4 mM Ca^{2+} to the assay medium increased specific [^{3}H]-serotonin binding, whereas monovalent cations apparently reduced the binding (reviewed in Leysen 1981). Ascorbic acid, commonly used in the binding assays as a reducing agent for the biogenic amines, was recently reported to cause lipid peroxidation of membranes in certain conditions, resulting in decreased binding. Moreover, the ascorbic acid effect is complex, displaying a so-called 'bell shaped' course: lipid peroxidation accompanied by a decrease in binding starts at 0.025 mM ascorbate, reaches a maximum at 0.5 mM (at which a four-fold reduction in binding is observed), then declines and is no longer observed above 6 mM ascorbate, probably because of an auto-inhibitory effect of ascorbate on the peroxidizing system (Muakkassah-Kelly, Andresen, Shih and Hochstein 1982). It was further shown that $CaCl_2$ (4 mM) could inhibit the ascorbate induced lipid peroxidation, which probably partly explains the effect of Ca^{2+} ions on the [³H]-serotonin binding (Muakkassah-Kelly et al. 1982, 1983). These problems particularly hampered quantitation of 5-HT_1 binding sites and assessment of the binding affinity of the [³H]-serotonin. Variations in K_D-values between 3 and 12 nM have been reported. There is, however, fair agreement about the general drug binding properties of 5-HT_1 sites labelled with [³H]-serotonin.

[³H]-Serotonin binding is potently inhibited by serotonin-like indole derivatives. Also most ergot derivatives, which have an indole moiety in their structure, bind with high affinity to the 5-HT_1 sites. Known serotonin antagonists, lacking the indole structure, show very low or no binding affinity for the sites, except for metitepine, a 6,7,6-membered tricyclic compound. However, metitepine is an extremely non-selective agent which binds potently to various different neurotransmitter receptor sites (see Table 4.1). Examples of inhibition curves of various serotonin agonists and antagonists are shown in Fig. 4.1.

Besides the fact that most antagonists only poorly inhibited the [³H]-serotonin binding, it was also observed that they produced shallow inhibition curves. This was interpreted by Pedigo, Yamamura and Nelson (1981) as being a possible indication for multiple binding sites. They introduced a subclassification of 5-HT_1 sites into 5-HT_{1A} sites and 5-HT_{1B} sites. 5-HT_{1A} sites, detected in the upper part of the antagonist inhibition curves, have been characterized by high affinity binding for both serotonin and serotonin antagonists. 5-HT_{1B} sites, represented in the lower part of the antagonist inhibition curves, should show high agonist but low antagonist binding affinity. Apart from the irregular binding curves there was no further identification of the binding site sub-types.

Probable heterogeneity of 5-HT_1 binding sites was also suggested by investigations with the new serotonin agonist denoted as 8-OH-DPAT (trivial name: 8-hydroxy-2(Di-n-propylamino)tetralin; chemical abstract name: 7-(dipropylamino)-5,6,7,8-tetrahydro-1-naphtolenol). 8-OH-DPAT (Fig. 10.5) proved to be the first compound without an indole

Table 4.1 Receptor binding profile of drugs used as radioligands for serotonin receptor sites

K_i values, nM

Receptor type [³H]-ligand*:	Serotonin 5-HT$_1$ [³H]-serotonin	Serotonin 5-HT$_2$ [³H]-ketanserin	Histamine-H$_1$ [³H]-mepyramine	Adrenergic-α$_1$ [³H]-WB4101	Adrenergic-α$_2$ [³H]-clonidine	Dopamine-D$_2$ [³H]-haloperidol
Serotonin	6	300	>>10 000	19 000	33 000	12 000
8-OH-DPAT	4	>1000	>1000	1030	480	2200
Ketanserin	>1000	0.39	10	10	>1000	220
Spiperone	100	0.53	>1000	10	>1000	0.16
Mianserin	1000	1.4	2.9	82	60	620
Metitepine	38	0.39	5.5	0.4	32	3.9
LSD	20	2.5	>1000	160	93	20
Metergoline	13	0.28	1100	38	380	23

*Methodological conditions see Leysen et al. (1981, 1982a).

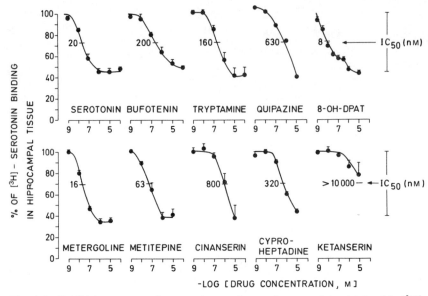

Fig. 4.1   Inhibition curves of serotonin agonists and antagonists obtained in [³H]-serotonin binding assays using rat hippocampal membranes. Assays were run at 37°C using 3 nM [³H]-serotonin in Tris–HCl buffer pH 7.6, containing 4 mM CaCl₂. Specific binding was defined as the portion of binding inhibited by 1 µM LSD. The rapid filtration procedure was used to harvest labelled membranes (full methodological description in Leysen 1981). IC₅₀-values indicate drug concentrations inhibiting 50% of the specific binding of the radioligand.

moiety which showed potent inhibition of [³H]-serotonin binding, but the inhibition curve for [³H]-serotonin binding revealed an intermediary plateau such as illustrated in Fig. 4.1. Middlemiss and Fozard (1983) used purportedly specific binding assays for measuring 5-HT$_{1A}$ sites (2 nM [³H]-serotonin, non-specific binding defined with 1 µM spiperone) and 5-HT$_{1B}$ sites (2 nM [³H]-serotonin in the presence of 1 µM spiperone throughout the buffer, non-specific binding defined with 10 µM 5-methoxytryptamine) and found high affinity binding of 8-OH-DPAT to 5-HT$_{1A}$ sites only. This led them to propose that the compound would be a useful tool to explore the significance of the 5-HT$_{1A}$ binding sites.

4.3.2   *Other labelled ligands for 5-HT$_1$ sites*

Theoretically, compounds with a nanomolar binding affinity for a particular binding site are potentially useful as labelled ligands to detect the receptor sites. From the data in Table 4.1, showing the receptor binding profile of several compounds used in serotonin receptor research, it can be inferred that besides serotonin, 8-OH-DPAT, LSD and metergoline are all radioligand candidates for the study of 5-HT_1 sites. Metitepine which also shows moderately high binding affinity for 5-HT_1 sites is a less suitable labelled ligand because of its multiple interactions with various different neurotransmitter receptor sites. [³H]-LSD has been reported to label 5-HT_1 sites in appropriate conditions (Peroutka and Snyder 1979). Since LSD binds also potently to 5-HT_2 sites, special precautions must be taken in order to avoid concomitant labelling of these sites. In whole cortex tissue, containing both 5-HT_1 and 5-HT_2 sites, a selective blocker of 5-HT_2 sites (e.g. spiperone) must be added to all assays. In hippocampal tissue, which is enriched in 5-HT_1 sites and virtually devoid of 5-HT_2 sites, [³H]-LSD labels predominantly 5-HT_1 sites. Binding affinities of drugs for the 5-HT_1 sites in hippocampal tissue, labelled with [³H]-serotonin and with [³H]-LSD were found to be highly significantly correlated (Leysen 1981). In contrast to [³H]-LSD, [¹²⁵I]-LSD was recently reported to label exclusively 5-HT_2 sites (Hartig, Kadan, Evans and Krohn 1983). [³H]-Metergoline appeared to be unsuitable for binding studies because of unacceptably high non-specific binding (Hamon, Mallat, Herbet, Nelson, Audinot, Pichat and Glowinski 1981).

As mentioned above, the new serotonin agonist 8-OH-DPAT appears an interesting compound for investigation of 5-HT_1 sites and its receptor profile reveals great selectivity for the 5-HT_1 sites. Unlike 5,6-dihydroxy- or 6,7-dihydroxy-2-aminotetralins and 5-, or 6-, or 7-monohydroxy-2-aminotetralins, which have dopamine agonist properties and bind to dopamine receptor sites (Cannon, Costall, Laduron, Leysen and Naylor 1978), 8-OH-DPAT was found to be a selective serotonin agonist (Hjorth, Carlsson, Lindberg, Sanchez, Wikström, Arvidsson, Hacksell and Nilsson 1982). The compound elicits a serotonin-like behavioural syndrome, but in contrast to the serotonin-induced behaviour (Jacobs 1976), 8-OH-DPAT-induced behaviour was reportedly not antagonizable by known serotonin agonists such as cyproheptadine and metergoline (Hjorth *et al.* 1982, and see Section 12.6). The first radioligand binding studies using [³H]-8-OH-DPAT were recently reported (Gozlan, El Mestikawy, Pichat, Glowinski and Hamon 1983; Marcinkiewicz, Vergé, Gozlan, Pichat and Hamon 1984). Sites labelled by [³H]-8-OH-DPAT in hippocampal and in striatal tissue were found to have somewhat different binding properties. In hippocampal tissue the K_D-value was about 4 nM, the maximal number of binding sites was about 600 fmoles per mg protein and non-specific binding defined in the presence of excess serotonin was rather low (less than 20 per cent of total binding up to 15 nM [³H]-8-OH-DPAT). In striatal tissue the K_D-value was higher (about 11 nM) and the B_{max} value lower (about 280 fmoles per mg protein). Unfortunately no information was provided on the non-specific binding of [³H]-8-OH-

86   Josée E. Leysen

DPAT in the striatal tissue. These are important missing data, since, as will
be illustrated in the following paragraph, [³H]-ligands can show considerable differences in non-specific binding in different brain areas which may
have impact on the specific binding of the [³H]-ligand. Most serotonin
agonists and antagonists revealed lower potencies for inhibition of [³H]-8-
OH-DPAT binding in the striatum than for inhibition of the binding in the
hippocampus; data are presented in Table 4.2. The drug binding properties
of sites labelled by 8-OH-DPAT in the hippocampus showed close similarities to the drug binding properties of 5-HT$_1$ sites labelled by serotonin
and there was a lower correlation coefficient with the drug binding properties of the striatal 8-OH-DPAT binding sites (see Table 4.4). Of note is the

Table 4.2   Inhibitory potencies ($-\log IC_{50}$, M) of various compounds in *in vitro*
binding assays using [³H]-8-OH-DPAT and [³H]-serotonin in rat hippocampus and
striatum

	Hippocampus		Striatum
	[³H]-8-OH-DPAT*	[³H]-serotonin†	[³H]-8-OH-DPAT*
Serotonin mimetics			
8-OH-DPAT	8.52	8.1	7.7
Serotonin	8.17	8.0	7.33
Bufotenin	7.67	7.1	5.99
5-Methoxy,N,N-dimethyl-tryptamine	7.40		6.54
α-Methyl-5-hydroxy-tryptamine	6.8		6.30
5-Methoxytryptamine	6.54		4.40
Tryptamine	5.43	6.6	4.63
Serotonin antagonists			
Metergoline	7.46	7.6	6.61
Methysergide	7.15	6.9	5.33
Metitepine	7.05	7.1	5.94
Cyproheptadine	6.01	6.1	5.28
Cinanserin	5.96	5.4	4.94
Pizotifen	5.66	5.7	5.21
Ketanserin	5.38	<5	4.87
Micellaneous			
Yohimbine	5.88	6.6	5.81
Haloperidol	5.45	5.0	5.38
Dopamine	4.77	4.6	3.29
Noradrenaline	3.77	3.6	3.26

Data from: *Gozlan *et al.* (1983), †Leysen *et al.* (1981).

very low binding affinity of most non-indole-like serotonin antagonists for the [^{3}H]-8-OH-DPAT binding sites. This contradicts the suggestion (Middlemiss and Fozard 1983; Marcinkiewicz *et al.* 1984) that [^{3}H]-8-OH-DPAT would label the 5-HT$_{1A}$ subclass of sites, which according to the original classification are characterized by a high binding affinity for serotonin antagonists (see above). Further extensive studies of [^{3}H]-8-OH-DPAT binding are required to characterize the binding sites in various brain areas and to investigate whether subclass(es) of 5-HT$_1$ sites are involved. Because of the particular pharmacological properties of the drug it is at the moment perhaps the most appropriate substance to trace the role of 5-HT$_1$ sites. Hitherto this has been a very controversial issue in the serotonin receptor binding research (see below, and Section 12.6). Binding studies are, however, liable to many artefacts and one should refrain from drawing premature conclusions or proposing loose hypotheses. Thorough investigations of binding properties and the demonstration of a direct relationship with a specific pharmacological or physiological effect are mandatory in order to substantiate that binding sites are receptor sites and to eventually allow subclassification of receptor sites.

4.4   Binding properties of 5-HT$_2$ sites

5-HT$_2$ sites were first identified using [^{3}H]-spiperone in frontal cortex tissue (Leysen and Laduron 1977; Leysen *et al.* 1978). The sites were characterized by a high binding affinity (of nanomolar order) for known serotonin antagonists belonging to different chemical classes. Serotonin itself shows micromolar binding affinity, but was found to bind at least 100-fold more potently to the sites than various other neurotransmitters. The use of [^{3}H]-spiperone in the 5-HT$_2$ receptor site studies suffered from the disadvantage that the ligand also binds potently to dopamine receptor sites. Detection of 5-HT$_2$ sites in the frontal cortex with [^{3}H]-spiperone was possible because of the predominant concentration of 5-HT$_2$ sites compared to very few dopamine sites in this area. However, in areas enriched in dopamine receptor sites, such as the nucleus accumbens and the tuberculum olfactorium, 5-HT$_2$ sites could only be detected with [^{3}H]-spiperone, when a dopamine receptor blocker (e.g. domperidone) was added to the assay medium (Leysen 1979). As mentioned above, [^{3}H]-LSD can label both 5-HT$_2$ and 5-HT$_1$ binding sites and labelling of dopamine receptor sites by this ligand has also been reported (reviewed in Leysen 1981). In the 5-HT$_2$ site enriched frontal cortical tissue, binding sites labelled by either [^{3}H]-LSD or [^{3}H]-spiperone show similar drug binding properties (Leysen 1981).

[^{125}I]-LSD was recently reported to be highly selective for 5-HT$_2$ sites (Hartig *et al.* 1983). Because of the very high specific radioactivity of this ligand it may be useful for autoradiographic studies. For current radio-ligand binding assays using crude membrane fractions the very highly labelled iodinated products are not of particular advantage, since labelling due to specific and non-specific binding increase to the same extent. More-over, if one uses the apparent advantage of reducing the tissue concentration with the highly labelled iodinated tracer, the variability of the assays increases.

Investigation of the binding properties of [^{3}H]-mianserin, a tetracyclic, mixed serotonin and histamine antagonist, has revealed concomitant labelling of 5-HT$_2$ and histamine-H$_1$ receptor sites (Peroutka and Snyder 1981), an observation which is in agreement with the receptor binding profile of the drug (see Table 4.1). To render this ligand suitable for 5-HT$_2$ receptor binding studies addition to all assays of a selective histamine-H$_1$ receptor blocker (e.g. triprolidine) is required.

Progress in 5-HT$_2$ binding site research was made by the introduction of [^{3}H]-ketanserin, a new serotonin antagonist which primarily binds to 5-HT$_2$ sites and lacks interaction with 5-HT$_1$ sites (Leysen, Awouters, Kennis, Laduron, Vandenberk and Janssen 1981; Leysen, Niemegeers, Van Nueten and Laduron 1982a). When used at nanomolar concentrations ($\leq$5 nM) this ligand was found to label solely 5-HT$_2$ sites. Histamine-H$_1$ and α_1-adrenergic sites, for which ketanserin shows moderate binding affinity (K_i=10 nM), are apparently not labelled at the low nanomolar [^{3}H]-ketanserin concentrations which are used in the 5-HT$_2$ binding assays.

[^{3}H]-Ketanserin revealed K_D-values (assayed at 37°C) for 5-HT$_2$ receptor sites of 0.4–1 nM. Specific [^{3}H]-ketanserin binding, defined by displacement with 1000-fold excess methysergide, displayed regular equilibrium binding kinetics, conform with a single site interaction accord-ing to the law of mass action. The binding affinity seemed to be somewhat higher in assays using frontal cortex tissue (K_D=0.4 nM) as compared to assays using striatal tissue (K_D=1.04 nM) or cat platelet membranes (K_D=1.02 nM) (Leysen, Gommeren and De Clerck 1983a). Investigations in frontal cortex tissue revealed that additives in the assay buffer influenced the [^{3}H]-ketanserin binding properties: ascorbic acid 0.1 per cent and also physiological ion concentrations (120 mM NaCl, 5 mM KCl, 2mM CaCl$_2$, 1 mM MgCl$_2$) slightly increased the K_D-value and reduced the B_{max}-value. The effect of the ions was apparently due to an increase in the dissociation rate of the [^{3}H]-ketanserin receptor complex from $t_{1/2}$ (37°C)=3.5 min in Tris-HCl buffer to $t_{1/2}$ (37°C)=1 min in Tris-HCl buffer containing the physiological ion concentrations (Leysen et al. 1982a). The reduction in binding in the presence of ascorbic acid could be caused by the lipid peroxydation induced by ascorbate (Muakkassah-Kelly et al. 1982). Figure 4.2 shows inhibition curves for [^{3}H]-ketanserin binding of serotonin

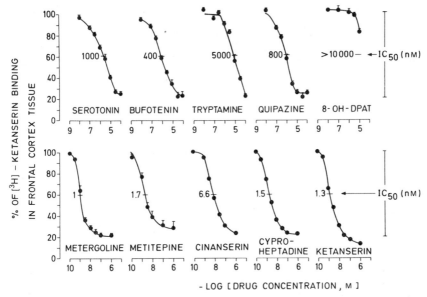

Fig. 4.2 Inhibition curves of serotonin agonists and antagonists obtained in [³H]-ketanserin binding assays using rat frontal cortex membranes. Assays were run at 37°C using 1 nM [³H]-ketanserin in Tris–HCl buffer pH 7.7. Specific binding was defined as the portion of binding inhibited by 1 μM methysergide. (Full methodological description in Leysen *et al.* 1982*a*)

agonists and antagonists from various chemical classes. Serotonin displays an $IC_{50}=1000$ nM ($K_i=300$ nM, calculated according to the Cheng and Prusoff equation), bufotenin is 2.5 times more active and tryptamine 5 times less active than serotonin, while 8-OH-DPAT shows no appreciable binding affinity for 5-HT$_2$ sites. All the serotonin antagonists have IC_{50}-values of nanomolar order. Comparison of the data in Figs. 4.1 and 4.2 clearly shows that 5-HT$_1$ and 5-HT$_2$ sites are unrelated. Otherwise, the binding affinities of drugs for 5-HT$_2$ sites labelled with [³H]-ketanserin in membrane preparation of different tissues, such as prefrontal cortex, striatum and cat platelets, or with [³H]-spiperone in membranes from frontal cortex or with [³H]-LSD in membrane preparation from human platelets were highly significantly correlated. K_i-values of a number of archetypal drugs measured in 5-HT$_2$ binding models using various labelled ligands and tissues are presented in Table 4.3. Table 4.4 shows the rank correlation coefficients between binding affinities of drugs measured in the various radioligand binding models for 5-HT$_1$ and 5-HT$_2$ sites.

Table 4.3 K_i-values (nM) of compounds for 5-HT$_2$ sites labelled with various [^{3}H]-ligands in various tissues from different species

	[^{3}H]-spiperone	[^{3}H]-ketanserin			[^{3}H]-LSD	
	Rat	Rat		Human	Cat	Human
	Frontal* cortex	Frontal† cortex	Striatum‡	Frontal** cortex	Platelets‡	Platelets§
Serotonin antagonist						
Ketanserin	2.1	0.39	0.57	2.28	0.60	6.95
Cyproheptadine	6.5	0.44	1.0	5.2	2.0	12.2
Methysergide	12	0.94	2.3	2.0	7.1	15.8
LSD	8.2	2.5		0.28		0.77
Pipamperone	5.3	0.78	2.7		4.0	
Dopamine antagonists						
Spiperone	1.2	0.53	0.45	0.92	1.1	6.79
Haloperidol	48	22	20	24.2	32	580
Domperidone	327	78	100		54	
Histamine antagonist						
Mepyramine	2600	>1000	1000	657	1140	1506
Adrenergic antagonist						
Prazosin		>1000	1800	>10 000	850	>10 000
Amine re-uptake blockers						
Amitriptyline	21	4.2	21		21	165
Desipramine		78	140		210	
Serotonin agonists						
Bufotenin	518	118	810	1144	72	
Serotonin	1033	296	2870		290	81.3
Tryptamine	5420	1482	8080		1070	
Various biogenic amines						
Dopamine	82 000	93 500	25 500	>10 000	20 000	>100 000
Norepinephrine	>100 000	>100 000	200 000	>10 000	53 200	>100 000
Histamine	>100 000	>100 000	102 000		42 500	>100 000

Table 4.4 Correlation between various radioligand binding models for 5-HT$_1$ and 5-HT$_2$ sites, using data from Tables 4.2 and 4.3

	I	II	III	IV	V	VI	VII	VIII	IX
	Spearman rank correlation coefficients								
5-HT$_1$ models									
(I) [³H]-Serotonin, rat hippocampus		r_s=0.912 p<0.001	r_s=0.869 p<0.001	r_s=−0.095 p>0.05	r_s=−0.017 p>0.05	r_s=−0.017 p>0.05	r_s=−0.154 p>0.05	r_s=−0.017 p>0.05	r_s=−0.018 p>0.05
(II) [³H]-8-OH-DPAT, rat hippocampus			r_s=0.860 p<0.001	r_s=0.214 p>0.05	r_s=0.400 p>0.05	r_s=0.400 p>0.05	r_s=0.564 p>0.05	r_s=0.400 p>0.05	r_s=0.378 p>0.05
(III) [³H]-8-OH-DPAT, rat striatum				r_s=0.143 p>0.05	r_s=0.417 p>0.05	r_s=0.417 p>0.05	r_s=0.564 p>0.05	r_s=0.417 p>0.05	r_s=0.270 p>0.05
5-HT$_2$ models									
(IV) [³H]-Spiperone, rat frontal cortex					r_s=0.960 p<0.001	r_s=0.967 p<0.001	r_s=0.924 p<0.001	r_s=0.984 p<0.001	r_s=0.907 p<0.001
(V) [³H]-Ketanserin, rat frontal cortex						r_s=0.949 p<0.001	r_s=0.887 p<0.001	r_s=0.979 p<0.001	r_s=0.879 p<0.001
(VI) [³H]-Ketanserin, rat striatum							r_s=0.890 p<0.001	r_s=0.973 p<0.001	r_s=0.893 p<0.001
(VII) [³H]-Ketanserin, human frontal cortex								r_s=0.890 p<0.001	r_s=0.953 p<0.001
(VIII) [³H]-Ketanserin, cat platelets									r_s=0.958 p<0.001
(IX) [³H]-LSD, human platelets									

4.5   Possible causes of anomalous binding data

The membrane preparations, which are used in the radioligand binding assay, consist of a micellar suspension of membrane fragments in the aqueous medium. Interactions of solutes with membrane micelles are complex, comprising different phenomena. The primary interactions between the solutes and the micelles are the surface effects. These can be of different nature (electrostatic, conformational rearrangements in the membranes, surface excess of drugs) and have impact on the overall binding process (Leysen and Gommeren 1981). Secondly, the several constituents of membranes (lipids, extrinsic and intrinsic proteins, carbohydrate chains, metals, etc. . .) are prone to many different interactions with ligands. The actual binding of a drug to the receptor protein, sometimes only represents a small component in the total binding. A number of possible interactions between drugs and membrane micelles are summarized in Table 4.5. Moreover, all the various interactions can be influenced by assay conditions (ions, ionic strength, pH, temperature) and can vary between different tissues. In view of the complexity of the binding process very careful identification of specific receptor binding is necessary and observations of irregular binding curves or small variations between drug binding properties in different tissues are not necessarily an indication of the involvement of multiple receptor sites. This will be illustrated in the following examples on possible interpretations of $5\text{-}HT_2$ binding data.

Data described in the above paragraph provide consistent evidence for

Table 4.5   Possible interactions between ligands and membrane micelles

| Type of interaction | Features | | | |
	Affinity	Saturating concentration	Inhibitory agents	Correlation
Binding to receptors	high	nanomolar	pharmacological congeners	pharmacological potencies
Binding to recognition sites of a chemical structure	high	nanomolar	chemical congeners	chemical structure
Metal chelation	high	micromolar	chelating agents	chelating potency
Electrostatic	low	no	reduced at high ionic strength	charge, dipole moment
Adsorption	low	no	—	lipophilicity
Surface effects	variable		detergents	physicochemical properties

similarities between 5-HT$_2$ sites labelled with various [^{3}H]-ligands in different tissues. Nevertheless the binding data derived from the different binding models are not completely identical, although this should theoretically be so when the law of mass action for single site interaction was strictly applicable. Using [^{3}H]-ketanserin in different tissues it appeared that the binding affinities of drugs were in general 3–5 times higher when measured in the frontal cortex as compared to the striatum or cat platelets, but the rank order of potency of the drugs in the three tissues were highly significantly correlated. Are these apparent variations in binding affinities necessarily indicative of differences in the specific binding sites in different tissues? Investigations of [^{3}H]-ketanserin binding revealed that in striatal and platelet tissue non-specific [^{3}H]-ketanserin binding represented a high proportion of total binding and the non-specific binding in these tissues was apparently of a different nature than that observed in cortex tissue. Both striatal and platelet membranes contained a high proportion of what is called displaceable non-specific binding sites (see Table 4.5). The various types of binding can be identified by careful investigation of inhibition of the binding by drugs of different chemical classes. Examples of inhibition curves obtained in [^{3}H]-ketanserin binding assays in striatal and platelet membranes are shown in Figs. 4.3 and 4.4. It should be noted that unlabelled ketanserin inhibits 70 to 80 per cent of the total [^{3}H]-ketanserin binding in both tissues, whereas, serotonin antagonists with a different chemical structure (spiperone, pipamperone, cyproheptadine, methysergide) or the serotonin agonist (bufotenin) reveal an inflection point or an intermediary plateau in their inhibition curves at the level of about 40 per cent inhibition of total binding. Only this portion of the binding involves specific binding to 5-HT$_2$ sites, the remaining binding is for about 40 per cent to displaceable non-specific sites and for 20 per cent to non-displaceable adsorption. Curves in Fig. 4.3 show that the chemical congener of ketanserin, R 51 230, which is a weak serotonin antagonist is able to displace [^{3}H]-ketanserin from the displaceable non-specific sites. Displaceable non-specific [^{3}H]-ligand binding is a commonly occurring, but not always recognized, phenomenon in radioligand binding studies (Leysen 1984). This type of binding does not seem to be related to a pharmacological or physiological function. It is, however, possible that through non-specific interactions the available free [^{3}H]-ligand at the receptor site is reduced or that the access of the [^{3}H]-ligand towards the specific binding sites is hindered. Non-specific binding to tissue must also occur with the non-labelled drugs, but for these quantitative data are missing. Because of possible effects of non-specific binding on specific binding, minor apparent variations in specific binding should not be regarded as a substantial indication for the involvement of different receptor sites.

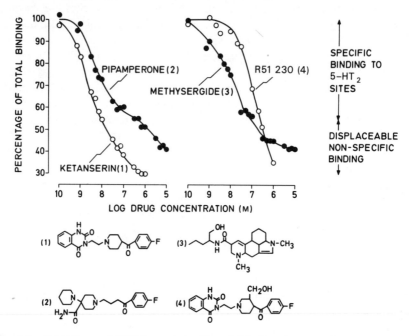

Fig. 4.3 Inhibition of [³H]-ketanserin (1 nM) binding in rat striatal membrane preparations by drugs of various chemical structures, (1) (2) (3) are potent serotonin antagonists, (4) is a chemical congener of ketanserin with very weak serotonin antagonistic properties.

A second apparent anomaly in the binding data described in the foregoing paragraph is the somewhat lower potency of drugs to inhibit [³H]-spiperone and [³H]-LSD binding compared to [³H]-ketanserin binding. This observation can probably be explained by the slow dissociation of spiperone and LSD from the receptor sites. Using a new technique, developed for measuring drug-receptor dissociation rates of unlabelled drugs (Leysen and Gommeren 1984), the following half-lifes of dissociation of drugs from the 5-HT$_2$ receptor sites were found: serotonin: 6.5±1.0 min (n=8), ketanserin: 11±1.0 min (n=10), mianserin: 15±3 min (n=5), spiperone: 30±4 min (n=11), LSD > 90 min. These values indicate the free dissociation of the unlabelled drug from drug-saturated receptor sites. With the slowly dissociating drugs, binding equilibrium will not be reached during relatively short incubation times (15 min at 37°C in assay with [³H]-ketanserin and [³H]-spiperone). Although in binding studies using [³H]-LSD in human platelets, incubation was run for four hours, this may still not be sufficiently long for reaching LSD binding equilibrium. The more slower the receptor

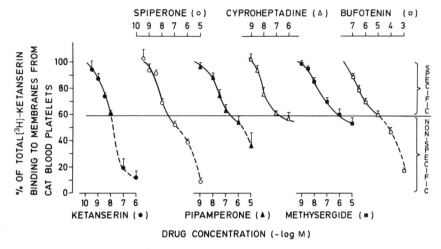

Fig. 4.4 Inhibition of [³H]-ketanserin binding in cat platelet membranes by serotonin antagonists of various chemical structures (quinazolinedione derivative: ketanserin; butyrophenones: spiperone, pipamperone; tricyclic derivative: cyproheptadine; ergoline derivative: methysergide) and an indole-like serotonin agonist (bufotenin) (full methodological description in Leysen *et al.* 1983*a*). (Figure taken from Leysen *et al.* 1983*a*, with permission from Elsevier Biomedical Press)

dissociation rate of the labelled ligand, the less readily will its binding be inhibited by non-labelled drugs, this conforms with the experimental findings presented in Table 4.3.

The fact that drugs show widely differing dissociation rates from receptor sites is a very new datum in receptor research and the implications for the interpretation of binding data is just starting to be recognized.

It can be concluded that data on specific binding are the eventual reflection of a number of complicated events which can be influenced by specific and non-specific interactions of the drugs with membranes and which involve a number of unknown and uncontrollable factors. Therefore binding data on themselves are not a reliable basis for hypothesizing the existence of receptors or receptor sub-types.

4.6   Serotonin-stimulated adenylate cyclase

Adenylate cyclase activity, stimulated by micromolar concentrations of serotonin was detected primarily in colliculi of new-born rats. Known serotonin antagonists were found to be poorly active or not active at the cyclase activity. The idea has arisen that binding sites which show decreased binding affinity for agonists in the presence of guanyl nucleotides are coupled to an adenylate cyclase (Rodbell 1980). Therefore a linkage of

5-HT$_1$ sites to the serotonin-stimulated adenylate cyclase was proposed (Peroutka, Lebovitz and Snyder 1979, 1981). However, no correlation was found between drug potencies to inhibit [^{3}H]-serotonin binding and to inhibit the serotonin-stimulated cAMP production. Moreover the [^{3}H]-serotonin binding sites and the serotonin-stimulated adenylate cyclase showed a distinct regional and subcellular distribution. This refuted a possible coupling between both systems (for reviews see Leysen 1981; Leysen and Tollenaere 1982). The physiological role of the cyclase is thus far unknown.

4.7   Localization of binding sites

The distribution of 5-HT$_1$ sites, labelled with [^{3}H]-serotonin, and of 5-HT$_2$ sites labelled with [^{3}H]-ketanserin, in rat brain, compared to the distribution of specific serotonin uptake is presented in Table 4.6. 5-HT$_1$ sites are mostly enriched in hippocampus and striatum and 5-HT$_2$ sites in the prefrontal cortex, the tuberculum olfactorium and the nucleus accumbens. Neither the distribution of 5-HT$_1$ or of 5-HT$_2$ sites corresponds to the regional brain distribution of specific serotonin uptake or to the distribution of serotonin content in the brain areas (Leysen and Tollenaere 1982; Leysen, Van Gompel, Verwimp and Niemegeers 1983b). This absence of correspondence between binding site density and degree of serotonergic innervation of brain areas is poorly understood and has occasionally been

Table 4.6 Proportional distribution in rat brain subcortical areas of 5-HT$_2$ and 5-HT$_1$ binding sites and serotonin uptake

	5-HT$_2$	5-HT$_1$	Uptake
Prefrontal cortex	100	100	100
Tuberculum olfactorium	56	67	156
Nucleus accumbens	40	—	92
Striatum	36	150	93
Substantia nigra	15	—	131
Thalamus	10	82	132
Hypothalamus	9	—	109
Medulla oblongata	10	—	87
Hippocampus	12	198	67
Spinal cord	1	—	48
Cerebellum	0	0	17

100% values in prefrontal cortex tissue represented: for 5-HT$_2$ site binding using 2 nM [^{3}H]-ketanserin: 21±6 fmols/mg wet weight tissue ($n=12$); for 5-HT$_1$ site binding using 3 nM [^{3}H]-serotonin: 13±1 fmols/mg wet weight tissue ($n=9$); for [^{3}H]-serotonin uptake using 10 nM [^{3}H]-serotonin: 150±30 fmols/mg wet weight tissue, 5 min.

raised as an argument against a neurotransmitter role of the sites (Barbaccia, Gandolfi, Chuang and Costa 1983). However, the ratio between functional and spare receptors and the relationship between receptor density, receptor function and intensity of elicited effect still are unknown factors. Otherwise 5-HT$_2$ receptor sites were found to be similarly distributed in brains of various mammalian species including man (Leysen et al. 1983b; Schotte, Maloteaux and Laduron 1983). Moreover, the ontogenetic development of brain 5-HT$_2$ sites, detectable from the 16th day after gestation and marked development occurring during the second postnatal week, was found to parallel that of various other neurotransmitter receptor sites in correspondence with neuronal development (Bruinink, Lichtensteiger and Schlumpf 1983). These observations support a probable neurotransmitter role for 5-HT$_2$ sites. In contrast, 5-HT$_1$ sites were found to be high during early embryonic development and decreased substantially during the first two weeks of life. Since a probable role of serotonin in cell differentiation had been suggested, the link was tentatively made between the decrease in 5-HT$_1$ sites shortly after birth and the likely postnatal decrease of serotonin sites involved in cell differentiation (Uzbekov, Murphy and Rose 1979).

Information on the cellular localization of the serotonergic binding sites in the brain was investigated in lesion studies in rat brain. Regarding 5-HT$_1$ sites, different findings have been made using different [^{3}H]-ligands. Destruction of catecholaminergic neurones following local injection of 6-hydroxydopamine, did not affect [^{3}H]-serotonin binding in the forebrain. This suggested that the 5-HT$_1$ sites either do not occur, or do so in an undetectable low amount, on dopaminergic or noradrenergic neurones. Lesioning of neuronal cell bodies in the striatum by local application of kainic acid reportedly resulted in a decrease of high affinity [^{3}H]-serotonin binding, but it did apparently not affect the striatal [^{3}H]-8-OH-DPAT binding. However, hippocampal [^{3}H]-8-OH-DPAT binding was found to decrease after injection of kainic acid in this brain area. Following destruction of serotonergic neurones by neurotoxins (5,6- or 5,7-DHT), forebrain [^{3}H]-serotonin binding was not altered and neither was [^{3}H]-8-OH-DPAT binding in the hippocampus, but [^{3}H]-8-OH-DPAT binding in the striatum seemed to be decreased. The lesion data were interpreted as follows. Some of the 5-HT$_1$ sites were localized postsynaptically (i.e. the binding sites affected by kainic acid lesions) whereas the striatal [^{3}H]-8-OH-DPAT sites were localized on serotonergic nerve terminals, and were hence called autoreceptor sites (Gozlan et al. 1983).

5-HT$_2$ sites, labelled by [^{3}H]-ketanserin in frontal cortical areas and striatal tissues were not affected by lesions of catecholaminergic and serotonergic neurones. High doses of kainic acid in the frontal cortex caused a decrease of 5-HT$_2$ sites which was found to be accompanied by a decrease in glutamic acid decarboxylase activity, the enzyme contained in GABA-ergic neurones. An association of 5-HT$_2$ sites with GABA-ergic neurones is therefore possible (Leysen, Geerts, Gommeren, Verwimp and Van Gompel 1982b; Leysen et al. 1983b). Apparent differences in effects of neuronal lesions on [^{3}H]-mianserin binding sites (which reportedly increased in hippocampus following destruction of serotonergic neurones) and [^{3}H]-ketanserin binding sites (not changed after the lesion) led Barbaccia et al. (1983) to suggest that [^{3}H]-mianserin and [^{3}H]-ketanserin recognition sites were not identical. The apparent absence of up-regulation of [^{3}H]-ketanserin binding sites following neuronal denervation was used as an argument to cast further doubt on the neurotransmitter receptor role of the sites (Barbaccia et al. 1983). It should be noted however that the conclusions of Barbaccia et al. (1983) are based on specific [^{3}H]-mianserin binding in the hippocampus, an area which is very poor in 5-HT$_2$ receptor sites. Moreover, since [^{3}H]-mianserin suffers from the disadvantage of labelling both histamine-H$_1$ and 5-HT$_2$ sites, the significance of these data remains in doubt.

In conclusion, binding site localization and binding data following neuronal lesions in themselves do not provide sufficient proof for either attributing or refuting a functional role of receptor binding sites. On one hand this is the consequence of our lack of information regarding the ratio between functionally active and so-called spare receptor sites. On the other hand it appears that binding studies still are too much prone to artefacts and findings need extremely critical evaluation, a point which is often neglected.

4.8   The role of binding sites

4.8.1   *The use of correlations for investigating functional roles of binding sites*

The crucial issue in receptor binding research is the demonstration of a role of the binding sites in pharmacological and physiological effects and the most readily available means is the investigation of a correlation between binding affinities of drugs and their potencies in functional tests.

The value of correlations has been much criticized, and correlations are often misused. Nevertheless, investigation of correlations can allow well-substantiated conclusions providing that a large series of drugs comprising compounds of different chemical classes and spanning a wide activity scale are considered. When *in vitro* binding data are compared to *in vivo* pharmacological potencies the problems of drug resorption, distribution and metabolism must be kept in mind and the deviation of a single compound from a well-established correlation line should not lead to receptor sub-types being hypothesized. When *in vitro* binding tests are compared to *in vitro* functional tests on isolated tissues, not only should rank order of potencies correspond, but binding and functional activities should occur at the same drug concentration.

According to these rules, investigation of the role of 5-HT$_2$ sites has been much more successful than investigation of the role of the 5-HT$_1$ sites. Indeed all known serotonin antagonists, belonging to various different chemical classes bind to 5-HT$_2$ sites and binding affinities of drugs vary from the sub-nanomolar to micromolar order of magnitude. In contrast, very few compounds potently bind to 5-HT$_1$ sites and they mostly are serotonin-mimetic indole derivatives. It is often alledged that the absence of potent and selective 5-HT$_1$ site blockers has hampered the investigation of its role. Preliminary behavioural data on this site are discussed later (sections 10.3, 11.3, 12.5 and 12.6).

4.8.2   *The binding affinity of the endogenous neurotransmitter*

The primary reason for allocating a receptor role to 5-HT$_1$ sites, resided in the nanomolar binding affinity of serotonin itself for the sites, whereas the micromolar affinity of serotonin for the 5-HT$_2$ sites was used as an argument against a functional role for the latter sites.

The idea that endogenous neurotransmitters must display nanomolar binding affinities for target receptor sites does not however conform to the concept of humoral neurotransmission. According to this concept, receptor sites are likely to be localized on the membranes forming the synaptic cleft, where high local concentrations of neurotransmitters are attained. The receptor binding affinity of the endogenous neurotransmitter must be of the order of magnitude of its local concentration, which presumably is of micromolar order. This requirement is fulfilled by 5-HT$_2$ sites, but not by 5-HT$_1$ sites. Nanomolar binding affinities of endogenous substances are to be reconciled with endocrine hormone actions or with a localization of 'neurotransmitter' receptor sites distant from synaptic regions of neurones. At present there are no tangible data on the functional role of such systems in the central nervous system.

4.8.3   *Role of binding sites in* in vivo *effects elicited by serotonin-mimetics*

Behavioural studies in rodents have revealed that administration of serotonin precursors or serotonin-mimetic agents elicites multiple effects, some of which are listed in Table 4.7 and are reviewed in detail in Chapter 12. Observed effects reflect behavioural excitation (head twitches, forepaw

100   Josée E. Leysen

Table 4.7 *In vivo* effects of serotonin-mimetics

Serotonin agonist treatment	Observation	Receptor site	Refs.
5-Hydroxytryptophan +monoamine oxidase inhibitor	serotonin syndrome in rats, mice: head twitches, forepaw treading, cyanosis, hindlimb abduction, . . . , gastric lesions, dead	multiple serotonin receptors	(1, 2)
5-Hydroxytryptophan	head twitches, rats, guinea pigs	5-HT$_2$	(3)
Mescaline	head twitches, rats, mice	5-HT$_2$	(4, 5)
Tryptamine	forepaw treading, rats	5-HT$_2$	(6)
5-Hydroxytryptophan	myoclonus, guinea pig	multiple serotonin receptors?	(7)
5-Hydroxytryptophan, LSD, quipazine	discriminative stimulus in rats	5-HT$_2$ component	(8, 9, 10, 11)
Compound 48/80	gastric lesions	5-HT$_2$	(12)
Serotonin	paw oedema	5-HT$_2$	(2)

(1) Chapter 12 (Section 12.2), (2) Ortman *et al.* 1982, (3) Peroutka *et al.* 1981, (4) Leysen *et al.* 1982a, (5) Niemegeers *et al.* 1983, (6) Leysen *et al.* 1978, (7) Chapter 12 (Section 12.4), (8) Colpaert *et al.* 1982, (9) Glennon *et al.* 1983, (10) Friedman *et al.* 1984, (11) Chapter 10 (Section 10.3), (12) Awouters *et al.* 1982.

treading, myoclonus, wet dog shakes) impairment of blood circulation (cyanosis, gastric lesions) and subjective perceptions of the drugs by the animals (discriminative stimulus). Several studies involving large series of serotonin antagonists from different chemical classes have shown that the binding affinities of the drugs for 5-HT$_2$ sites are highly significantly correlated with the potencies of the drugs to antagonize serotonin-induced behavioural excitation, measured in different ways (see Table 4.7). *In vitro* drug binding affinities are similarly correlated with the drug potencies to antagonize serotonin-induced gastric lesions (Awouters, Leysen, De Clerck and Van Nueten 1982), or serotonin-induced paw oedema (Ortmann, Bischoff, Radeke, Buech and Delini-Stula 1982). From these studies a role of 5-HT$_2$ sites in serotonin-mediated behavioural excitation, impairment of blood circulation and inflammation is clearly apparent. Other studies, on a limited number of drugs have suggested involvement of multiple, not further identified, serotonin receptors in 5-hydroxytryptophan induced myoclonus in guinea pigs (see Section 12.4). In studies on serotonin agonist (5-HTP, LSD, quipazine) induced discriminative stimulus in rats, certain authors recognized a 5-HT$_2$ receptor site component (Colpaert, Niemegeers and Janssen 1982; Glennon, Young and Rosecrans 1983; Friedman, Barrett and Sanders-Bush 1984 and see Section 10.3). This would indicate that subjective feelings (presumably anxiety, neurotic depression, tension) which are perceived by animals following treatment with serotonin agonists are partially mediated by 5-HT$_2$ sites. Besides the effects mentioned above, serotonin has been suggested to mediate other behaviours such as sexual behaviour (Ahlenius, Larsson, Svensson, Hjorth, Carlsson, Lindberg, Wikström, Sanchez, Arvidsson, Hacksell and Nilsson 1981), sleep (Harmann 1974), and pain perception (Hylden and Wilcox 1983). In general, studies related to these effects have been restricted to the use of only a few serotonin agonists and antagonists, which does not allow a proper identification and characterization of the type of receptor sites which are involved. No study could thus far provide reliable proof for a role of 5-HT$_1$ binding sites, in any behavioural effect induced by serotonin although there are now a few interesting suggestions (see Sections 12.5 and 12.6).

4.8.4   *Role of binding sites in effects of serotonin on isolated organs*

The majority (90 per cent) of whole body serotonin is found in the periphery where it occurs in the enterochromaffin cells in the gut and in blood platelets. Serotonin was first discovered in these peripheral cells and the early investigations with the endogenous amine concerned its action on peripheral tissues. *In vitro* pharmacological studies, demonstrated multiple effects of serotonin on smooth muscle tissue of various organs; examples are given in Table 4.8. The early defined 'M' and 'D' receptors have not yet found a corresponding radioligand binding model (Leysen *et al.* 1982*a*; Leysen 1983). However, a very close correspondence is found between binding affinities of drugs for 5-HT$_2$ sites (either measured using [^{3}H]-spiperone (Leysen 1981) or using [^{3}H]-ketanserin (Leysen *et al.* 1982*a*)) and drug potencies to antagonize serotonin-induced contractions in rat caudal artery (correlation shown in Fig. 4.5) in rabbit femoral artery (Van Nueten, De Ridder and Vanhoutte 1982*a*) and in dog basilar artery (Müller-Schweinitzer and Engel 1983; Van Nueten, De Clerck and Vanhoutte 1984). Moreover, to elicit vessel contraction, serotonin concentrations of $10^{-7}-10^{-6}$ M are required (Van Nueten 1983), which corresponds to the binding affinity of serotonin for the 5-HT$_2$ sites (see Fig. 4.2). Hence the functional role of 5-HT$_2$ sites in the vascular smooth muscle is well documented. The failure to directly label 5-HT$_2$ sites in vascular tissue homogenates is probably due to their having a low receptor density but high functional activity in peripheral tissue. It would appear that a direct relationship between total receptor density and function is not necessary. A controversial issue concerns the effect of serotonin in the basilary artery. Peroutka, Noguchi, Tolner and Allen (1983) suggested that these effects were mediated by 5-HT$_1$ sites. However, antagonism of the

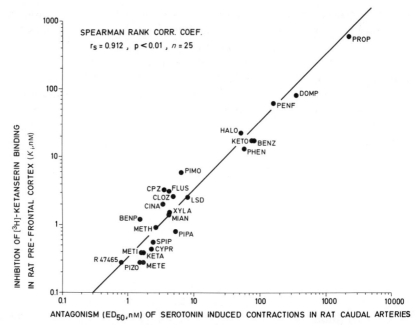

Fig. 4.5 Correlation between *in vitro* measured binding affinities of drugs for 5-HT$_2$ sites (K_i values for inhibition of [^{3}H]-ketanserin binding in rat frontal cortex tissue) and the potencies of drugs to antagonize serotonin-induced contractions in isolated rat caudal arteries. (Figure taken from Leysen *et al.* 1982*a*, with permission from the American Society for Pharmacology and Experimental Therapeutics)

PIZO: pizotifen
METE: metergoline
METI: metitepine
KETA: ketanserin
CYPR: cyproheptadine
SPIP: spiperone
METH: methysergide
PIPA: pipamperone
BENP: benperidol
MIAN: mianserin
XYLA: xylamidine
CINA: cinanserin

CLOZ: clozapine
CPZ: chlorpromazine
FLUS: fluspirilene
PIMO: pimozide
PHEN: phentolamine
KETO: ketotifen
BENZ: benzotropine
HALO: haloperidol
PENF: penfluridol
DOMP: domperidone
PROP: propanolol

serotonin induced contraction in the basilary artery appear to be complex and not of a fully competitive nature (Van Nueten *et al.* 1984). The conclusions of the study by Peroutka *et al.* (1983) seem to be based on a misinterpretation of the *in vitro* pharmacological findings. By more thorough data analysis other investigators have proved that 5-HT$_2$ sites, and not 5-HT$_1$ sites are involved in the serotonin-induced basilary artery constriction (Müller-Schweinitzer and Engel 1983; van Nueten *et al.* 1984). Besides its direct action on serotonin receptors, serotonin at higher

Table 4.8   Serotonin-induced effects on isolated tissues

Tissue and serotonin-induced effect	Typical antagonists	Receptor site	Ref.
Gastrointestinal smooth muscle			
contraction of ileum	morphine	'M'-receptor	(1)
contraction of ileum, fundus	phenoxybenzamine	'D'-receptor	(1, 2)
Vascular smooth muscle			
contraction of { rat caudal artery, rabbit femoral artery, dog basilar artery }	{ ketanserin, cyproheptadine, methysergide }	5-HT$_2$	(3, 4) (5, 6, 7)
contraction of { dog saphenous vein, rabbit ear artery }	{ phentolamine, prazosin }	α_1-adrenergic	(3)
relaxation of contracted vessels dog coronary artery	metitepine methysergide	?	(6, 7)
Trachea, guinea pig	{ ketanserin, cyproheptadine, methysergide }	5-HT$_2$	(8)

(1) Gaddum and Picarelli 1957, (2) Vane 1957, (3) Van Nueten *et al.* 1981, (4) Leysen *et al.* 1982*a*, (5) Van Nueten *et al.* 1982*a*, (6) Müller-Schweinitzer and Engel 1983, (7) Van Nueten *et al.* 1984, (8) Van Nueten *et al.* 1982*b*.

concentrations can cross-react with α_1-adrenergic receptor in certain vascular tissues (see Table 4.8). Antagonism of these effects is only accomplished by α_1-adrenergic antagonists. Finally, serotonin was found to induce relaxation of contracted vessels and the effect is antagonizable by metitepine and methysergide (Van Nueten *et al.* 1984). Too few substances have been found active in this test to allow receptor identification.

4.8.5   *Mediation by serotonin of neurotransmitter release*

A role for 5-HT$_1$ sites has most often been proposed in the apparent attenuation by serotonin of its own release in brain tissue, i.e. a so-called autoreceptor function (Martin and Sanders-Bush 1982; Engel, Göthert, Müller-Schweinitzer, Schlicker, Sistonen and Stadler 1983). Also serotonin regulation of dopamine release such as has been measured in striatal slices (Ennis, Kemp and Cox 1981). The regulation of noradrenaline from sympathetic nerve endings measured in canine saphenous vein segments (Engel *et al.* 1983) has also been investigated for a possible role of 5-HT$_1$ sites.

The release experiments consist of pre-loading the tissue slices or crude synaptosome preparations with the tritiated neurotransmitter, followed by stimulation of the [^{3}H]-amine release by potassium or electrical impulses. A general problem with the *in vitro* release studies is the uncertainty of whether the labelled neurotransmitters are stored in the same compartment and released in the same way as the endogenous amine.

It was first hypothesized that $5-HT_1$ sites would mediate serotonergic regulation of potassium evoked [^{3}H]-dopamine release from striatal slices (Ennis *et al.* 1981). The evidence advanced was an apparent correlation between the rank order of potencies of five serotonin antagonists in the binding and release test. The hypothesis has been refuted because of a disagreement between binding data used by Ennis *et al.* (1981) and those reported by several other groups (Peroutka and Snyder 1979; Nelson, Herbet, Enjalbert, Bockaert and Hamon 1980; Leysen and Tollenaere 1982).

The hypothesis on an autoreceptor role for $5-HT_1$ sites originated from an observed correlation between the binding affinities of a rather limited number of serotonin agonists and the potencies of the agonists to inhibit potassium-stimulated release of [^{3}H]-serotonin from hypothalamic synaptosomes (Martin and Sanders-Bush 1982). But the relative binding affinity of serotonin antagonists (metergoline > metitepine > methysergide) did not correspond to their potency in the serotonin release test (metitepine $\gg$ metergoline $\geq$ methysergide). Engel *et al.* (1983) supported the $5-HT_1$ autoreceptor hypothesis based on their finding of a rank correlation between binding affinities of indoleamines for $5-HT_1$ sites in cortex tissue and inhibition of electrically-evoked [^{3}H]-serotonin release from cortex slices. They provided, however, no adequate explanation for the 100-fold weaker potency of most (but not all) serotonin agonists for inhibiting serotonin release in their cortex slice model as compared to the hypothalamic synaptosome model used by Martin and Sanders-Bush (1982). In the study of Engel *et al.* (1983) the potencies of serotonin antagonists did not fit the correlation between the drug potencies in the binding and release test. Ennis and Cox (Ennis and Cox 1982; Cox and Ennis 1982) did not share the viewpoint that serotonin autoreceptors would be of the $5-HT_1$ binding site type. They supplied apparent evidence that dendritic serotonin autoreceptors in the raphe, which in their hands showed some functional similarities with autoreceptors on serotonergic nerve endings in the hypothalamus, were different from $5-HT_1$ binding sites. Engel tried to reconcile the contradictory findings by making the rather awkward suggestion that different types of autoreceptors may exist on the same neurone! Finally, a distinct hypothesis accruing from the study of Engel, concerned the probable role of $5-HT_1$ sites in the serotonin-induced attenuation of noradrenaline release from sympathetic nerves in the dog saphenous vein. The hypothesis was solely based on investigations with some serotonin agonists; it awaits confirmation by studies using antagonists!

Despite many discrepancies and criticisms, an argument found in nearly all release studies, is that the lack of effect of neuronal lesions (destruction of serotonergic or catecholaminergic neurones) on $5\text{-}HT_1$ binding sites (see Section 4.7) cannot adequately refute a localization of the sites on the destroyed neurones. It is presumed that the portion of presynaptic binding sites may be very small and beyond the detection level of binding studies. What these authors disregard as evidence, represented the main conclusive force to attribute an autoreceptor function to striatal [³H]-8-OH-DPAT binding sites (Gozlan et al. 1983) (see Section 4.7). The functional evidence for the presumed autoreceptor role of the striatal [³H]-8-OH-DPAT binding sites only concerned the greater binding affinity of 5-MeODMT as compared to bufotenin (see Table 4.2), which was found to correspond to the potencies of the compounds for inhibiting [³H]-serotonin release in the study of Martin and Sanders-Bush (1982). However, Gozlan et al. (1983) failed to mention the fact that tryptamine, which bound 80 times weaker than 5-MeODMT, was 15 times more potent than the latter compound in the serotonin release test. Upon re-analysis of entire published data we did not find a significant correlation between either agonist or antagonist binding affinities for striatal [³H]-8-OH-DPAT binding sites (Gozlan et al. 1983) and the drug potencies in release tests (Martin and Sanders-Bush 1982; Engel et al. 1983).

It can be concluded that the evidence for an autoreceptor function for $5\text{-}HT_1$ sites or for a sub-population of [³H]-8-OH-DPAT binding sites is far from convincing. This conclusion has also been reached by Moret (Section 2.5) in her review of this area. That leaves us with a population of binding sites, allegedly consisting of multiple subclasses, for which a function is still missing.

4.8.6 *Mediation by serotonin of neuronal firing*

Serotonin has been found to exert both excitatory and inhibitory effects on neuronal firing. Electrophysiological studies led to a particular subclassification of serotonin receptor sites (see Rogawski and Aghajanian 1981 and Section 7.1.2), distinct from the classification based on binding studies. Peroutka, Lebovitz and Snyder (1981) made the assumption that the electrophysiologically defined excitatory serotonin receptor would be akin to $5\text{-}HT_2$ binding sites. However, the authors apparently confused excitatory behaviour with neuronal excitation; the relationship between both has not been demonstrated. Moreover, Peroutka et al. (1981) further suggested, without any foundation, that neuronal inhibition would be mediated by $5\text{-}HT_1$ sites. Neither allegation is substantiated by direct comparison of drug potencies in the binding tests and in the electrophysiological tests. From the limited data available little correspondence is apparent between the electrophysiologically defined receptors and receptor binding sites.

4.9   The platelet: a model for characterization of 5-HT$_2$ receptors?

Serotonin is known to affect platelet function. In some species (such as man) this is reflected in a mild response (reversible platelet shape change), whereas in other species (such as the cat) serotonin stimulation of platelets produces a marked aggregation response which becomes irreversible when high serotonin concentrations are used (De Clerck and Herman 1983; De Clerck, Xhonneux, Leysen and Janssen 1984). The finding that ketanserin was particularly effective in inhibiting serotonin-induced platelet function tentatively suggested a role for 5-HT$_2$ sites (De Clerck and Herman 1983). The first successes in demonstrating 5-HT$_2$ receptor binding sites on platelet membranes with radioligand binding techniques were achieved using [^{3}H]-ketanserin and cat platelet membrane preparations (Leysen *et al.* 1983*a*) and shortly thereafter using [^{3}H]-LSD and human platelet membranes (Geaney, Schächter, Elliott and Grahame-Smith 1984). As described earlier in Fig. 4.4 and Tables 4.3 and 4.4 the binding site on platelet membranes display characteristics similar to 5-HT$_2$ sites characterized in brain tissue. Several lines of evidence were obtained demonstrating that serotonin-induced functional responses in platelets were mediated by 5-HT$_2$ receptor sites. Studies using cat platelets showed that the potency of serotonin to induce platelet aggregation expressed by its K_m-value$=6.10^{-7}$M (calculated using the initial slope of aggregation induced by serotonin) matched the K_i-value$=2.9.10^{-7}$M of serotonin for inhibition of [^{3}H]-ketanserin binding to the cat platelet membranes. Moreover, such as illustrated in Fig. 4.6 the potencies of antagonists to inhibit the serotonin-induced aggregation closely correspond to the 5-HT$_2$ binding affinities of the drugs (De Clerck *et al.* 1984). The fact that drugs belonging to other pharmacological classes (selective dopamine antagonists, histamine antagonists, α_1-adrenergic receptor blockers) were poorly or not active, corroborated the specificity of the 5-HT$_2$ binding sites and the aggregation response to serotonin. The inactivity of selective serotonin uptake blockers in these tests demonstrated that the effects were unrelated to the uptake of serotonin in platelets (Leysen *et al.* 1983*a*; De Clerck *et al.* 1984). It was similarly found that binding affinities of drugs for 5-HT$_2$ sites detected with [^{3}H]-LSD in human platelet membranes were highly significantly correlated with the potencies of the compounds to antagonize serotonin-induced shape change in human platelets (Geaney *et al.* 1984).

Besides the fact that the platelet model allowed investigation of the relationship between binding and function in the same tissue, it also proved to be suitable for the study of the biochemical response coupled to serotonin receptor activation. Recently it was found that stimulation of intact

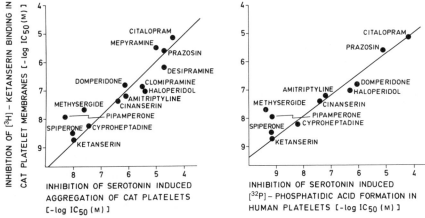

Fig. 4.6 Correlation between *in vitro* measured binding affinities of drugs for 5-HT$_2$ sites on cat platelet membranes (Leysen *et al.* 1983*a*) and the drug potencies to inhibit serotonin-induced aggregation of cat platelets (De Clerck *et al.* 1984) or to inhibit serotonin-induced [^{32}P]-phosphatidic acid formation in human platelets (De Chaffroy de Courcelles *et al.* 1984*b*).

human platelets with serotonin causes a rapid, initially linear increase in ^{32}P-incorporation in phosphatidic acid in the platelet membranes, a measurement which seems to be a reliable estimate of phosphatidyl inositol turnover (de Chaffoy de Courcelles, Leysen, Van Belle, and Janssen 1984*b*). Serotonin concentration dependent changes in [^{32}P]-phosphatidic acid formation ran entirely parallel with changes in the slope of serotonin-induced platelet aggregation. The serotonin concentration producing half-maximal stimulation of [^{32}P]-phosphatidic acid formation ($=2\times10^{-7}$M) also matched the K_i-value ($=2.9\times10^{-7}$M) of serotonin for binding to 5-HT$_2$ receptor sites. Antagonists revealed the same potencies for inhibition of the biochemical response (serotonin-stimulated [^{32}P]-phosphatidic acid formation), the functional test (serotonin-stimulated platelet shape change) and [^{3}H]-ligand binding to 5-HT$_2$ receptor sites ([^{3}H]-ketanserin binding in rat brain tissue or cat platelets, [^{3}H]-LSD binding in human platelets) (de Chaffoy de Courcelles *et al.* 1984*b*; Geaney *et al.* 1984). The highly significant intercorrelations between drug potencies measured in the various tests are illustrated in Fig. 4.6. Hence, ample evidence was obtained indicating that [^{32}P]-phosphatidic acid formation is a primary response directly coupled to 5-HT$_2$ receptor stimulation in human platelets.

Various hypotheses have been put forward regarding the sequence of events taking place in membranes where phosphatidylinositol turnover seems to be coupled to receptor stimulation (see Chapter 5). For the 5-HT$_2$

receptor system, de Chaffoy de Courcelles *et al.* (1984*b*) suggested that the initiating factor following serotonin receptor occupation is activation of phosphatidylinositol-4′,5′-biphosphate phosphomonoesterase. This enzyme hydrolyses PIP$_2$ (phosphatidylinositol-4′,5′-biphosphate) (which is an extremely strong Ca-chelator and is likely to be a Ca-storage site within the membrane) into PIP (phosphatidylinositol-4′-phosphate). Through that reaction sufficient Ca is released locally (Erne, Bühler, Affolter and Bürgisser 1983) necessary for the activation of phospholipase C. The latter enzyme generated diacylglycerol from PIP and this reaction product triggers on its turn protein kinase C activity (de Chaffoy de Courcelles, Roevens and Van Belle 1984*a*). Activation of this enzyme has been proposed as a key event in transmission of receptor stimuli (Nishizuka 1983). Current research is directed towards providing experimental proof for the sequence of events involved in phosphatidylinositol turnover coupled signal transduction and to examine whether Ca possibly fulfils the role of second messenger in these reactions. The platelet model represented the first tissue in which substantial indications were found about the signal transducing system coupled to 5-HT$_2$ receptor sites. It must now be confirmed whether the same coupling between 5-HT$_2$ receptor sites and phospholipid turnover exists in brain areas and various peripheral tissues where the receptor sites were found to have a role.

Finally apart from the importance of platelets for the fundamental research towards receptor coupled events, platelets represent an easily accessible and promising tool for the investigation of the role of 5-HT$_2$ receptor sites in disease states in humans. Of particular note is the suggested involvement of 5-HT$_2$ receptor sites in depressive illnesses (see Sections 6.4 and 6.7). In addition receptor alteration upon chronic treatment of man with drugs affecting the serotonergic system can now be closely followed. This should eventually lead to a better understanding of the hypothesized adaptive mechanism of serotonin receptors. A whole new field of human clinical research remains to be explored here.

4.10   Receptor solubilization and purification

Isolation and purification of receptor molecules is an ambitious goal of biochemical receptor research which eventually could resolve many issues. It will allow investigation of the nature and composition of the receptive macromolecule(s) and reconstitution of the receptor complex. It can provide a final answer to the existence of receptor sub-types: whether they represent distinct macromolecular entities, whether various drugs interact with different regions of the receptor complex or whether apparent differences in responses or binding properties found in different tissues or due to minor changes in the micro-environment of the receptor. Otherwise receptor purification can be used for raising monoclonal antibodies which in turn can be applied for further receptor purification and for receptor localization at the subcellular level. However, receptor isolation studies are still in a very early stage. The only reported achievements were the solubilization of active ligand binding macromolecular complexes.

Ilien, Gorissen and Laduron (1982a) described solubilization of 5-HT$_2$ receptor sites from rat and dog brain using lysolecithin as a detergent. The solubilized receptor sites could be labelled by [^{3}H]-spiperone, [^{3}H]-LSD and [^{3}H]-ketanserin and apparently retained the same drug binding properties as the receptor sites in the membranes (Ilien $et\ al.$ 1982a; Ilien, Schotte and Laduron 1982b). Similar results were recently obtained using human brain tissue (Schotte, Maloteaux and Laduron 1984). Chan and Madras (1982, 1983) reported solubilization of [^{3}H]-mianserin binding sites from dog frontal cortex using the detergent digitonin, and obtained partial purification by gel filtration and isoelectric focusing.

Solubilization of 5-HT$_1$ sites has also been attempted (VandenBerg, Allgren, Todd and Ciaranello 1983). By using a detergent mixture, high affinity [^{3}H]-serotonin binding activity was solubilized from bovine cerebral cortex. These authors particularly investigated the presence of high and low affinity binding components in the solubilized extract. In the presence of detergent, only a single binding component was found in the solubilized extract, but two components became apparent upon removal of the detergent. The two binding components were not separable by Sephacryl column chromatography and glycerol gradient sedimentation. The aim of the study was to obtain information on possible multiple receptor sites and several sophisticated interpretation of the experimental data were discussed: (i) interconvertible forms of a single receptor protein, (ii) different binding sites on a single protein, (iii) physically independent sites, and (iv) aggregates of receptor units interacting in a negatively co-operative manner. However, the likely explanation was not considered that curved Scatchard plots may be a result of surface effects on membranes (see Section 4.5, and Leysen and Gommeren 1981). The presence of detergents may reduce these surface effects and removal of detergents probably results in a re-aggregation of solubilized membrane components through which microscopic membrane vesicles are reformed.

Receptor purification is tedious, very slow in progress, and often disappointing, but the eventual goals should encourage biochemists to continue.

4.11 General conclusions

The serotonin 5-HT$_1$ and 5-HT$_2$ sites represent two distinct types of binding sites, with different drug binding properties and distinct distribution in the brain and peripheral tissues. Attention has been paid in this chapter to the fact that radioligand binding is prone to many artefacts and binding data in themselves are not a sufficient indication for attributing a receptor role to binding sites or for receptor site subclassification. A careful evaluation of the relationship between binding sites and pharmacological or physiological effects is mandatory. Only for the 5-HT$_2$ sites have multiple roles been demonstrated: they seem to mediate serotonin-induced behavioural excitation in rodents, serotonin-impaired blood circulation, serotonin-provoked inflammation, serotonin-induced tracheal and vascular smooth muscle contraction and serotonin-induced platelet function. Moreover, it has been suggested that 5-HT$_2$ type receptors have a role in the perception of subjective feelings (such as anxiety, neurotic depression, tension) induced by serotonin-mimetic agents in laboratory animals. A finding which can tentatively be extended to a proposed involvement of 5-HT$_2$ sites in the ethiology of depression and anxiety in humans.

In contrast a role for 5-HT$_1$ sites labelled by [^{3}H]-serotonin or for alledged sub-types labelled by [^{3}H]-8-OH-DPAT has not been irrefutably demonstrated. Until this has been achieved these sites should be considered as mere binding sites.

Notwithstanding the diversity of functions mediated by 5-HT$_2$ sites and the absence of confirmed functions for 5-HT$_1$ sites, many other functional effects of serotonin remain for which a corresponding radioligand binding model has not yet been identified. Broad investigation and precise characterization of several serotonin-mediated functions is hindered by the lack of potent and selective antagonists.

The platelet appears to be an interesting integral model for the study of 5-HT$_2$ receptors. 5-HT$_2$ binding sites can be labelled on platelet membranes. Evidence has been obtained that phosphatidic acid acid–phosphatidylinositol turnover is a primary biochemical response following 5-HT$_2$ receptor stimulation; this represents the first indication regarding a signal transducing system coupled to the 5-HT$_2$ receptor. Both the binding and the biochemical response were shown to be directly related to serotonin-induced platelet function. Moreover the platelets provide a tool for the study of 5-HT$_2$ receptors in relation with human diseases. A whole field of difficult, but challenging research towards serotonin receptor isolation and molecular identification remains to be explored.

References

Ahlenius, S., Larsson, K., Svensson, L., Hjorth, S., Carlsson, A., Lindberg, P., Wikström, H., Sanchez, D., Arvidsson, L.-E., Hacksell, U. and Nilsson, J. L. G. (1981). Effects of a new type of 5-HT receptor agonist on male rat sexual behavior. *Pharmac. Biochem. Behav.* **15**, 785.

Awouters, F., Leysen, J. E., De Clerck, F. and Van Nueten, J. M. (1982). General Pharmacological profile of ketanserin (R 41 468), a selective 5-HT$_2$ receptor antagonist. In *5-Hydroxytryptamine in peripheral reactions* (eds. F. De Clerck and P. M. Vanhoutte) p. 193. Raven Press, New York.

Barbaccia, M. L., Gandolfi, O., Chuang, D.-M. and Costa, E. (1983). Differences in the regulatory adaptation of the 5-HT$_2$ recognition sites labelled by ^{3}H-mianserin and ^{3}H-ketanserin. *Neuropharmacology* **22**, 123.

Bennett, J. P. Jr. and Snyder, S. H. (1975). Stereospecific binding of *d*-lysergic acid diethylamide (LSD) to brain membranes: relationship to serotonin receptors. *Brain Res.* **94**, 523.

—— and Snyder, S. H. (1976). Serotonin and lysergic acid diethylamide binding in rat brain membranes: relationship to post-synaptic serotonin receptors. *Molec. Pharmac.* **12**, 373.

Bruinink, A., Lichtensteiger, W. and Schlumpf, M. (1983). Pre- and post-natal ontogeny and characterization of dopaminergic D$_2$, serotonergic S$_2$, and spirodecanone binding sites in rat forebrain. *J. Neurochem.* **40**, 1227.

Cannon, J. S., Costall, B., Laduron, P. M., Leysen, J. E. and Naylor, P. J. (1978). Effects of some derivatives of 2-aminotetralin on dopamine-sensitive adenylate cyclase and on the binding of [^{3}H]haloperidol to neuroleptic receptors in the rat striatum. *Biochem. Pharmac.* **27**, 1417.

Chan, B. and Madras, B. K. (1982). [^{3}H]Mianserin binding to solubilized membranes of frontal cortex. *Eur. J. Pharmac.* **83**, 1.

—— and Madras, B. K. (1983). Partial purification of [^{3}H]mianserin binding sites. *Eur. J. Pharmac.* **87**, 357.

Cheng, Y. C. and Prusoff, W. H. (1973). Relationship between the inhibition constant (K$_i$) and the concentration of inhibitor which causes 50 per cent inhibition (I$_{50}$) of an enzymatic reaction. *Biochem. Pharmac.* **22**, 3099.

Colpaert, F. C., Niemegeers, C. J. E. and Janssen, P. A. J. (1982). A drug discrimination analysis of lysergic acid diethylamide (LSD): In vivo agonist and antagonist effects of purported 5-hydroxytryptamine antagonist and of pirenperone, a LSD-antagonist. *J. Pharmac. exp. Ther.* **221**, 206.

Cox, B. and Ennis, C. (1982). Characterization of 5-hydroxytryptaminergic autoreceptors in the rat hypothalamus. *J. Pharm. Pharmac.* **34**, 438.

De Chaffoy de Courcelles, D., Roevens, P. and Van Belle, H. (1984*a*). Stimulation by serotonin of 40 K and 20 K protein phosphorylation in human platelets. *FEBS Lett.* **171**, 289.

—— De Clerck, F., Leysen, J. E., Van Belle, H. and Janssen, P. A. J. (1984*b*). Evidence that phospholipid turnover is the signal transducing system coupled to serotonin-S_2 receptor sites. *J. Biol. Chem.* (in press).

De Clerck, F. F. and Herman, A. G. (1983). 5-Hydroxytryptamine and platelet aggregation. *Fedn. Proc.* **42**, 228.

—— Xhonneux, B., Leysen, J. and Janssen, P. A. J. (1984). The involvement of 5-HT_2-receptor sites in the activation of cat platelets. *Thromb. Res.* **33**, 305.

De Lean, A., Munson, P. J. and Rodbard, D. (1978). Simultaneous analysis of families of sigmoidal curves: application to bioassay, radioligand assay, and physiological dose-response curves. *Am. J. Physiol.* **235**, E97.

—— Munson, P. J. and Rodbard, D. (1979). Multi-subsite receptors for multivalent ligands. Application to drugs, hormones and neurotransmitters. *Molec. Pharmac.* **15**, 60.

Engel, G., Göthert, M., Müller-Schweinitzer, E., Schlicker, E., Sistonen, L. and Stadler, P. A. (1983). Evidence for common pharmacological properties of [^{3}H]5-hydroxytryptamine binding sites, presynaptic 5-hydroxytryptamine autoreceptors in CNS and inhibitory presynaptic 5-hydroxytryptamine receptors on sympathetic nerves. *Naunyn-Schmiedebergs Arch. Pharmac.* **324**, 116.

Ennis, C. and Cox, B. (1982). Pharmacological evidence for the existence of two distinct serotonin receptors in rat brain. *Neuropharmacology* **21**, 41.

—— Kemp, J. D. and Cox, B. (1981). Characterisation of inhibitory 5-hydroxytryptamine receptors that modulate dopamine release in the striatum. *J. Neurochem.* **36**, 1515.

Erne, P., Bühler, F. R., Affolter, H. and Bürgisser, E. (1983). Excitatory and inhibitory modulation of intracellular free calcium in human platelets by hormones and drugs. *Eur. J. Pharmac.* **91**, 331.

Farrow, J. T. and Van Vunakis, H. (1972). Binding of *d*-lysergic acid diethylamide to subcellular fractions from rat brain. *Nature* **237**, 164.

Friedman, R., Barrett, R. J. and Sanders-Bush, E. (1984). Discriminative stimulus properties of quipazine: mediation by serotonin$_2$ binding sites. *J. Pharmac. Exp. Ther.* **228**, 628.

Gaddum, J. H. and Picarelli, Z. P. (1957). Two kinds of tryptamine receptors. *Br. J. Pharmac.* **12**, 323.

Geaney, D. P., Schächter, M., Elliott, J. M. and Grahame-Smith, D. G. (1984). Characterisation of [^{3}H]lysergic acid diethylamide binding to a 5-hydroxytryptamine receptor on human platelet membranes. *Eur. J. Pharmac.* **97**, 87.

Glennon, R. A., Young, R. and Rosecrans, J. A. (1983). Antagonism of the effects of the hallucinogen DOM and the purported 5-HT agonist quipazine by 5-HT_2 antagonists. *Eur. J. Pharmac.* **91**, 197.

Gozlan, H., El Mestikawy, S., Pichat, L., Glowinski, J. and Hamon, M. (1983). Identification of presynaptic serotonin autoreceptors using a new ligand: ^{3}H-PAT. *Nature* **305**, 140.

Hamon, M., Mallat, M., Herbet, A., Nelson, D. L., Audinot, M., Pichat, L. and Glowinski, J. (1981). ^{3}H-Metergoline: A new ligand of 5-HT receptors in rat brain. *J. Neurochem.* **36**, 613.

Hartig, P. R., Kadan, M. J., Evans, M. J. and Krohn, A. M. (1983). ^{125}I-LSD: A high sensitivity ligand for serotonin receptors. *Eur. J. Pharmac.* **89**, 321.

Hartman, E. (1974). L-Tryptophan: A possible natural hypnotic substance. *J. Am. Med. Assoc.* **230**, 1680.

Hjorth, S., Carlsson, A., Lindberg, P., Sanchez, D., Wikström, H., Arvidsson, L.-E., Hacksell, U. and Nilsson, J. L. G. (1982). 8-Hydroxy-2-(di-*n*-propylamino)tetralin, 8-OH-DPAT, a potent and selective simplified ergot congener with central 5-HT-receptor stimulating activity. *J. Neural. Transm.* **55**, 169.

Hylden, J. L. K. and Wilcox, G. L. (1983). Intrathecal serotonin in mice: Analgesia and inhibition of a spinal action of substance P. *Life Sci.* **33**, 789.

Ilien, B., Gorissen, H. and Laduron, P. M. (1982*a*). Characterization of solubilized serotonin (S$_2$) receptors in rat brain. *Mol. Pharmac.* **22**, 243.

—— Schotte, A. and Laduron, P. M. (1982*b*). Solubilized serotonin receptors from rat and dog brain. *FEBS Lett.* **138**, 311.

Langley, (1878). On the physiology of the salivary secretion. *J. Physiol.* **1**, 339.

Leysen, J. E. (1979). Different neuroleptic receptors in various rat brain areas. In *Catecholamines: basic and clinical frontiers* (eds. F. Usdin, I. J. Kopin and J. Barchas) p. 556. Pergamon Press, New York.

—— (1981). Serotoninergic receptors in brain tissue: Properties and identification of various [^{3}H]ligand binding sites in vitro. *J. Physiol.* **77**, 351.

—— (1983). Rationale and importance of binding studies in respect to characterization of serotonin receptors. In *Vascular neuroeffector mechanisms* (eds. J. A. Bevan *et al.*) p. 259. Raven Press, New York.

—— (1984). Problems in in vitro receptor binding studies and identification and role of serotonin receptor sites. *Neuropharmacology* **23**, 247.

—— and Gommeren, W. (1981). Optimal conditions for [^{3}H]apomorphine binding and anomalous equilibrium binding of [^{3}H]apomorphine and [^{3}H]spiperone to rat striatal membranes: Involvement of surface phenomena versus multiple binding sites. *J. Neurochem.* **36**, 201.

—— and Gommeren, W. (1984). The dissociation rate of unlabelled dopamine antagonists and agonists from the dopamine-D$_2$ receptor, application of an original filter method. *J. Recept. Res.*, (in press).

—— and Laduron, P. M. (1977). A serotonergic component of neuroleptic receptors. *Archs. int. Pharmacodyn. Thér.* **230**, 337.

114   Josée E. Leysen

——— and Tollenaere, J. P. (1982). Chapter 1. Biochemical models for serotonin receptors. *A. med. chem. Rep.* **17**, 1.

——— Gommeren, W. and De Clerck, F. (1983*a*). Demonstration of S_2-receptor binding sites on cat blood platelets using [^{3}H]ketanserin. *Eur. J. Pharmac.* **88**, 125.

——— Niemegeers, C. J. E., Tollenaere, J. P. and Laduron, P. M. (1978). Serotonergic component of neuroleptic receptors. *Nature* **272**, 168.

——— Niemegeers, C. J. E., Van Nueten, J. M. and Laduron, P. M. (1982*a*). [^{3}H]Ketanserin (R 41 468), a selective ^{3}H-ligand for serotonin$_2$ receptor binding sites. *Molec. Pharmac.* **21**, 301.

——— Van Gompel, P., Verwimp, M. and Niemegeers, C. J. E. (1983*b*). Role and localization of serotonin$_2$ (S$_2$-receptor-binding sites: Effects of neuronal lesions. In *CNS receptors. From molecular pharmacology to behavior* (eds. P. Mandel and F. V. DeFeudis) p. 373. Raven Press, New York.

——— Awouters, F., Kennis, L., Laduron, P. M., Vandenberk, J. and Janssen, P. A. J. (1981). Receptor binding profile of R 41 468, a novel antagonist at 5-HT$_2$ receptors. *Life Sci.* **28**, 1015.

———Geerts, R., Gommeren, W., Verwimp, M. and Van Gompel, P. (1982*b*). Regional distribution of serotonin-2 receptor binding sites in the brain and effects of neuronal lesions. *Archs. int. Pharmacodyn. Thér.* **256**, 301.

Marcinkiewicz, M., Vergé, D., Gozlan, H., Pichat, L. and Hamon, M. (1984). Autoradiographic evidence for the heterogeneity of 5-HT$_1$ sites in the rat brain. *Brain Res.* **291**, 159.

Martin, L. L. and Sanders-Bush, E. (1982). Comparison of the pharmacological characteristics of 5-HT$_1$ and 5-HT$_2$ binding sites with those of serotonin autoreceptors which modulate serotonin release. *Naunyn-Schmiedebergs Arch. Pharmac.* **321**, 165.

Middlemiss, D. N. and Fozard, J. R. (1983). 8-Hydroxy-2-(di-*n*-propylamino)-tetralin discriminates between subtypes of the 5-HT$_1$ recognition site. *Eur. J. Pharmac.* **90**, 151.

Muakkassah-Kelly, S. F., Andresen, J. W., Shih, J. C. and Hochstein, P. (1982). Decreased [^{3}H]-serotonin and [^{3}H]-spiperone binding consequent to lipid peroxidation in rat cortical membranes. *Biochem. biophys. Res. Commun.* **104**, 1003.

——— ——— ——— ——— (1983). Dual effects of ascorbate on serotonin and spiperone binding in rat cortical membranes. *J. Neurochem.* **41**, 1429.

Müller-Schweinitzer, E. and Engel, G. (1983). Evidence for mediation by 5-HT$_2$ receptors of 5-hydroxytryptamine-induced contraction of canine basilar artery. *Naunyn-Schmiedeberg's Arch. Pharmac.* **324**, 287.

Nahorski, S. R. and Willcocks, A. L. (1983). Interactions of β-adrenoceptor antagonists with 5-hydroxytryptamine subtypes in rat cerebral cortex. *Br. J. Pharmac.* **78**, 107P.

Nelson, D. L., Herbet, A., Enjalbert, A., Bockaert, J. and Hamon, M. (1980). Serotonin-sensitive adenylate cyclase and [^{3}H]-serotonin binding sites in the CNS of the rat. I. Kinetic parameters and pharmacological properties. *Biochem. Pharmac.* **29**, 2455.

Niemegeers, C. J. E., Colpaert, F. C., Leysen, J. E., Awouters, F. and Janssen, P. A. J. (1983). Mescaline-induced head-twitches in the rat: an *in vivo* method to evaluate serotonin-S$_2$ antagonists. *Drug. Dev. Res.* **3**, 123.

Nishizuka, Y. (1983). Phospholipid degradation and signal translation for protein phosphorylation. *Trends Biochem. Sci.* **8**, 13.

Ortmann, R., Bischoff, S., Radeke, E., Buech, O. and Delini-Stula, A. (1982). Correlations between different measures of antiserotonin activity of drugs. Study with neuroleptics and serotonin receptor blockers. *Naunyn-Schmiedebergs Arch. Pharmac.* **321**, 265.

Pedigo, N. W., Yamamura, H. I. and Nelson, D. L. (1981). Discrimination of multiple [^{3}H]5-hydroxytryptamine binding sites by the neuroleptic spiperone in rat brain. *J. Neurochem.* **36**, 220.

Peroutka, S. J. and Snyder, S. H. (1979). Multiple serotonin receptors differential binding of [^{3}H]-5-hydroxytryptamine, [^{3}H]-lysergic and diethylamide, [^{3}H]-spiroperidol. *Molec. Pharmac.* **16**, 687.

—— and Snyder, S. H. (1981). [^{3}H]Mianserin: Differential labelling of serotonin$_2$ and histamine$_1$ receptors in rat brain. *J. Pharmac. exp. Ther.* **216**, 142.

—— Lebovitz, R. M. and Snyder, S. H. (1979). Serotonin receptor binding sites affected differentially by guanine nucleotides. *Molec. Pharmac.* **16**, 700.

—— Lebovitz, R. M. and Snyder, S. H. (1981). Two distinct central serotonin receptors with different physiological functions. *Science* **212**, 827.

—— Noguchi, M., Tolner, D. J. and Allen, G. S. (1983). Serotonin-induced contraction of canine basilar artery: Mediation by 5-HT$_1$ receptors. *Brain Res.* **259**, 327.

Rodbell, M. (1980). The role of hormone receptors and GTP-regulatory proteins in membrane transduction. *Nature* **284**, 17.

Rogawski, M. A. and Aghajanian, G. K. (1981). Serotonin autoreceptors on dorsal raphe neurons: Structure-activity relationships of tryptamine analogs. *J. Neurosci.* **1**, 1148.

Scatchard, G. (1949). The attractions of proteins for small molecules and ions. *Annls. N. Y. Acad. Sci.* **51**, 660.

Schotte, A., Maloteaux, J. M. and Laduron, P. M. (1983). Characterization and regional distribution of serotonin S$_2$-receptors in human brain. *Brain Res.* **276**, 231.

—— Maloteaux, J. M. and Laduron, P. M. (1984). Solubilization of serotonin S_2-receptors from human brain. *Eur. J. pharmac.* **100**, 329.

Segal, I. H. (1975). *Enzyme kinetics. Behavior and analysis of rapid equilibrium and steady-state enzyme systems.* J. Wiley & Sons, New York.

Uzbekov, M. G., Murphy, S. and Rose, S. P. R. (1979). Ontogenesis of serotonin 'receptors' in different regions of rat brain. *Brain Res.* **168**, 195.

VandenBerg, S. R., Allgren, R. L., Todd, R. D. and Ciaranello, R. D. (1983). Solubilization and characterization of high-affinity [^{3}H]serotonin binding sites from bovine cortical membranes. *Proc. natn. Acad. Sci. USA* **80**, 3508.

Vane, J. R. (1957). A sensitive method for the assay of 5-hydroxytryptamine. *Br. J. Pharmac.* **12**, 344.

Van Nueten, J. M. (1983). 5-Hydroxytryptamine and precapillary vessels. *Fedn. Proc.* **42**, 223.

——— Janssen, P. A. J., De Ridder, W. and Vanhoutte, P. M. (1982*a*). Interaction between 5-hydroxytryptamine and other vasoconstrictor substances in the isolated femoral artery of the rabbit effect of ketanserin (R 41 468). *Eur. J. Pharmac.* **77**, 281.

——— Janssen, P. A. J., Van Beek, J., Xhonneux, R., Verbeuren, T. J. and Vanhoutte, P. M. (1981). Vascular effects of ketanserin (R 41 468), a novel antagonist of 5-HT$_2$ serotonergic receptors. *J. Pharmac. exp. Ther.* **218**, 217.

——— Leysen, J. E., De Clerck, F. and Vanhoutte, P. M. (1984). Serotonergic receptor subtypes and vascular reactivity. *J. cardiovasc. Pharmac.*, (in press).

——— Leysen, J. E., Vanhoutte, P. M. and Janssen, P. A. J. (1982*b*). Serotonergic responses in vascular and non-vascular tissues. *Archs. Int. Pharmacodyn. Thér.* **256**, 331.

5

Inositol phospholipid breakdown as an index of serotonin receptor function

MICHAEL C. W. MINCHIN

5.1 Introduction

Our understanding of serotonin pharmacology has been much increased by the development of microiontophoretic and ligand binding techniques for the study of discrete receptor interactions (see Chapters 4, 6 and 7). However, until recently there have been no functional measures of serotonin receptor activation other than the membrane conductance changes recorded by microelectrodes and, meanwhile, awareness of the limitations of ligand binding experiments has been increasing. In particular, binding sites may not necessarily be coupled to an effector mechanism on a one-to-one basis; indeed, the concept of a receptor reserve has been known for some time and there has been a growing requirement for some functional biochemical measure of the consequences of the interaction of serotonin with its receptors. In some tissues, for example, the blowfly salivary gland, about which more will be said later, serotonin generates second messengers such as Ca^{2+} and cAMP (Berridge 1981*a*). In the brain, it appears that in some regions similar phenomena may exist; for example, in the rat hippocampus serotonin stimulates the production of cAMP (Barbaccia, Brunello, Chuang and Costa (1983) in a rather similar fashion to the well-known second messenger effects of noradrenaline and dopamine. However, the generation of cyclic nucleotides by serotonin is not common within the CNS and in some cases can only be observed in tissues from neonatal animals (see, for example, Mallat and Hamon 1982). More recently a presynaptic serotonin receptor has been described that regulates the release of serotonin from nerve terminals and hence has been termed an autoreceptor. The response of this receptor may easily be measured by the inhibition of release that it evokes and thereby provides a good functional model for this particular type of receptor (see Section 2.2.2).

This chapter is concerned with a newly-discovered membrane function of serotonin, that of enhancing the breakdown of inositol phospholipids and with the possible consequences of this for the study of serotonin pharmacology.

5.2   The phosphatidylinositol cycle

Hawthorne and Pickard (1979) estimated that in the brain phosphatidylinositol (PI) accounts for 3 per cent of the total phospholipid; the structure of PI is shown in Fig. 5.1. It is of interest to note, and is certainly of considerable functional importance from the point of view of prostanoid biosynthesis, that position 2 on the glycerol skeleton of PI is enriched with arachidonate in mammalian cells. The lower half of Fig. 5.1 depicts the cycle of events involved in the turnover of PI. Two other inositol phospholipids are in fairly rapid equilibrium with PI; these are phosphatidylinositol-4-phosphate (PIP) and phosphatidylinositol 4, 5-bisphosphate (PIP_2) and together these two phosphorylated derivatives of PI account for a similar quantity of phospholipid as PI itself in nervous tissue. Breakdown of these three inositol phospholipids occurs by the action of a phosphodiesterase of the phospholipase C type, though whether

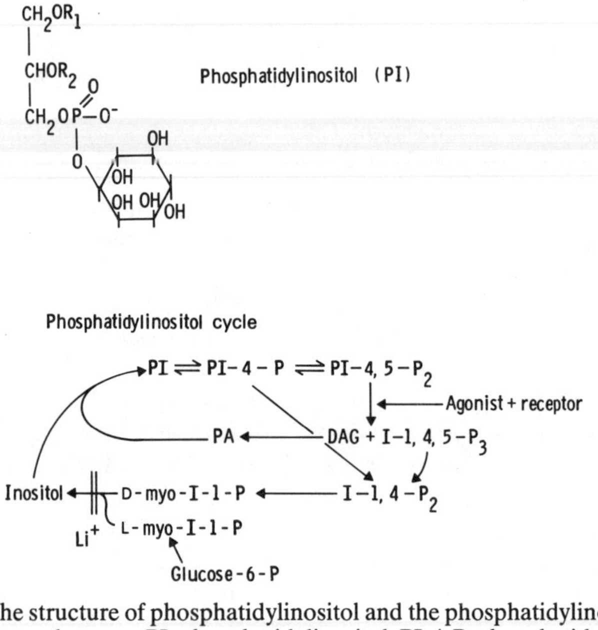

Fig. 5.1   The structure of phosphatidylinositol and the phosphatidylinositol cycle. R_1 and R_2 are acyl groups. PI, phosphatidylinositol; PI-4-P, phosphatidylinositol-4-phosphate; PI-4-,5-P_2, phosphatidylinositol-4,5-bisphosphate; DAG, diacylglycerol; PA, phosphatidic acid; I-1,4,5,-P_3, inositol 1,4,5-trisphosphate; I-1,4,-P_2, inositol 1,4-bisphosphate; I-1-P, inositol 1-phosphate.

there are specific enzymes for the three substrates is not known at present. It is possible that control of the hydrolysis of these membrane-bound phospholipids by the phosphodiesterase, which is soluble (Irvine and Dawson 1978), is achieved by alterations in membrane structure (induced by ligands for example) which expose the substrate to or protect it from the enzyme (Irvine, Letcher and Dawson 1984). The products of this cleavage are diacylglycerol and the appropriate inositol phosphate; the former is first phosphorylated, giving phosphatidate, and subsequently activated by CTP, giving a CDP-diacylglycerol moiety, then recombined with free inositol to give PI again, whilst the latter is hydrolysed by phosphomono-esterase action to yield, ultimately, free inositol which is then available for the resynthesis of PI. Much of the inositol required for PI synthesis in the brain derives from this recycling process since inositol crosses the blood–brain barrier extremely poorly (Margolis, Press, Altzuler and Stewart 1971; Spector and Lorenzo 1975) and the *de novo* synthesis from glucose-6-phosphate is very slow. This may prove to be some significance when we come to discuss the effects of lithium.

In 1955 Hokin and Hokin described the stimulation of the PI cycle in brain slices by acetylcholine and since then many others have reported similar findings. In a wide variety of tissues it was noted that agonist-induced stimulation of PI turnover was often linked with a rise in cytosolic Ca^{2+} and this prompted Michell (1975) to propose a formal link between these two phenomena. He suggested that receptor-stimulated inositol lipid metabolism was essential for receptor-stimulated mobilization of intracellular Ca^{2+} to occur. For some time it was believed that the initiating event was the stimulation of phosphodiesterase attack on PI but recently it has become clear that in many tissues the initial reaction is the hydrolysis of PIP and/or PIP_2 and that the observed increased turnover of PI occurs subsequent to this, probably serving to replenish the expended polyphosphoinositides.

For more detailed coverage of the arguments involved in the evolution of the hypothesis relating phosphoinositide metabolism and intracellular Ca^{2+} mobilization the reader is referred to the many excellent recent reviews of this topic (Michell, Kirk, Jones, Downes and Creba 1981; Berridge 1981b; Cockcroft 1981; Putney 1981; Michell 1982, 1983).

5.3   Serotonin-stimulated inositol phospholipid metabolism in the periphery

There are few peripheral systems in which serotonin has been shown to alter inositol phospholipid metabolism. Undoubtedly the most extensively studied is the blowfly salivary gland, in which serotonin regulates fluid secretion by two mechanisms, one involving elevated intracellular cAMP and the other raised intracellular free calcium levels. It is the second of these which is thought to be associated with phospholipid metabolism, as has been demonstrated by Berridge and his coworkers. For example, stimulation of the glands with cAMP led to fluid secretion but not $^{45}Ca^{2+}$ transport or [^{3}H]-inositol release from pre-labelled PI. On the other hand, stimulation with serotonin not only increased the rate of fluid secretion but also promoted $^{45}Ca^{2+}$ transport into the lumen and [^{3}H]-inositol release (Fain and Berridge 1979a). The threshold concentrations of serotonin necessary for fluid secretion, calcium transport and [^{3}H]-inositol release were very similar in the salivary gland. However, the dose-response curve of serotonin for [^{3}H]-inositol release was shallower and displaced by more than two orders of magnitude to the right of the dose-response curve for fluid secretion, suggesting the presence of spare receptors (Fain and Berridge 1979a). In addition, release of [^{3}H]-inositol induced by serotonin was shown not to depend upon the intracellular Ca^{2+} concentration. When pre-labelled glands were stimulated by serotonin in the presence of EGTA until the rate of secretion was reduced to zero, [^{3}H]-inositol release remained at maximum. Conversely, the calcium ionophore A23187 promoted Ca^{2+} transport into the saliva and increased the rate of fluid secretion but failed to increase [^{3}H]-inositol release (Fain and Berridge 1979a). These experiments support the suggestion (Michell 1975) that PI breakdown is responsible for the increase in intracellular Ca^{2+} concentration and not a consequence of it. Interestingly, only a small fraction of the total PI in the salivary gland appeared to be involved in intracellular Ca^{2+} regulation. Of the 135 pmol of PI in each gland only about 9 pmol was necessary to maintain serotonin-induced Ca^{2+} transport in experiments in which glands were depleted of PI by prolonged exposure to serotonin and then reactivated by incubation with myo-inositol (Fain and Berridge 1979b).

Recently, Berridge and his coworkers have investigated the nature of the inositol phospholipid hydrolysis induced by serotonin in the blowfly salivary gland in greater detail. At short stimulation times, around 60–90 seconds, there was a large increase in the amount of [^{3}H]-inositol-1,4-bisphosphate (IP$_2$) in the gland extracts with only small increases in [^{3}H]-inositol-1-phosphate (IP) and [^{3}H]-inositol-1,4,5-trisphosphate (IP$_3$) (Berridge, Dawson, Downes, Heslop and Irvine 1983). However, at very short stimulation times, up to 4 seconds, an increase in IP$_3$ could be detected which declined whilst an increase in IP$_2$ occurred which continued for up to 8 seconds (Berridge 1983). These experiments suggest that the initial response to serotonin is the hydrolysis of PIP$_2$.

An important methodological breakthrough was achieved when Berridge, Downes and Hanley (1982) made use of an earlier observation by Hallcher and Sherman (1980) that Li$^+$ inhibited myo-inositol-

1-phosphatase. By labelling inositol phospholipids with [³H]-inositol and stimulating the tissue with a suitable agonist in the presence of a few milli-molar Li⁺ large increases in the level of [³H]-IP could be generated. This occurs because Li⁺ does not inhibit inositol-1,4-phosphatase or inositol-1,4,5-phosphatase and so the [³H]-products of the hydrolysis of PI, PIP and PIP₂ accumulate as [³H]-IP which increases with duration of stimulation. The amplification of inositol phospholipid responses afforded by this technique has greatly facilitated the study of receptors that function by hydrolysing inositol phospholipids.

In mammals, there is evidence linking two peripheral serotonin receptors with inositol phospholipid metabolism. The serotonin receptors that mediate the contraction of the longitudinal smooth muscle of the guinea pig ileum do so, it is believed, by increasing the intracellular Ca^{2+} concentration. Jafferji and Michell (1976) have shown that serotonin increased the incorporation of ^{32}P into PI in the muscle and that this effect was abolished by methysergide, suggesting that this receptor is linked to inositol phospholipid hydrolysis.

The human blood platelet is weakly activated by serotonin, both shape change and, in some individuals, aggregation being induced by the monoamine. Recently, the platelet receptor responsible for these effects has been identified and characterized by [³H]-LSD binding (Geaney, Schächter, Elliot and Grahame-Smith 1984). In common with many substances which activate platelets serotonin has now also been shown to stimulate the breakdown of inositol phospholipids (Section 4.9; Brydon, Drummond, Kirkpatrick, MacIntyre, Pollock and Shaw 1984; Minchin and Schächter, unpublished). This phenomenon is accompanied by an increase in intracellular free Ca^{2+} concentration (Erne, Bühler, Affolter and Bürgisser 1983; Brydon et al. 1984) and may also be linked to the phos-phorylation of a 40 000 MW protein via a diacylglycerol stimulated, Ca^{2+}-dependent protein kinase—the C-kinase described by Nishizuka and his colleagues (Takai, Kishimoto, Iwasa, Kawahara, Mori and Nishizuka 1979; Kawahara, Takai, Minakuchi, Sano and Nishizuka 1980). This triad of events—phospholipid hydrolysis, elevation of intracellular free Ca^{2+} and protein phosphorylation may prove to be crucial in the development of the platelet response to a variety of activators. The response includes shape change, secretion and aggregation and it is possible that protein phos-phorylation resulting from the action of the C-kinase is involved in the mechanism of secretion (Sano, Takai, Yamanishi and Nishizuka 1983). Furthermore, serotonin-mediated biochemical responses in the platelet may provide a peripheral model for their central nervous system counter-parts and thus provide a valuable tool in the characterization of abnormalities of serotonin function in disease and in the assessment of drug action in man.

5.4   Serotonin-induced stimulation of inositol phospholipid metabolism in the central nervous system

Early experiments by Hokin (1970) established that in guinea pig cerebral cortex serotonin stimulated the incorporation of ^{32}P into phosphatidic acid. However, the concentrations required were high, equal to or greater than 1 mM, and the pharmacological nature of this response was not investigated. A number of other brain regions were examined, including striatum, hypothalamus, olfactory bulbs and cerebellar cortex and all demonstrated an increased labelling of phosphatidic acid at a concentration of serotonin of 10 mM. The magnitude of the response in the different regions does not correlate with the respective number of [^{3}H]-ketanserin (5-HT$_2$) binding sites in guinea pig brain tissue (Leysen, Niemegeers, Van Neuten and Laduron 1982) although the correlation is somewhat better in the case of 5-HT$_1$ receptor binding sites (Peroutka and Snyder 1981), although this is based on an uncomfortably small number of regions. This same high concentration of serotonin was additionally found to increase the labelling of PI and PIP (Hokin 1970) and the increased labelling of phosphatidic acid and PI was later confirmed by Abdel-Latif, Yau and Smith (1974) using 10 mM serotonin. Subsequently, Reddy and Sastry (1979) demonstrated increased ^{32}P labelling of inositol phosphatides and phosphatidic acid in cerebral cortex suspensions incubated with serotonin. This could be detected at a serotonin concentration of 10 μM and increased up to 10 mM. The effect of serotonin also increased between 7 and 21 days of age. In contrast with the study of Hokin (1970), however, serotonin had a much greater effect on ^{32}P incorporation in the cerebellum than in the cerebral cortex.

Since, in these experiments, the measurement being made was the incorporation of ^{32}P into inositol phosphatides and phosphatidic acid it is not possible to deduce from them whether alterations in the synthesis and/or in the breakdown of the labelled metabolites was responsible for the observed effects. The same could be said for the recent experiments of Jolles Van Dongen, Ten Haaf and Gispen (1982). These authors reported that serotonin, at concentrations above 1 μM, inhibited the transfer of ^{32}P from [γ-^{32}P]-ATP into PIP, PIP$_2$ and phosphatidic acid in a lysed crude mitochondrial/synaptosomal preparation. The relevance of these effects to receptor-mediated events is uncertain.

In the rat pineal gland serotonin (100 μM) was shown to increase ^{32}P incorporation into PI and phosphatidylethanolamine (Muraki 1972) whilst the calf pineal showed an increased incorporation of ^{32}P into PI but a decreased incorporation into phosphatidylethanolamine at a serotonin concentration of 10 mM, but not 10 μM (Basinska, Sastry and Stancer

(1973). Again, the physiological relevance of these findings is at present unclear.

As with the blowfly salivary gland, the investigation of phospholipid metabolism in the brain has been greatly facilitated by the use of Li^+ to inhibit inositol-1-phosphatase. By labelling brain slices with [³H]-inositol and subsequently stimulating them with neurotransmitters in the presence of Li^+ a measure of the breakdown of inositol phospholipids may be obtained from the increased levels of [³H]-inositol-1-phosphate, which can be separated from free [³H]-inositol on ion exchange columns (Berridge *et al.* 1982). Since under these conditions [³H]-inositol-1-phosphate accumulates with time as long as the effector molecule is present, a considerable amplification of the responses may be obtained and Berridge *et al.* (1982) were able to detect considerable increases in [³H]-inositol-1-phosphate levels in cerebral cortical slices exposed to various neurotransmitters, including serotonin. In our laboratory we have been able to confirm this finding and show that the dose-response curve for serotonin is sigmoid in shape, reaching a maximum at a serotonin concentration of between 30 and 100 μM (Fig. 5.2). In addition, a number of other substances with serotonin

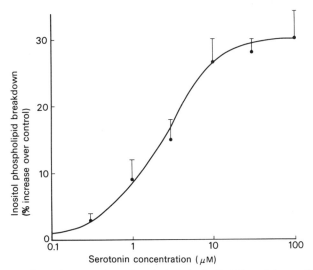

Fig. 5.2   The effect of serotonin on inositol phospholipid breakdown. Rat cerebral cortex slices were incubated with [³H]-inositol in the presence of 10 mM Li^+ for 30 min and stimulated with various concentrations of serotonin. After a further 45 min the incubation was terminated and [³H]-inositol phosphates separated from free [³H]-inositol by ion exchange chromatography. Results are expressed as the per cent increase in [³H]-inositol phosphates compared to control incubations performed at the same time in the absence of serotonin. Each value is the mean ±S.E.M. of 3–7 experiments, each performed in triplicate.

124   Michael C. W. Minchin

agonist properties also promoted the breakdown of inositol phospholipids
(Table 5.1). These included one, RU 24969, which is thought to have some
specificity for 5-HT$_1$ receptors (Hunt and Oberlander 1981). On the other
hand, studies with antagonists show that drugs that are potent inhibitors of
[^{3}H]-ketanserin binding to 5-HT$_2$ receptors, such as ketanserin itself, metergoline and pizotifen (Leysen *et al.* 1982) are also effective inhibitors of the
serotonin-mediated breakdown of inositol phospholipids (Table 5.2).
These findings are in agreement with those of Conn and Sanders-Bush
(1984) who have recently demonstrated that ketanserin competitively inhibited the accumulation of [^{3}H]-inositol phosphates induced by serotonin.
The serotonin-mediated response was uneffected, however, by the hista-
mine-H$_1$ blocker mepyramine, the muscarinic antagonist atropine and the
α_1-adrenoceptor antagonist prazosin (Table 5.2, and Brown, Kendall and
Nahorski 1984) at concentrations that were effective in antagonizing the
inositol phospholipid breakdown induced by activation of their respective
receptors. Nor was it inhibited by antidepressant drugs (Table 5.2) which

Table 5.1   The effects of compounds related to serotonin on inositol phospholipid
breakdown

	Agonist conc. (μM)	Inositol phospholipid breakdown (% increase over controls)
Quipazine	100	38± 6 (6)
LSD	30	16± 0.2 (3)
5-Methoxytryptamine	100	52±12 (8)
RU 24969	100	21± 6 (11)
5-Methoxy-$N'N'$-dimethyltryptamine	100	22 (2)

Tryptamine and melatonin were inactive at 100 μM. Each value is the mean ±S.E.M. of the
number of determinations (in brackets) from 1–4 experiments.

Table 5.2   The effects of some serotonin antagonists upon the inositol phospholipid breakdown induced by serotonin (100 μM)

Antagonist (1 μM)	% of control
Metergoline	54±18 (4)
Ketanserin	53±19 (3)
Methiothepin	95±31 (3)
Methysergide	75± 9 (4)
Pirenperone	31±25 (4)
Pizotifen	56± 9 (5)

Each value is the mean ±S.E.M. of the number of experiments in brackets, each determined in
triplicate. Inactive at 1 μM: mepyramine, atropine, tetrodotoxin. Inactive at 5 μM: amitriptyline, desipramine, iprindole.

have been shown to be potent antagonists at the 'glycogenolytic serotonin receptor' described by Quach, Rose, Duchemin and Schwartz (1982). Thus it would appear that the serotonin-induced breakdown of inositol phospholipids has characteristics associated with $5\text{-}HT_2$ receptors although a possible involvement of $5\text{-}HT_1$ receptors (suggested by the weak response to RU 24969) cannot at present be excluded.

It has been suggested, on the basis of inhibitor studies, that the $5\text{-}HT_1$ receptor corresponds to that mediating serotonin stimulation of adenylate cyclase (Peroutka, Lebovitz and Snyder 1981) although it has also been demonstrated that $5\text{-}HT_1$ binding sites have a pharmacological profile similar to that of the presynaptic serotonin autoreceptor (Martin and Sanders-Bush 1982, and Section 2.5). The second messenger, if any, associated with $5\text{-}HT_2$ receptors, however, is unknown, although given the effect of serotonin upon Ca^{2+} flux in the blowfly salivary gland it is conceivable that Ca^{2+} may prove to be involved in the sequelae of $5\text{-}HT_2$ receptor activation in the brain. Furthermore, it is unlikely that an adenylate cyclase linked receptor would also precipitate inositol phospholipid hydrolysis. In the blowfly salivary gland, which possesses two serotonin receptors, one linked to adenylate cyclase and one mediating increased Ca^{2+} mobilization, it has been shown that cAMP has no effect upon inositol phospholipid hydrolysis or Ca^{2+} transport (Fain and Berridge 1979a). Therefore, if the $5\text{-}HT_2$ receptor promotes inositol phospholipid breakdown which itself triggers Ca^{2+} mobilization, as suggested by Michell (1975), the phospholipid hydrolysis should be relatively resistant to removal of extracellular Ca^{2+}. Recent evidence suggests that this indeed may be the case; the serotonin-induced accumulation of $[^3H]$-inositol phosphates in pre-labelled cerebral cortex slices was only partially suppressed by omission of Ca^{2+} from the medium, although vigorous chelation of Ca^{2+} with EGTA completely inhibited the response (Kendall and Nahorski 1984). It seems, therefore, that a small amount of (probably intracellular) Ca^{2+} is necessary for the maximum response to occur; this has also been noted for other neurotransmitters (see Downes, 1982; Kendall and Nahorski 1984).

5.5   Conclusions

Clearly, the study of the interrelationships between receptor activation and inositol phospholipid metabolism is still in its infancy. Many important questions remain to be answered. For example, because the serotonin-mediated breakdown of inositol phospholipids in the brain is a rather weak response compared to, say, the muscarinic cholinergic response, it has been difficult to establish the pattern of events occurring at very short times after addition of serotonin to the preparation. Both in the case of acetylcholine in the brain and serotonin in the blowfly salivary gland the initial event appears to be hydrolysis of polyphosphoinositides (Berridge 1983; Berridge *et al.* 1983) and it is important to establish whether this characteristic extends to serotonin receptors in the brain, since it appears that it is the breakdown of polyphosphoinositides, and not PI, which triggers Ca^{2+} mobilization (Streb, Irvine, Berridge and Schulz 1983). It may also be possible to gain more direct information on this point by using the fluorescent intracellular Ca^{2+} indicator quin 2 (Tsien, Pozzan and Rink 1982). Synaptosome preparations appear to contain, in addition to nerve terminals, postsynaptic structures or dendrosomes. Evidence in favour of this comes from experiments by Agranoff and his colleagues (Fisher, Boast and Agranoff 1980) in which hippocampal synaptosomes prepared from hippocampi in which all the cholinergic terminals had been destroyed by lesioning of the fornix, displayed a normal increase in PI labelling following stimulation with muscarinic agonists. It may, therefore, be possible to load such dendrosomes with quin 2 and correlate the time course of any increase in intracellular Ca^{2+} with serotonin-induced inositol phospholipid hydrolysis. This assumes that the localization of the serotonin response is postsynaptic; clearly it will be necessary to make lesions of serotonergic pathways to establish this point. Lesioning experiments may also reveal adaptive processes involving inositol phospholipid metabolism. For example, sympathetic denervation of the rat parotid gland did not alter receptor-stimulated phospholipase C activity but it did depress the re-synthesis of PI following receptor stimulation (Downes, Dibner and Hanley 1983). Such changes may contribute to long-term adaptations of receptor function.

Finally, lithium may exert some of its therapeutic effect by influencing phosphoinositide metabolism (Berridge *et al.* 1982). *In vivo,* Li^+ decreases inositol levels and increases inositol-1-phosphate levels in the brain (Allison and Stewart 1971; Allison, Blisner, Holland, Hipps and Sherman 1976). Since the inositol isomer involved is the D-isomer (Sherman, Leavitt, Honchar, Hallcher and Phillips 1981) the changes are probably due to phosphoinositide metabolism, particularly since Li^+ has been shown to inhibit inositol-1-phosphatase with an IC_{50} value of 0.8 mM (Hallcher and Sherman 1980). Furthermore, as phosphoinositide turnover is slow in the absence of receptor stimulation, at least *in vitro* (Berridge *et al.* 1982), depletion of free inositol at active synapses in the presence of Li^+ at therapeutic concentrations may eventually depress function. *In vivo* experiments have shown that atropine abolishes the changes in inositol and inositol-1-phosphate levels caused by Li^+ in the CNS, suggesting that these changes occur at muscarinic receptors (Allison *et al.* 1976) but it is possible that a quantitatively small influence of Li^+ at other strategic synapses may be present. Were this to include effects at central serotonin receptors, the function of which may be involved in antidepressant treatments (Van Praag and De Haan 1980; Green and Nutt 1983), at least part of the therapeutic action of Li^+ may begin to receive a rational explanation.

It is clear that further study of the involvement of inositol phospholipid metabolism in central receptor-mediated events will yield much of value in our understanding, not only of the biochemical substrates of receptor function, but also of the pharmaco-dynamics of drug action.

References

Abdel-Latif, A. A., Yau, S-J. and Smith, J. P. (1974). Effect of neurotransmitters on phospholipid metabolism in rat cerebral-cortex slices—cellular and subcellular distribution. *J. Neurochem.* **22,** 383.

Allison, J. H. and Stewart, M. A. (1971). Reduced brain inositol in lithium-treated rats. *Nature (New Biol).* **233,** 267.

——Blisner, M. E., Holland, W. H., Hipps, P. P. and Sherman, W. R. (1976). Increased brain myo-inositol-1-phosphate in lithium-treated rats. *Biochem. Biophys, Res. Commun.* **71,** 664.

Barbaccia, M. L., Brunello, N., Chuang, D-M. and Costa, E. (1983). Serotonin-elicited amplification of adenylate cyclase activity in hippocampal membranes from adult rat. *J. Neurochem.* **40,** 1671.

Basinska, J., Sastry, P. S. and Stancer, H. C. (1973). Incorporation of ^{32}Pi orthophosphate into phospholipids of calf pineal slices in the presence and absence of neurotransmitters. *Endocrinology* **92,** 1588.

Berridge, M. J. (1981*a*). Phosphatidylinositol metabolism and calcium gating in a 5-HT receptor system. In *Drug receptors and their effectors* (ed. N. J. M. Birdsall) p. 75. Macmillan, London.

—— (1981*b*). Phosphatidylinositol hydrolysis: a multifunctional transducing mechanism. *Mol. Cell. Endocrinol.* **24,** 115.

—— (1983). Rapid accumulation of inositol trisphosphate reveals that agonists hydrolyse polyphosphoinositides instead of phosphatidylinositol. *Biochem. J.* **212,** 849.

—— Downes, C. P. and Hanley, M. R. (1982). Lithium amplifies agonist-dependent phosphatidylinositol responses in brain and salivary glands. *Biochem. J.* **206,** 587.

—— Dawson, R. M. C., Downes, C. P., Heslop, J. P. and Irvine, R. F. (1983). Changes in the levels of inositol phosphates after agonist-dependent hydrolysis of membrane phosphoinositides. *Biochem. J.* **212,** 473.

Brown, E., Kendall, D. A. and Nahorski, S. R. (1984). Inositol phospholipid hydrolysis in rat cerebral cortex slices: 1. Receptor characterization. *J. Neurochem.* **42,** 1379.

Brydon, L. J., Drummond, A. H., Kirkpatrick, K. A., MacIntyre, D. E., Pollock, W. K. and Shaw, A. M. (1984). Agonist-induced inositol phospholipid turnover and calcium flux in human platelet activation. *Br. J. Pharmac.* **81**, 187P.

Cockcroft, S. (1981). Does phosphatidylinositol breakdown control the Ca^{2+}-gating mechanism? *Trends Pharmac. Sci.* **2**, 340.

Conn, P. J. and Sanders-Bush, E. (1984). Selective 5-HT$_2$ antagonists inhibit serotonin stimulated phosphatidylinositol metabolism in cerebral cortex. *Neuropharmacology* **23**, 993.

Downes, C. P. (1982). Receptor-stimulated inositol phospholipid metabolism in the central nervous system. *Cell Calcium* **3**, 413.

——— Dibner, M. D. and Hanley, M. R. (1983). Sympathetic denervation impairs agonist-stimulated phosphatidylinositol metabolism in rat parotid glands. *Biochem. J.* **214**, 865.

Erne, P., Bühler, F. R., Affolter, H. and Bürgisser, E. (1983). Excitatory and inhibitory modulation of intracellular free calcium in human platelets by hormones and drugs. *Eur. J. Pharmac.* **91**, 331.

Fain, J. N. and Berridge, M. J. (1979*a*). Relationship between hormonal activation of phosphatidylinositol hydrolysis, fluid secretion and calcium flux in the blowfly salivary gland. *Biochem. J.* **178**, 45.

——— ——— (1979*b*). Relationship between phosphatidylinositol synthesis and recovery of 5-hydroxytryptamine-responsive Ca^{2+} flux in blowfly salivary glands. *Biochem. J.* **180**, 655.

Fisher, S. K., Boast, C. A. and Agranoff, B. W. (1980). The muscarinic stimulation of phospholipid labelling in hippocampus is independent of its cholinergic input. *Brain Res.* **189**, 284.

Geaney, D. P., Schächter, M., Elliott, J. M. and Grahame-Smith, D. G. (1984). Characterization of (^{3}H)lysergic acid diethylamide binding to a 5-hydroxytryptamine receptor on human platelet membranes. *Eur. J. Pharmac.* **97**, 87.

Green, A. R. and Nutt, D. J. (1983). Antidepressants. In *Psychopharmacology* (eds. D. G. Grahame-Smith and P. J. Cowen) Vol. 1, Part 1, p. 1. Excerpta Medica, Amsterdam.

Hallcher, L. M. and Sherman, W. R. (1980). The effects of lithium ion and other agents on the activity of myo-inositol-1-phosphatase from bovine brain. *J. Biol. Chem.* **255**, 10896.

Hawthorne, J. N. and Pickard, M. R. (1979). Phospholipids in synaptic function. *J. Neurochem.* **32**, 5.

Hokin, M. R. (1970). Effects of dopamine, gamma-aminobutyric acid and 5-hydroxytryptamine on incorporation of ^{32}P into phosphatides in slices from the guinea pig brain. *J. Neurochem.* **17**, 357.

Hokin, L. E. and Hokin, M. R. (1955). Effects of acetylcholine on the turnover of phosphoryl units in individual phospholipids of pancreas slices and brain cortex slices. *Biochim. Biophys. Acta.* **18**, 102.

Hunt, P. and Oberlander, C. (1981). The interaction of indole derivatives with the serotonin receptor and non-dopaminergic circling behaviour. In *Serotonin—current aspects of neurochemistry and function* (ed. B. Haber) p. 547. Plenum Press, New York.

Irvine, R. F. and Dawson, R. M. C. (1978). The distribution of calcium-dependent phosphatidylinositol-specific phosphodiesterase in rat brain. *J. Neurochem.* **31**, 1427.

——Letcher, A. J. and Dawson, R. M. C. (1984). Phosphatidylinositol-4,5 bisphosphate phosphodiesterase and phosphomonoesterase activities of rat brain: some properties and possible control mechanisms. *Biochem. J.* **218**, 177.

Jafferji, S. S. and Michell, R. H. (1976). Stimulation of phosphatidylinositol turnover by histamine, 5-hydroxytryptamine and adrenaline in the longitudinal smooth muscle of guinea pig ileum. *Biochem. Pharmac.* **25**, 1429.

Jolles, J., Van Dongen, C. J., Ten Haaf, J. and Gispen, W. H. (1982). Polyphosphoinositide metabolism in rat brain: effects of neuropeptides, neurotransmitters and cyclic nucleotides. *Peptides* **3**, 709.

Kawahara, Y., Takai, Y., Minakuchi, R., Sano, K. and Nishizuka, Y. (1980). Possible involvement of Ca^{2+}-activated, phospholipid-dependent protein kinase in platelet activation. *J. Biochem. (Tokyo)* **88**, 913.

Kendall, D. A. and Nahorski, S. R. (1984). Inositol phospholipid hydrolysis in rat cerebral cortical slices: II. Calcium requirements. *J. Neurochem.* **42**, 1388.

Leysen, J. E., Niemegeers, C. J. E., Van Neuten, J. M. and Laduron, P. M. (1982). (^{3}H)ketanserin (R 41 468), a selective (^{3}H)-ligand for serotonin$_2$ receptor binding sites. *Mol. Pharmac.* **21**, 301.

Mallat, M. and Hamon, M. (1982). Ca^{2+}-guanine nucleotide interactions in brain membranes. 1. Modulation of central 5-hydroxytryptamine receptors in the rat. *J. Neurochem.* **38**, 151.

Margolis, R. V., Press, R., Altzuler, N. and Stewart, M. A. (1971). Inositol production by the brain in normal and alloxan-diabetic dogs. *Brain Res.* **28**, 535.

Martin, L. L. and Sanders-Bush, E. (1982). Comparison of the pharmacological characteristics of 5-HT$_1$ and 5-HT$_2$ binding sites with those of serotonin autoreceptors which modulate serotonin release. *Naunyn-Schmiedeberg's Arch. Pharmac.* **321**, 165.

Michell, R. H. (1975). Inositol phospholipids and cell surface receptor function. *Biochim. Biophys. Acta.* **415**, 81.

—— (1982). *Cell Calcium,* (special issue on Inositol lipids and cell calcium) **3**, 285.

—— (1983). Polyphosphoinositide breakdown as the initiating reaction in receptor-stimulated inositol phospholipid metabolism. *Life Sci.* **32**, 2083.

—— Kirk, C. J., Jones, L. M., Downes, C. P. and Creba, J. A. (1981). The stimulation of inositol lipid metabolism that accompanies calcium mobilization in stimulated cells: defined characteristics and unanswered questions. *Phil. Trans. R. Soc. Lond. B.* **296**, 123.

Muraki, T. (1972). Effects of drugs on the phospholipid metabolism of the pineal body of rats. *Biochem. Pharmac.* **21**, 2536.

Peroutka, S. J. and Snyder, S. H. (1981). Two distinct serotonin receptors: regional variations in receptor binding in mammalian brain. *Brain Res.* **208**, 339.

—— Lebovitz, R. M. and Snyder, S. H. (1981). Two distinct central serotonin receptors with different physiological functions. *Science* **212**, 827.

Putney, J. W. (1981). Recent hypotheses regarding the phosphatidylinositol effect. *Life Sci.* **29**, 1183.

Quach, T. T., Rose, C., Duchemin, A. M. and Schwartz, J. C. (1982). Glycogenolysis induced by serotonin in brain: identification of a new class of receptor. *Nature* **298**, 373.

Reddy, P. V. and Sastry, P. S. (1979). Studies on neurotransmitter-stimulated phospholipid metabolism with cerebral tissue suspensions: a possible biochemical correlate of synaptogenesis in normal and under-nourished rats. *Brain Res.* **168**, 287.

Sano, K., Takai, Y., Yamanishi, J. and Nishizuka, Y. (1983). A role of calcium-activated phospholipid-dependent protein kinase in human platelet activation. *J. Biol. Chem.* **258**, 2010.

Sherman, W. R., Leavitt, A. L., Honchar, M. P., Hallcher, L. M. and Phillips, B. E. (1981). Evidence that lithium alters phosphoinositide metabolism: chronic administration elevates primarily D-myo-inositol-1-phosphate in cerebral cortex of the rat. *J. Neurochem.* **36**, 1947.

Spector, R. and Lorenzo, A. V. (1975). Myo-inositol transport in the central nervous system. *Am. J. Physiol.* **228**, 1510.

Streb, H., Irvine, R. F., Berridge, M. J. and Schulz, I. (1983). Release of Ca^{2+} from a nonmitochondrial intracellular store in pancreatic acinar cells by inositol-1,4,5-trisphosphate. *Nature* **306**, 67.

Takai, Y., Kishimoto, A., Iwasa, Y., Kawahara, Y., Mori, T. and Nishizuka, Y. (1979). Calcium-dependent activation of a multifunctional protein kinase by membrane phospholipids. *J. Biol. Chem.* **254**, 3692.

Tsien, R. Y., Pozzan, T. and Rink, T. J. (1982). Calcium homeostasis in intact lymphocytes: cytoplasmic free calcium monitored with a new, intracellularly trapped fluorescent indicator. *J. Cell. Biol.* **94**, 325.

Van Praag, H. and De Haan, S. (1980). Depression vulnerability and 5-hydroxytryptophan prophylaxis. *Psychiat. Res.* **3**, 75.

6

Effects of antidepressant drugs on serotonin receptor mechanisms

SVEN-OVE ÖGREN AND KJELL FUXE

6.1 Introduction

Much evidence implicates brain serotonin as an important neurotrans-mitter involved in affective disorders and in the clinical effect of antidepressant treatments (Coppen 1967; van Praag 1982). The experi-mental basis for a serotonergic involvement in affective disorders (the indoleamine hypothesis of depression) was to a large extent generated by discoveries in the 1960s. The mapping and histochemical demonstration of the mesolimbic and mesocortical serotonin pathways (Dahlström and Fuxe 1964; Andén, Dahlström, Fuxe, Larsson, Olson and Ungerstedt 1966) innervating large parts of the limbic system and the neocortex, provided a rational basis for investigation of the role of serotonin in the mediation of mood and emotional behaviours (see Fuxe, Hökfelt and Ungerstedt 1970). In addition reserpine which depletes brain serotonin stores was found to produce a state of 'behavioural depression' in animals which was attenu-ated or reversed by antidepressant drugs. The most pivotal discovery was the demonstration that central serotonin neurones contain an active uptake mechanism operating at the neuronal membrane. The observation that tricyclic antidepressants (TCAs) can block this mechanism indicated that they enhanced the availability of the transmitter at the synaptic cleft (Carlsson, Corrodi, Fuxe and Hökfelt 1969a; Ross and Renyi 1969). Moreover, the monoamine oxidase inhibitors (MAOIs) were found to decrease the brain serotonin catabolism and thereby elevate brain sero-tonin content (Spector 1963). Thus, the two major classes of drugs effective in depression were postulated to enhance the functional activity of brain serotonin (Lapin and Oxenkrug 1969; Carlsson et al. 1969a) by increasing the synaptic concentration of serotonin and thereby alleviate a hypothesized transmitter deficiency in depressed patients.

However, despite significant progress the clinical significance of the acute pharmacological action of antidepressant drugs has remained elusive. Antidepressant drugs have been shown to cause manifold inter-actions with serotonin and other neurotransmitter systems (Hall and Ögren 1981; Tang and Seeman 1980) but it is not clear which actions are associated with the clinical effect.

The accumulated evidence from this type of study suggests that the inter-action with adrenergic, histaminergic or muscarinic receptors is better correlated to the side effect profile than to therapeutic efficacy (Hall and Ögren 1981). It is also now known that different types of antidepressant drugs differ widely in their acute action on serotonergic neuronal mechanisms. Since the effects of antidepressant drugs on the various aspects of the overall regulation of serotonin neurotransmission has been comprehensively dealt with in several reviews (Langer, Zarifian, Briley, Raisman and Sechter 1981; Ögren, Ross, Hall and Archer 1983a; Sugrue 1983; Svensson 1983), it is here only briefly summarized to provide a necessary background. Table 6.1 summarizes the number of potential sites of action by which antidepressant drugs can modify serotonin neurotrans-mission. Different types of antidepressant drugs have been shown to affect

Table 6.1   Sites of action of antidepressant drugs on serotonergic neuronal mechanisms

---

1. Effects on the high affinity serotonin uptake mechanism in serotonin neurones
2. Action on [^{3}II] imipramine high affinity binding sites at the neuronal membrane and raphe cell bodies
3. Effects on serotonin neuronal firing, turnover, synthesis and release possibly associated with an action on the uptake mechanism
4. Direct action on postsynaptic recognition sites for serotonin receptors
5. Adaptive alterations in postsynaptic serotonin receptors
6. Possible modulation of co-transmitter release

---

each of the receptor mechanisms and regulatory neuronal mechanisms to a varying degree. For instance, the relative potency of the antidepressants including the TCAs on presynaptic neuronal uptake varies considerably and most of the TCAs are more effective on noradrenaline (NA) than on serotonin uptake *in vivo* (Ross and Renyi 1975a, b) which is reflected in their relative potency on neuronal processes in serotonin neurones including actions on synthesis, release and neuronal activity (Ögren *et al.* 1983a; Sugrue 1983). The action on the serotonin uptake mechanism is related to the binding of several tricyclic antidepressant drugs to the high affinity binding site of [^{3}H]-imipramine located in the serotonin nerve membrane (Langer and Briley 1981; Chapter 3) or on serotonin cell bodies of the median and dorsal raphe nuclei in the midbrain (Fuxe, Ögren, Agnati, Benfenati, Caricchioli, Fredholm, Andersson, Farabegoli and Eneroth 1983*a*). Interestingly, new types of selective serotonin uptake blockers also have a high affinity for the [^{3}H]-imipramine binding sites. The recent discovery that selective serotonin re-uptake blockers such as zimeldine (Ögren, Ross, Hall, Holm and Renyi 1981*a*) possess antidepressant activity (Åberg-Wistedt 1982; Gershon, Georgotas, Newton and Bush 1982) could indicate an important role of the serotonin transport mechanism for the antidepressant action. However, an action at the serotonin uptake site or a high affinity for the [^{3}H]-imipramine binding sites is not a prerequisite for an antidepressant effect and appears not to be a common mechanism of action for all antidepressant drugs. This conclusion is based on the introduction of antidepressant drugs with weak or no inhibitory actions on the serotonin re-uptake site (such as buprobrion, mianserin and iprindole) and with different pharmacological profiles (Table 6.2). These compounds have also a negligible affinity for the [^{3}H]-imipramine binding site (Langer and Briley 1981). It should also be pointed out that no correlation has been established between clinical potency and serotonin uptake inhibition *in vitro* and *in vivo* (Ögren, Fuxe, Agnati, Gustafsson, Jonsson and Holm 1979; Hall and Ögren 1981).

Due to the apparent failure to define any common site of action for antidepressant drugs at presynaptic serotonin mechanisms the recent focus of research has been directed to studies on postsynaptic serotonergic receptor mechanisms. The availability of radiolabelled ligands for studies of these receptors has made this development possible. The discovery that several antidepressant drugs have a high affinity *in vitro* for the [^{3}H]-*d*-LSD binding sites in the cerebral cortex (Fuxe, Ögren, Agnati, Gustafsson and Jonsson 1977; Ögren *et al.* 1979) suggested that antidepressant drugs may have a direct action on postsynaptic receptors. In addition, it has also recently been demonstrated that chronic antidepressant treatment with several types of antidepressant drugs can induce adaptive modifications in serotonergic receptor mechanisms (de Montigny and Aghajanian 1978; Peroutka and Snyder 1980*a*; Friedmann and Dallob 1979; Fuxe, Ögren and Agnati 1979; Fuxe, Ögren, Agnati, Eneroth, Holm and Andersson 1981; Savage, Frazer and Mendels 1979 and see Section 12.7) similar to the reductions observed in β-adrenoceptor binding and noradrenaline sensitive adenylate cyclase in brain homogenates (Sulser 1979). The major focus of this chapter is to summarize and consider the available data on the acute and chronic effects of antidepressant drugs on postsynaptic serotonergic receptor mechanisms monitored in particular by receptor binding techniques.

Table 6.2 Uptake inhibitory potencies of newer antidepressants

	Noradrenaline	Serotonin	Dopamine	Ref.
Alaproclate	–	++	–	Lindberg et al. (1978)
Amineptine	–	–	–	Unpublished product manual
Amoxapine	++	–	–	Coupet et al. (1979)
Bupropion	–	–	++	Cooper et al. (1980)
Citalopram	–	+++	–	Hyttel (1982)
Doxepin	++	+	–	Pinder et al. (1977)
Femoxetine	–	++	–	Buus-Lassen et al. (1975)
Fluoxetine	–	++	–	Wong et al. (1974)
Fluvoxamine	–	++	–	Claassen et al. (1977)
Mianserin	++	–	–	Baumann and Maitre (1977)
Norzimeldine	++	+++	–	Ross and Renyi (1977)
Trazodone	–	+	–	Clements-Jewery et al. (1980)
Viloxazine	+	–	–	Greenwood (1982)
Zimeldine	–	++	–	Ross et al. (1976)

–, no effect; +, weak effect; ++, strong effect; +++, very strong effect.

6.2   Acute effect of antidepressant drugs on serotonin receptors

6.2.1   *Effects on 5-HT$_1$ binding sites*

Multiple types of serotonin receptors have been implicated from electro-physiological (Aghajanian and Wang 1978; Section 7.1.2) and receptor binding studies (Leysen, Niemegeers, Tollenaere and Laduron 1978; Peroutka and Snyder 1979; Whitaker and Seeman 1978; Chapter 4). Several different radioligands have been used to study the receptors that mediate the action of serotonin in the brain (see Chapter 4). [^{3}H]-Serotonin binds with a high affinity constant (K_D 1–3 nM) to synaptic membrane preparations from various brain regions (Bennett and Snyder 1976) According to Peroutka and Snyder (1979) serotonin binding sites may be classified as 5-HT$_1$ or 5-HT$_2$. The 5-HT$_1$ binding site has a high affinity for [^{3}H]-serotonin and [^{3}H]-d-LSD (nanomolar K_D) and a low affinity for spiperone binding sites (micromolar K_D). The characteristics of the high affinity [^{3}H]-serotonin binding site is consistent with the labelling of a sero-tonin receptor (Bennett and Snyder 1976) which most likely has a postsynaptic location (Nelson, Herbet, Bourgoin, Glowinski and Hamon 1978). The [^{3}H]-serotonin binding site may represent a postsynaptic sero-tonin receptor system linked to a serotonin sensitive adenylate cyclase system (G. Fillion, Beaudoin, Rousselle, Denian, M. P. Fillion, Dray and Jacob 1979).

Table 6.3 summarizes data from different laboratories and different membrane preparations on the displacement of [^{3}H]-serotonin binding *in vitro*. Of particular interest is the results with new types of antidepressant drugs with different pharmacological profiles (Table 6.2). Although there are some notable exceptions, the accumulated evidence from these studies, indicates that most types of antidepressant drugs including serotonin uptake blockers (see Wong, Bymaster, Reid and Threlkeld 1983) have a low affinity for the 5-HT$_1$ binding sites as tested in preparations from various regions of the rat brain. Most antidepressant drugs inhibit [^{3}H]-serotonin binding in the micromolar or higher range in the rat brain. It is notable, however, that tricyclic drugs and mianserin have a considerable affinity for [^{3}H]-serotonin binding sites in the grey matter of the calf frontal lobe (Tang and Seeman 1980). Whether there exists species differences in [^{3}H]-serotonin binding, remains to be determined.

In contrast to the action on the recognition site of [^{3}H]-serotonin in micromolar concentrations, antidepressants can modulate [^{3}H]-serotonin binding affinity at nanomolar concentrations. Thus, pre-exposure of synaptic membranes to various types of antidepressant drugs such as the tricyclics as well as mianserin and iprindole, similar to incubation with serotonin has been shown to increase the affinity of the serotonin binding

Table 6.3   Effects of antidepressant drugs on serotonergic receptors *in vitro*

	[3H]-mianserin IC$_{50}$ (μM) or K$_i$ (μM)	[3H]-serotonin IC$_{50}$ (μM)		[3H]-d-LSD IC$_{50}$ (μM) or K$_i$ (μM)		[3H]-spiperone IC$_{50}$ (μM) or K$_i$ (μM)	[3H]-ketanserin K$_i$ (μM)
NA uptake blockers							
Desipramine	0.180[g]	3.07[c]	16.1[a]	3.45[a]	0.710[d]		0.078[e]
Doxepine	0.030[g]	0.24[c]	1.81[b]		0.550[d]	0.093[b]	
Nomifensine	5.0[g]	1.37[c]	9.88[a]	3.47[a]	1.0[d]		0.372[e]
Nortriptyline	0.028[h]	0.38[c]	1.00[a]	0.302[a]			
Maprotiline	0.0386[h]	0.22[c]	15.8[a]	3.05[a]	0.066[d]		
NA and serotonin uptake blockers							
Amitriptyline	0.014[g]	0.24[c]	1.52[a]	0.150[a]	0.048	0.021[e]	0.0042[e]
Imipramine	0.070[g]	1.08[c]	24.6[a]	1.35[a]	1.5[d]	0.164[b]	
Trimipramine	0.060[g]	0.32[c]		0.300[c]			0.037[e]
Serotonin uptake blockers							
Alaproclate		100[a]		100[a]		0.0033[b]	
Amoxapine		1.21[b]				8.76[b]	
Citalopram	1.20[g]	>100[b]					1.66[e]
Clomipramine	0.0072[h]	21.2[a]		0.917[a]		0.082[e]	0.0093[e]
Femoxetine		12.0[b]			2.1[d]	0.710[b]	
Fluoxetine	2.50[g]	100[b]			23[d]	0.93[b]	
Fluvoxamine		38.8[b]		14.7[a]		13.9[b]	
Norzimeldine		63.1[a]				3[f]	
Trazodone	0.0167[h]	0.68[b]				0.014[b]	
Zimeldine		2.5[c]	33.2[a]	10.9[a]		1.89[b]	
Weak uptake blockers							
Amineptine		100[b]					
Bupropion		100[b]				74[b]	
Iprindole	0.500[g]	6.0[c]	15.2[a]	5.80[a]	19[d]	>100[b]	
Mianserin	+MIA 0.004, −MIA 0.018[g]	0.09[c]	1.21[a]	0.097[a]	0.19[d]	0.0072[b]	0.0014[e]
Viloxazine		16.7[b]				33.3[b]	

[a]Hall and Ögren 1981 (rat cortex), [b]Hall *et al.* 1984a (rat cortex), [c]Tang and Seeman 1980 (calf frontal cortex, grey matter), [d]Ögren *et al.* 1979 (rat cortex), [e]Leysen *et al.* 1981 (rat prefrontal cortex), [f]Ögren *et al.* 1982 (rat cortex), [g]Dumbrille-Ross *et al.* 1980 (rat cortex), [h]Whitaker and Cross 1980 (Calf n. caudate).

site for [³H]-serotonin (Fillion and Fillion 1981) from an active state of 'low affinity' and a desensitized 'high affinity' state. At the 'high affinity' state of serotonin binding, activation of serotonin dependent adenylate cyclase activity is blocked. Thus, the 'high affinity' state, in which the receptors has a high affinity for the agonist, would correspond to a desensitized state (Fillion 1983). Whether this phenomenon occurs *in vivo* is not known but it could suggest a mechanism by which serotonin neurotransmission is stabilized (see Fuxe, Ögren, Agnati, Benfenati, Fredholm, Andersson, Zini and Eneroth 1983*b*), since serotonin neurotransmission is operating now via a serotonin site in the higher nanomolar range (5–10 nM) compared with a lower nanomolar range in the untreated animals.

6.2.2 Effects on *[³H]*-d-LSD binding sites

[³H]-*d*-LSD was the first radioligand to be available for the analysis of 5-HT$_2$ receptors (Bennett and Snyder 1976). Most studies indicate that [³H]-*d*-LSD labels two distinct recognition sites in the frontal cortex, the [³H]-spiperone 5-HT$_2$ site as well as the [³H]-serotonin site (Peroutka and Snyder 1979, Seeman, Westman, Coscina and Warsh 1980). Several well-known antidepressant drugs such as amitriptyline, nortriptyline and mianserin were originally found to be potent inhibitors of those serotonin receptors which are labelled by [³H]-*d*-LSD in the dorsal cortex of the rat (Fuxe *et al.* 1977; Fuxe, Ögren, Everitt, Agnati, Eneroth, Gustafsson, Jonsson, Skett and Holm 1978). The same compounds were also found to potently block serotonin-mediated responses (Fuxe *et al.* 1977). Subsequent studies (Ögren *et al.* 1979) confirmed this finding (see Table 6.3). The results indicate that the affinities for the [³H]-*d*-LSD binding site is higher than for that of [³H]-serotonin for most antidepressant drugs which is consistent with the relative affinity of [³H]-*d*-LSD for the 5-HT$_1$ and 5-HT$_2$ binding sites (Peroutka and Snyder 1979).

This analysis provided the first evidence for the existence of two types of serotonin receptors within the brain. Thus, no correlation was found to exist between the affinity of various types of antidepressant drugs for the [³H]-serotonin v. the [³H]-*d*-LSD binding sites or the high affinity serotonin uptake site (Ögren *et al.* 1979). Moreover, a high correlation was demonstrated between the affinity of a number of antidepressant drugs for the [³H]-*d*-LSD binding sites in the dorsal cortex and their ability to block serotonin-mediated behaviour (head twitch behaviour) induced by L-5-HTP and *d*-LSD in mice (Ögren *et al.* 1979; Fig. 6.1). No significant correlation was found between the affinity of these antidepressant drugs for [³H]-serotonin binding and their ability to block the serotonin-dependent behaviours. These results, therefore indicated the existence of two types of serotonin receptors and also suggested a functional correlate

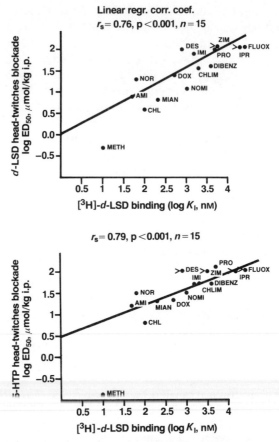

Fig. 6.1   Correlation between affinities of antidepressant drugs for the $[^3H]$-d-LSD binding site in the rat frontal cortex and inhibition of serotonin-mediated behaviour (head twitches) in mice. The test drugs were injected i.p. 60 min prior to the injection of d-LSD (0.5 mg/kg s.c.) or dl-5-HTP (90 mg/kg i.v.). ED_{50} values were calculated by probit analysis and fitted by linear regression analysis. (Data modified from Ögren *et al.* 1979)

to the serotonin receptors labelled by $[^3H]$-d-LSD but not to the ones labelled by $[^3H]$-serotonin.

6.2.3   *Effects on [³H]-spiperone binding sites*

Several different ligands including $[^3H]$-spiperone and $[^3H]$-ketanserin have been used to characterize the serotonin binding site classified as 5-HT$_2$. It has been shown that $[^3H]$-spiperone labels in the frontal cortex a receptor site which is predominantly serotonergic and identical to that labelled by $[^3H]$-d-LSD (Leysen *et al.* 1978; Peroutka and Snyder 1979) although this claim has been dismissed by one group based on the low affinity of serotonin and serotonin agonists for the [^{3}H]-spiperone binding site in the frontal cortex (Middlemiss, Carroll, Fisher and Mounsey 1980). In view of the low affinity of the [^{3}H]-spiperone labelled site for serotonin it is possible that the [^{3}H]-spiperone binding site is not a proper serotonin receptor but a regulatory site at several types of serotonin receptors (see Fuxe, Ögren, Agnati, Andersson and Eneroth 1982a, b). Autoradiographic studies have also shown that at least in some cortical areas the [^{3}H]-spiperone binding sites are distinct from the 5-HT$_1$ receptors. In addition, they show a poor correlation with serotonin innervation as shown by immunocytochemical studies (Köhler 1984). It must be pointed out that in most brain regions with the exception of the cortex, [^{3}H]-spiperone labels dopamine receptors of the D$_2$ type as well as the spirodecanone binding site, the transmitter of which still is unknown (Seeman 1980). Recent studies indicate that [^{3}H]-spiperone may bind to two sites in mouse cortical membranes, one is the authentic 5-HT$_2$ binding site and the other component corresponds to the α_1-adrenoceptor (Morgan, Marcusson and Finch 1984). The α_1-adrenoceptor component can be eliminated by the antagonists prazosin (30 nM). It is possible that the studies using [^{3}H]-spiperone, [^{3}H]-ketanserin or [^{3}H]-d-LSD have included the α_1-binding component, since most antagonists with the exception of methysergide fail to distinguish between the serotonergic and adrenergic component (see Leysen, Niemegeers, van Nueten and Laduron 1982 and Section 4.4).

The studies using labelled [^{3}H]-spiperone have shown a considerable potency of tricyclic antidepressant drugs to displace [^{3}H]-spiperone from its binding sites (Peroutka and Snyder 1980b; Fuxe et al. 1982a, b; Ögren, Fuxe, Archer, Johansson and Holm 1982; Table 6.3). Among the weak uptake blockers amoxapine (IC$_{50}$=3.3 nM) and trazodone (IC$_{50}$=14 nM) are very potent while all selective uptake blockers have a low potency (Hall, Sällenmark and Wedel 1984a). Imipramine, amitryptyline, clomipramine and mianserin displace [^{3}H]-spiperone with IC$_{50}$ values ranging from 10–160 nM (Table 6.3). In agreement with the findings with d-LSD it was shown that the ability of antidepressant compounds to block serotonin-mediated behaviours such as head twitch behaviour correlated to their potency to displace [^{3}H]-spiperone from its binding sites (Ögren et al. 1982; Peroutka, Lebovitz and Snyder 1981). These results indicate that several types of antidepressant compounds have the ability to block 5-HT$_2$ receptors in the brain.

6.2.4   Effects on [^{3}H]-ketanserin binding sites

The high affinity binding observed with [^{3}H]-ketanserin has resulted in ketanserin being considered a highly selective ligand for 5-HT$_2$ receptors in the CNS (Leysen *et al.* 1982). About 70% of [³H]-ketanserin binding (2 nM) in the prefrontal cortex is to 5-HT₂ serotonin binding sites as defined by methysergide (1 μM) inhibition. In agreement with the previous studies using [³H]-*d*-LSD and [³H]-spiperone binding several different types of antidepressant compounds have a very high affinity for the [³H]-ketanserin binding sites in the prefrontal cortex (Leysen *et al.* 1982; Table 6.3). The K_i values usually range from 5–80 nM for the tricyclic drugs. Mianserin is extremely potent with a K_i of about 1 nM, while nomifensine and the serotonin uptake blocker citalopram have a low potency. These results further underline the observations that many types of antidepressant drugs bind with a high affinity to receptors in the brain which has characteristics of a 5-HT₂ receptor.

6.2.5   Effects on [³H]-mianserin binding sites

Mianserin is a tetracyclic antidepressant compound which is a potent serotonin receptor antagonist in the brain and periphery as shown in functional and behavioural studies (Maj, Sowinska, Baran, Gancarczyk and Rawlow 1978; Fuxe *et al.* 1977). Unlike the tricyclic antidepressant drugs, it has a weak inhibitory action on monoamine uptake (Goodlet, Mireyless and Sugrue 1977). Specific high affinity binding sites for [³H]-mianserin have been described in rat cortical membranes (Dumbrille-Ross, Tang and Seeman 1980) and in the caudate nucleus of the calf (Whitaker and Cross 1980). In these studies unspecific binding has been defined in the presence of 0.5 and 1 μM of non radioactive mianserin, respectively. Characterization of the [³H]-mianserin recognition site suggests a postsynaptic location at serotonin and histamine receptors (Dumbrille Ross *et al.* 1980; Peroutka and Snyder 1981; Whitaker and Cross 1980). Of substantial interest is that [³H]-mianserin binding is not associated with the high affinity binding site for [³H]-imipramine (Brunello, Chuang and Costa 1982*a*), which is closely related to the uptake site for serotonin (Section 3.4). The compound has a substantial affinity for the histamine-H₁ recognition site (Peroutka and Snyder 1981) and to some extent also for α_1- and α_2-adrenoceptors (Hall and Ögren 1981) which has to be recognized in the interpretation of the binding studies. Scatchard analysis of the specific binding, however, indicated a single population of binding sites in the cortex (see Dumbrille-Ross *et al.* 1980).

The specific binding of [³H]-mianserin in the rat cortex has been shown to be potently displaced by low concentrations of tricyclic antidepressants (10–180 nM) and at higher concentrations of the atypical antidepressant drugs and the selective uptake blockers (Dumbrille-Ross *et al.* 1980; Table 6.3). The results with the calf caudate (Whitaker and Cross 1980) are largely in agreement with the results in the rat. However, the drugs are generally more potent in the calf preparation consistent with previous data for [^{3}H]-spiperone (Tang and Seeman 1980). The present results illustrate the high potency of a number of antidepressants on putative serotonergic post-synaptic receptors. It is apparent that although the results with [^{3}H]-mianserin are roughly similar to those obtained with [^{3}H]-spiperone the effects of antidepressants on the two ligands are not identical. These results support the view that the recognition site labelled by mianserin is not identical to the one labelled by [^{3}H]-spiperone (see Whitaker and Cross, 1980 and also Section 4.7 for further discussion of [^{3}H]-mianserin binding).

6.2.6   Functional importance of the action of antidepressant drugs on serotonin receptors

Table 6.4 summarizes data with a number of antidepressant drugs on pre- and postsynaptic serotonin receptors. It is apparent that several anti-depressant drugs have affinities for the 5-HT$_2$ receptor binding sites *in vitro* at concentrations in the same range or even lower than those affecting amine uptake. The functional significance of these receptor affinities has been examined in behavioural studies. As already pointed out, a striking correlation between the affinity of a number of antidepressant drugs for the [^{3}H]-*d*-LSD binding site in the dorsal cortex and their ability to block serotonin-mediated behaviour (head twitches) induced by L-5-HTP and *d*-LSD in mice was demonstrated (Ögren *et al.* 1979). The subsequent observation that the potency of these drugs to block serotonin-mediated behaviour (head twitches) correlates with their potency to displace [^{3}H]-spiperone *in vitro* indicates that they possess 5-HT$_2$ receptor blocking activity in the brain (Fuxe *et al.* 1982 *a*, *b*; Ögren *et al.* 1982; Snyder and Peroutka 1982). In addition, the potencies of antidepressant drugs to block mescaline-induced head twitches in rats are correlated with their inhibition of [^{3}H]-ketanserin binding (Leysen *et al.* 1982). Thus, several antidepressant drugs can be considered as 5-HT$_2$ receptor blockers and inhibit serotonin-mediated effects (see also Section 12.7).

An important question is whether the antidepressant drugs block sero-tonin receptors at doses which may be of clinical relevance. Since tricyclic drugs such as imipramine and amitriptyline reach concentrations in the rat brain of around 1–10 μM following acute and repeated administration at 'therapeutic' dose levels (Daniel, Adamus, Melzacka, Szymmura and Vetulani 1981; Glotzbach and Preskorn 1982) direct interference with postsynaptic serotonin receptors is likely. Much evidence suggests that antidepressant drugs may block 5-HT$_2$ receptors *in vivo* in doses which are lower than those affecting serotonin uptake and in the same dose range as those blocking noradrenaline uptake (Ögren *et al.* 1982). Since the action

Table 6.4   Effects of antidepressant drugs on pre- and postsynaptic serotonin receptor sites

| | Uptake inhibition in vitro | Inhibition of serotonin receptor binding | | | |
	[³H]-Serotonin IC₅₀ (nM)	[³H]-Serotonin (5-HT₁) IC₅₀ (nM)	[³H]-d-Lysergide (5-HT₁+5-HT₂) Kᵢ (nM)	[³H]-Spiperone (5-HT₂) IC₅₀ (nM) or Kᵢ (nM)	[³H]-Ketanserin (5-HT₂) Kᵢ (nM)
Zimeldine	240	33 200	10 900	1890	—
Amitriptyline	180	1520	48	21	4
Clomipramine	18	21 200	2100	82	9
Imipramine	140	24 600	1500	260	37
Mianserin	1200	1210	190	13	1.4
Trazodone	580	680	—	14	—
Citalopram	1.8	>100 000	—	8760	1660

Data from: Ross and Renyi (1977), Hyttel (1982), Hall and Ögren (1981), Hall et al. (1984a), Ögren et al. (1979, 1982), Leysen et al. (1982).

on the uptake mechanism and the blockade of postsynaptic receptors will result in functionally opposite effects in acute studies, the question as to which effect predominates is important. The studies presented above clearly indicate that receptor blockade is an important property of several tricyclic antidepressant drugs (amitriptyline, desipramine, nortriptyline, imipramine, dibenzepine). Studies by Maj and coworkers have also clearly established that antidepressant drugs such as doxepin, mianserin and trazodone are potent blockers of serotonin-mediated responses *in vivo* (Maj, Gancarczyk, Gorszczyk and Rawlow 1977; Maj, Palider and Rawlow 1979*a*; Maj, Lewandowska and Rawlow 1979*b*; Maj *et al.* 1978). In contrast, serotonin uptake blockers such as zimeldine, norzimeldine and fluoxetine have a very low affinity for the $[^3H]$-*d*-LSD and $[^3H]$-spiperone binding sites. These compounds enhance and do not block serotonin-mediated behaviours (Ögren *et al.* 1979).

6.3   Effects of chronic treatment with antidepressants on serotonergic receptor mechanisms: 5-HT$_1$ receptors

6.3.1   *Studies on uptake blockers*

It is generally accepted that treatment of depressed patients with antidepressant compounds requires at least 2–3 weeks of treatment until a significant therapeutic response is observed (Oswald, Brezinora and Dunleavy 1972). An explanation for the therapeutic delay might be that most compounds, e.g. tricyclic drugs cannot be given at therapeutic dose levels initially due to side effects. Thus, the therapeutic delay reflects the time-course of the step-wise increment of dose levels to obtain steady state levels of the compound and a significant effect of amine uptake. However, this explanation appears debatable since the selective uptake blocker zimeldine can be given at full dose from day 1. Interestingly, the time-course for the effect of zimeldine on symptoms of depression is similar to that of imipramine and amitriptyline (Åberg-Wistedt 1982; Gershon *et al.* 1982).

Recent evidence suggests that the adaptive changes at postsynaptic receptor mechanisms which occur following repeated treatment with antidepressant drugs may be more relevant for their therapeutic action than the acute pharmacological effect. However, the results have been inconsistent; probably reflecting the large variability in performing the binding assay, the choice of brain area as well as route and duration of administration. A typical feature of most of these studies is the employment of relatively high i.p. doses given to rats at which the specificity of the drugs may be low. Oral administration may be critical since most antidepressants are subject to extensive demethylation in both rats and human subjects which changes their relative potency to inhibit noradrenaline and serotonin uptake (see Ross and Renyi 1975 b). Little data also exists on sex- and strain-dependent variables. Since the inconsistencies reported may partly depend on methodology (Fuxe et al. 1983b) some further methodo- logical considerations are required. Most studies of [³H]-serotonin binding have followed the procedure by Bennett and Snyder (1976) as modified by Nelson et al. (1978) to remove endogenous serotonin. Using this method most investigators have reported the existence of only one high affinity binding site for [³H]-serotonin in membranes prepared from various rat brain areas. Some authors have distinguished two binding sites for [³H]- serotonin with a high and a low affinity ($K_D \sim 12$ nM) in the rat brain (Fillion 1983; Segawa, Mizuta and Nomura 1979). In crude bovine cortical mem- branes also a higher affinity ($K_D \sim 1-3$ nM) and lower affinity ($K_D \sim 10-30$ nM) component of [³H]-serotonin binding were distinguished as defined from curvilinear Scatchard plots (VandenBerg, Allgren, Todd and Ciaranello 1983). Segawa and coworkers demonstrated two binding sites using crude membrane fractions from the whole rat brain. However, the binding assay was performed at 0°C in contrast to 25°C used in most other studies. However, it is not clear whether the low affinity binding site for [³H]- serotonin is associated with neuronal postsynaptic receptors for serotonin (see Fillion 1983). In addition, it is possible that different affinity states of [³H]-serotonin binding may exist depending on temperature (see Blaschuk and Tang 1983) and incubation or exposure time to the agonist (Fillion 1983). Switches from low to high affinity states of the 5-HT$_1$ receptor may reflect (see Fillion 1983) different conformational states of the same receptor-protein (see also Section 4.3.1).

The studies by Segawa and coworkers (1979) indicated changes in 5-HT$_1$ receptor mechanisms in membranes from whole rat brains following chronic antidepressant treatment. Thus, three weeks of treatment with imipramine, desipramine and amitriptyline (10 mg/kg i.p. twice daily) reduced the B_{max}-values for the low and the high affinity component of the [³H]-serotonin binding. High i.p. doses of mianserin (2×15 mg/kg) were also found to reduce the B_{max}-values (25–30%) of [³H]-serotonin binding sites with both low and high affinity while iprindole and nomifensine failed to change [³H]-serotonin binding (Segawa, Mizuta, and Uehara 1982). None of these drugs were found to affect the uptake of [³H]-serotonin into brain synaptosomes. Interestingly, methiothepin (a preferential antagonist of the serotonin autoreceptors) accelerated the decrease in [³H]-serotonin binding induced by repeated imipramine treatment suggesting that elevation of intrasynaptic serotonin may be involved.

Repeated treatment with desipramine, imipramine, zimeldine and the putative antidepressant drug alaproclate at low oral doses (2×10 µmol/kg, p.o. for 14 days) induced two binding sites for [³H]-serotonin (a high and low affinity component) in the dorsal part of the cerebral cortex (Fuxe *et al.* 1979, 1981, 1982*a, b,* 1983*b*). The assay was performed at room temperature. The saturation analysis using Scatchard plots indicates that the best fit was no longer to a straight line but to a curve (see Fig. 6.2). A high affinity and a low affinity component were distinguished with one low (~1 nM) and a higher K_D value (~6–8 nM) which is considerably higher than the corresponding value in the control group. Of particular interest is the observation that the density of the high affinity component was low. Interestingly fluoxetine (2×10 µmol/kg p.o.) only produced a slight change in affinity but no change in B_{max}-values. Thus, these results similar to the data of Segawa indicate a down-regulation of the high affinity binding site for [^{3}H]-serotonin. These changes have so far mainly been observed in the

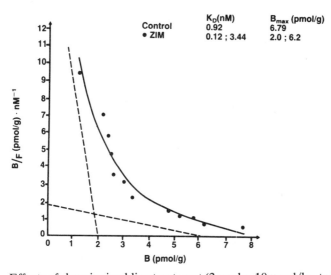

Fig. 6.2 Effects of chronic zimeldine treatment (2 weeks, 10 µmol/kg, twice daily, p.o., the last dose given 24 hours before killing) on the binding characteristics of [^{3}H]-serotonin binding sites in the dorsal part of the cerebral cortex (mainly parietal cortex) in the male rat. Saturation analysis has been performed using 10 concentrations of [^{3}H]-serotonin ranging from 0.1 to 10 mM with three replicates per concentration. Unspecific binding was defined as the binding in the presence of serotonin (10 µM). The binding procedure was that of Bennett and Snyder (1976) as modified by Nelson *et al.* (1978). The Scatchard plots are shown. In the case of the saline treated gruup, the best fit line is a straight line and it was calculated using standard parametrical procedures. Following zimeldine treatment the best fit line was a curve and the two binding lines were calculated by a modification of Hart's procedure (1965). From the two binding lines the theoretical curve also shown was obtained using the procedure of Rosenthal (see Fuxe *et al.* 1981). (Data from Fuxe *et al.* 1983*a*, with permission)

dorsal part of the cerebral cortex 24 hours following the last drug treatment. The brain area may be important since a nonsignificant reduction of [^{3}H]-serotonin binding was observed in the whole cerebral cortex following oral zimeldine treatment (5, 12.5 and 25 µmol/kg) for 14 days (Ross et al. 1981). It is possible that the effects of chronic antidepressant treatment may be related to an action on the processes regulating the affinity state of the serotonin receptor (see above). The variability in the stability of the high affinity component may explain why several groups in the literature have failed to observe the switch from one to two binding site appearances of [^{3}H]-serotonin binding in cortical membranes upon chronic antidepressant treatment.

Taken together, the results of Segawa and coworkers suggest a downregulation of 5-HT$_1$ receptor mechanisms since a reduction in the B_{max}-values of both the low affinity component was found at 0°C. Furthermore, the appearance of a very high and low affinity component in [^{3}H]-serotonin binding sites of the dorsal cerebral cortex (and a decrease in the high and an increase in the B_{max}-values of the low affinity component) could indicate a mechanism by which serotonin neurotransmission is stabilized (Fuxe, Ögren, Agnati and Calza 1982c; Fuxe et al. 1983b). Thus the action of antidepressants at 5-HT$_1$ receptors may lead to a stabilization of serotonin synaptic transmission via a decreased ratio: change in serotonin transmission/change in serotonin release (Fuxe et al. 1982c) since the very high affinity component of the [^{3}H]-serotonin binding site may become uncoupled from its biological effector upon chronic treatment with antidepressant drugs (Fillion and Fillion 1981). Instead, the decoding of 5-HT$_1$ transmission may be performed by an increased number of 5-HT$_1$ receptors of a fairly low affinity ($K_D \sim 6$–8 nM).

However, administration of the serotonin uptake blocker fluoxetine (Wong, Horng, Bymaster, Hauser and Molloy 1974) to mice (10 mg/kg i.p. daily for 14 days) which produced evidence for decreased brain serotonin synthesis, failed to alter [^{3}H]-serotonin binding in the whole brain (Hwang, Magnusson and van Woert 1980). Interestingly, a more prolonged administration to rats with fluoxetine caused changes in [^{3}H]-serotonin binding. In the study by Wong and Bymaster (1981) a high and a low affinity component of [^{3}H]-serotonin binding was also distinguished (see Segawa et al. 1979). Both in the group fed the fluoxetine diet for 2, 4 and 6 weeks or injected with 10 mg/kg i.p. daily for 27 or 46 days the number of high affinity sites (B_{max}-values) was consistently reduced with a maximum reduction of 47% after oral treatment. Interestingly, the reduction in B_{max}-values was considerably greater after 6 than after 2 weeks of treatment.

On the other hand, several workers studying the modulation of 5-HT$_1$ receptors by chronic antidepressant treatment have failed to demonstrate any significant changes in the characteristics of the [^{3}H]-serotonin binding sites in synaptic membranes with different types of antidepressant compounds. for instance, four weeks of treatment with clomipramine (20 mg/kg) failed to modify [^{3}H]-serotonin binding (Wirz-Justice, Krauchi, Lichtsteiner and Feer 1978). The receptor binding was also not effected by chronic L-5-HTP (100 mg/kg i.p. for 4 weeks) or metergoline (2 mg/kg daily for 3 weeks). Using a different binding procedure than Segawa *et al.* (1979) Peroutka and Snyder (1980*a*) reported a 20 per cent reduction in [^{3}H]-serotonin binding during a three-week treatment (10 mg/kg i.p. daily) with imipramine but not with amitriptyline, desipramine, iprindole, fluoxetine, chlorpromazine, haloperidol and methysergide. In a similar study clomipramine (2×10 mg/kg i.p.) and fluoxetine (2×10 mg/kg i.p.) as well as amitriptyline (2×10 mg/kg i.p.) failed to change [^{3}H]-serotonin binding in the cerebral cortex and hippocampus following 16 days of treatment (Savage *et al.* 1979; Savage, Mendels and Frazer 1980*a*). In a subsequent study amitriptyline, imipramine, fluoxetine, clomipramine (all 15 mg/kg twice daily for 14 days) as well as metergoline (2 mg/kg) failed to affect [^{3}H]-serotonin binding in the frontal cortex (Stolz, Marsden and Middlemiss 1983).

Administration of imipramine (10 mg/kg for 21 days) was shown to cause a 60 per cent reduction of [^{3}H]-serotonin binding in the hippocampus and striatum and a 33 per cent reduction in the frontal cortex (Maggi, U'Pritchard and Enna 1980). Also desipramine, but not fluoxetine or the noradrenaline uptake blocker nisoxetine (10 mg/kg), caused a significant reduction of [^{3}H]-serotonin binding in the frontal cortex but not in the hippocampus suggesting that the frontal cortex may be more sensitive to this effect. Scatchard analysis indicated that the reduction in binding was due to a reduction in receptor sites (Maggi *et al.* 1980). None of these treatments changed muscarinic receptor binding. Interestingly, the combined treatment with nisoxetine and fluoxetine reduced [^{3}H]-serotonin binding suggesting that serotonin receptor changes require inhibition of both noradrenaline and serotonin uptake. Desipramine has also been reported to cause a slight but significant reduction in serotonin receptor binding in the rat cerebral cortex following 2 but not 1, 4, 6, or 12 weeks of treatment (Bergstrom and Kellar 1979). Mianserin (10 mg/kg i.p. for 14 days) did not change [^{3}H]-serotonin binding in the frontal cortex, hippocampus and pons-medulla (Blackshear and Sanders-Bush 1982).

Thus, at the present time no general agreement exists as to the involvement of 5-HT$_1$ receptors in the mechanism of action of antidepressant compounds. Some of these results may depend on the experimental conditions, e.g. the brain area analysed as well as treatment time and dose. Thus, the serotonin-releasing drug fenfluramine caused a time dependent reduction in [^{3}H]-serotonin binding only in the diencephalon and cortex. Moreover, the antagonist metergoline only after 28 but not 14 days increased [^{3}H]-serotonin binding in the striatum, hippocampus and cortex but not in the diencephalon and brain stem (Samanin, Mennini, Ferraris, Bendotti and Borsini 1980). It must be realized that antidepressant drugs can both increase and decrease (lowering of firing rate in serotonin nerve cells) serotonin neurotransmission. Therefore, the response at the 5-HT$_1$ receptor level in different brain areas will vary depending on whether the action is one of increase or decrease or no change in 5-HT$_1$ receptor activity. Taken together, the available data show that changes in central serotonin receptor number (and sometimes affinity) may occur after repeated treatment with drugs acting preferentially on brain serotonin (see Section 7.4).

6.3.2   Monoamine oxidase inhibitors

Chronic administration of unselective monoamine oxidase inhibitors such as nialamide and pargyline appears to produce a reduction in the number of 5-HT$_1$ binding sites. Changes of affinity of the 5-HT$_1$ binding sites could not be demonstrated. Pargyline (25 mg/kg i.p. for 3 weeks) produced a marked decrease in [^{3}H]-serotonin binding (Peroutka and Snyder 1980a). The changes in serotonin receptors caused by the MAO-inhibitors seem to be rather specific since no change was observed in muscarinic, α-adrenoceptor or dopamine receptors (Peroutka and Snyder 1980a). It is notable that the decrease in β-adrenoceptor density following repeated pargyline is of a lesser magnitude than the changes in 5-HT$_1$ receptor binding.

Also nialamide (40 mg/kg i.p.) caused a marked, significant decrease in the density of [^{3}H]-serotonin binding in the hippocampus and cerebral cortex after only 4 days of treatment accompanied by an elevation of endogenous serotonin levels (Savage et $al.$ 1980a). Tranylcypromine (5 mg/kg), also reduced [^{3}H]-serotonin binding in the cortex (Savage et $al.$ 1980a) with no reduction in binding being observed in rats given a single i.p. injection, indicating that the results are not due to direct effects of the drugs in the binding assay. In a similar study repeated nialamide (40 mg/kg i.p. twice daily for 7 days) but not acute treatment markedly reduced [^{3}H]-serotonin binding in the brain stem and spinal cord (Lucki and Frazer 1982). The nialamide treatment did not significantly affect the dissociation constant (K_D) but only the maximum number of binding sites (B_{max}) was decreased. Also phenelzine (10 mg/kg i.p. once daily for 7 days) markedly reduced [^{3}H]-serotonin binding in the thoracic-lumbar region of the spinal cord (Lucki and Frazer 1982).

The role of the A and B form of monoamine oxidase has also been investigated. Nialamide and tranylcypromine inhibit both the A-type and the B-type of MAO (Maxwell and White 1978). Interestingly, only the selective A-type MAOI clorgyline (Johnston 1968) but not the B-type MAOI deprenyl (Knoll and Magyar 1972) nor pargyline given at a low dose (0.5 mg/kg) which is probably selective for the B-form of MAO (Fuller and Roush 1972) reduced [^{3}H]-serotonin binding in the cerebral cortex (Savage *et al.* 1979, 1980*a*). Since serotonin is a preferential substrate for type A MAO (see Johnston 1968), these results suggest that changes in serotonin activity may be responsible for a down-regulation of the [^{3}H]-serotonin binding site, e.g. the ability to increase serotonin concentrations following chronic MAOI treatment. This hypothesis has received further support since depletion of brain serotonin with the synthesis inhibitor, PCPA blocked the clorgyline induced reduction in [^{3}H]-serotonin binding (Savage, Mendels and Frazer 1980*b*). In contrast, depletion of NA and DA by α-methyl-paratyrosine failed to effect the action of repeated clorgyline treatment. Interestingly, the putative serotonin agonists trifluoromethyl-phenylpiperazine and quipazine also significantly reduced [^{3}H]-serotonin binding upon repeated treatment (Savage *et al.* 1980*b*). Thus, an increase of serotonergic activity seems to be a prerequisite for the ability of monoamine oxidase inhibitors to down-regulate 5-HT$_1$ receptors seen as a reduction in the B_{max} values. It remains to be determined whether these results may indicate differential effects of tricyclic antidepressants, serotonin uptake blockers and monoamine oxidase inhibitors in their actions on 5-HT$_1$ receptor mechanisms (Lucki and Frazer 1982; Peroutka and Snyder 1980*a*; Savage *et al.* 1979, 1980*a*).

6.4 Effects of chronic treatment with antidepressants on serotonergic receptor mechanisms: 5-HT$_2$ receptors

6.4.1 *Studies on uptake blockers*

The effects of repeated antidepressant treatment on 5-HT$_2$-labelled [^{3}H]-spiperone binding sites have been examined by several investigators using different areas of the brain and definition of specific binding. Most investigators have used variations of the method of Peroutka and Snyder (1979) or Leysen *et al.* (1978). In these methods specific binding is defined as that measure in the presence of either *d*-LSD (1 μM) or (+)-butaclamol (10 μM). However, some investigators have also employed high concentrations of serotonin (1–10 μM).

Peroutka and Snyder (1980*a*) were the first to demonstrate that long-term treatment with antidepressant drugs can produce adaptive changes in the 5-HT$_2$ receptors of the rat frontal cortex using [^{3}H]-spiperone as the radioligand. Administration of the tricyclic drugs desipramine, imipramine, amitriptyline and the atypical antidepressant iprindole was reported to decrease the number of [^{3}H]-spiperone binding sites by about 20–40%

without changing the apparent affinity (Peroutka and Snyder 1980*a*). These alterations were found to develop within 2–3 weeks of treatment at a daily dose of 10 mg/kg i.p. However, a single injection of 10 mg/kg of amitriptyline caused a marked reduction in [³H]-spiperone binding which was more pronounced than the effect on serotonin uptake. The effects of amitriptyline were found to be dose-related and significant at a dose of 2.5 mg/kg i.p. after 21 days of administration (Peroutka and Snyder 1980*b*). Three weeks administration of amitriptyline and iprindole (10 mg/kg) was also found to reduce the binding of [³H]-spiperone by 38–40% in the whole cerebral cortex (Kellar, Cascio, Butler and Kurtzke 1981) and a similar treatment with imipramine (10 mg/kg) reduced the binding of [³H]-spiperone by 35–50% in both the cortex and the hippocampus (Kendall, Stancel and Enna 1981). In a separate study amitriptyline and desipramine (10 mg/kg/day i.p. for 21 days) treatment was associated with a marked decrease in specific binding of [³H]-spiperone and [³H]-mianserin (see below) in the whole cortex (Dumbrille-Ross, Tang and Coscina 1982). No change was found in the binding of [³H]-spiperone in the striatum (a marker for dopamine D_2 receptors). These results indicate that the reduction in the density of [³H]-spiperone may represent downregulation of serotonin receptors due to an enhancement of postsynaptic serotonin activity (Peroutka and Snyder 1980*a*).

However, other groups have failed to demonstrate any reduction in the number of [³H]-spiperone binding sites in the frontal cortex upon chronic treatment with the tricyclic drugs amitriptyline and imipramine given at a dose of 2×15 mg/kg for a period of 14 days (Stoltz *et al.* 1983). On the other hand, fluoxetine and clomipramine in contrast to previous data (at a dose of 2×15 mg/kg) significantly reduced [³H]-spiperone binding. Interestingly, this study differs from the previous ones since non-specific binding was defined as that remaining in the presence of 100 μM non-radioactive serotonin and the assays were performed 72 hours following drug withdrawal. This study illustrates the possibility that the definition of binding specificity may be critical. Since the results are not based on Scatchard analysis and only one concentration of the ligand was employed the results must, however, be interpreted with caution. In contrast to most other studies, repeated desipramine treatment (5 mg/kg i.p. twice daily for 14 days) was reported to increase the number of [³H]-spiperone binding sites in the rat frontal cortex (Green, Heal, Johnson, Laurence and Nimgaonkar 1983*a*). The discrepancy between this finding and other data is not apparent but may reflect differences in methodology and a decrease has been found by the group following repeated desipramine administration to mice (Goodwin, Green and Johnson 1984).

Rüdeberg (1983) reported that treatment with amitriptyline (10 mg/kg i.p. for 10 days) reduced, while desipramine (10 mg/kg i.p.) did not show any effect, contrary to the results of Peroutka and Snyder (1980*a*). On the other hand, Rüdeberg (1983) reported that desipramine (2×7 mg/kg) for a period of 5 days caused a reduction. The most rapid and potent effects were found with mianserin. Even a single injection (7 mg/kg i.p.) decreased the specific binding for at least 66 hours, whereas the maximal effect, already obtained after 5 days of treatment, lasted for more than 90 hours (Fig. 6.3). Since the K_D values were unaffected 24 hours after injection the long-term reduction in [³H]-spiperone binding is probably not related to a direct interference between the test-drug and the receptor. These results may, however, be explained by an irreversible attachment of mianserin to the 5-HT$_2$ receptors followed by a restoration in binding capacity due to newly synthesized 5-HT$_2$ receptors. A minor degree of irreversible blockade may also be caused by those antidepressant drugs which acutely reduce the B_{max} values of [³H]-spiperone binding (see also Section 12.7).

The studies by Rüdeberg (1983) indicate that a treatment time of 5 days is sufficient for several drugs when given at a dose of 7–10 mg/kg i.p. twice daily. If the same dose is given only once a day, a longer treatment time is required. Not all the tested antidepressant drugs were found to induce a decrease in specific [³H]-spiperone binding; clomipramine, nomifensine

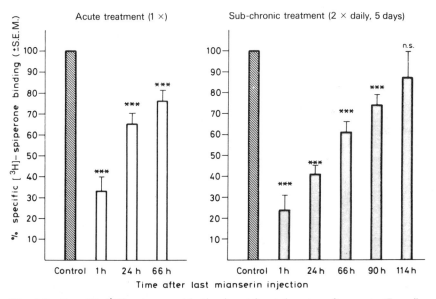

Fig. 6.3 Specific [³H]-spiperone binding in rat frontal cortex after acute (7 mg/kg i.p.) and sub-chronic (2×7 mg/kg i.p. for 5 days) treatment with mianserin. Controls were given corresponding saline treatment (5 ml/kg i.p. striped columns). Each column represents mean ±S.E.M. of 6 animals in per cent of the respective control group. ***p <0.001; n.s., not significantly different from control. (Data taken from Rüdeberg 1983, with permission.)

and zimeldine failed to show any effect while desipramine was only weakly active. Interestingly, the atypical antidepressant drugs iprindole, amoxapine, dibenzepine and mianserin were clearly active. In the case of the active drugs, Eadie-Hofstee analyses indicated that the decreases in binding depended on a reduction of 5-HT$_2$ receptor densities and not on changes in affinity.

Long-term treatment with mianserin (10 mg/kg daily for 14 days) caused a pronounced decrease in [^{3}H]-spiperone binding in the frontal cortex with only minor reductions in the hippocampus and pons-medulla (Blackshear and Sanders-Bush 1982). Thus, these results further emphasize the possible regional nature of the modulation of [^{3}H]-spiperone binding (see below). The effects of mianserin on [^{3}H]-spiperone were clearly dose-related in the range 0.05–1.0 mg/kg with half maximum inhibition between 0.05 and 0.125 mg/kg (Blackshear and Sanders-Bush 1982). The reduction of [^{3}H]-spiperone binding sites was accompanied by a blockade of serotonin-induced head twitches in mice which is compatible with a down-regulation of 5-HT$_2$ receptors (see below and Section 12.7).

Interestingly, the time-course for recovery of the number of [^{3}H]-spiperone binding sites appeared to be about the same for both mianserin and the tricyclic drugs. The B_{max} values of [^{3}H]-spiperone were 'normalized' within 7–10 days after discontinuation of repeated treatment with amitriptyline (10 mg/kg) (Peroutka and Snyder 1980b) and mianserin (Fig. 6.3; Blackshear and Sanders-Bush 1982; Rüdeberg 1983) which may reflect the lag time in the synthesis of new 5-HT$_2$ receptors (see above). This explanation also offers a way to understand why 5-HT$_2$ receptor blocking agents can reduce and not increase the number of its own receptors.

Also, long-term oral treatment with desipramine and imipramine in low doses (2×10 µmol/kg p.o.) was found to produce a reduction in the number of 5-HT$_2$ binding sites in the frontal cortex in order of 20–40% while the K_D values were unchanged (Fuxe et al. 1982a, b). In these studies d-LSD was used to define specific binding. Oral treatment with zimeldine (2×20 µmol/kg), a selective serotonin uptake blocker, produced also a reduction in the density of the [^{3}H]-spiperone sites in the frontal cortex (Fuxe et al. 1982a, b) or whole cerebral cortex (Hall, Ross and Ögren 1982). In a more recent study imipramine treatment (2×10 µmol/kg daily, for 14 days) was shown to produce a 20 per cent reduction in the number of [^{3}H]-spiperone binding sites in the frontal cortex as well as in the dorsal cortex, while the K_D values were not altered (Ögren, Fuxe, Agnati and Celani 1984). The [^{3}H]-spiperone binding sites probably represents 5-HT$_2$ receptors, since ketanserin (100 nM) was used to define nonspecific binding (Leysen et al. 1982). A similar repeated treatment with the selective serotonin uptake blockers zimeldine (Fuxe et al. 1983a) and alaproclate, but not the NA uptake blocker maprotiline (Ögren et al. 1984) produced a

15–20% reduction in the number of 5-HT$_2$ receptors in the frontal cortex of the male rat. Similar to previous i.p. studies, the effects of oral treatment on [^{3}H]-spiperone binding was shown to be time dependent and a reduction in [^{3}H]-spiperone binding in the dorsal cerebral cortex (mainly parietal cortex) was observed after one week of treatment (Fuxe *et al.* 1982*a*, *b*).

However, the effects of oral zimeldine on the [^{3}H]-spiperone-labelled sites were found to depend on the cortical area analysed. Thus repeated treatment with zimeldine induced an increase in the number of [^{3}H]-spiperone binding sites in the dorsal part of the cerebral cortex, while a reduction was observed in the frontal cortex. Interestingly, the effect on [^{3}H]-spiperone binding in the dorsal part of the cortex is not produced by alaproclate (Fuxe *et al.* 1983*b*) or fluoxetine, while alaproclate similar to zimeldine reduced the density of [^{3}H]-spiperone in the prefrontal cortex (Ögren *et al.* 1984). Thus, these data suggest that there may exist regional heterogeneities in the responses to different antidepressant drugs with regard to [^{3}H]-spiperone binding.

Somewhat unexpectedly not all serotonin uptake blockers produced the same effect on [^{3}H]-spiperone binding as did zimeldine. Thus, fluoxetine 10 mg/kg i.p. daily for 3 weeks (Peroutka and Snyder 1980*a*) or 2×10 µmol/kg p.o. failed to change the [^{3}H]-spiperone binding in the prefrontal cortex or dorsal cerebral cortex (Fuxe *et al.* 1983*b*). The non-tricyclic antidepressant trazodone, which is a relatively weak serotonin uptake blocker *in vivo* but a potent serotonin antagonist (Maj *et al.* 1979*a*; Riblet, Gatewood and Mayol 1979) also did not change [^{3}H]-spiperone binding even when given in a high dose (40 mg/kg i.p. for 4 days) (Taylor, Allen, Ashworth, Becker, Hyslop and Riblet 1981). Surprisingly, concurrent administration of phenoxybenzamine and trazodone resulted in a decreased 5-HT$_2$ receptor binding, indicating that blockade of central α-receptors may accelerate the down-regulation of 5-HT$_2$ receptors (see below). Citalopram is the most selective serotonin uptake inhibitor developed so far (see Hyttel 1982). Double blind studies (Kragh-Sörensen 1983) suggest that citalopram has antidepressant properties. In contrast to zimeldine, repeated treatment of citalopram (40 mg/kg daily for 13 days or 10 mg/kg daily for 21 days) failed to significantly change [^{3}H]-spiperone binding in the frontal cortex or in various parts of the cortex including the occipital plus temporal cortex or the whole cortex (Hyttel, Fredricson, Overo and Arnt 1984; Buckett, Strange, Stuart and Thomas 1983). It should be noted, however, that specific binding in one study (Hyttel *et al.* 1984), was examined in the presence of mianserin (1 µM) while in the other study (Buckett *et al.* 1983) *d*-LSD (1 µM) was used to define specific binding. The accumulated evidence suggests that although citalopram, alaproclate fluoxetine and zimeldine are relatively similar with regard to effects on the serotonin uptake mechanisms they differ with regard to the effects on 5-HT$_2$

receptors labelled by [³H]-spiperone. Thus, citalopram and fluoxetine resemble each other while zimeldine and alaproclate to some extent resemble the tricyclic drugs. However, there also exist other critical differences between these compounds. Zimeldine down-regulates NA-stimulated adenylate cyclase activity with no changes in the isoprenaline-stimulated adenylate cyclase activity or in β-adrenoceptor number (Mishra, Janowsky and Sulser 1980). Some studies, however, have indicated a slight decrease in β-receptor number (Sethy and Harris 1981; Ross *et al.* 1981). Fluoxetine neither influences NA-stimulated adenylate cyclase activity (Mishra *et al.* 1980) nor β-receptor number (Peroutka and Snyder 1980*a*). Alaproclate failed to change the number of β-adrenoceptor sites while reducing the [³H]-spiperone sites (Fuxe *et al.* 1983*b*).

6.4.2 *Monoamine oxidase inhibitors*

Three weeks administration of tranylcypromine (5 mg/kg) reduced the binding of [³H]-spiperone by 45 per cent (Kellar *et al.* 1981). In a similar design pargyline (25 mg/kg i.p. daily) reduced the [³H]-spiperone and [³H]-

Table 6.5 Effects of antidepressant drugs on 5-HT$_2$ receptors labelled by [³H]-spiperone

Noradrenaline and serotonin uptake blockers	
Amitriptyline	+
Imipramine	+
Dibenzepine	+
Noradrenaline uptake blockers	
Desipramine	+
Maprotiline	0
Nomifensine	0
Selective serotonin uptake blockers	
Alaproclate	+
Amoxapine	+
Zimeldine	+
Citalopram	0
Fluoxetine	0
Clomipramine	0, +
Trazodone	0
Weak uptake blockers	
Iprindole	+
Mianserin	+
Monoamine oxidase inhibitors	
Pargyline	+
Tranylcypromine	+
Amiflamine	0

+, effect; 0, no effect.

d-LSD binding by 35 per cent (Peroutka and Snyder 1980a). On the other hand, the selective A-type inhibitor amiflamine (Fig. 1.2) which produces a selective increase of serotonin levels in the brain (Ögren, Ask, Holm, Florwall, Lindbom, Lundström and Ross 1981b) did not alter [^{3}H]-spiperone binding in the frontal and dorsal cortex of the rat (Ögren et $al.$ 1984).

Taken together (see Table 6.5) a great number of antidepressant drugs of different structural types have been shown to down-regulate the [^{3}H]-spiperone binding sites. The available data suggest that no simple relationship exists between the pharmacological properties of the compounds and the effects on [^{3}H]-spiperone binding (see also Section 12.7).

6.5   Mechanisms of 5-HT$_2$ receptor regulation by antidepressant drugs

6.5.1   Role of pre- and postsynaptic serotonin

Early experimental evidence suggested that the down-regulation of the 5-HT$_2$ receptor in cortex labelled by [^{3}H]-spiperone was mediated by a presynaptic action (Peroutka and Snyder 1980a). The down-regulation by the serotonin uptake inhibitor zimeldine is consistent with this view. However, the reduction following desipramine (a weak serotonin uptake blocker, Carlsson et $al.$ 1969a, b) or iprindole (a drug which does not inhibit uptake) and the failure of citalopram and fluoxetine (two other selective serotonin uptake blockers) to cause down-regulation are not consistent with the serotonin availability hypothesis. Moreover, the relative magnitude of the reduction of the density of [^{3}H]-spiperone sites is not in agreement with the predicted changes in serotonin availability by various drugs. This conclusion is also in agreement with lesion studies. Chronic treatment with amitriptyline (20 mg/kg) resulted in a reduced number of cerebral 5-HT$_2$ receptors ([^{3}H]-spiperone binding) which was only partially blocked by pretreatment with the serotonin neurotoxin p-chloroamphetamine (Clements-Jewery and Robson 1982). PCA treatment was found to slightly attenuate the reduction in [^{3}H]-spiperone binding sites induced by repeated zimeldine treatment (10 mg/kg i.p. for 3 weeks), while the reductions of [^{3}H]-spiperone binding by repeated mianserin (10 mg/kg) or desipramine (5 mg/kg) treatment were not affected (Hall, Ross, and Sällemark 1984b).

In another study the midbrain dorsal and median raphe nuclei were subjected to radiofrequency heat lesions (Dumbrille-Ross et $al.$ 1982) which produced a marked impairment of serotonin uptake in the forebrain. Both amitriptyline and desipramine (both given at 10 mg/kg i.p. for 21 days) decreased the density of specific binding of both [^{3}H]-spiperone and [^{3}H]-mianserin in the cortex with about the same extent in both raphe-lesioned and intact rats. Moreover, raphe-lesions which diminished [^{3}H]-imipramine high affinity binding did not alter the imipramine induced decrease in 5-HT$_2$ receptor. Therefore, tricyclic antidepressant drugs may be acting on a common site other than the high affinity uptake site or [^{3}H]-imipramine binding sites to affect 5-HT$_2$ receptor numbers. Interestingly, some evidence indicates that in contrast to the down regulation of β-adrenergic receptor coupled adenylate cyclase system (Janowsky, Sternaka, Gillespie and Sulser 1982) intact NA input is not essential for down regulation of 5-HT$_2$ receptors (Dumbrille-Ross and Tang 1983). Thus, lesions of NA neurones by 6-hydroxydopamine failed to alter the effects of imipramine on [^{3}H]-spiperone binding (Dumbrille-Ross and Tang 1983). These studies indicate that intact presynaptic serotonergic terminals are not necessary for the inductions of 5-HT$_2$ down-regulation and further emphasize a post-synaptic locus for the action of several antidepressant drugs (see Ögren et al. 1979).

Changes in 5-HT$_2$ binding sites appear not to be directly related to action on dopaminergic or α-adrenoceptor mechanisms, since the selective dopamine (DA) antagonist haloperidol (10 mg/kg) (Peroutka and Snyder 1980a; Rüdeberg 1983), the α-adrenoceptor antagonist phenoxybenzamine (10 mg/kg, for four days Taylor et al. 1981) and the selective α$_1$-adrenoceptor antagonist prazosin (Rüdeberg 1983) failed to influence [^{3}H]-spiperone binding. Recent studies, however, have shown that blockade of central α$_2$-receptors can accelerate the down-regulation of 5-HT$_2$ receptors by antidepressants (Crews, Scott and Shorstein 1983) suggesting a possible contribution of NA mechanisms. However, the precise pharmacological mechanisms underlying changes in 5-HT$_2$ sites is not clear. Thus, reductions in [^{3}H]-spiperone binding have also been reported following repeated treatment with non-established antidepressant drugs. Thioridazine (32 mg/kg p.o.) markedly reduced specific [^{3}H]-spiperone binding after 10 days of treatment while a similar treatment with chlorpromazine (10 mg/kg) failed to influence [^{3}H]-spiperone binding (Rüdeberg 1983; Peroutka and Snyder 1980a). However, a more recent study showed that chlorpromazine (5 mg/kg twice daily for 14 days) reduced [^{3}H]-spiperone binding (Mikuni and Meltzer 1984).

Interestingly, the combination of imipramine and chlorpromazine produced a greater reduction in 5-HT$_2$ binding sites than either treatment alone (Mikuni and Meltzer 1984). This finding is of interest, since both thioridazine and chlorpromazine are quite potent serotonin receptor antagonists (Leysen, Awouters, Kennis, Laduron, Vandenberk and Janssen 1981; Leysen et al. 1982). A similar selective action of various serotonin antagonists is also reported. In view of the similarity in the action of methysergide, mianserin, pizotifen and cinanserin on 5-HT$_1$ and 5-HT$_2$ receptors (Leysen et al. 1981, 1982) it is surprising that cinanserin (Rüdeberg 1983) and methysergide (Peroutka and Snyder 1980a) unlike the two other compounds failed to affect [^{3}H]-spiperone binding. Interestingly, pizotifen unlike cinanserine produced after both acute and repeated administration similar effects as mianserin with a marked decrease in the B_{max} values (-34 per cent) (Rüdeberg 1983).

As already discussed the changes in [^{3}H]-spiperone binding in the frontal cortex are, however, difficult to explain on the basis of 5-HT$_2$ receptor blocking activity, since a blocking action at a receptor should theoretically result in an increase or up-regulation of the number of transmitter receptors. Besides an irreversible attachment to the receptor by antidepressant drugs an alternative explanation is the possibility of modulation via other receptor mechanisms controlling the 5-HT$_2$ receptor. It has been suggested that tricyclic antidepressant drugs can modulate the secretion of serotonin co-modulators which may contribute to the adaptive changes demonstrated (Fuxe et al. 1983b). Some recent evidence also suggests that release of co-modulators, which are of a peptidergic nature [TRH and substances P (SP)], may regulate the properties of serotonin receptors (see Fuxe et al. 1983b).

6.5.2 Hormonal regulation of 5-HT$_2$ receptors

Some data have also suggested that the receptor responses to some antidepressant drugs may depend on the hormonal state of the animal. Thus, the decrease in 5-HT$_2$ receptors in the cortex and hippocampus produced by long-term administration of imipramine was blocked in ovariectomized rats. Administration of oestradiol or progesterone separately or in combination re-established the effect on [^{3}H]-spiperone binding (Kendall et al. 1981). Castration was also found to block the decrease of 5-HT$_2$ but not β-adrenoceptors caused by chronic administration of imipramine and iprindole (Kendall, Stancel and Enna 1982). In contrast, the receptor responses to trazodone, mianserin and pargyline were not influenced by castration (Kendall et al. 1982). It is notable that a later study have shown that the reduction of [^{3}H]-spiperone by imipramine and mianserin in the cortex was uneffected by ovariectomy and hypophysectomy (Buckett et al. 1983).

6.5.3 Comparison of the actions on β-adrenoceptors and on 5-HT$_2$ receptors

It is well-established that repeated treatment with antidepressant drugs including uptake inhibitors, atypical drugs and monoamine oxidase inhibitors causes decrease in the number of β-adrenoceptors (Banerjee, Kung, Riggi and Chanda 1977; Maggi *et al.* 1980). However, changes in serotonergic receptor function seem to be more sensitive to chronic treatment than corresponding changes in β-adrenoceptors. Thus, although there is a significant decrease in both [³H]-spiperone and β-adrenoceptors one week following amitriptyline (10 mg/kg i.p. daily) the maximal reduction of [³H]-spiperone is much greater (Peroutka and Snyder 1980*b*; Snyder and Peroutka 1982). Interestingly, a maximal reduction of the binding of both ligands requires three weeks of treatment. Similar results have also been obtained following oral treatment. Thus, alaproclate and zimeldine, which affect serotonin binding at low oral doses (2×10 μmol/ kg) failed to affect the β-adrenoceptors in the cortex (Fuxe *et al.* 1983*b*).

6.5.4 *Effects of antidepressant drugs on 5-HT₂ receptors as examined by different types of 5-HT₂ ligands*

Consistent with the results on [³H]-spiperone binding amitriptyline, imipramine, desipramine, iprindole but not fluoxetine produced a reduction of [³H]-*d*-LSD binding following 3 weeks of treatment at a dose of 10 mg/kg i.p. daily. Again, methysergide, chlorpromazine and haloperidol were ineffective (Peroutka and Snyder 1980*a*). A few preliminary studies have also been performed using the radioligand [³H]-ketanserin as a marker for 5-HT₂ receptors. Repeated imipramine treatment (10 mg/kg i.p. for 21 days) decreased specific binding of [³H]-ketanserin in the hippocampus (Barbaccia, Gandolfi, Chuang and Costa 1983). Chronic oral zimeldine treatment (2×10 mmol/kg) was found to produce a small but significant reduction in the B_{max} values of [³H]-ketanserin binding sites in the dorsal cortex of the rat (Ögren *et al.* 1984).

[³H]-Mianserin has also been used as a radioligand for 5-HT₂ receptors in chronic studies with antidepressant drugs. As already indicated [³H]-mianserin was found to be displaced by compounds with affinity for the 5-HT₂ recognition sites. A common site of action between the tricyclic drugs and mianserin could be important for the antidepressant action of mianserin (Peroutka and Snyder 1981; Blackshear and Sanders-Bush 1982). Recent data do not accord with this hypothesis since repeated daily injections of mianserin and desipramine (10 mg/kg i.p. for 20 days) did not change the binding characteristics of [³H]-mianserin to cortical membranes (Brunello *et al.* 1982*b*), while a similar treatment is known to decrease the number of cortical 5-HT₂ recognition sites labelled by [³H]-spiperone (Blackshear and Sanders-Bush 1982). Interestingly, repeated imipramine (10 mg/kg i.p. for 21 days) also failed to alter specific [³H]-mianserin binding in the hippocampus, while a similar treatment decreased [³H]-ketanserin binding (Barbaccia *et al.* 1983). These data suggest that

[^{3}H]-mianserin and [^{3}H]-ketanserin do not label the same recognition site and that antidepressant drugs do not down regulate the [^{3}H]-mianserin site. In contrast Dumbrille-Ross *et al.* (1982) demonstrated that chronic desipramine and amitriptyline treatment (10 mg/kg i.p. for 21 days) could produce a down regulation of 5-HT$_2$ receptors in the frontal lobe as indicated by a reduction in the B_{max} values of [^{3}H]-mianserin. However, in this study non-specific binding was defined as that remaining in the presence of 80 nM spiperone while in the studies of Costa and coworkers cold mianserin (1 μM) was used to determine non-specific binding. It is likely, therefore, that the sub-population of 5-HT$_2$ receptors labelled by [^{3}H]-mianserin (defined by cold spiperone) is down-regulated by chronic mianserin treatment.

6.6 Functional significance of alterations in serotonin receptor binding

There are several problems in correlating the serotonergic receptor alterations with functional or behavioural changes. The major problem is that the relationship between the 5-HT$_1$ and 5-HT$_2$ binding sites and the physiological receptor is not clear. Based on electrophysiological investigations both inhibitory and excitatory roles of brain serotonin have been identified (Aghajanian and Wang 1978, and Section 7.1) suggests that the 5-HT$_1$ receptor identified in binding assays may mediate inhibitory responses while the 5-HT$_2$ site would trigger excitatory behaviours induced by serotonin receptor activation (McCall and Aghajanian 1979; Peroutka *et al.* 1981). However, recent studies suggest that the functional role of 5-HT$_1$ and 5-HT$_2$ receptors may require re-assessment (see Middlemiss 1982, and Sections 12.5 and 12.6). Another major problem is the selectivity and relevance of behavioural models presumed to reflect serotonergic receptor function. Several different 'serotonin models' have been employed in which the synaptic activity of serotonin probably differs (see Ögren, Fuxe, Berge and Agnati 1983*b*) and some of the behavioural models such as the 'serotonin syndrome' (Jacobs 1976; Green and Grahame-Smith 1976) induced by serotonin precursors involve to a large extent brain DA (Andrews, Fernando and Curzon 1982). It is also notable that serotonin neurones in the brain stem and spinal cord are believed to mediate the motor effects of serotonin receptor stimulation (McCall and Aghajanian 1979) such as the serotonin motor syndrome (Jacobs 1976) and head twitches and head shakes in rats (Corne, Pickering and Warner 1963; Jacobs 1976). However, most of the serotonin receptors in the brain stem and spinal cord are of the 5-HT$_1$ sub-type (Blackshear, Steranka and Sanders-Bush 1981) while the motor effects of serotonin agonists may be associated with the 5-HT$_2$ rather than the 5-HT$_1$ site (Peroutka *et al.* 1981).

These problems and the behavioural models are reviewed in detail in Chapter 12.

One of the most important considerations in interpreting the data from binding studies is whether differences in ligand binding reflects alterations in the functional state of the receptor. Behavioural and neurophysiological studies indicate that antidepressant treatment can induce adaptive changes in serotonin synapses which can be interpreted as either super- or sub-sensitivity development at the receptor (Fuxe *et al.* 1979; de Montigny and Aghajanian 1978; Ögren *et al.* 1982). When considering the reduction in the number of 5-HT$_2$ receptors observed after antidepressant treatment in the frontal lobe, it has usually been assumed that these results represent biochemical signs of 5-HT$_2$ receptor down-regulation. However, some investigators have found evidence for increased serotonin-mediated behaviour following chronic treatment with antidepressant drugs (Friedmann and Dallob, 1979; Green *et al.* 1983*a*; Jones 1980). Other evidence indicates a decreased serotonergic function following repeated antidepressant treatment (Fuxe *et al.* 1979; Ögren *et al.* 1982, 1983*b*; Stoltz *et al.* 1983). The head twitch responses induced in rats by a high dose of the agonist 5-methoxy-N,N-dimethyltryptamine (5-MeODMT, 4 mg/kg

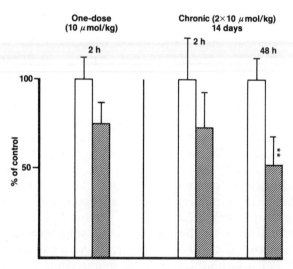

Fig. 6.4 Effects of repeated oral imipramine treatment on 5-MeODMT induced head twitches in rats. Imipramine was given in a single oral dose (10 μmol/kg) or in repeated oral doses (10 μmol/kg) twice daily for 2 weeks. 5-MeODMT (4 mg/kg) was injected s.c. at different times following the last treatment. Means ±S.E.M. are shown for 8 rats. **p <0.01 v. control (Tukey's t-test). (Data taken from Ögren *et al.* 1983*b*, with permission)

s.c.) (Fuxe, Holmstedt and Jonsson 1972) and which are believed to be mediated by 5-HT$_2$ receptor activation (Peroutka *et al.* 1981) are significantly reduced 48 hours following repeated administration of imipramine in clinically relevant doses (2×10 μmol/kg p.o.) (Fig. 6.4). Similar results have been observed with desipramine and zimeldine (Ögren *et al.* 1983*b*) and with clomipramine and fluoxetine (Stoltz *et al.* 1983) as well as amitriptyline (Pawlowski and Melzacka 1983). However, chronic imipramine treatment performed in the same way as described above had no effects on the forepaw treading (part of the serotonin syndrome) induced by 5-MeODT (4 mg/kg, s.c.), indicating a selectivity of chronic antidepressant treatment to down-regulate different types of serotonin receptors (Ögren *et al.* 1984). In contrast to the uptake blockers repeated administration of monoamineoxidase inhibitors has been reported to markedly reduce the serotonin syndrome in the rat induced by serotonin agonists (Lucki and Frazer 1982).

Interestingly, behavioural signs of serotonin receptor super-sensitivity have also been observed following repeated desipramine and imipramine treatment as evaluated in studies on head twitch behaviour. Thus, following a small dose of 5-MeODMT (1 mg/kg) enhanced numbers of head twitches were found 48 hours following chronic desipramine and imipramine treatment. Similar results were observed by Stoltz and co-workers (1983) showing that chronic amitriptyline and imipramine treatment 72 hours following the last injection potentiated the serotonin syndrome induced by 5-MeODMT (2.5 mg/kg i.p.). Also in mice the enhanced responsiveness of 5-MeODMT was first observed 24 hours after cessation of chronic treatment with the tricyclic drugs iprindole and trazodone (Friedman, Cooper and Dallob 1983). It should be pointed out that imipramine, desipramine, amitriptyline, iprindole, clomipramine, trazodone as well as fluoxetine all reduced the 5-MeODMT induced responses following chronic treatment when 5-MeODMT was given 30 minutes or 2 hours after the last treatment (Friedman *et al.* 1983; Ögren *et al.* 1983*b*; Stoltz *et al.* 1983). In view of the serotonin receptor blocking activity of tricyclic drugs (Ögren *et al.* 1979, 1982) these results may be explained on the basis of serotonin receptor blockade (see Fuxe *et al.* 1978) leading to super-sensitivity development. Interestingly, chonic zimeldine treatment does not produce this type of behavioural serotonin receptor super-sensitivity upon withdrawal which may be related to its inability acutely and following chronic administration to block serotonin-dependent behaviours such as head twitch behaviour (Ögren *et al.* 1982, and Section 12.7).

Other evidence indicates also a down regulation of serotonin receptor activity. The deficit in avoidance induced by 5-MeODMT is significantly attenuated in rats treated chronically with zimeldine or imipramine (Ölgren *et al.* 1982). There is evidence for a serotonin system in the forebrain which exerts an inhibitory role in active avoidance learning probably by modulation of storage and/or retrieval processes (Ögren *et al.* 1982). Thus, serotonergic receptor sub-sensitivity seems to exist in the forebrain. Interestingly, evidence for development of serotonin receptor sub-sensitivity upon chronic treatment has also been observed in spinal mechanisms of analgesia studied by the tail-flick test (Ögren *et al.* 1983*b*). The descending bulbospinal serotonin neurone systems innervating the dorsal horn, especially the substantia gelatinosa, exert an inhibitory influence on pain transmission at spinal cord level (Basbaum 1981). Following chronic zimeldine treatment, the pain sensitivity was enhanced as evaluated in the tail-flick test, while acute treatment with zimeldine reduced pain sensitivity in this model (Ögren and Holm 1980). The results suggest a presynaptic down-regulation of serotonin activity probably via reduced serotonin release. However, this down-regulation seems to result in a postsynaptic receptor super-sensitivity since the agonist 5-MeODMT, at a postsynaptic dose, produced an enhanced reduction in pain sensitivity following chronic zimeldine treatment (Fig. 6.5).

Recent findings also suggest a heterogeneity in the response of the neuroendocrine system to chronic antidepressant treatment. In the rat there is evidence for facilitatory serotonin mechanisms located in the hypothalamus and involved in the regulation of corticosterone and prolactin

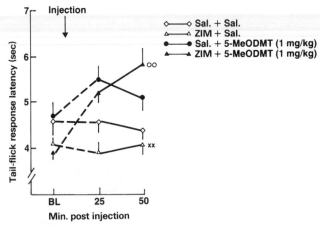

Fig. 6.5 Effects of long-term oral zimelidine (ZIM) treatment on tail-flick responses (pain sensitivity) in the rat. ZIM was given twice daily in a dose of 10 μmol/kg orally. Tail-flick baseline latency (BL) was examined 48 h following withdrawal. Tail-flick latency was also examined 25 and 50 min following injection of 5-MeODMT (1 mg/kg s.c.) (see Fig. 6.4). *p <0.01 ZIM v. saline; **p <0.01 ZIM/5-MeODMT v. saline/5-MeODMT. (Data taken from Ögren *et al.* 1983*b*, with permission)

secretion (Kordon, Héry, Fzafarczk, Ixart and Assenmacher 1981). Chronic zimeldine treatment which acutely increases corticosterone was found to reduce corticosterone secretion 2 hours following the last dose (Fig. 6.6), while a hypersecretion was observed 24 hours after dosing (Fuxe *et al.* 1983*a*). An opposite result was observed with prolactin. These results imply a down-regulation of the serotonin receptors controlling ACTH secretion during chronic zimeldine treatment. The rapid rebound is probably related to an up-regulation of the facilitatory serotonergic mechanisms, participating in the regulation of ACTH secretion. In the case of prolactin, the serotonin receptors controlling release of this hormone obviously can still respond to the treatment with no evidence for down-regulation of the receptor response.

In agreement with some of these results there also exists neurophysiological evidence that chronic treatment with desipramine and imipramine can lead to serotonin receptor super-sensitivity as demonstrated in the hippocampus and in the nuc. motorius nervi facialis, while chronic zimeldine treatment failed to induce neurophysiological signs of receptor

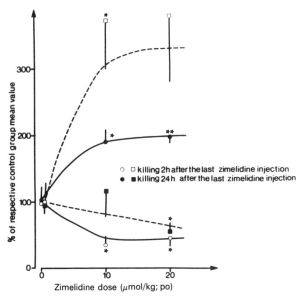

Fig. 6.6   The effects of oral, long-term zimelidine treatment (2 weeks, 10 or 20 μmol/kg twice daily) on the serum prolactin levels (□, ■) and the serum corticosterone levels (O, ●) 2 h and 24 h following the last dose of zimelidine. Means ±S.E.M. are shown for 6 rats. The respective hormone levels are given in per cent of the saline treated group mean value. Statistical analysis was performed using Wilcoxon test: one-way classification comparing all possible pairs of treatment. *p <0.05 v. saline; *p <0.01 v. saline. (Data taken from Fuxe *et al.* 1983*b*, with permission)

super-sensitivity, at least in the areas analysed. (de Montigny and Aghajanian 1978; Menkes, Aghjanian and McCall 1980; de Montigny *et al.* 1981). On the other hand, Olpe and Schellenberg (1981) failed to observe any changes in the cortical nerve cell responsitivity to iontotheoretically applied serotonin following chronic treatment with tricyclic drugs while chronic clorgyline produced a marked desensitization to serotonin. However, other neurophysiological data suggest sub-sensitivity development in the cortex. Serotonin applied onto the occipital cortex reduced the amplitude of the secondary component of the light-evoked response and this action was counteracted by chronic treatment with amitriptyline and zimeldine (Pawlowski, Stach and Kacz 1982).

These results are obviously difficult to reconcile with the biochemical findings reported above, but may suggest differences in the adaptive changes occuring in different types of serotonin receptor populations in various brain areas. In view of the serotonin receptor blocking action of several antidepressant drugs (Ögren *et al.* 1979) it should be considered that some of the neurophysiological responses observed (de Montigny and Aghajanian 1978) may reflect the development of a super-sensitivity response. It must be noted that zimeldine which lacks receptor blocking activity does not induce sensitization of serotonin receptors (de Montigny *et al.* 1981). These results emphasize the importance of performing detailed regional analysis of the adaptive changes taking place in serotonin receptors upon chronic treatment with different types of drugs. The biochemical mechanism underlining the dose-dependent differences in the responses to serotonin agonists upon antidepressant treatment is presently unknown since serotonin receptor mechanisms in the relevant brain regions have not been analysed.

Taken together, the studies with head twitches responses suggest reduced responsiveness of serotonin following chronic antidepressant treatment followed by signs of serotonin receptor super-sensitivity or sub-sensitivity development upon drug withdrawal. These results suggest that chronic antidepressant treatment in some brain areas can narrow the range of serotonin receptor activity (possibly $5-HT_2$ activity) which may represent one mechanism for stabilization of neurotransmission, thereby avoiding wide fluctuations in serotonin receptor activity.

6.7 Discussion

The findings of this review indicate that postsynaptic alterations in serotonin receptor functions are caused by a great number of clinically established antidepressant drugs. However, there is no single common change in brain serotonergic receptors after chronic antidepressant treatment. The most consistent finding is that chronic treatment with a great number of antidepressant drugs (amitriptyline, imipramine, desipramine, iprindole, mianserin, pargyline, zimeldine, alaproclate) produce a decrease in the density of the $[^3H]$-spiperone-labelled serotonin receptors in the frontal cortex of the rat. However, the mechanism behind the decrease in 5-HT$_2$ receptor density is at present not clear. Although several antidepressants have a high affinity for the 5-HT$_2$ recognition sites *in vitro* and *in vivo* the long-term adaptive changes in 5-HT$_2$ receptors do not directly correlate with the affinity for the recognition site or changes in serotonin availability. Some data suggest that the hormonal state of the animal is critical for down-regulation of 5-HT$_2$ receptors (see Kendall *et al.* 1982).

Although, most of the drugs found to decrease $[^3H]$-spiperone binding have been shown to have a therapeutic effect in depression the evidence presented here indicates that down-regulation of 5-HT$_2$ receptors may not be a common denominator for treatment having antidepressant activity. In contrast to the effects of antidepressant drugs repeated, but not acute ECS, increases the binding of $[^3H]$-spiperone in cerebral cortical membranes (Kellar *et al.* 1981; Vetulani, Lebrecht and Pilc 1981; Green, Johnson and Nimgoankar 1983 *b*, and see Section 12.7.

Moreover, chronic treatment with lithium in clinically relevant doses induced in the rat an increase in the number of 5-HT$_2$ receptors using $[^3H]$-spiperone as radioligand in the cerebral cortex (Kellar *et al.* 1981). However, also drugs with a non-established use as antidepressants (chlorpromazine, thioridazine, pizotifen) can reduce the density of 5-HT$_2$ binding sites although it should be noted that thioridazine, pizotifen as well as chlorpromazine have been suggested to possess antidepressant properties (Standal 1977; de Jonghe, Van der Helm, Schalkan and Thiel 1973). The ability of chlorpromazine to decrease 5-HT$_2$ binding and to enhance the down-regulation caused by imipramine is of interest in view of the observation that chlorpromazine may have an antidepressant action (see Mikuni and Meltzer 1984).

Concerning the adaptive changes which occur in 5-HT$_1$ receptors following chronic antidepressant treatment limited data are at present available. Several investigators have reported alterations in the high affinity binding sites with impramine, selective uptake blockers such as zimeldine and with the monoamine oxidase inhibitors. Whether the changes in $[^3H]$-serotonin binding sites only relate to drugs which predominately effect the serotonin system remains to be further investigated. It should be noted that long-term lithium administration reduces $[^3H]$-serotonin binding in the hippocampus but not in the cortex accompanied by an increased basal release of serotonin (Treiser, Cascio, O'Donohue, Thoa, Jacobowitz and Kellar 1981). There is an intriguing possibility that antidepressant drugs can induce a very high affinity binding site for $[^3H]$-serotonin both *in vitro* (Fillion and Fillion 1981) and *in vivo* (Fuxe *et al.* 1983*b*) which is uncoupled from the biological effector, a serotonin-dependent cyclase. These effects could suggest a mechanism by which serotonin neurotransmission is stabilized since transmission will operate through a 'low affinity' state of the receptor (see above and Fuxe *et al.* 1983*b*).

More extended clinical studies are required to assess the significance of serotonin receptor changes. Some data suggest a differential susceptibility to different antidepressants suggesting that the pathophsiology of patient sub-groups is critical. It should be pointed out that patients with 'endogenous' forms of depression may respond better to tricyclic drugs than patients with atypical forms of depression (Bielski and Friedel 1976). Interestingly, a recent study also showed that patients with several depressive epidsodes responded better to the serotonin uptake blocker zimeldine than to the NA uptake blocker maprotiline (Hällström and Nyström 1983). Patients with few episodes responded better to the latter drug. Moreover, atypical depressives are reported to respond favourably to MAOIs (Quitkin, Rifkin and Klein 1979). It will therefore be relevant to compare the responses in patient groups with different clinical features using compounds with different effects on serotonin receptor systems.

The functional and behavioural paradigms designed to assess serotonin receptor function have not clearly identified whether the changes observed correlate with the alteration observed in receptor binding assays. The results underline the heterogeneity of the central serotonin neurone systems in their responses to chronic antidepressant treatment. Whether there exist specific changes in postsynaptic receptor sensitivity which may be linked to the antidepressant action following long-term administration of antidepressant drugs is, thus, not known. Interestingly, the neuroendocrine paradigm designed to assess postsynaptic serotonin function indicates a down-regulation of those mechanisms involved in corticosterone secretion following chronic zimeldine treatment (Fuxe *et al.* 1983*a*). It is notable that the serum cortisol responses to 5-HTP in unmedicated depressed and manic patients were both significantly greater than that of normal controls (Meltzer, Uberkoman-Wiita, Robertson, Tricou and Lowy 1983). These results suggest that at least some serotonin receptors may be super-sensitive in some depressed patients.

The observation that both acute and long-term antidepressant treatment produces effects on serotonin receptor systems raises the question whether changes in receptors may underly some of the clinical manifestations of depression (see Charney, Menkes, Phil and Heninger 1981). Some preliminary data suggest that the number of $5\text{-}HT_2$ receptors are increased in the frontal cortex of suicide victims (Stanley and Mann 1983). If this finding can be confirmed in large samples, it is consistent with the down-regulation of $5\text{-}HT_2$ receptors observed with many antidepressant drugs. The findings that antidepressant drugs can exert a direct action on some serotonin receptors in the rat brain are of particular interest in view of recent theories on the mechanism of action of antidepressant drugs. Based on receptor ligand studies and on functional studies it was suggested that antidepressant compounds may in part act via blockade of certain types of serotonin receptors within the brain or via down-regulation or stabilization of functional serotonin activity (Ögren *et al.* 1979). These ideas are in line with other present theories indicating overfunction of serotonin neuronal activity in depressive states (Aprison and Hingten 1981; Takahashi, Tateishi, Yoshida, Nagayama and Tachiki 1981) due to the development of hypersensitive serotonin receptors.

Considerably more information is required before the significance of the serotonin receptor interaction of antidepressant drugs can be clearly evaluated. It is notable that a recent study has also shown antidepressant properties by cyproheptadine, a serotonin receptor antagonist (Bansal and Brown 1983). Since concentrations of antidepressant drugs in the rat brain following chronic treatment is normally in the range of 1 μM or higher a direct action on postsynaptic serotonin receptos in the human brain *in vivo* can not be excluded.

Acknowledgements

This work has been supported by a grant (MH 25504) from the National Institute of Mental Health, NIH, Bethesda, Maryland, USA. We gratefully acknowledge the excellent assistance of Mrs Eva Sepa and B. A. C. Johansson. We thank Drs H. Hall, J. Hyttel and J. Marcusson for allowing us to use unpublished papers.

References

Åberg-Wistedt, A. (1982). A double-blind study of zimelidine, a serotonin uptake inhibitor, and desipramine, a noradrenergic uptake inhibitor, in endogenous depression. *Acta Psychiat. Scand.* **66**, 50.

Aghajanian, G. K. and Wang, R. Y. (1978). Physiology and pharmacolcoy of central serotonin neurones. In *Psychopharmacology: A generation of progress* (eds. M. A. Lipton, A. Di Mascio and K. F. Killam) p. 171. Raven Press, New York.

Andén, N.-E., Dahlström, A., Fuxe, K., Larsson, K., Olson, L. and Ungerstedt, U. (1966). Ascending monoamines neurons to the telencephalon and diencephalon. *Acta Physiol. Scand.* **67**, 313.

Andrews, C. D., Fernando, J. C. R. and Curzon, G. (1982). Differential involvement of dopamine-containing tracts in 5-hydroxytryptamine-dependent behaviours caused by amphetamine in large doses. *Neuropharmacology* **21**, 63.

Aprison, M. H. and Hingten, J. N. (1981). Hypersensitive serotonergic receptors: A new hypothesis for one subgroup of unipolar depression derived from an animal model. In *Serotonin. Current aspects of neurochemistry and function* (eds. B. Haber, S. Gabay, M. R. Issidorides and S. G. A. Alivisatos) p. 627. Plenum Press, New York.

Åsberg, M., Thorén, P., Träskman, L., Bertilsson, L. and Ringberger, V. (1976). Serotonin depression—a biochemical subgroup within the affective disorders? *Science* **191**, 478.

Banerjee, S. P., Kung, L. S., Riggi, S. J. and Chanda, S. K. (1977). Development of β-adrenergic receptor subsensitivity by antidepressants. *Nature* **268**, 455.

Bansal, S. and Brown, W. A. (1983). Cyproheptadine in depression. *Lancet* **i**, 803.

Barbaccia, M. L., Gandolfi, O., Chuang, D. M. and Costa, E. (1983). Differences in the regulatory adaption of the 5-HT₂ recognition sites labelled by ³H-mianserin or ³H-ketanserin. *Neuropharmacology* **22**, 123.

Basbaum, A. I. (1981). Descending control of pain transmission: Possible serotonergic enkephalinergic interactions. In *Serotonin. Current aspects of neurochemistry and function.* (eds. B. Harber, S. Gabay, M. R. Issidorides and S. G. A. Alivisatos) p. 177. Plenum Press, New York.

Bauman, P. A. and Maitre, L. (1977). Blockade of preysynaptic α-receptors and of amine uptake in the rat brain by the antidepressant mianserin. *Naunyn-Schmiedeberg's Arch. Pharmac.* **300**, 31.

Bennett, J. P. and Snyder, S. H. (1976). Serotonin and lysergic acid diethylamide binding in rat brain membranes: Relationship to postsynaptic serotonin receptors. *Mol. Pharmac.* **12**, 373.

Bergstrom, D. A. and Kellar, K. J. (1979). Adrenergic and serotonin receptor binding in rat brain after chronic desmethylimipramine treatment. *J. Pharmac. Exp. Ther.* **209**, 256.

Bielski, R. J. and Friedel, R. O. (1976). Prediction of tricyclic antidepressant response. *Arch. Gen. Psychiat.* **33**, 1479.

Blackshear, M. A., Steranka, L. R. and Sanders-Bush, E. (1981). Multiple serotonin receptors: Regional distribution and effect of raphe lesions. *Eur. J. Pharmac.* **76**, 325.

Blackshear, M. A. and Sanders-Bush, E. (1982). Serotonon receptor sensitivity after acute and chronic treatment with mianserin. *J. Pharmac. Exp. Ther.* **221**, 303.

Blaschuk, K. and Tang, S. W. (1983). Temperature sensitive reversible loss of ³H-serotonin binding sites. Abstract, *Soc. Neurosci.* **9(1)**, 175.

Brunello, N., Chuang, D. M. and Costa, E. (1982*a*). Different synaptic locations of mianserin and imipramine binding sites. *Science* **215**, 1112.
—————— (1982*b*). Characterization of typical and atypical antidepres-

5 Effects of antidepressant drugs**169**

sant recognition sites in rat brain. In *Typical and atypical antidepressants: molecular mechanisms* (eds. E. Costa and G. Racagni) p. 179. Raven Press, New York.

Buckett, W. R., Strange, P. G., Stuart, E. M. and Thomas, P. C. (1983). Chronic antidepressant treatment, hormonal manipulation and cortical serotonin S_2 receptors. *Br. J. Pharmac.* **79**, 297.

Buus-Lassen, J., Squires, R. F., Christensen, J. A. and Molander, L. (1975). Neurochemical and pharmacological studies on a new 5-HT uptake inhibitor, FG 4963, with potential antidepressant properties. *Psychopharmacologia* **42**, 21.

Carlsson, A., Corrodi, H., Fuxe, K. and Hökfelt, T. (1969a). Effect of antidepressant drugs on the depletion of intraneuronal brain 5-hydroxytryptamine stores caused by 4-methyl-α-ethyl-meta-tyramine. *Eur. J. Pharmac.* **5**, 357.

—— —— —— —— —— (1969b). Effect of some antidepressant drugs on the depletion of intraneuronal brain catecholamine stores caused by 4,α-dimethyl-meta-tyramine. *Eur. J. Pharmac.* **5**, 367.

Charney, D. S., Menkes, D. B., Phil, M. and Heninger, G. R. (1981). Receptor sensitivity and the mechanism of action of antidepressant treatment. *Arch. Gen. Psychiat.* **38**, 1160.

Claassen, V., Davies, J. E., Hertting, G. and Placheta, P. (1977). Fluvoxamine, a specific 5-hydroxytryptamine uptake inhibitor. *Br. J. Pharmac.* **60**, 505.

Clements-Jewery, S. and Robson, P. A. (1982). Intact 5-HT neuroterminals are not required for 5-HT$_2$ receptor down-regulation by amitriptyline. *Neuropharmacology* **21**, 725.

—— —— and Chidley, L. J. (1980). Biochemical investigations into the mode of action of trazodone. *Neuropharmacology* **19**, 1165.

Cooper, B. R., Hester, T. J. and Maxwell, R. A. (1980). Behavioural and biochemical effects of the antidepressant bupropion (Wellbutrin): Evidence for selective blockade of dopamine uptake *in vivo. J. Pharmac. Exp. Ther.* **215**, 127.

Coppen, A. (1967). The biochemistry of affective disorders. *Br. J. Psychiat.* **113**, 1237.

Corne, S. J., Pickering, R. W. and Warner, B. T. (1963). A method for assessing the effects of drugs on the central actions of 5-hydroxytryptamine. *Br. J. Pharmac.* **20**, 106.

Coupet, J., Rauh, C. E., Szues-Myers, V. A. and Yunger, L. M. (1979). 2-Chloro-11-(piperazinyl)dibenz(b.f.)(1,4)oxazepine (amoxapine), an antidepressant with antipsychotic properties—a possible role for 7-hydroxyamoxapine. *Biochem. Pharmac.* **28**, 2514.

Crews, F. T., Scott, J. A. and Shorstein, N. H. (1983). Rapid down-regulation of serotonin$_2$ receptor binding during combined administration of tricyclic antidepressant drugs and α_2 antagonists. *Neuropharmacology* **22**, 1203.

Dahlström, A. and Fuxe, K. (1964). Evidence for the existence of mono-amine containing neurons in the central nervous system. I. Demonstration of monoamines in the cell bodies of brain stem neurons. *Acta Physiol. Scand.* **62**, Suppl. 232, 1.

de Jonghe, F. E. R. E. R., van der Helm, H. J., Schalken, H. F. A. and Thiel, J. H. (1973). Therapeutic effect and plasma level of thioridazine. *Acta Phychiat. Scand.* **49**, 535.

Daniel, W., Adamus, A., Melzacka, M., Szymmura, J. and Vetulani, J. (1981). Cerebral pharmacokinetics of imipramine in rats after single and multiple dosages. *Nauyn-Schmiedeberg's Arch. Pharmac.* **317**, 209.

de Montigny, C. and Aghajanian, G. K. (1978). Tricyclic antidepressants: Long-term treatment increases responsivity of rat brain forebrain neurons to serotonin. *Science* **202**, 1303.

—— Blier, P., Caillé. G. and Kouassi, E. (1981). Pre- and postsynaptic effects of zimeldine and norzimeldine on the serotonergic system: single cell studies in the rat. *Acta Psychiat. Scand.* **63**, suppl. **290**, 79.

Dumbrille-Ross, A. and Tang, S. W. (1983). Noradrenergic and sero-tonergic input necessary for imipramine-induced changes in beta but not S_2 receptor densities. *Psychiat. Res.* **9**, 207.

—— —— and Coscina, V. (1982). Lack of effect of raphe lesions on serotonin S_2 receptor changes induced by amitriptyline and desmethyl-imipramine. *Psychiat. Res.* **7**, 145.

—— —— and Seeman, P. (1980). High-affinity binding of [^{3}H]-mianserin to rat cerebral cortex. *Eur. J. Pharmac.* **68**, 395.

Fillion, G. (1983). Central serotonin receptors: Regulation mechanisms at the molecular level. In *Molecular pharmacology of neurotransmitter receptors* (eds. T. Segawa, H. I. Yamamura and K. Kuriyama) p. 115. Raven Press, New York.

—— and Fillion, M. P. (1981). Modulation of affinity of postsynaptic serotonin receptors by antidepressant drugs. *Nature* **292**, 349.

—— Beaudoin, D., Rouselle, J. C., Denian, J. M., Fillion, M. P., Dray, F. and Jacob, J. (1979). Decrease of [^{3}H]-5-HT high affinity binding and 5-HT adenylate cyclase activation after kainic acid lesions in rat brain striatum. *J. Neurochem.* **33**, 567.

Friedman, E. and Dallob, A. (1979). Enhanced serotonin receptor activity after chronic treatment with imipramine or amitriptyline. *Commun. Psychopharmac.* **3**, 89.

Friedman, E., Cooper, T. B. and Dallob, A. (1983). Effects of chronic antidepressant treatment on serotonin receptor activity in mice. *Eur. J. Pharmac.* **89**, 69.

Fuller, R. W. and Roush, B. W. (1972). Substrate-selective and tissue-selective inhibition of monoamine oxidase. *Arch. Int. Pharmacodyn. Ther.* **198,** 270.

Fuxe, K., Hökfelt, T. and Ungerstedt, U. (1970). Morphological and functional aspects of central monoamine neurons. In *International Review of Neurobiology,* Vol. 13, p. 93. Academic Press, New York.

—— Holmstedt, B. and Jonsson, G. (1972). Effects of 5-methoxy-N,N-dimethyltryptamine on central monoamine neurons. *Eur. J. Pharmac.* **19,** 25.

—— Ögren, S. O. and Agnati, L. F. (1979). The effects of chronic treatment with the 5-hydroxytryptamine uptake blocker zimeldine on central 5-hydroxytryptamine mechanisms. Evidence for the induction of a low affinity binding site for 5-hydroxytryptamine. *Neurosci. Lett.* **13,** 307.

—— —— —— and Calza, I. (1982*c*). Evidence for stabilization of cortical 5-HT neurotransmission by chronic treatment with antidepressant drugs: induction of a high and a low affinity component in ^{3}H-5-HT binding sites. *Acta Physiol. Scand.* **114,** 477.

—— —— —— Andersson, K. and Eneroth, P. (1982*a*). On the mechanism of action of antidepressant drugs: Indications of reductions in 5-HT neurotransmission in some brain regions upon subchronic treatment. In *New vistas in depression,* (eds. S. Z. Langer, R. Takahashi, T. Segawa and M. Briley) p. 49. Pergamon Press, New York.

—— —— —— —— and Eneroth, P. (1982*b*). Effects of subchronic antidepressant drug treatment on central serotonergic mechanism in the male rat. In *Typical and atypical antidepressants: molecular mechanisms* (eds. E. Costa and G. Racagni) p. 91. Raven Press, New York.

—— —— —— Gustafsson, J. Å. and Jonsson, G. (1977). On the mechanism of action of the antidepressant drugs amitriptyline and nortriptyline. Evidence for 5-hydroxytryptamine receptor blocking activity. *Neurosci. Lett.* **6,** 339.

—— —— —— Eneroth, P., Holm, A.-C. and Andersson, K. (1981). Long-term treatment with zimelidine leads to a reduction in 5-hydroxytryptamine neurotransmission within the central nervous system of the mouse and rat. *Neurosci. Lett.* **21,** 57.

—— —— —— Benfenati, F., Fredholm, B., Andersson, K., Zini, I. and Eneroth, P. (1983*b*). Chronic antidepressant treatment and central 5-HT synapses. *Neuropharmacology* **22,** 389.

—— —— —— —— Cavicchioli, L., Fredholm, B., Andersson, K., Farabegoli, C. and Eneroth, P. (1983*a*). Regional variations in 5-HT receptor populations and in ^{3}H-imipramine binding sites in their responses to chronic antidepressant treatment. In *Frontiers in neuropsychiatric research,* (eds. E. Usdin, M. Goldstein, A. Friedhoff and A. Georgotas) p. 33. Macmillan, London.

—— —— Everitt, B. J., Agnati, L. F., Eneroth, P., Gustafsson, J. Å.,

Jonsson, G., Skett, P. and Holm, A.-C. (1978). The effect of antidepressant drugs of the imipramine type on various monoamine systems and their relation to changes in behaviour and neuroendocrine function. In *Symposium Medicum Hoechst on depressive disorders* (ed. S. Garattini) p. 69. F. K. Schattauer, Stuttgart.

Gershon, S., Georgotas, A., Newton, R. and Bush, D. (1982). Clinical evaluation of two antidepressants. *Adv. Biochem. Psychopharmac.* **32,** 57.

Glotzback, R. K. and Preskorn, S. H. (1982). Brain concentrations of tricyclic antidepressants: Single dose kinetics and relationship to plasma concentrations in chronically dosed rats. *Psychopharmacology* **78,** 25.

Goodlet, I., Mireyless, S. E. and Sugrue, M. R. (1977). Effects of mianserin, a new antidepressant, on the in vitro and in vivo uptake of monoamines. *Br. J. Pharmac.* **61,** 307.

Goodwin, G. M., Green, A. R. and Johnson, P. (1984). 5-HT$_2$ receptor characteristics in frontal cortex and 5-HT$_2$ receptor-mediated head-twitch behaviour following antidepressant treatment to mice. *Br. J. Pharmac.* **83,** 235.

Green, A. R. and Grahame-Smith, D. G. (1976). Effects of drugs on the processes regulating the functional activity of brain 5-hydroxytryptamine. *Nature,* **260,** 487.

—— Johnson, P. and Nimgaonkar, V. L. (1983 b). Increased 5-HT$_2$ receptor number in brain as a probable explanation for the enhanced 5-hydroxytryptamine-mediated behaviour following repeated electroconvulsive shock administration to rats. *Br. J. Pharmac.* **80,** 173.

—— Heal, D. J., Johnson, P., Laurence, B. E. and Nimgaonkar, V. L. (1983 a). Antidepressant treatments: effects in rodents on dose-response curves of 5-hydroxytryptamine- and dopamine-mediated behaviours and 5-HT$_2$ receptor number in frontal cortex. *Br. J. Pharmac.* **80,** 377.

Greenwood, D. T. (1982). Viloxazine and neurotransmitter function. In *Typical and atypical antidepressants: molecular mechanisms* (eds. E. Costa and G. Racagni) p. 287. Raven Press, New York.

Hall, H. and Ögren, S. O. (1981). Effects of antidepressant drugs on different receptors in the brain. *Eur. J. Pharmac.* **70,** 393.

—— Ross, S. B. and Ögren, S. O. (1982). Effects of zimelidine on various transmitter systems in the brain. In *Typical and atypical antidepressants: molecular mechanisms* (eds. E. Costa and G. Racagni) p. 321. Raven Press, New York.

—— —— and Sällemark, M. (1984 b). Effect of destruction of central noradrenergic and serotonergic nerve teminals by systemic neurotoxins on the long-term effects of antidepressants on β-adrenoceptors and 5-HT$_2$ binding sites in the rat cerebral cortex. *J. Neural Transm.* **59,** 9.

—— Sällemark, M. and Wedel, I. (1984 a). Acute effects of atypical antidepressants on various receptors in the rat brain. *Acta Pharmac. Toxicol.* **54**, 379.

Hällström, T. and Nyström, C. (1983). A double blind efficacy comparison between zimelidine and maprotiline in the treatment of depressed outpatients. *Nord. Psykiatr. Tidskr.* **1**, 97.

Hart, H. E. (1965). Determination of equilibrium constants and maximum binding capacities in complex in vitro systems. *Bull. Math. Biophys.* **27**, 87.

Hwang, E. C., Magnussen, I. U. and van Woert, M. H. (1980). Effects of chronic fluoxetine administration on serotonin metabolism. *Res. Comm. Chem. Pathol. Pharmac.* **29**, 79.

Hyttel, J. (1982). Citalopram-pharmacological profile of a specific serotonin uptake inhibitor with antidepressant activity. *Prog. Neuro-Psychopharmac. Biol. Psychiat.* **6**, 277.

—— Fredricson Overo, K. and Arnt, J. (1984). Biochemical effects and drug levels in rats after long-term treatment with the specific 5-HT uptake inhibitor, citalopram. *Psychopharmacology* **83**, 20.

Janowsky, A., Steranka, L. R., Gillespie, D. D. and Sulser, F. (1982). Role of neuronal signal input in the down-regulation of central noradrenergic receptor function by antidepressant drugs. *J. Neurochem.* **39**, 290.

Jacobs, B. L. (1976). An animal behaviour model for studying central serotonergic synapses. *Life Sci.* **19**, 777.

Johnston, J. P. (1968). Some observations upon a new inhibitor of monoamine oxidase in brain tissue. *Biochem. Pharmac.* **17**, 1285.

Jones, R. S. G. (1980). Enhancement of 5-hydroxytryptamine-induced behavioural effects following chronic administration of antidepressant drugs. *Psychopharmacology* **69**, 307.

Kellar, K. J., Cascio, C. S., Butler, J. A. and Kurtzke, R. N. (1981). Differential effects of electroconvulsive shock and antidepressant drugs on serotonin-2 receptors in rat brain. *Eur. J. Pharmac.* **69**, 515.

Kendall, D. A., Stancel, G. M. and Enna, S. J. (1981). Imipramine: Effect of ovarian steroids on modifications in serotonin receptor binding. *Science* **211**, 1183.

—— —— —— (1982). The influence of sex hormones on antidepressant-induced alterations in neurotransmitter receptor binding. *J. Neuroscience* **2**, 354.

Knoll, J. and Magyar, K. (1972). Some puzzling pharmacological effects of monoamine oxidase inhibitors. *Adv. Biochem. Psychopharmacol.* **5**, 393.

Köhler, C. (1984). An autoradiographic study of serotonin receptor distribution in the hippocampal region in the rat brain. *Neuroscience.* (in press).

Kordon, C., Héry, M., Fzafarczyk, A., Ixart, G. and Assenmacher, I. (1981).

Serotonin and the regulation of pituitary hormone secretion and of neuroendocrine rhythms. *J. Physiol. (Paris)* **77**, 489.

Kragh-Sörensen, P. (1983). Abstract, *VII World Congress of Psychiatry, Vienna*, 11–16 July, p. 179.

Langer, S. Z. and Briley, M. (1981). High affinity ^{3}H-imipramine binding: A new biological tool for studies in depression. *Trends Neurosci.* **4**, 28.

—— Zarifian, E., Briley, M., Raisman, R. and Sechter, D. (1981). High affinity binding of ^{3}H-imipramine in brain and platelets and its relevance to the biochemistry of affective disorders. *Life Sci.* **29**, 211.

Lapin, I. P. and Oxenkrug, G. F. (1969). Intensification of the central serotonergic process as a possible determinant of the thymoleptic effect. *Lancet* i, 132.

Leysen, J. E., Niemegeers, C. J. E., Tollenaere, J. P. and Laduron, P. M. (1978). Serotonergic component of neuroleptic receptors. *Nature* **272**, 168.

—— —— van Nueten, J. M. and Laduron, P. M. (1982). [^{3}H]-ketanserin (R41468), a selective ^{3}H-ligand for serotonin$_2$ receptor binding sites. Binding properties, brain distribution and functional role. *Mol. Pharmac.* **21**, 301.

—— Awouters, F., Kennis, L., Laduron, P. M., Vandenberk, J. and Janssen, P. A. J. (1981). Receptor binding profile of R 41468, a novel antagonist at 5-HT$_2$ receptors. *Life Sci.* **28**, 1015.

Lindberg, U. H., Thorberg, S. O., Bengtsson, S., Renyi, A. L., Ross, S. B. and Ögren, S. O. (1978). Inhibitors of neuronal monoamine uptake. 2. Selective inhibition of 5-hydroxytryptamine uptake by α-amino acid esters of phenethyl alcohols. *J. Med. Chem.* **21**, 448.

Lucki, I. and Frazer, A. (1982). Prevention of the serotonin syndrome in rats by repeated administration of monoamine oxidase inhibitors but not tricyclic antidepressants. *Psychopharmacology* **77**, 205.

McCall, R. B. and Aghajanian, G. K. (1979). Serotonergic facilitation of facial motoneuron excitation. *Brain Res.* **169**, 11.

Maj, J., Lewandowska, A. and Rawlow, A. (1979*b*). Central antiserotonin action of amitriptylin. *Pharmakopsychiat. Neuropsychopharmac.* **12**, 281.

—— Palider, W. and Rawlow, A. (1979*a*). Trazodone, a central serotonin antagonist and agonist. *J. Neural. Transm.* **44**, 237.

—— Gancarczyk, L., Gorszczyk, E. and Rawlow, A. (1977). Doxepin as a blocker of central serotonin receptors. *Pharmakopsychiat. Neuropsychopharmac.* **10**, 310.

—— Sowinska, H., Baran, L., Gancarczyk, L. and Rawlow, A. (1978). The central antiserotonergic action of mianserin. *Psychopharmacology* **59**, 79.

Maggi, A., U'Prichard, D. C. and Enna, S. J. (1980). Differential effects of antidepressant treatment on brain monoaminergic receptors. *Eur. J. Pharmac.* **61**, 91.

Maxwell, R. A. and White, H. L. (1978). Tricyclic and monoamine oxidase inhibitor antidepressants: Structure activity relationships. In *Handbook of psychopharmacology* (eds. L. L. Iversen, S. D. Iversen and S. H. Snyder) Vol. 14, p. 83. Plenum Press, New York.

Meltzer, H. Y., Uberkoman-Wiita, B., Robertson, A., Tricou, B. J. and Lowy, M. (1983). Enhanced serum cortisol response to 5-hydroxy-tryptophan in depression and mania. *Life Sci.* **33**, 2541.

Menkes, D. B., Aghajanian, G. K. and McCall, R. B. (1980). Chronic antidepressant treatment enhances α-adrenergic and serotonergic responses in the facial nucleus. *Life Sci.* **27**, 45.

Middlemiss, D. N. (1982). Multiple 5-hydrotryptamine receptors in the central nervous system of the rat. In *Presynaptic receptors* (L. De Belleroche) p. 46. Ellis Horwood, Chichester.

—— Carroll, J. A., Fisher, R. W. and Mounsey, I. J. (1980). Does [³H]-spiroperidol label a 5-HT receptor in the frontal cortex of the rat? *Eur. J. Pharmac.* **66**, 253.

Mikuni, M. and Meltzer, H. (1984). Reduction of serotonin-2 receptors in rat cerebral cortex after subchronic administration of imipramine, chloropromazine and the combination thereof. *Life Sci.* **34**, 87.

Mishra, R., Janowsky, A. and Sulser, F. (1980). Action of mianserin and zimeldine on the norepinephrine receptor coupled adenylate cyclase system in brain: Subsensitivity without reduction in β-adrenergic receptor binding. *Neuropharmacology* **19**, 983.

Morgan, D. G., Marcusson, J. O. and Finch, C. E. (1984). Contamination of serotonin-2 binding sites by an alpha-1 adrenergic component in assays with (³H)-spiperone. *Life Sci.* **34**, 2507.

Nelson, D. L., Herbet, A., Bourgoin, S., Glowinski, J. and Hamon, M. (1978). Characteristics of central 5-HT receptors and their adaptive changes following intracerebral 5,7-dihydroxytryptamine administration in the rat. *Mol. Pharmac.* **14**, 983.

Ögren, S. O. (1982). Forebrain serotonin and avoidance learning: Behavioural and biochemical studies on the acute effect of p-chloro-amphetamine on one-way active avoidance learning in the male rat. *Pharmac. Biochem. Behav.* **16**, 881.

—— and Holm, A.-C. (1980). Test-specific effects of the 5-HT reuptake inhibitors alaproclate and zimeldine on pain sensitivity and morphine analgesia. *J. Neural. Transm.* **47**, 253.

—— Fuxe, K., Agnati, L. F. and Celani, M. F. (1984). Effects of antidepressant drugs on cerebral serotonin receptor mechanisms. *Acta Pharmac. Toxicol.* (in press).

—— —— Berge, O. G. and Agnati, L. F. (1983b). Effects of chronic administration of antidepressant drugs on central serotonergic receptor mechanisms. In *Frontiers in neuropsychiatric research* (eds. E. Usdin, M. Goldstein, A. J. Friedhoff and A. Georgotas) p. 93. Macmillan, London.

—— Ross, S. B., Hall, H. and Archer, T. (1983a). Biochemical and behavioural effects of antidepressant drugs. In *Antidepressants,* (eds. G. D. Burrows, T. R. Norman and B. Davies) p. 13. Elsevier, Amsterdam.

—— Fuxe, K., Archer, T., Johansson, G. and Holm, A.-C. (1982). Behavioural and biochemical studies on the effects of acute and chronic administration of antidepressant drugs on central serotonergic receptor mechanisms. In *New vistas in depression.* (eds. S. Z. Langer, R. Takahashi, T. Segawa and M. Briley) p. 11. Pergamon Press, New York.

—— Ross, S. B., Hall, H., Holm, A.-C. and Renyi, A. L. (1981a). The pharmacology of zimelidine: A 5-HT selective reuptake inhibitor. *Acta Psychiat. Scand.* **63,** suppl. 290, 127.

—— Fuxe, K., Agnati, L. F., Gustafsson, J.-Å., Jonsson, G. and Holm, A.-C. (1979). Reevaluation of the indoleamine hypothesis of depression. Evidence for a reduction of functional activity of central 5-HT systems by antidepressant drugs. *J. Neural. Transm.* **46,** 85.

—— Ask, A. L., Holm, A. C., Florwall, L., Lindbom, L. O., Lundström, J. and Ross, S. B. (1981b). Biochemical and pharmacological properties of a new selective and reversible monoamine oxidase inhibitor, FLA 336(+) In *Monoamine oxidase inhibitors: the state of the art* (eds. M. B. H. Youdim and E. S. Paykel) p. 103. John Wiley, London.

Olpe, H. R. and Schellenberg, A. (1981). The sensitivity of cortical neurons to serotonin: Effect of chronic treatment with antidepressants, serotonin-uptake inhibitors and monoamine-oxidase-blocking drugs. *J. Neural. Transm.* **51,** 233.

Oswald, I., Brezinora, V. and Dunleavy, D. L. F. (1972). On the slowness of action of tricyclic antidepressant drugs. *Br. J. Psychiat.* **120,** 673.

Pawlowski, L. and Melzacka, M. (1983). Inhibition of head twitch response to quipazine in rats by chronic amitriptyline but not fluvoxamine or citalopram. *8th Congress of the Polish Pharmacological Society,* September 26–28, Warsaw (Abstract).

—— Stach, R. and Kacz, D. (1982). Chronic treatment with amitriptyline and zimelidine: Attenuation of serotonin-induced changes in light-evoked responses from the occipital cortex. In *New vistas in depression* (eds. S. Z. Langer, R. Takahashi, T. Segawa and M. Briley) p. 73. Pergamon Press, New York.

Peroutka, S. J. and Snyder, S. H. (1979). Multiple serotonin receptors: Differential binding of [^{3}H]-serotonin, [^{3}H]-lysergic acid diethylamide and [^{3}H]-spiroperidol. *Mol. Pharmac.* **16,** 687.

—— —— (1980a). Long-term antidepressant treatment decreases spiro-peridol-labelled serotonin receptor binding. *Science* **210,** 88.

—— —— (1980*b*). Regulation of serotonin₂ (5-HT-2) receptors labelled with ³H-spiroperidol by chronic treatment with the antidepressant amitriptyline. *J. Pharmac. Exp. Ther.* **215,** 582.

—— —— (1981). ³H-mianserin: Differential labelling of serotonin₂ and histamine₁ receptors in rat brain. *J. Pharmac. Exp. Ther.* **216,** 142.

—— Lebovitz, R. M. and Snyder, S. H. (1981). Two distinct central serotonin receptors with different physiological functions. *Science* **212,** 827.

Pinder, R. M., Brogden, R. N., Speight, T. M. and Avery, G. S. (1977). Doxepin up to date: a review of its pharmacological properites and therapeutic efficacy with particular reference to depression. *Drugs* **13,** 161.

van Praag, H. M. (1982). Neurotransmitters and CNS disease. Depression. *Lancet* **ii,** 1259.

Quitkin, F., Rifkin, A. and Klein, D. F. (1979). Monoamine oxidase inhibitors. *Arch. Gen. Psychiat.* **36,** 749.

Raisman, R., Briley, M. S. and Langer, S. Z. (1980). Specific tricyclic antidepressant binding sites in rat brain charcterized by high-affinity ³H-imipramine binding. *Eur. J. Pharmac.* **61,** 373.

Riblet, L. A., Gatewood, C. F. and Mayol, R. F. (1979). Comparative effects of trazodone and tricyclic antidepressants on uptake of selected neurotransmitters by isolated rat brain synaptosomes. *Psychopharmacology* **63,** 99.

Ross, S. B. and Renyi, A. L. (1969). Inhibition of the uptake of tritiated 5-hydroxytryptamine in brain slices. *Life Sci.* **6,** 1407.

—— —— (1975*a*). Tricyclic antidepressants agents. I. Comparison of the inhibition of the uptake of ³H-noradrenaline and ¹⁴-C-5-hydroxytryptamine in slices and crude synaptosome preparations of the midbrain-hypothalamus region of the rat brain. *Acta Pharmac. Toxicol.* **36,** 382.

—— —— (1975*b*). Tricyclic antidepressants agents. II. Effects of oral administration on the uptake of ³H-noradrenaline and ¹⁴-C-5-hydroxytryptamine in slices from the midbrain-hypothalamus region of the rat brain. *Acta Pharmac. Toxicol.* **36,** 395.

—— —— (1977). Inhibition of the neuronal uptake of 5-hydroxytryptamine and noradrenaline in rat brain by (Z)-and(E)-3-(4-bromophenyl)-*N,N*-dimethyl-3-(3-pyridyl)allylamines and their secondary analogues. *Neuropharmacology* **16,** 57.

—— Ögren, S. O. and Renyi, A. L. (1976). Z-dimethylamino-1-(4-bromophenyl)-1-(3-pyridyl)propene (H102/09) a new selective inhibitor of the neuronal 5-hydroxytryptamine uptake. *Acta Pharmac. Toxicol.* **39,** 152.

—— Hall, H., Renyi, A. L. and Westerlund, D. (1981). Effects of zimelidine on serotonergic and noradrenergic neurons after repeated administration in the rat. *Psychopharmacology* **72**, 219.

Rüdeberg, C. (1983). Effects of single and multiple doses of antidepressant drugs on the ^{3}H-spiperone-labelled serotonin receptors in the frontal cortex of the rat. In *Frontiers in neuropsychiatric research* (eds. E. Usdin, M. Goldstein, A. J. Friedhoff and A. Georgotas) p. 135. Macmillan, London.

Samanin, R., Mennini, T., Ferraris, A., Bendotti, C. and Borsini, F. (1980). Repeated treatment with *d*-fenfluramine or metergoline alters cortex binding of ^{3}H-serotonin and serotonergic sensitivity in rats. *Eur. J. Pharmac.* **61**, 203.

Savage, D. D., Frazer, A. and Mendels, J. (1979). Differential effects of monoamine oxidase inhibitors and serotonin reuptake inhibitors on ^{3}H-serotonin receptor binding in rat brain. *Eur. J. Pharmac.* **58**, 87.

—— Mendels, J. and Frazer, A. (1980*a*). Monoamine oxidase inhbitors and serotonin uptake inhibitors: Differential effects on ^{3}H-serotonin binding sites in rat brain. *J. Pharmac. Exp. Ther.* **212**, 259.

—— —— —— (1980*b*). Decrease in ^{3}H-serotonin binding in rat brain produced by the repeated administration of either monoamine oxidase inhibitors or centrally acting serotonin agonists. *Neuropharmacology* **19**, 1063.

Seeman, P. (1980). Brain dopamine receptors. *Pharmacological Reviews* **32**, 229.

—— Westman., K., Coscina, D. and Warsch, J. J. (1980). Serotonin receptors in hippocampus and frontal cortex. *Eur. J. Pharmac.* **66**, 179.

Segawa, T., Mizuta, T. and Nomura, Y. (1979). Modifications of central 5-hydroxytryptamine binding sites in synaptic membranes from rat brain after long-term administration of tricyclic antidepressants. *Eur. J. Pharmac.* **58**, 75.

—— —— and Uehara, M. (1982). Role of central serotonergic system as related to the pathogenesis of depression: Effect of antidepressants on rat central serotonergic activity. In *New vistas in depression* (eds. S. Z. Langer, R. Takahashi, T. Segawa and M. Briley) p. 3. Pergamon Press, New York.

Sethy, V. H. and Harris, D. W. (1981). Effect of norepinephrine uptake blocker on β-adrenergic receptors of the rat cerebral cortex. *Eur. J. Pharmac.* **75**, 53.

Snyder, S. H. and Peroutka, S. J. (1982). A possible role of serotonin receptors in antidepressant drug action. *Pharmacopsychiat.* **15**, 131.

Spector, S. (1963). Monoamine oxidase in control of brain serotonin and norepinephrine content. *Ann. N.Y. Acad. Sci.* **107**, 856.

Standal, J. E. (1977). Pizotifen as an antidepressant. *Acta Psychiat. Scand.* **56**, 276.

Stanley, M. and Mann, J. J. (1983). Increased serotonin-2 binding sites in frontal cortex of suicide victims. *Lancet* **i,** 214.

Stolz, J. F., Marsden, C. A. and Middlemiss, D. N. (1983). Effect of chronic antidepressant treatment and subsequent withdrawal on [³H]-5-hydroxytryptamine and [³H]-spiperone binding in rat frontal cortex and serotonin receptor mediated behaviour. *Psychopharmacology* **80,** 150.

Sugrue, M. F. (1983). Do antidepressants possess a common mechanism of action? *Biochem. Pharmac.* **32,** 1811.

Sulser, F. (1979). New perspectives on the mode of action of antidepressant drugs. *Trends Pharmac. Sci.* **1,** 92.

Svensson, T. H. (1983). Mode of action of antidepressant agents and ECT-adaptive changes after subchronic treatment. In *The origins of depression: current concepts and approaches.* (ed. J. Angst) p. 367. Springer Verlag, Berlin.

Takahashi, R., Tateishi, T., Yoshida, H., Nagayama, H. and Tachiki, L. (1981). Serotonin metabolism of animal model of depression. In *Serotonin Current aspects of neurochemistry and function* (eds. B. Haber, S. Gabay, M. R. Issidorides and S. G. A. Alivisatos) p. 603. Plenum Press, New York.

Tang, S. W. and Seeman, P. (1980). Effect of antidepressant drugs on serotonergic and adrenergic receptors. *Naunyn-Schmiedeberg's Arch. Pharmac.* **311,** 255.

Taylor, D. P., Allen, L. E., Ashworth, E. M., Becker, J. A., Hyslop, D. K. and Riblet, L. A. (1981). Treatment with trazodone plus phenoxybenzamine accelerates development of decreased type 2 serotonin binding in rat cortex. *Neuropharmacology* **20,** 573.

Treiser, S. L., Cascio, C. S., O'Donohue, T. L., Thoa, N. B., Jacobowitz, D. M. and Kellar, K. J. (1981). Lithium increases serotonin release and decreases serotonin receptors in the hippocampus. *Science* **213,** 1529.

VandenBerg, S. R., Allgren, R. L., Todd, R. D. and Ciaranello, R. D. (1983). Solubilization and characterization of high-affinity [³H]-serotonin binding sites from bovine cortical membranes. *Proc. Natl. Acad. Sci.* **80,** 3508.

Vetulani, J., Lebrecht, V. and Pilc, A. (1981). Enhancement of responsiveness of the central serotonergic system and serotonin-2 receptor density in rat frontal cortex by electroconvulsive treatment. *Eur. J. Pharmac.* **76,** 81.

Whitaker, P. M. and Seeman, P. (1978). Selective labeling of serotonin receptors by *d*-[³H]-lysergic acid diethylamide in calf caudate. *Proc. Natl. Acad. Sci.* **75,** 5783.

Whitaker, P. M. and Cross, A. J. (1980). ³H-Mianserin binding in calf caudate: Possible involvement of serotonin receptors in antidepressant drug action. *Biochem. Pharmac.* **29,** 2709.

Wirz-Justice, A., Krauchi, K., Lichtsteiner, M. and Feer, H. (1978). Is it possible to modify serotonin receptor sensitivity? *Life Sci.* **23,** 1249.

Wong, D. T. and Bymaster, F. P. (1981). Subsensitivity of serotonin receptors after long-term treatment of rats with fluoxetine. *Res. Comm. Chem. Pathol. Pharmac.* **32,** 41.

—— —— Reid, L. R. and Threlkeld, P. G. (1983). Fluoxetine and two other serotonin inhibitors without affinity for neuronal receptors. *Biochem. Pharmac.* **32,** 1287.

—— Horng, J. S., Bymaster, F. P., Hauser, K. L. and Molloy, B. B. (1974). A specific serotonin uptake inhibitor: Lilly 110140, 3-(β-trifluoromethyl-phenoxy)-*N*-methyl-3-phenylpropylamine. *Life Sci.* **15,** 471.

7

Electrophysiological aspects of serotonin neuropharmacology: implications for antidepressant treatments

CLAUDE DE MONTIGNY AND PIERRE BLIER

7.1 Introduction

7.1.1 *The distinction between the aetiological and the therapeutic aspect of monoaminergic hypotheses*

The notion that serotonin might be involved either in the pathogenesis of depression or in the therapeutic effect of antidepressant drugs, or both, is a long-standing one (Coppen 1967; Lapin and Oxenkrug 1969). From the early enthusiasm there succeeded a disenchantment phase brought about by conflicting clinical data. It is only in the last few years that novel electrophysiological, neuropharmacological, and clinical psychopharmacological evidence has revived interest in the notion that serotonergic neurotransmission might play a key role in mediating the therapeutic effect of antidepressant treatments.

The monoaminergic hypotheses of major affective disorders have been plagued with a confusion between their aetiological and therapeutic aspects. This confusion arose from the origin of these hypotheses. Indeed, it is from therapeutic observations [i.e. the efficacy of monoamine oxidase inhibitor (MAOI) and tricyclic antidepressant (TCA) drugs] that it was postulated that a deficiency of serotonin or noradrenaline might underlly major depression. It must be borne in mind that the aetiological and the therapeutic aspects of these hypotheses are totally distinct. For instance, should it be demonstrated that antidepressant treatments exert their therapeutic effect via the serotonin system, this would by no means constitute an indication of a dysfunction of this system in major depression. Indeed, medical pharmacology is full of examples of the fact that acting on an intact system can compensate for the dysfunction of another totally distinct system. The present chapter will focus exclusively on the involvement of the serotonin system in mediating the therapeutic effect of antidepressant treatments. As for a possible aetiological role of serotonin in major depression, the reader is referred to a recent review by Siever, Guttmacher and Murphy (1984).

7.1.2 *The electrophysiologically-defined serotonin receptors*

The development of the technique of radioligand binding has given rise to such an immense proliferation of biochemical binding studies that, for many, the term 'receptor' has become synonymous with 'binding site'. It is true that for several neuronal systems, there is a very close correspondence between biochemical and physiological data, but this is clearly not the case as yet for the serotonin system. Biochemical studies have identified at least two distinct postsynaptic serotonin binding sites, denoted 5-HT$_1$ and 5-HT$_2$ (Peroutka and Snyder 1979); however, it is probable that the 5-HT$_1$ sites might represent several separate populations (Nelson, Weck and Taylor 1983, and see Chapter 4). Recently, Gozlan, El Mestikawy, Pichat, Glowinski and Hamon (1983), using [^{3}H]-8,hydroxy-2-(di-1-propylamino) tetralin (8-OH-DPAT) have identified one presynaptic serotonin binding site.

Aghajanian (1981) proposed the designations S$_1$, S$_2$ and S$_3$ for the three types of serotonin receptors identified electrophysiologically by their response to serotonin agonists in the mammalian brain. The S$_1$ receptor, present *inter alia* on motoneurones, amplifies the neuronal response to excitatory amino acids (McCall and Aghajanian 1979). This receptor is readily blocked by the classical serotonin antagonists such as cyproheptadine, metergoline and methysergide (McCall and Aghajanian 1980). The S$_2$ receptor is an autoreceptor, the pharmacology of which has been studied electrophysiologically at the level of the soma of serotonin neurones in the nucleus raphe dorsalis. Since its activation decreases the firing activity of serotonin neurones, it is believed to play an important role in the autoregulation of the serotonin system (Aghajanian 1978). This autoreceptor is characterized by its extreme sensitivity to LSD whereas the classical serotonin antagonists are completely ineffective (Aghajanian 1981). The S$_3$ receptor is that found in most forebrain regions receiving a dense serotonergic input. Its activation produces a depression of the firing rate. LSD is a weak agonist at this site and the classical serotonin antagonists are not consistently active at this site (Haigler and Aghajanian 1974; Aghajanian 1981). The discovery of an effective antagonist for this serotonin receptor would undoubtedly contribute a major tool for understanding its role.

It must be emphasized that, as of now, there is no correspondence between the *postsynaptic* binding sites identified biochemically and the S_1 and S_3 receptors (see Aghajanian 1981). However, it would seem from preliminary data that the *presynaptic* binding site identified by Gozlan *et al.* (1983) using $[^3H]$-8-OH-DPAT may correspond to the S_2 autoreceptor, since this drug is extremely powerful (even more so than LSD) in depressing the firing activity of dorsal raphe serotonin neurones (de Montigny *et al.*, unpublished observations).

7.2   Tricyclic antidepressant drugs

Following the demonstration by Glowinski and Axelrod (1964) of the monoaminergic re-uptake blocking property of amitriptyline and impramine, it has long been assumed that this property might be responsible for their therapeutic activity, However, several observations failed to support this hypothesis. For instance, iprindole, a TCA drug devoid of any re-uptake blocking activity, is an effective antidepressant drug (see de Montigny 1982). However, there was the crucial observation of Shopsin, Gershon, Goldstein, Friedman and Wilk (1975) that the administration of the serotonin synthesis inhibitor (PCPA) reversed the therapeutic effect of imipramine in major depression. This constitutes strong evidence for the mediating role of serotonin in the antidepressant effect of TCA drugs.

It was this apparent paradox which prompted us to study the effect of long-term TCA drug administration on the responsiveness of rat forebrain postsynaptic neurones to microiontophoretically applied serotonin. We found that long-term, but not acute, administration of different types of TCA drugs (including iprindole) induced an increased responsiveness of ventral lateral geniculate and hippocampal neurones to serotonin whereas the responsiveness to GABA was unchanged (de Montigny and Aghajanian 1978). Several reports have substantiated and extended this finding to other CNS regions (Gallager and Bunney 1979; Jones 1980; Menkes and Aghajanian 1981; Menkes, Aghajanian and McCall 1980; Wang and Aghajanian 1980). This sensitization of the postsynaptic target neurones does not result in a decreased activity of serotonin neurones (Blier and de Montigny 1980).

Several groups have studied the effect of long-term TCA drug treatment on serotonin binding sites. Whereas their effect on the 5-HT_1 site is variable, they generally decrease the number of 5-HT_2 binding sites (Maggi, U'Prichard and Enna 1980; Peroutka and Snyder 1980; see also Sections 6.4 and 12.7). Obviously, the first explanation that comes to mind for the discrepancy between the data of these radioligand binding studies and those of the electrophysiological studies is the fact mentioned above that binding sites do not appear to correspond to electrophysiologically-defined receptors. However, it is highly interesting that both the S_1- and the S_3-mediated responses are enhanced by long-term TCA drug response. Moreover, in the facial motor nucleus and in the lateral geniculate nucleus, there is a parallel increase of the responsiveness to noradrenaline mediated by an α_1-adrenoceptor (Menkes and Aghajanian 1981; Menkes *et al.* 1980). Hence, we must come to terms with the fact that the responses mediated by three distinct receptors (S_1, S_3 and α_1) are enhanced in a parallel manner by long term treatment with different types of TCA drugs. This, we believe, constitutes a strong indication that the molecular modification underlying the enhancement of the responsiveness to serotonin by TCA drugs might not occur at the level of the membrane receptor but rather at the 'post-receptor' level, such as the coupling between the receptor and the effector mechanisms. Assuming that the nature of this coupling is the same for the S_1, S_3 and α_1 receptors, this would account for the similar effects TCA have on all three responses, and, at the same time, unravel the apparently paradoxical biochemical and electrophysiological findings.

7.3   Tetracyclic antidepressant drugs and ECS

Heninger, Charney and Sternberg (1983) reported that lithium addition potentiated the antidepressant effect of mianserin, a tetracyclic antidepressant drug, as it had been reported for TCA drugs (de Montigny, Mayer and Deschênes 1981*b*, and see Section 7.6). This prompted us to study the effect of long-term mianserin administration on the responsiveness of rat forebrain neurones to serotonin. Comparing the effects of saline, imipramine and mianserin, we found that mianserin was as effective as imipramine in inducing a sensitization of hippocampal pyramidal neurones to the amine (Blier, de Montigny and Tardif 1984).

Since electroconvulsive shock treatment (ECS) remains the most effective treatment for major depression (Royal College of Psychiatrists 1977; Brandon, Cowley, McDonald, Neville, Palmer and Wellstood-Eason 1984) we thought it would be interesting to see if ECS would also affect neuronal sensitivity to serotonin all the more as Costain, Green and Grahame-Smith (1979) had reported an increased serotonin-mediated behavioural response in the rat following repeated ECS (see Section 12.7.2). Consistent with this finding, we observed in the rat hippocampus that repeated, but not a single, ECS markedly enhanced the effects of microiontophoretically-applied serotonin and 5-MeODMT, a serotonin receptor agonist (de Montigny and Aghajanian 1977), whereas the responses to GABA and noradrenaline were unchanged (de Montigny 1984).

It might appear at first sight surprising that drugs and an electrical current can induce a similar neurobiological modification. However, it is a long-standing clinical notion that ECS and TCA drugs have much in common. For instance, the profile of the 'good responder' to TCA drugs is almost superimposable on that of the 'good responder' to ECS, and the ECS-induced remission can be maintained with TCA drugs (Carney, Roth and Garside 1965; Royal College of Psychiatrists 1977). Even though the absence of a direct comparison of ECS and TCA drugs in the same experimental series precludes a definite conclusion, it is tempting to speculate that the apparently greater sensitization induced by ECS might account for their greater clinical efficacy.

7.4   Antidepressant serotonin re-uptake blockers

During the last five years, several selective serotonin uptake blockers have been shown to be effective antidepressant drugs. We have studied electrophysiologically zimelidine (zimeldine) and indalpine (Blier and de Montigny 1983; Blier *et al.* 1984; de Montigny, Blier, Caillé and Kouassi 1981*a*). Long-term treatment with either drug did not affect the responsiveness of rat hippocampal pyramidal neurones to microiontoporetic applications of serotonin or GABA (Table 7.1). However, the response of the same neurones to the electrical stimulation of the ascending serotonin pathway was increased (Fig. 7.1) indicating an enhancement of serotonin neurotransmission due to a presynaptic modification, presumably inhibition of re-uptake.

These observations raised a troublesome question. If these drugs were acting through blocking serotonin re-uptake, why would they require long-term administration to produce their full antidepressant effect as the

Table 7.1   Responsiveness of CA_3 hippocampal pyramidal neurones to microiontophoretic applications of serotonin and GABA in control and zimeldine-treated rats*

	Control	Zimeldine
Serotonin	168±11 (28)†	185±11 (33)
GABA	200±35 (15)	169±24 (17)

*Zimeldine was administered at a daily dose of 5 mg/kg i.p., for 14 days; values are expressed as mean $I \cdot T_{50}$ ±S.E.M. $I \cdot T_{50}$ values for serotonin and GABA in zimeldine-pretreated rats were not significantly different (p <0.05) from control values. †Numbers in parentheses: number of units tested. (Reproduced from Blier and de Montigny 1983, with permission from the Society of Neuroscience)

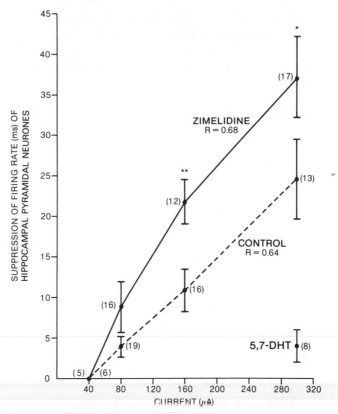

Fig. 7.1 Relationship between the duration of suppression of CA_3 hippocampal pyramidal neurone firing and the intensity of the current used to stimulate the ascending serotonin pathway in control and zimeldine-pre-treated rats (5 mg/kg, i.p., q.d.×14 days). Values are expressed as means ±S.E.M.; the number of cells tested is indicated in parentheses. *p <0.05, **p <0.001. (Student's t-test, comparing control and zimelidine groups.) (Reproduced from Blier and de Montigny 1983, with permission from the Society for Neuroscience)

re-uptake process is blocked in a matter of minutes or hours? In order to address this question, the activity of serotonin neurones in the raphe dorsalis was determined during the repeated administration of these drugs. After 2 days of treatment, there was a drastic reduction of the firing activity of serotonin neurones; after 7 days of treatment, there was a 75 per cent recovery; and after 14 days of treatment, their activity was back to normal (Fig. 7.2).

The reduction of the activity of serotonin neurones by a short-term treatment with a selective uptake blocker was not unexpected since it has been shown by Sheard, Zolovick and Aghajanian (1972) and Scuvée-Moreau

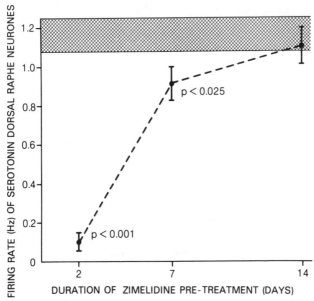

Fig. 7.2   Graph showing the mean firing rate (±S.E.M.) of serotonin neurones in rats treated for 2, 7 and 14 days with zimeldine (5 mg/kg, i.p., q.d.). The shaded zone represents the range (S.E.M.×2) of the firing frequency of serotonin neurones recorded from the same area in control rats. (Reproduced from Blier and de Montigny 1983, with permission from the Society for Neuroscience)

and Dresse (1979) that acute administration of serotonin uptake blockers depressed the firing rate of serotonin neurones, presumably by increasing the amount of amine available to the autoreceptor of these neurones. Thus, it was the recovery of the activity, as the treatment was pursued, which called for an explanation. To this end, LSD, an agonist at the autoreceptor (Aghajanian 1974), was used to assess the sensitivity of the autoreceptor. It was found that, in rats treated for 14 days with either zimeldine or indalpine, there was a 2- to 3-fold increase of the ED_{50} of LSD, indicating a drastic desensitization of the autoreceptor Fig. 7.3. Thus, the progressive desensitization of the autoreceptors frees the serotonin neurones from the potent autoregulatory process, allowing them to resume normal firing activity despite the higher concentration of serotonin in their vicinity. Hence, this provided an outlet from the theoretical impasse, despite their immediate action on re-uptake, these drugs require a long-term administration before enhanced serotonergic neurotransmission is attained.

The evidence that the somatic serotonin autoreceptor can desensitize raises a potentially important issue. Cerrito and Raiteri (1979) and Sanders-Bush (1982) have demonstrated the major role played by the serotonin autoreceptor at the level of the terminal boutons in controlling the

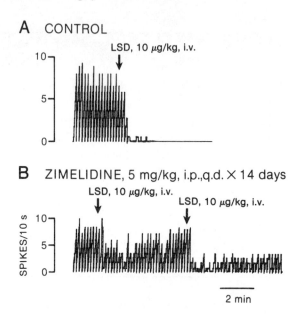

Fig. 7.3 Integrated firing rate histogram of a single serotonin neurone recorded from the dorsal mesencephalic raphe nucleus in a control rat (A) and a rat treated with zimelidine (5 mg/kg, i.p., q.d.) for 14 days (B). LSD was injected in the tail vein at times marked by arrows. Time base applies to both traces. (Reproduced from de Montigny and Blier 1984, with permission from Raven Press)

amount of monoamine released. This raises the possibility that this auto-receptor, which appears pharmacologically very similar, if not identical, to the somatic autoreceptor (Sanders-Bush 1982), might also desensitize under chronic serotonin re-uptake blockade. If this were the case, it would mean that more serotonin molecules per impulse would enter the synaptic cleft. Hence, the augmented effect of the stimulation of the ascending serotonin pathway on the target neurones might be due more to the desensitization than to the re-uptake blockade itself! This hypothesis is currently being tested in our laboratory.

7.5 Monoamine oxidase inhibitors

Monoamine oxidase inhibitors were among the first drugs shown to be effective for the treatment of major depression (Loomer, Saunders and Kline 1957). It has long been assumed that the blockade of the degradation of either serotonin or noradrenaline or both was the basis of their clinical effect. This has received firm support by the demonstration of the effectiveness of clorgyline (Herd 1969; Wheatley 1970), a selective type A MAOI and of the ineffectiveness of deprenyl (selegiline) when administered at doses selective for the type B MAO (Mendis, Pare, Sadler, Glover and Stern 1981).

As was mentioned in the preceding section on serotonin re-uptake blockers, there is also, in the case of MAOIs, the question of time lag between the rapid biochemical effect of MAOIs and their delayed clinical effect. Second, there is the perennial question of whether the serotonin system or the noradrenaline system or both mediate their therapeutic effect.

In an attempt to address these two questions, we recently carried out the assessment of serotonin and noradrenaline neuronal activity under repeated administration of clorgyline and deprenyl, administered at a selective regimen, and of phenelzine, a non-specific MAOI (Blier and de Montigny 1984).

As could be expected from its selectivity for MAO-B, depenyl failed to alter either serotonin or noradrenaline neuronal firing activity as well as the sensitivity of their respective autoreceptors. However, both clorgyline and phenelzine were found to affect differentially serotonin and noradrenaline neurones. Dorsal raphe serotonin neurones (as was the case after serotonin re-uptake blockers) showed a marked decrease in the firing rate during the first week of treatment with a complete recovery after 21 days of treatment. This recovery was accompanied by a marked desensitization of the auto-receptor indicated by the decreased effectiveness of LSD. However, the story was different for noradrenaline neurones in the locus coeruleus. They showed, with both MAOIs, a 50 per cent decrease of the firing rate after 2 days of treatment but failed to show any recovery after 7 and 21 days of treatment. In keeping with this striking observation, the sensitivity of their α_2-adrenoceptor autoreceptor, assessed by their neuronal responsiveness to clonidine, remained unchanged.

The main information provided by these data is that serotonin neurones possess a plasticity which noradrenaline neurones lack. In other words, under long-term MAOI treatment, serotonergic neurones show an adaptation whereas noradrenergic neurones do not. This fundamental difference between these two monoaminergic systems is undoubtly crucial in under-standing their respective roles in the control of affective behaviour and, even possibly, in the response to stressful environmental conditions.

As far as the two initial questions are concerned, these data provide clues indicating an answer. The time lag between the biochemical and the clinical effects of MAOIs can now be accounted for by the long time constant of the recovery of firing activity of serotonin neurones. Hence, a significant increase in serotonin neurotransmission would be achieved only long after the effective inhibition of the enzyme. As for the question of whether the serotonin or noradrenergic systems mediate the therapeutic effect of MAOIs, the answer is suggested by following the same line of thought. If the action of MAOIs on noradrenaline neurones were responsible for their antidepressant effect, then a therapeutic response ought to be occurring after 2 days of treatment since the modification of noradrenergic neurone activity is the same after 2 days and 21 days of treatment. Thus, these data constitute further evidence that it is the delayed augmentation of sero-tonergic neurotransmission which best accounts for their therapeutic effect. This conclusion must however, be at the present time, considered as tentative; it must await the direct assessment of synaptic transmission efficacy using a paradigm where the effects of the activation of the whole systems on their target neurones are assessed. These experiments are currently under way in our laboratory.

7.6   The rapid antidepressant effect of lithium addition in tricyclic antidepressant-refractory depression

The animal electrophysiological data reviewed in previous sections converge on the notion that the TCA, ECS, specific serotonin re-uptake blockers and antidepressant MAOIs may achieve their clinical effect through a common final pathway that of enhancing serotonin neurotrans-mission. However, as attractive as pre-clinical animal data may be, the conclusions drawn from them must be regarded as tentative, until confirmed by evidence obtained in patients.

Thus, the question we had in mind was: 'How would it be possible to show that TCA drugs induce in depressed patients a neuronal sensitization to serotonin and that this might be related to their antidepressant effect?' We imagined that studying depressed patients treated with, but not responding to, a TCA drug might prove the best approach. This reasoning was based on the assumption that if a sensitization of postsynaptic neurones had been induced by the TCA treatment, then, increasing the activity of serotonin neurones should bring about a rapid improvement of depression.

Among the several means of enhancing the action of serotonergic neurones on their target neurones, we chose lithium for the following reasons. First, Grahame-Smith and Green (1974) had elegantly demon-strated that lithium enhances serotonin neurotransmission. Secondly, the enhancement is obtained within 36 hours (Grahame-Smith and Green 1974). Thirdly, very low concentrations of the ion are required to enhance serotonin synthesis (Knapp and Mandell 1975).

In the first publication, we reported the marked improvement brought about by lithium addition in eight patients presenting a major depression resistant to a TCA treatment of at least three weeks (de Montigny et al. 1981b). Subsequently, we reported 42 observations of lithium addition in TCA-resistant patients. In 31 of them (74 per cent), lithium addition brought about within 48 hours a greater than 50 per cent improvement as measured on the Hamilton Rating Scale for Depression (de Montigny, Cournoyer, Morrisette, Langlois and Caillé 1983). Heninger *et al.* (1983) confirmed the augmentation of the antidepressant effect of TCA drugs, by lithium addition in a double-blind, placebo-controlled study and we have just completed a double-blind cross-over study of 12 TCA-refractory patients who received lithium and placebo for 48 hours at an interval of 1 week. Lithium, but not placebo, produced a marked amelioration (Cournoyer, de Montigny, Ouellette, Leblanc, Langlois and Elie 1984).

That the serotonin system might mediate the therapeutic effect of lithium addition in TCA-refractory depression is suggested by the congruence of several aspects of the clinical phenomenon with neurobiological data. First, the time constants of the neuropharmacological effects of TCA drugs and of lithium on the serotonin system are consistent with the clinical phenomenology: the postsynaptic sensitization to the amine is obtained only after a long-term treatment with a TCA drugs, whereas the effect of lithium on serotonin neurones is very rapid. It is noteworthy in this respect that lithium addition exerts the same rapid antidepressant effect in patients not responding to long-term treatment with iprindole (de Montigny, Elie and Caillé 1985), a TCA drug which is devoid of any effect on serotonin re-uptake but which, as with other TCA drugs, induces a postsynaptic sensitization to serotonin (de Montigny and Aghajanian 1978; Menkes *et al.* 1980; Wang and Aghajanian 1980). Secondly, marked improvements were obtained with lithium blood levels as low as 0.4 mEq/l. This is consistent with the fact that low concentrations of lithium are sufficient to stimulate serotonin synthesis (Broderik and Lynch 1982). Thirdly, we found a significant positive correlation between the effects of lithium addition on clinical status and on plasma prolactin (de Montigny and Cournoyer, unpublished observations). Since prolactin release is under the stimulatory control of the serotonin system (Clemens, Roush and Fuller 1978; Ferrari, Caldara, Rumussi, Rampini, Telloli, Zaatar and Curtatelli 1978) this observation constitutes strong evidence for the involvement of serotonin neurotransmission in mediating the rapid antidepressant effect of lithium addition in TCA-resistant depression. Finally, it is striking that the clinical phenomenon observed with lithium addition is exactly the mirror image of that reported by Shopsin *et al.* (1975) that is, adding a serotonin synthesis inhibitor reversed the therapeutic effect of the TCA drug.

7.7 Conclusion

The novel electrophysiological and clinical data reviewed in this chapter have revived the notion that the serotonin system might play a pivotal role in mediating the effect of antidepressant treatments. However, we believe that beyond that, this story constitutes an example of the possible contribution of basic pharmacology to the advancement of medical therapeutics.

Acknowledgement

Supported by Grants MT-6444 and MA-7649 from the Medical Research Council of Canada, by Grant 810462 from the Fonds de la Recherche en Santé du Québec (FRSQ) and by a Scholarship to C. de M. and a Studentship to P.B. from the FRSQ.

References

Aghajanian, G. K. (1978). Feedback regulation of central monoaminergic neurons: evidence from single cell recording studies. In *Essays in neurochemistry and neuropharmacology* (eds. M. B. H. Youdim and W. Lovenberg) Vol. 3, p. 1. John Wiley, Chichester.

—— (1981). The modulatory role of serotonin of multiple receptors in brain. In *Serotonin neurotransmission and behavior* (eds. B. L. Jacobs and A. Gelperin) p. 156. MIT Press, Cambridge, Massachusetts.

—— and Haigler, H. J. (1974). Mode of action of LSD on serotonergic neurons. *Adv. Biochem. Psychopharmac.* **10**, 167.

Blier, P. and de Montigny, C. (1980). Effect of chronic tricyclic antidepressant treatment on the serotonergic autoreceptor. A microiontophoretic study in the rat. *Naunyn Schmiedeberg's Arch. Pharmac.* **314**, 123.

—— —— (1983). Electrophysiological investigations on the effect of repeated zimeldine administration on serotonergic neurotransmission in the rat. *J. Neurosci.* **3**, 1270.

—— —— (1984*a*). The effect of repeated administration of monoamine oxidase inhibitors on the firing activity of serotoninergic and noradrenergic neurons. *Coll. Int. neuropsychopharmac (Florence)* **14**, F-35.

—— —— and Tardif, D. (1984). Effects of two antidepressant drugs, mianserin and indalpine on the serotoninergic system: single cell studies in the rat. *Psychopharmacology* **844**, 242.

Brandon, S., Cowley, P., McDonald, C., Neville, P., Palmer, R. and Wellstood-Eason, B. (1984). Electroconvulsive therapy: results in depressive illness from the Leicestershire trial. *Br. med. J.* **288**, 22.

Broderick, P. and Lynch, V. (1982). Behavioral and biochemical changes induced by lithium and L-tryptophan in muricidal rats. *Neuropharmacology* **21**, 671.

Carney, M. W. P., Roth, M. and Garside, R. F. (1965). The diagnosis of depressive syndromes and the prediction of E.C.T. response. *Br. J. Psychiat.* **111**, 659.

Cerrito, F. and Raiteri, M. (1979). Serotonin release is modulated by pre-synaptic autoreceptors. *Eur. J. Pharmac.* **57**, 427.

Clemens, J. A., Roush, M. E. and Fuller, R. W. (1978). Evidence that serotonin neurons stimulate secretion of prolactin releasing factor. *Life Sci.* **22**, 2209.

Coppen, A. (1967). Biochemistry of affective disorders. *Br. J. Psychiat.* **113**, 1237.

Costain, D. W., Green, A. R. and Grahame-Smith, D. G. (1979). Enhanced 5-hydroxytryptamine-mediated behavioural responses in rats following repeated electroconvulsive shock: relevance to the mechanism of the antidepressive effect of electroconvulsive therapy. *Psychopharmacology* **61**, 167.

Cournoyer, C., de Montigny, C., Ouellette, J., Leblanc, G., Langlois, R. and Elie, R. (1984). Lithium addition in tricyclic-resistant unipolar depression: a placebo-controlled study. *Coll. Int. Neuropsychopharmac. (Florence)* **14**, F-177.

de Montigny, C. (1982). Iprindole: a cornerstone in the neurobiological investigation of antidepressant treatments. *Mod. Probl. Pharmaco-psychiat.* **18**, 102.

—— (1984). Electroconvulsive shock treatments enhance responsiveness of forebrain neurons to serotonin. *J. Pharmac. Exp. Ther.* **228**, 230.

—— and Aghajanian, G. K. (1977). Preferential action of 5-methoxydi-methyltryptamine and 5-methoxytryptamine on presynaptic serotonin receptors: a comparative iontophoretic study with LSD and serotonin. *Neuropharmacology* **16**, 811.

—— —— (1978). Tricyclic antidepressants: long term treatment increases responsivity of rat forebrain neurons to serotonin. *Science* **202**, 1303.

—— and Blier, P. (1984). Effects of antidepressant treatments on 5-HT neurotransmission: electrophysiological and clinical studies. In *Frontiers in biochemical and pharmacological research in depression*. (eds. E. Usdin, M, Åsberg, L. Bertilsson and L. Sjöqvist) p. 223. Raven Press, New York.

—— —— Caillé, G. and Kouassi, E. (1981*a*). Pre- and postsynaptic-effects of zimeldine and norzimelidine on the serotonergic system: single cell studies in the rat. *Acta Psychiatr. Scand.* **63** (Supple. 290), 79.

—— Elie, R. and Caillé, G. (1985). Rapid response to lithium addition in iprindole-resistant unipolar depression. *Am. J. Psychiat.* (in press).

—— Cournoyer, G., Morisette, R., Langlois, R. and Caillé, G. (1983). Further studies on the antidepressant effect of lithium addition in patients not responding to a tricyclic antidepressant drugs. *Arch. Gen. Psychiat.* **40**, 1327.

—— Grunberg, P., Mayer, A. and Deschênes, J. P. (1981*b*). Lithium induces a rapid relief of depression in tricyclic antidepressant drug nonresponders. *Br. J. Psychiat.* **138,** 252.

Ferrari, C., Caldara, R., Romussi, M., Rampini, P., Telloli, P. Zaatar, S. and Curtatelli, G. (1978). Prolactin suppression by serotonin antagonists in man: further evidence for serotoninergic control of prolactin secretion. *Neuroendocrinology* **25,** 319.

Gallager, D. W. and Bunney, W. E., Jr. (1979). Failure of chronic lithium treatment to block tricyclic antidepressant-induced 5-HT supersensitivity. *Naunyn Schmiedeberg's Arch. Pharmac.* **307,** 129.

Glowinski, J. and Axelrod, J. (1964). Inhibition of uptake of tritiated noradrenaline in the intact rat brain by imipramine and structurally related compounds. *Nature* **204,** 1318.

Gozlan, H., El Mestikawy, S., Pichat, L., Glowinski, J. and Hamon, M. (1983). Identification of presynaptic serotonin autoreceptor using a new ligand: [^{3}H]PAT. *Nature* **305,** 140.

Grahame-Smith, D. G. and Green, A. R. (1974). The role of brain 5-hydroxytryptamine in the hyperactivity produced in rats by lithium and monoamine oxidase inhibition. *Br. J. Pharmac.* **52,** 19.

Haigler, H. J. and Aghajanian, G. K. (1974). Peripheral serotonin antagonists: failure to antagonize serotonin in brain areas receiving a prominent serotonergic input. *J. Neural. Transm.* **35,** 257.

Heninger, G. R., Charney, D. S. and Sternberg, D. E. (1983). Lithium carbonate augmentation of antidepressant treatments. *Arch. Gen. Psychiat.* **40,** 1335.

Herd, J. A. (1969). A new antidepressant—M and B 9302. A pilot study and a double-blind controlled trial. *Clin. trial* **6,** 219.

Jones, R. S. G. (1980). Long-term administration of atropine, imipramine, and viloxazine alters responsiveness of rat cortical neurones to acetylcholine. *Can. J. Physiol. pharmac.* **58,** 531.

Knapp, S. and Mandell. A. J. (1975). Effects of lithium chloride on parameters of biosynthetic capacity for 5-hydroxytryptamine in rat brain. *J. Pharmac. Exp. Ther.* **193,** 812.

Lapin, J. P. and Oxenkrug, G. F. (1969). Intensification of the central serotonergic processes as a possible determinant of the thymoleptic effect. *Lancet* **i,** 132.

Loomer, H. P., Saunders, J. C. and Kline, N. S. (1957). A clinical and pharmaco dynamic evaluation of iproniazid as a psychic energizer. *Psychiat. Res. Rep. Amer. Psychiat. Ass.* **8,** 129.

McCall, R. B. and Aghajanian, G. K. (1979). Serotonergic facilitation of facial motoneuron excitation. *Brain Res.* **169,** 11.

—— —— (1980). Pharmacological characterization of serotonin receptors in the facial motor nucleus: a microiontophoretic study. *Eur. J. Pharmac.* **65,** 175.

Maggi, A., U'Prichard, D. C. and Enna, S. J. (1980). Differential effects of antidepressant treatment on brain monoaminergic receptors. *Eur. J. Pharmac.* **61,** 91.

Mendis, N., Pare, C. M. B., Sadler, M., Glover, V. and Stern, G. M. (1981). Is the failure of (−)deprenyl, a selective monoamine oxidase B inhibitor, to alleviate depression related to freedom from the cheese effect? *Psychopharmacology* **73,** 87.

Menkes, D. B. and Aghajanian, G. K. (1981). α_1-Adrenoceptor-mediated responses in the lateral geniculate nucleus are enhanced by chronic antidepressant treatment. *Eur. J. Pharmac.* **74,** 27.

—— —— McCall, R. B. (1980). Chronic antidepressant treatment enhances adrenergic and serotonergic responses in the facial nucleus. *Life Sci.* **27,** 45.

Nelson, Dl., Weck, B., and Taylor, W. (1983). Discrimination of heterogeneous serotonin$_1$ receptors by indolealkylamines. In *CNS receptors—from molecular pharmacology to behavior* (eds. P. Mandel and F. V. De Feudis) p. 337. Raven Press, New York.

Peroutka, S. J. and Snyder, S. H. (1979). Multiple serotonin receptors: differential binding of [^{3}H]5-hydroxytryptamine, [^{3}H]lysergic acid diethylamide and [^{3}H]spiroperidol. *Mol. Pharmac.* **16,** 687.

—— —— (1980). Long-term antidepressant treatment decreases spiroperidol-labelled serotonin receptor binding. *Science* **210,** 88.

Royal College of Psychiatrists (1977). Memorandum on the use of electroconvulsive therapy. *Br. J. Psychiat.* **131,** 261.

Sanders-Bush, E. (1982). Regulation of serotonin storage and release. In *Serotonin in biological psychiatry* (eds. B. T. Ho, J. C. Schoolar and E. Usdin) p. 17. Raven Press, New York.

Scuvée-Moreau, J. and Dresse, A. R. (1979). Effect of various antidepressant drugs on the spontaneous firing rate of locus coeruleus and dorsal raphe neurons of the rat. *Eur. J. Pharmac.* **57,** 219.

Sheard, M. H., Zolovick, A. and Aghajanian, G. K. (1972). Raphe neurons: Effect of tricyclic antidepressant drugs. *Brain Res.* **43,** 690.

Shopsin, B., Gershon, M., Goldstein, M., Friedman, E. and Wilk, S. (1975). Use of synthesis inhibitors in defining a role for biogenic amines during imipramine treatment in depressed patients. *Psychopharmac. Commun.* **1,** 239.

Siever, L. J., Guttmacher, L. B. and Murphy, D. L. (1984). Serotonergic receptors: Evaluation of their possible role in the affective disorders. In *Neurobiology of mood disorders* (eds. R. M. Post and J. C. Ballenger) p. 587. Williams and Wilkins, London.

Wang, R. Y. and Aghajanian, G. K. (1980). Enhanced sensitivity of amygdaloid neurons to serotonin and norepinephrine after chronic antidepressant treatment. *Commun. Psychopharmac.* **4,** 83.

Wheatley, D. (1970). Comparative trial of a new monoamine oxidase inhibitor in depression. *Br. J. Psychiat.* **117,** 573.

8

An overview of brain serotonergic unit
activity and its relevance to the
neuropharmacology of serotonin

BARRY L. JACOBS

8.1 Introduction

Serotonergic neurones comprise the site at which many psychoactive drugs
exert a significant portion of their action. Thus, they constitute the essential
interface between such drugs and their functional effects. Basic compre-
hension of the variables that control and influence the activity of brain
serotonergic neurones is therefore a prerequisite to fully understanding the
action of drugs upon the central serotonergic system. In turn, this is critical
for understanding the effect of these drugs upon physiology and behaviour

Since other contributors to this volume have provided detailed descriptions of various drug effects on the brain serotonergic system, I will focus
my discussion on the physiological and behavioural variables that control,
or correlate with, the electrophysiological activity of serotonergic
neurones. I will, however, discuss drug effects when they provide information that has been helpful in understanding the basic processes that control
serotonergic unit activity. A case in point is the study of hallucinogenic
drugs which has proven useful in analysing various types of serotonergic
receptors. Hallucinogenic drugs will also be discussed in their own right
since electrophysiological studies of serotonergic neurones have contributed importantly to our understanding the mechanism of action of these
compounds (see also Chapter 10).

Over the past 15 years, electrophysiological studies of serotonergic
neurones have been carried out at a variety of different levels of analysis,
including *in vivo* and *in vitro* experiments utilizing both intracellular and
extracellular recordings, and studies conducted in both anaesthetized and
unanaesthetized animals. Each of these experimental approaches provides
inherent advantages for better understanding the basic neurophysiology of

serotonergic neurones and, consequently, the influence of drugs upon serotonergic neurotransmission. Therefore, I will review the experimental findings that have been gathered from all of these approaches. However, since much of the interest in drugs affecting the serotonergic system derives from the fact that many of them are psychoactive, I will place a major emphasis on electrophysiological studies of serotonergic neurones in unanaesthetized and unrestrained (freely moving) animals. I will confine the discussion to studies conducted in mammals, and almost exclusively to those in rats and cats, since virtually all electrophysiological studies of serotonergic neurones have been carried out in these two species (see chapters in Jacobs and Gelperin 1981, and Osborne 1982 for electrophysiological studies of serotonergic neurotransmission in invertebrates).

Prior to beginning, it may be useful to review some basic issues of relevance to understanding drug action on the brain serotonergic system. First, although it may be a simple matter to specify that a particular drug influences serotonergic neurotransmission, it is typically very difficult to define precisely the neuropharmacological mechanism involved and the site of action on the serotonergic neorone-target neurone complex. Another complication is the fact that there is now clearcut evidence for several different types of serotonin receptors, each with its own distinct set of functional properties (see Chapters 2, 4, 6 and 12). Finally, drugs with relatively complex molecular structures, such as LSD, may exert important effects on several different neurotransmitter systems, and, in addition, may exert more than one effect (e.g. agonist and antagonist) on a given neurotransmitter system (see Section 7.1.2).

8.2   Activity of serotonergic neurones in anaesthetized and/or immobilized rats

In the mid-1960s the organization of the brain serotonergic system was mapped by the use of fluorescence histochemistry (Dahlstrom and Fuxe 1964). The pattern of localization of serotonergic neurones within the brain appears to be virtually constant across all the vertebrate species (Parent 1981). They are confined almost exclusively to the midline area of the brainstem (from medulla to mesencephalon), largely in the classically defined raphe nuclei (Taber, Brodal and Walberg 1961). Their axon terminal distribution is diffuse and projects widely, to virtually all portions of the central nervous system (Fuxe 1965). The densest aggregation of serotonergic cell bodies, at least in the mammalian brain, is found in the dorsal and median raphe nuclei (DRN and MRN) (Dahlstrom and Fuxe 1964). In an attempt to record the activity of serotonergic neurones, Aghajanian, Foote and Sheard (1968, 1970) placed microelectrodes into the area of the DRN and MRN of chloral hydrate anaesthetized rats. They consistently found cells with a slow (1–2 spikes/s) and highly regular discharge pattern whose firing rate was completely suppressed by systemic administration of low doses of LSD (fig. 8.1). Based on these initial results, they hypothesized that these slow and regular firing cells were serotonergic since this type of activity was encountered only in brain areas known to contain serotonergic cells, and because only this type of neurone showed a consistent depressant response to low doses of LSD (an effect predicted on the basis of previous neurochemical studies). The assumption that these were serotonergic neurones has now been supported by a large body of experimental evidence (reviewed in Jacobs, Heym and Steinfels 1984).

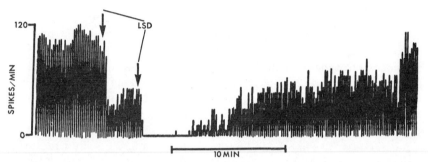

Fig. 8.1   Typical response of a raphe unit to LSD. This unit, which was situated in the dorsal raphe, had a spontaneous rate of about 120 spikes/min. Within 30 s after an i.v. injection of 10 μg/kg of LSD, this unit slowed; after a second dose of 5 μg/kg it ceased firing entirely. Recovery took place gradually over a period of about 20 min. The record consists of consecutive 10 s samples of the analogue output of a counter triggered by the unit spikes. (Reproduced from Aghajanian *et al.* (1968) with permission from American Society of Experimental Pharmacology and Therapeutics)

Because of the slow discharge rate and extraordinary regularity of firing of serotonergic neurones, it has often been speculated that these neurones may be endogenously active or autoactive. Thus, rather than being driven by cells presynaptic to them, their slow firing rate and regular discharge pattern may be characteristics intrinsic to the individual serotonergic neurone itself. Our belief that serotonergic neurones in the DRN might be autoactive was strengthened by an experiment in our laboratory in which rats were prepared with a complete transection of the neuraxis, immediately rostral to the DRN (Mosko and Jacobs 1977). This cut isolated this nucleus from the influence of the entire forebrain. Somewhat surprisingly, the neurones in the DRN of these brain transected animals were virtually indistinguishable from those of intact animals. Although their discharge rate was approximately 30 per cent higher than a corresponding group of neurones in intact animals, these DRN cells still displayed the slow rhtymic discharge pattern that characterizes serotonergic unit activity. In an attempt to analyse this issue more directly, we examined the activity of serotonergic neurones in the rat DRN *in vitro* (Mosko and Jacobs 1976). By cutting 400 μm thick sections through the rat brainstem, we effectively isolated the serotonergic neurones in this tissue slab from the influence of neurones in the remaining forebrain and brainstem. Influences intrinsic to the tissue slab, of course, remained. *In vitro,* most DRN neurones displayed discharge patterns indistinguishable from those observed *in vivo.* DRN neurones *in vitro* uniformly fired with a regular pattern. Some of these units fired at a rate substantially higher than that observed *in vivo,* but over 60 per cent of the cells had a mean discharge rate between 0.8 and 2.5 spikes/ s. Non-serotonergic neurones in the area of the DRN did not display these slow, regular discharge characteristics. Taking our initial results one step further, VanderMaelen and Aghajanian (1983) recently reported the removal of calcium ions from the bath, with a concomitant increase in magnesium ions, had little effect upon the slow and rhythmic discharge of the cells. This is important because these manipulations of the ionic con- stituency of the bath are assumed to block all synaptic transmission. Finally, a report by Aghajanian and VanderMaelen (1982) provides perhaps the strongest evidence for autoactivity of brain serotonergic neurones. Intracellular recordings from serotonergic neurones in the DRN of the chloral hydrate anaesthetized rat revealed that these cells display a post-spike hyperpolarization followed by a gradual depolarization, with- out evidence of EPSPs, leading to the succeeding spike. This neuro-physiological pattern was also found in subsequent intracellular analyses of serotonergic unit activity recorded *in vitro* from tissue slices (VanderMaelen and Aghajanian 1983).

A major variable that controls the activity of brain serotonergic neurones is negative neuronal feedback. Thus, as the level of brain serotonin increases, for example, following administration of L-tryptophan or L-5- hydroxytryptophan, the activity of brain serotonergic neurones correspondingly decreases (Aghajanian 1972; Trulson and Jacobs 1976) (Fig. 8.2). It is presumed that this neuronal decrease is a homeostatic response which acts to compensate for the pharmacologically induced increase in synaptic levels of serotonin. During the past several years, a good deal of evidence has been accumulated to indicate that the nature of this feedback is short-loop, or local, rather than long-loop, involving the response of postsynaptic target neurones. Increased levels of synaptic serotonin appear to influence serotonergic neurones directly, either as it is released from axons or dendrites of neighbouring neurones, or via axon collateral feedback. This is based on the following evidence. First, studies employing iontophoresis show that serotonergic neurones are directly responsive to the application of serotonin. Aghajanian, Haigler and Bloom

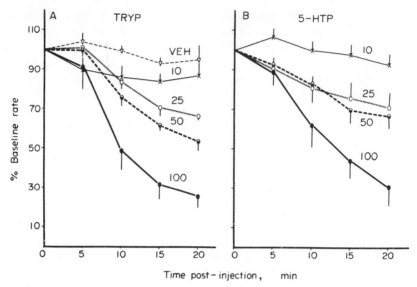

Fig. 8.2   Effects of systemic administration of L-tryptophan (TRYP) and vehicle (VEH) (A) and L-5-HTP (B) on raphe unit activity over time. Each drug was administered intra-peritoneally in doses of 10, 25, 50 and 100 mg/kg. Each point represents the mean for 4 to 6 cells, and vertical bars represent standard errors. (Reproduced from Trulson and Jacobs 1976, with permission from Pergamon Press)

(1972) reported that the ejection of serotonin onto serotonergic neurones in the DRN strongly depressed their rate of firing. Secondly, experiments in our laboratory mentioned above, demonstrated that the ability of drugs that release the endogenous stores of serotonin, or block its re-uptake, to depress serotonergic unit activity in the DRN was independent of any feedback from the forebrain, since the effectiveness of these compounds was undiminished in animals whose neuraxes were transected immediately rostral to the nucleus (Mosko and Jacobs 1977). When we coupled these neurophysiological findings with anatomical evidence that serotonergic neurones make synaptic connections with serotonergic neurones in the same, as well as other, raphe nuclei (Mosko, Haubrich and Jacobs 1977), it is clear that these cells are under local, rather than long-loop, feedback control. Thirdly, Aghajanian has further refined and elaborated this latter concept by providing physiological evidence that serotonergic neurones are under the influence of axon collateral inhibition (Wang and Aghajanian 1977), and anatomical evidence that neighbouring serotonergic neurones make dendro-dendritic and dendro-somatic connections (Aghajanian 1981).

The activity of serotonergic neurones is also influenced by neurochemically identifiable afferent inputs other than the serotonergic ones already discussed. There is evidence for afferents from GABAergic interneurones in the area of the DRN (Gottesfeld, Hoover, Muth and Jacobowitz 1978; Belin, Aguera, Tappaz, McRae-Degueurce, Bobillier and Pujol 1979), from catecholaminergic neurones (Fuxe 1965; Chu and Bloom 1974; Baraban and Aghajanian 1981), and from histaminergic neurones (Taylor, Gfeller and Snyder 1972). Although a large number of other afferent inputs to raphe nuclei have been demonstrated through the use of retrograde tracing techniques, they will not be discussed here since their lack of neurochemical identity precludes their manipulation by pharmacological means. Wang, Gallager and Aghajanian (1976) reported that stimulation of the pontine reticular formation produced a short latency suppression of serotonergic unit activity in the DRN. The effects were blocked by systemic administration of the GABA antagonist, picrotoxin, but not by the glycine antagonist, strychnine. There was some specificity to this action, since stimulation of the locus coeruleus did not produce these suppressive effects (Wang et al. 1976; Anderson, Pasquier, Forbes and Morgane 1977). Based on anatomical studies indicating that the lateral habenula constitutes the densest input to the DRN, Wang and Aghajanian (1977) and Stern, Johnson, Bronzino and Morgane (1979) examined the effects of electrical stimulation of this region in chloral hydrate anaesthetized rats. Once again, a strong suppressive effect (latency of 15 ms and duration of 50–400 ms) upon serotonergic neurones was seen. This effect was also blocked by picrotoxin but not by strychnine (Wang and Aghajanian 1977). On the basis of these data, the latter authors hypothesized that the habenulo-raphe pathway was GABAergic in nature. More recent evidence, however, indicates that the GABAergic neurones involved in this response are not long-axoned neurones located in the habenula, but are small interneurones found in the area of the DRN (Gottesfeld et al. 1978; Belin et al. 1979). Finally, Gallardo and Pasquier (1980) have shown that habenula lesions increase choline acetyltransferase activity in the DRN, thus providing evidence that the habenulo-raphe pathway must be cholinergic.

The issue of neurochemical identity of afferent inputs to serotonergic neurones in the DRN has also been addressed by means of studies employing microiontophoresis. Application of either GABA or glycine suppressed the activity of serotonergic neurones, and these effects were blocked by their respective antagonists, picrotoxin and strychnine (Gallager and Aghajanian 1976b). Recently, Lakoski and Aghajanian (1982) reported that iontophoretic application of histamine exerted a depressant effect on serotonergic unit activity in the DRN. Iontophoretic application of noradrenaline had no effect upon serotonergic unit activity (Svensson, Bunney and Aghajanian 1975; Gallager and Aghajanian

1976a), however the application of noradrenergic antagonists strongly suppressed the activity of these neurones, and this effect could be reversed by the iontophoretic application of noradrenaline (Baraban and Aghajanian 1980). These data imply that adrenergic neurones provide a tonic excitatory input to serotonergic neurones in the DRN. Since systemic or iontophoretic application of a variety of α-adrenoceptor antagonist drugs (also, reserpine and low doses of clonidine) can completely suppress the activity of serotonergic neurones in the DRN of chloral hydrate anaesthetized rats, Aghajanian and his colleagues hyptothesized that the activity of serotonergic neurones was dependent upon a continued adrenergic input (Svensson et al. 1975; Gallager and Aghajanian 1976a; Baraban, Wang and Aghajanian 1978; Baraban and Aghajanian 1980). We have conducted similar experiments in freely moving cats (described below). While we agree that an excitatory adrenergic input to the DRN influences the activity of serotonergic neurones, we do not believe that, under physiological conditions, their continued activity is dependent upon this tonic input.

To summarize, serotonin, GABA, and histamine exert depressive effects upon serotonergic neurones. Evidence also indicates that glycine may be an inhibitory neurotransmitter on these neurones. Noradrenaline is the only transmitter known to exert an excitatory influence on serotonergic neurones. This noradrenergic input seems to be a tonic one operating at a near maximal level since the iontophoretic application of noradrenaline on spontaneously active cells has little further effect. These data are of obvious relevance to pharmacological studies, since it can reasonabltt be assumed that drugs known to affect serotonergic, GABAergic, histaminergic, and noradrenergic neurotransmission will also engage or disengage the activity of brain serotonergic neurones. Most of the data described above were taken from studies on the DRN because very little neuropharmacological work has been devoted to any of the other groups of serotonergic neurones. Recently, however, this has changed with the growing interest of the role, in analgesia, of serotonergic neurones in the area of nucleus raphe magnus (see below).

Serotonergic unit activity in anaesthetized animals (primarily rats) has been studied in conjunction with the manipulation of several different physiological systems (e.g. body temperature and blood pressure), several pituitary-adrenal hormones (e.g. ACTH and corticosterone), and with stimuli from several sensory modalities (e.g. light flash and electrical shocks to peripheral nerves). In general these studies have revealed that manipulating these variables produces either not effect on serotonergic unit activity, or, the results of several studies on the same issue have been contradictory (for a detailed description of these experiments see a review by Jacobs et al. 1984). Largely because of this lack of response of serotonergic neurones to a variety of manipulations, and the equivocal nature of these findings, several years ago we decided to begin studying the activity of serotonergic neurones, in unanaesthetized, freely moving animals. Our assumption was that these neurones might manifest a greater degree of responsivity in the unanaesthetized, awake animal. In a concrete example of the advantages of this approach, we have found that chloral hydrate anaesthesia can *categorically* alter the response of brain serotonergic neurones to sensory stimuli, brain stimulation, and systemically administered drugs (Heym, Steinfels and Jacobs 1984). Of course, we were also interested in this approach because it would allow us to relate changes in serotonergic unit activity directly to changes in behaviour, including those produced by drugs.

8.3   Activity of serotonergic neurones in freely moving cats

To date, we have studied the activity of serotonergic neurones in freely moving cats within the four largest clusters of these cells in the mammalian brain: DRN; nucleus centralis superior (NCS) (homologous to MRN in the rat); nucleus raphe magnus (NRM); and nucleus raphe pallidus (NRP). I will focus my discussion of serotonergic neurones in freely moving cats on our studies of the DRN. I will also briefly mention our recent studies of serotonergic neurones in the NRM and their relation to nociception and analgesia. Detailed descriptions of our studies on the activity of serotonergic neurones in NCS and NRP are contained in two of our recent publications (Heym *et al.* 1982*a*; Rasmussen, Heym and Jacobs 1984).

Our studies utilize a method that employs bundles of chronically implanted flexible insulated microwires (32 or 64 µm in diameter) that can be moved through the brain by means of an attached mechanical microdrive. At the time of surgery, gross electrodes are also implanted for recording eye movements, EEG, and nuchal EMG (depending on the particular study, electrodes may also be implanted for brain stimulation, recording PGO waves, recording heart rate, etc.). All potentials are led from the animal by means of a counterweighted low noise cable. The experiments are conducted in an electrically shielded chamber (65×65×95 cm) with a plexiglass wall which permits continuous visual monitoring on a t.v. screen. Complete methodological details have been published elsewhere (Jacobs 1982; Jacobs *et al.* 1984).

Our initial objective was to characterize the activity of serotonergic neurones in the DRN across the sleep-wake-arousal cycle (Trulson and Jacobs 1979*a*). This analysis would then provide a firm data base for use in future experiments. Serotonergic neurones were identified on-line by the characteristic slow and rhythmic activity described above. Further support for their neurochemical identity was provided by the fact that their spontaneous neuronal activity was strongly suppressed by low doses of the

204   Barry L. Jacobs

serotonergic agonist drugs 5-MeODMT and LSD. Furthermore, histo-
logical examination revealed that neurones with these characteristics were
localized precisely and exclusively in those brain areas where serotonergic
neurones are found. Although the activity of DRN serotonergic neurones
is slow and regular when the cat is in a quiet waking state, this activity shows
dramatic changes across the sleep-waking-arousal cycle (McGinty and
Harper 1976; Trulson and Jacobs 1979a). From a quiet waking rate of
approximately 3 spikes/s, the activity of these neurones will often physically increase to 6 spikes/s in response to an arousing stimulus.
Reciprocally, the activity of these neurones shows a gradual decline as the
cat becomes drowsy and enters slow wave sleep. A decrease in the regularity of firing accompanies this overall slowing of neuronal activity. During
the middle and latter stages of slow wave sleep, the activity of DRN
serotonergic neurones displays a strong inverse relationship with two of the
electrophysiological events associated with slow wave sleep and REM
sleep, namely slow wave sleep spindles and PGO waves. Finally, the
culmination of this state-dependent decrease in unit activity occurs as that
cat enters REM sleep. During REM sleep the activity of these neurones
typically becomes completely quiescent (Fig. 8.3). Thus, although serotonergic neurones display a slow, clock-like activity under quiet waking

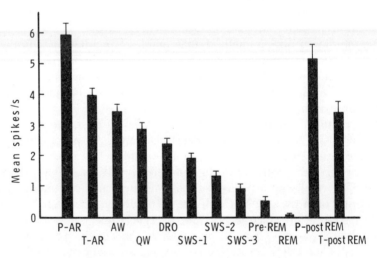

Fig. 8.3 Mean discharge rate of dorsal raphe neurons across sleep–wakefulness–
arousal states. Abbreviations: P-AR, phasic arousal; T-AR, tonic arousal; AW, active
waking; QW, quiet waking; DRO, drowsy; SWS-1, early slow wave sleep; SWS-2;
middle slow wave sleep; SWS-3, late slow wave sleep; P-PostREM, phasic post-
REM; T-PostREM, tonic post-REM. Vertical bars represent S.E.M. (Redrawn from
Trulson and Jacobs 1979)

conditions, this activity can be drastically altered in conjunction with changes in sleep-wake-arousal state.

Because brain serotonin metabolism is strongly influenced by the light-dark cycle (Quay and Meyer 1978) (true circadian rhythmicity has not been closely examined), we decided to examine wether the light-dark cycle would influence the activity of DRN serotonergic neurones (Trulson and Jacobs 1983). When we held state constant, that is, studied the activity of a given neurone, e.g. during REM sleep or quiet waking, at various times during the light-dark cycle, we found the activity to be completely stable. Thus, the activity of these cells is not influenced by the light-dark cycle, nor, by inference, the circadian cycle. Why then have a number of investigators reported that serotonin metabolism varies as a function of the light-dark cycle? There are two obvious, non-mutually exclusive, possibilities. First, since neurotransmitter metabolism occurs predominantly at the axon terminal, local conditions at these sites may modulate metabolism somewhat independently from a constant spike discharge rate being generated at the cell body. Secondly, the variation in serotonin metabolism across the light-dark cycle may simply be secondary to the well-known distribution of sleep and waking across this cycle. For example, rats are known to spend most of their time in the dark, awake and active. The activity of their brain serotonergic neurones would also be active, and metabolism would be elevated (which it is). By contrast, rats spend most of their time in the light, sleeping and quiescent. During this period, their serotonergic neurones would be inactive, and metabolism would be decreased (which it is). Regardless of the true explanation for the variation in serotonin metabolism across the light-dark cycle, its existence introduces an additional variable into any pharmacological studies on the brain serotonergic system.

Repetitive presentation (once every 2 s) of phasic auditory (click) or visual stimuli (flash) both produce a similar effect on DRN serotonergic unit activity of excitation followed by inhibition (Heym, Trulson and Jacobs 1982b). (The effects of these stimuli in two different modalities are so similar, in fact, that we proposed that they are mediated by a common mechanism.) The excitation has a latency of approximately 40–50 ms and a duration of 60–80 ms. The inhibitory period which follows the excitation has a mean duration of approximately 250 ms and corresponds to the normal interspike interval that typically follows the discharge of these highly regular firing neurones. No evidence of habituation of these responses of DRN serotonergic neurones was seen, even after thousands of stimulus presentations. By contrast, serotonergic neurones in the NRM show only a modest excitation-inhibition response to these same auditory and visual stimuli (Fornal, Auerbach, Heym and Jacobs 1983). Perhaps of greater interest, however, is our failure to observe any specific response of

NRM serotonergic neurones to various types of painful or noxious stimuli, or to the systemic administration of morphine, in a dose that produces analgesia (Auerbach, Fornal, Heym and Jacobs 1983). This is somewhat surprising in light of the large literature which implicates the NRM in nociception and analgesia. It is also worth pointing out that the high degree of responsivity of DRN serotonergic neurones in freely moving cats is in contrast to previous studies conducted in chloral hydrate anaesthetized animals. Under these latter conditions, neuronal excitability in response to external inputs is quite damped. By studying the same cell under both conditions (unanaesthetized v. anaesthetized), we have shown that this difference is specifically attributable to the presence of the anaesthetic (Heym *et al.* 1984).

The sensory stimulation data just presented are also of interest because of their parallel to the response of brain noradrenergic neurones to the same types of stimuli (Aston-Jones and Bloom 1981). One possibility is that noradrenergic neurones in the locus coeruleus or in other nuclei drive the activity of serotonergic neurones. Neuro-anatomical data showing catecholamine terminals in the DRN are consistent with this idea. Furthermore, as described above, there is evidence from studies done in chloral hydrate anaesthetized rats which demonstrates that DRN unit activity is dependent on tonic noradrenergic input (Baraban and Aghajanian 1980; Aghajanian 1981). In order to test these hypotheses regarding the influence of noradrenergic neurones upon serotonergic unit activity, we administered a variety of drugs that decrease noradrenergic neurotransmission, and examined the effects upon spontaneous, as well as sensory evoked, activity of DRN serotonergic neurones in freely moving cats (Heym, Trulson and Jacobs 1981). The most impressive effect of WB-4101, prazosin, phenoxybenzamine, and a low dose of clonidine was strong sedation accompanied by catalepsy and high voltage slow waves in the EEG. At this time, many of the cells studied showed a significant depression of activity. However, when a sensory stimulus was presented, or when the behavioural state of the animal was equated pre- and post-drug, there was no significant effect upon unit activity. Once again, it seems that serotonergic neurones are responding in a state-dependent manner. While we believe that brain noradrenergic neurones may influence the activity of DRN neurones, we do not feel that the activity of these serotonergic neurones is critically *dependent* on such an input. Chloral hydrate anaesthesia appears to make the critical difference in these two sets of studies. When we administered WB-4101 to a cat that had been anaesthetized with chloral hydrate, the activity of DRN units was rapidly and completely suppressed (Heym *et al.* 1981, 1984). This study underscores the utility of conducting neurophysiological studies of drug action in unanaesthetized animals.

One of the issues that currently concerns us most is trying to determine the major variable(s) that control the activity of brain serotonergic neurones. An important initial clue was provided by our analysis of serotonergic unit activity during REM sleep. As described above, the activity of DRN neurones becomes virtually quiescent during REM sleep periods. Since it is well known that a major feature of REM sleep is centrally induced atonia, we hypothesized that this might be importantly related to the decrease in DRN serotonergic unit activity. We have now tested this hypothesis in several different ways, but only for the DRN group of serotonergic neurones. In the first series of studies, we lesioned a small area of the pons whose destruction in cats was known to result in REM sleep without atonia. An animal with such a lesion has periods which, by all criteria, look like REM sleep, except that antigravity mucle tonus is present and the animals are therefore capable of engaging in complex species-specific behaviours, including locomotion. Consistent with our hypothesis, we found that the activity of DRN neurones was significantly increased during REM sleep periods in these animals (Trulson, Jacobs and Morrison 1981 *a*, Fig. 8.4). In fact, the level of neuronal activity was directly related to the level of motor activity. A second series of experiments examined this issue in a somewhat reciprocal manner (Steinfels, Heym, Strecker and Jacobs 1983). When the cholinergic agonist drug carbachol is directly injected into specific areas of the pons, it can produce atonia in an otherwise awake cat (the awake animal is capable of tracking a moving object in its visual field). When the carbachol injection induced atonia, we observed a complete suppression of DRN serotonergic unit activity, whereas injections in nearby areas, which failed to induce atonia, had no effect on unit activity. Several additional experiments in this series led us to conclude that the critical variable was centrally induced atonia, rather than muscle paralysis *per se*. When we administered dantrolene, a peripherally acting muscle relaxant, or succinylcholine, an antagonist at the neuromuscular junction, we observed no change in DRN serotonergic unit activity in spite of a profound loss of muscle tonus (with succinylcholine, respiration was suppressed and the animals were artificially respired). These experiments also excluded, incidentally, peripheral afferent feedback from muscles as being important in influencing DRN serotonergic unit activity. When we administered mephenesin, a centrally acting muscle relaxant, we again observed a large decrease in serotonergic unit activity.

We also have a number of other, somewhat anecdotal, pieces of evidence supporting the hypothesis that the activity of DRN neurones is related to some aspect of central motor activity (Jacobs *et al.* 1984). Although the activity of serotonergic neurones is unrelated to movement *per se,* they often will change dramatically during specific behavioural acts. For example, when cats are lifted by the scruff of the neck, they typically

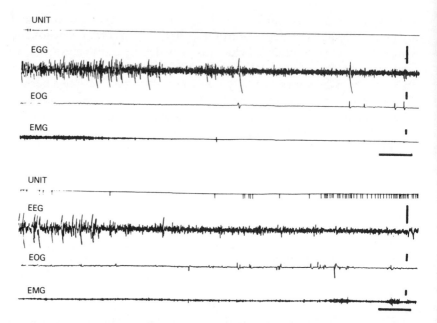

Fig. 8.4 Polygraph records displaying onset of REM sleep in a normal cat (top) and pontine lesioned cat (bottom). Note the increased EMG activity about one minute after REM sleep onset in the bottom record and the correlated increase in unit activity. Abbreviations: EEG, electroencephalogram; EOG, electrooculogram; EMG, electromyogram. Calibrations: EEG, 100 μV, EOG, 500 μV; EMG, 100 μV; time, 10 s. (Reproduced from Trulson *et al.* 1981*a*, with permission from Elsevier)

become limp or immobilized. During this time, we typically see a large decrease in DRN serotonergic unit activity, which quickly returns to baseline when the animal is released. Similarly, we have observed decreases in DRN serotonergic unit activity during defecation, a period when the animal assumes an immobilized, fixed position. Finally, we have also seen decreased unit activity during the conditioned production of the sensorimotor rhythm, a cortical EEG pattern associated with decreased motor activity or behavioural inhibition.

Thus, under a number of conditions, we have observed an inverse relationship between central motor outflow and the activity of serotonergic unit activity in the DRN. We are currently attempting to determine whether this is a relationship with activity in central motor pathways or possibly due to a variable which is strongly, and possibly inextricably, linked to central motor activity. A prime candidate for such a concomitant of motor activity is sympathetic tone, or its central representation.

8.4   Effects of hallucinogenic drugs on serotonergic unit activity

Evidence from a variety of different experimental approaches leads to the conclusion that an alteration(s) in serotonergic neurotransmission plays a critical role in the action of 'LSD-like' hallucinogenic drugs. Electrophysiological studies of serotonergic neurones have contributed more to our understanding of the mechanism of action of hallucinogenic drugs than to any other class of psycho-active drug. Before briefly reviewing these studies it may be worthwhile to consider a few of the complexities regarding the relationship between hallucinogenic drugs and brain serotonergic neurotransmission. Receptor binding studies indicate that there are at least two different types of serotonergic receptors, 5-HT_1 and 5-HT_2 (Peroutka and Snyder 1979; Chapter 4). Furthermore, recent studies indicate that there may be two sub-types of the 5-HT_1 receptor (Pedigo, Yamamura and Nelson 1981). On the other hand, neurophysiological analyses indicate that there may be three different types of serotonergic receptors, S_1–S_3 (Aghajanian 1981). It is not yet clear what relationship, if any, the two classifications have with each other. LSD, the prototypic hallucinogen, not only has multiple effects on the brain serotonergic system, but is also known to interact with several other neurotransmitter systems (Jacobs 1983). Detailed reviews of this literature have been recently published (Jacobs 1983, 1984).

  Based on neurochemical data indicating that LSD and related hallucinogens might act by decreasing the turnover of serotonin, it was hypothesized that this might be mediated by a decrease in unit activity of serotonergic neurones (Aghajanian et al. 1968). Aghajanian and co-workers tested this by recording the activity of mesencephalic serotonergic neurones in chloral hydrate anaesthetized rats and they systematically administering low doses of LSD (Aghajanian et al. 1968, 1970). As predicted, LSD and several other hallucinogens produced a complete and immediate suppression of the discharge of these neurones. The next issue was to try to determine how this was mediated. Was this action of hallucinogenic drugs upon serotonergic neurones a direct one, or was it mediated by some type of feedback loop? In the most direct examination of this issue, application of LSD directly onto serotonergic neurones, by means of iontophoresis, produced a profound suppression of serotonergic unit activity at low ejection currents (Haigler and Aghajanian 1973). In a subsequent experiment, they reported that although iontophoretic application of serotonin was equally potent in suppressing the activity of serotonergic neurones and their target neurones in various forebrain areas, the action of LSD was preferential to serotonergic neurones (Haigler and Aghajanian 1974). Thus, it was hypothesized, when LSD is administered systemically, it depresses serotonergic unit activity, while having little direct effect on target neurones. Because much of serotonin's synaptic effect in the forebrain is inhibitory, this produces a dramatic disinhibition of forebrain target neurones. Not only did this provide a model for explaining the action of hallucinogenic drugs, but it indicated that there were two different types of serotonergic receptors, both of which mediated an inhibitory response to serotonin: one of the serotonergic neurone itself that was sensitive to LSD (a presynaptic receptor); and one on forebrain target neurones that was relatively insensitive to LSD (a postsynaptic receptor). Subsequently, McCall and Aghajanian (1979) described a third type of receptor in the brainstem where serotonin exerts little direct effect on neuronal activity, but appears to amplify excitatory inputs from other sources (modulatory receptor). At this site, hallucinogens such as LSD potentiate this modulatory effect of serotonin (McCall and Aghajanian 1980).

We were interested in testing the hypothesis that the hallucinogens exerted their behavioural and psychological effects by preferentially depressing the activity of brain serotonergic neurones, so as to produce a disinhibition of various forebrain neurones. We did this by examining the activity of serotonergic units in the DRN of feely moving cats. This would allow us to examine the behavioural and neurophysiological effects of drugs simultaneously. In our first experiment, we administered 5-MeODMT and found that the onset, peak and offset of behavioural effects of the drug were strongly correlated with the onset, peak, and offset of changes in serotonergic unit activity (Trulson and Jacobs 1979b) (Fig. 8.5). These data strongly supported the 'presynaptic serotonin hypothesis' for the action of hallucinogenic drugs. We then extended this approach to the study of LSD, and although the results were generally similar to those for 5-MeODMT, there were two critical differences (Trulson and Jacobs 1979c). First, LSD produced a depression of serotonergic unit activity lasting 3–4 hours, while the behavioural effects lasted for 6–8 hours. Secondly, when LSD was readministered 24 hours later, it produced little or no behavioural effect (tolerance), but the effect on serotonergic unit activity was the same as that observed on the previous day. Based upon those two dissociations of unit activity and behaviour, we began to question the validity of the presynaptic hypothesis. Our scepticism was reinforced by later experiments with 2,5-dimethoxy-4-methylamphetamine (DOM), mescaline, and psilocin (Trulson, Heym and Jacobs 1981b). With all three drugs we observed the same temporal dissociations of serotonergic unit activity and behaviour as observed with LSD. In addition, low doses of DOM and psilocin produced no significant decrease in serotonergic unit activity in any of the cells tested, however, they did produce significant behavioural effects. Similarly, mescaline produced no overall significant change in unit activity, but did produce large behavioural effects.

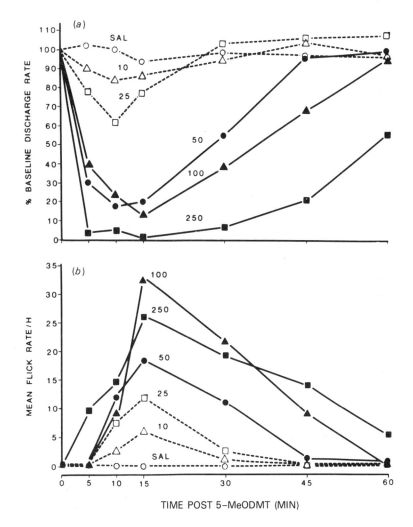

Fig. 8.5 Time course for the effects of saline (SAL, O– – – – –O) and 5-MeODT
(10 △– – – – –△, 25 □– – – – –□, 50 ●– – – – –●, 100 ▲– –.– – –▲ and 250 µg/kg
■– – – – –■) on the activity of dorsal raphe neurones (a) and the rate of occurrence
of limb flicks (b). (Reproduced from Trulson and Jacobs 1979b, with permission
from Elsevier)

Finally, I will describe a recent study from my laboratory which provides
perhaps the most damaging evidence to the presynaptic serotonergic hypothesis of hallucinogenic drug action (Heym, Rasmussen and Jacobs,
submitted for publication). When cats are given any of a variety of serotonin antagonist drugs (e.g. ketanserin, mianserin, or metergoline) prior to

being administered LSD, the behavioural effects of LSD are blocked in a dose-dependent manner. This was not attributable to non-specific sedative or cataleptic actions of these drugs since, when give alone, they did not significantly depress spontaneous locomotor activity. Furthermore, they did not block the behavioural effects of the non-hallucinogens, lisuride or apomorphine, but they did block the behavioural effects of DOM. The main goal of this experiment, however, was to examine whether pre-treatment with the serotonin antagonist mianserin would also block the depressant effects of LSD upon serotonergic neurones in the DRN or NCS of the cat. If it did, these combined behavioural and electrophysiological data would support the presynaptic hypothesis. The results were unequivocal. Pre-treatment, with a dose of mianserin that produced a complete blockade of the behavioural effects of LSD, exerted no statistically significant blocking effect on the typical suppression of DRN serotonergic unit activity induced by LSD. Similar results were obtained in experiments utilizing the serotonin antagonist ketanserin. These data argue in favour of a postsynaptic serotonergic action as being important in mediating the effects of LSD and related hallucinogens. This is also consistent with a large body of behavioural pharmacology experiments (for reviews see Jacobs 1983, and Chapter 10).

8.5   Conclusion

Serotonergic neurones in the DRN comprise a homogeneous group of cells, with many shared characteristics. In fact, this homogeneity extends, to a large degree, to serotonergic neurones in various other nuclei (e.g. NCS, NRM and NRP—see Jacobs et al. 1984 for a reiew of these data). Because of the similarity of the response of different serotonergic neurones to various stimuli, drugs, or physiological variables, it seems appropriate to consider brain serotonergic neurones as a 'system' whose components work in concert to produce integrated functional effects. This is important from a pharmacological point of view since it can be assumed, initially, that particular drugs, by engaging or disengaging the brain serotonergic system, will produce somewhat global effects.

Serotonergic neurones in the DRN manifest a slow and highly regular pattern of firing which appears to be autonomous. This neuronal activity is locally modulated by inputs from neighbouring serotonergic neurones. In addition, these neurones receive neurochemically identifiable inputs from GABAergic, noradrenergic, and histaminergic neurones. At least in some instances, the response of serotonergic neurones to various types of inputs appears to be dramatically altered by anaesthesia. Their spontaneous activity is strongly state dependent, manifesting an almost linear positive relationship to level of arousal. Serotonergic neuronal dischatge rate does not vary as a function of phase of the light-dark cycle, however, serotonin metabolism at the level of the axon terminal may be independently modulated by this cycle. The activity of serotonergic neurones can be phasically activated by sensory stimuli, and their tonic discharge level may be dependent on level of neuronal activity in central motor or autonomic systems. These data are of obvious relevance to interpreting the behavioural and physiological effects of drugs acting upon the brain serotonergic system. They also provide important information for interpreting brain neurochemical changes induced by drug administration.

Finally, turning to a specific class of drug, it seems clear that a major site of action of the hallucinogens is on the brain serotonergic neurone. Our most recent data indicate that an important aspect of the action of hallucinogenic drugs is as a direct serotonin agonist. However, many issues regarding the action of hallucinogenic drugs still remain unanswered: at what serotonin receptor is this action exerted; what is the importance of the action of these drugs directly on serotonergic neurone cell bodies and terminals (the presynaptic effect); is the well-known serotonin antagonist effects of some hallucinogens important; and what, if any, is the significance of the actions of hallucinogens on other neurotransmitter systems?

Acknowledgements

The author's work described in this chapter was supported by a grant from the N.I.M.H. (MH-23433). I wish to thank Drs James Heym and George Steinfels, and Mr Kurt Rasmussen for their contributions.

References

Aghajanian, G. K. (1972). Chemical-feedback regulation of serotonin-containing neurons in brain. *Ann. N. Y. Acad. Sci.* **193**, 86.

—— (1981). The modulatory role of serotonin at multiple receptors in brain. In *Serotonin neurotransmission and behaviour* (eds. B. L. Jacobs and A. Gelperin) p. 156. MIT Press, Cambridge, Mass.

—— and VanderMaelen, C. P. (1982). Intracellular recordings from serotonergic dorsal raphe neurons: pacemaker potentials and the effect of LSD. *Brain Res.* **238**, 463.

—— Foote, W. E. and Sheard, M. H. (1968). Lysergic acid diethlamide: Sensitive neuronal units in the midbrain raphe. *Science* **161**, 706.

—— —— —— (1970). Action of psychotogenic drugs on single midbrain raphe neurons. *J. Pharmac. Exp. Ther.* **171**, 178.

——Haigler, H. J. and Bloom, F. E. (1972). Lysergic acid diethylamide and serotonin: direct actions on serotonin containing neurons. *Life Sci.* **11**, 615.

Anderson, C. D., Pasquier, D. A., Forbes, W. B. and Morgane, P. J. (1977). Locus coeruleus-to-dorsal raphe input examined by electrophysiological and morphological methods. *Brain Res. Bull.* **2**, 209.

Aston-Jones, G. and Bloom, F. E. (1981). Norepinephrine-containing locus coeruleus neurons in behaving rats exhibit pronounced responses to non-noxious environmental stimuli. *J. Neurosci.* **1**, 887.

Auerbach, S., Fornal, C., Heym, J. and Jacobs, B. L. (1983). Unit activity of serotonergic neurons in n. raphe magnus of freely moving cats: response to morphine and noxious stimuli. *Soc. Neurosci. Abst.* **9**, 553.

Baraban, J. M. and Aghajanian, G. K. (1980). Suppression of firing activity of 5-HT neurons in the dorsal raphe by alpha-adrenoceptor antagonists. *Neuropharmacology* **19**, 355.

—— —— (1981). Noradrenergic innervation of serotonergic neurons in the dorsal raphe: demonstration by electron microscopic autoradiography. *Brain Res.* **204**, 1.

—— Wang, R. Y. and Aghajanian, G. K. (1978). Reserpine suppression of dorsal raphe neuronal firing: mediation by adrenergic system. *Eur. J. Pharmac.* **52**, 27.

Belin, M. F., Aguera, M., Tappaz, M., McRae-Degueurce, A., Bobillier, P. and Pujol, J. F. (1979). GABA-accumulating neurons in the nucleus raphe dorsalis and periaqueductal gray in the rat: a biochemical and autoradiographic study. *Brain Res.* **170**, 279.

Chu, N. S. and Bloom, F. E. (1974). The catecholamine-containing neurons in the cat dorso-lateral pontine tegmentum: distribution of the cell bodies and some axonal projections. *Brain Res.* **66**, 1.

Dahlstrom, A. and Fuxe, K. (1964). Evidence for the existence of monoamine-containing neurons in the central nervous system: I. Demonstration of monoamines in the cell bodies of brainstem neurons. *Acta Physiol. Scand.* **62** (Suppl. 232) 1.

Fornal, C., Auerbach, S., Heym, J. and Jacobs, B. L. (1983). Unit activity of serotonergic neurons in n. raphe magnus in freely moving cats. *Soc. Neurosci. Abstr.* **9**, 553.

Fuxe, K. (1965). Evidence for the existence of monoamine neurons in the central nervous system. IV. The distribution of monoamine terminals in the central nervous system. *Acta Physiol. Scand.* **64** (Suppl. 247) 41.

Gallager, D. W. and Aghajanian, G. K. (1976a). Effect of antipsychotic drugs on the firing of dorsal raphe cells. I. Role of adrenergic system. *Eur. J. Pharmac.* **39**, 341.

—— —— (1976b). Effect of antipsychotic drugs on the firing of dorsal raphe cells. II. Reversal by picrotoxin. *Eur. J. Pharmac.* **39**, 357.

Gallardo, M. R. G. P. and Pasquier, D. A. (1980). Increase in activity of choline acetyltransferase in the dorsal raphe nucleus following habenular deafferentation, *Brain Res.* **194**, 578.

Gottesfield, Z., Hoover, D. B., Muth, E. A. and Jacobowitz, D. M. (1978). Lack of biochemical evidence for a direct habenulo-raphe GABAergic pathway. *Brain Res.* **141**, 353.

Haigler, H. J. and Aghajanian, G. K. (1973). Mescaline and LSD: Direct and indirect effects on serotonin-containing neurons in brain. *Eur. J. Pharmac.* **21**, 53.

——— (1974). Lysergic acid diethylamide and serotonin: A comparison of effects on serotonergic neurons and neurons receiving serotonergic input. *J. Pharm. Exp. Ther.* **188**, 688.

Heym, J., Steinfels, G. F. and Jacobs, B. L. (1982*a*). Activity of serotonin-containing neurons in the nucleus raphe pallidus of freely moving cats. *Brain Res.* **251**, 259.

——— ——— ——— (1984). Chloral hydrate anesthesia alters the responsiveness of central serotonergic neurons in the cat. *Brain Res.* **291**, 63.

—— Trulson, M. E. and Jacobs, B. L. (1981). Effects of adrenergic drugs on raphe unit activity in freely moving cats. *Eur. J. Pharmac.* **74**, 117.

——— ——— ——— (1982*b*). Raphe unit activity in freely moving cats: effects of phasic auditory and visual stimuli. *Brain Res.* **232**, 29.

Jacobs, B. L. (1982). Recording serotonergic unit activity in the brains of freely moving cats. *J. Histochem. Cystochem.* **8**, 815.

—— (1983). Mechanism of action of hallucinogenic drugs: focus upon postsynaptic serotonergic receptors. In *Psychopharmacology 1 Part 1: Preclinical psychopharmacology* (ed. D. G. Grahame-Smith) p. 344. Excerpta Medica, Amsterdam.

—— (1984). (ed.) *Hallucinogens: neurochemical, behavioral and clinical perspectives.* Raven Press, New York.

—— and Gelperin, A. (1981). (eds.) *Serotonin neurotransmission and behavior.* MIT Press, Cambridge, Mass.

—— Heym, J. and Steinfels, G. F. (1984). Physiological and behavioral analysis of raphe unit activity. In *Handbook of psychopharmacology* (eds. L.L. Iversen, S.D. Iversen, and S.H. Snyder) Vol. 18, p. 343. Plenum Press, New York. (In press).

Lakoski, J. M. and Aghajanian, G. K. (1982). Effect of histamine on the activity of serotonin-containing neurons in the dorsal raphe. *Soc. Neurosci. Abstr.* **8**, 276.

McCall, R. B. and Aghajanian, G. K. (1979). Serotonergic facilitation of facial motoneuron excitation. *Brain Res.* **169**, 11.

—— —— (1980). Hallucinogens potentiate responses to serotonin and norpinephrine in the facial motor nucleus. *Life Sciences* **26**, 1149.

McGinty, D. J. and Harper, R. M. (1976). Dorsal raphe neurons: Depression of firing during sleep in cats. *Brain Res.* **101**, 569.

Mosko, S. S., Haubrich, D. and Jacobs, B. L. (1977). Serotonergic afferents to the dorsal raphe nucleus: Evidence from HRP and synaptosomal uptake studies. *Brain Res.* **119,** 269.

—— and Jacobs, B. L. (1976). Recording of dorsal raphe unit activity in vitro. *Neurosci. Lett.* **2,** 195.

—— —— (1977). Electrophysiological evidence against negative neuronal feedback from the forebrain controlling midbrain raphe unit activity. *Brain Res.* **119,** 291.

Osborne, N. N. (1982). (ed.) *Biology of serotonergic transmission.* John Wiley, New York.

Parent, A. (1981). The anatomy of serotonin-containing neurons across phylogeny. In *Serotonin neurotransmission and behavior* (eds. B. L. Jacobs and A. Gelperin) p. 3. MIT Press, Cambridge, Mass.

Pedigo, N. W., Yamamura, H. I. and Nelson, D. L. (1981). Discrimination of multiple [^{3}H]5-hydroxytryptamine binding sites by the neuroleptic spiperone in rat brain. *J. Neurochem.* **36,** 220.

Peroutka, S. J. and Snyder, S. H. (1979). Multiple serotonin receptors: differential binding of [^{3}H]5-hydroxytryptamine, [^{3}H]lysergic acid diethylamide and [^{3}H] spiroperidol. *Mol. Pharmac.* **16,** 687.

Quay, W. B. and Meyer, D. C. (1978). Rhythmicity and periodic functions of the central nervous system and serotonin. In *Serotonin in health and disease. Physiological regulation and pharmacological action* (ed. W. B. Essman) Vol. 2, p. 159. Spectrum Publications, New York.

Rasmussen, K., Heym, J. and Jacobs, B. L. (1984). Activity of serotonin-containing neurons in nucleus centralis superior of freely moving cats. *Exper. Neurol.* **83,** 302.

Steinfels, G. F., Heym, J., Strecker, R. E. and Jacobs, B. L. (1983). Raphe unit activity in freely moving cats is altered by manipulations of central but not peripheral motor systems. *Brain Res.* **279,** 77.

Stern, W. C., Johnson, A., Bronzino, J. D. and Morgane, P. J. (1979). Effects of electrical stimulation of the lateral habenula on single-unit activity of raphe neurons. *Exper. Neurol.* **65,** 326.

Svensson, T. H., Bunney, B. S. and Aghajanian, G. K. (1975). Inhibition of both noradrenergic and serotonergic neurons in brain by the α-adrenergic agonist clonidine. *Brain Res.* **92,** 291.

Taber, E., Brodal, A. and Walberg, F. (1961). The raphe nuclei of the brain stem in the cat. I. Normal topography and cytoarchitecture and general discussion. *J. Comp. Neurol.* **116,** 161.

Taylor, K. M., Gfeller, E. and Snyder, S. H. (1972). Regional localization of histamine and histidine in the brain of the rhesus monkey. *Brain Res.* **41,** 171.

Trulson, M. E. and Jacobs, B. L. (1976). Dose-response relationships between systemically administered L-tryptophan or L-5 hydroxy-tryptophan and raphe unit activity in the rat. *Neuropharmacology* **15,** 339.

—— —— (1979a). Raphe unit activity in freely moving cats: Correlation with level of behavioral arousal. *Brain Res.* **163,** 135.

—— —— (1979*b*). Effects of 5-methoxy-*N,N*-dimethyl-tryptamine on behavior and raphe unit activity in freely-moving cats. *Eur. J. Pharmac.* **54,** 43.

—— —— (1979*c*). Dissociations between the effects of LSD on behavior and raphe unit activity in freely moving cats. *Science* **205,** 515.

—— —— (1983). Raphe unit activity in freely moving cats: lack of diurnal variation. *Neurosci. Lett.* **36,** 285.

—— —— and Morrison, A. R. (1981*a*). Raphe unit activity during REM sleep in normal cats and in pontine lesioned cats displaying REM sleep without atonia. *Brain Res.* **226,** 75.

—— Heym, J. and Jacobs, B. L. (1981*b*). Dissociations between the effects of hallucinogenic drugs on behaviour and raphe unit activity in freely moving cats. *Brain Res.* **215,** 275.

VanderMaelen, C. P. and Aghajanian, G. K. (1983). Evidence for a calcium -activated potassium conductance in serotonergic dorsal raphe neurons. *Soc. Neurosci. Abstr.* **9,** 500.

Wang, R. Y. and Aghajanian, G. K. (1977). Antidromically identified serotonergic neurons in the rat midbrain raphe: evidence for collateral inhibition. *Brain Res.* **132,** 186.

—— Gallager, D. W. and Aghajanian, G. K. (1976). Stimulation of pontine reticular formation suppresses firing of serotonergic neurones in the dorsal raphe. *Nature* **264,** 365.

9

In vivo monitoring of pharmacological and physiological changes in endogenous serotonin release and metabolism

CHARLES A. MARSDEN

9.1 Introduction

One of the criteria for establishing a chemical as a neurotransmitter is the demonstration of its calcium dependent stimulation-induced released from nerve endings into the extracellular space. With the CNS the majority of studies attempting to show this depend on *in vitro* preparations. Neuro-scientists however, are not only interested in establishing that a particular chemical is a neurotransmitter, but they also wish to understand the functional importance of particular neurotransmitters and how individual neurotransmitter pathways interact with others. Such studies need techniques that can be used *in vivo,* preferably in the freely moving animal, to monitor changes in transmitter release and metabolism and correlate these to behavioural, physiological and pharmacological events. The present chapter will briefly describe three possible approaches to the problem of monitoring *endogenous* serotonin release and metabolism and demonstrate how such methods can provide new information about the role of serotonin neuronal systems in the CNS. The techniques discussed fall into two categories. The first, and most commonly used, group includes the intracerebral perfusion techniques in which brain perfusate samples, containing serotonin, and its metabolite 5-HIAA, are collected and assayed by specific and highly sensitive analytical techniques. The second category attempts to continuously monitor changes in transmitter release and metabolism using an electrochemical detector electrode implanted in brain (Conti, Strope, Adams and Marsden 1978; Marsden, Conti, Strope, Curzon and Adams 1979; Adams and Marsden 1982).

In vivo Electrochemistry differs from the other methods for monitoring endogenous levels in that the 'assay' is performed *in situ* without any collection of perfusates and subsequent separation and measurement of the amines and metabolites in the perfusates. There are advantages and disadvantages to this approach—on the one hand *in situ* measurement allows rapid and frequent sampling but the lack of an external separation step imposes important restraints on the interpretation of the signals obtained which will be discussed in this chapter.

The theoretical and practical aspects of all techniques are discussed in detail elsewhere (e.g. see Marsden 1984).

9.2   Perfusion and collection techniques

The increasing interest in the possibility of monitoring changes in *endogenous* amine release and metabolism clearly relates to the marked improvement in the sensitivity of the assays that are now available to measure serotonin, its precursor 5-HTP and 5-HIAA. In particular, the development of assays based on high performance liquid chromatography combined with electrochemical detection (LCEC) (see Adams and Marsden 1982) has allowed the separation and detection of amines and their metabolites in small (e.g. 10 µl) samples with minimal preparation (Mefford 1981). Similarly, the development of other sensitive assay methods for other compounds of interest to the neuroscientist has prompted the increased use of the perfusion methods; these assays include radioimmuno, radioenzymatic, gas chromatographic and mass fragmentographic assays.

Earlier studies, although dependent on less sensitive assays, were able to demonstrate endogenous serotonin release *in vivo* using both anaesthetized (Holman and Vogt 1972; Ternaux, Boireau, Bourgoin, Hamon, Héry and Glowinski 1976) unanaesthetized (Mayers and Beleslin 1971*a*) and encephale isole (Ternaux, Héry, Hamon, Bourgoin and Glowinski 1977) preparations. These techniques used either cortical cups or push-pull perfusion; the latter approach was originally described by Gaddum (1961). Most push-pull perfusion studies however, do not measure endogenous amines but efflux of exogenously administered radiolabelled transmitter (e.g. Holloway 1975; Héry, Simonnet, Bourgoin, Soubrié, Artaud, Hamon and Glowinski 1979; Reisine, Soubrié, Holloway 1975; Artaud and Glowinski 1982). While studies using labelled transmitters have an important place in the investigation of transmitter interactions in the CNS, the methodological aspects will not be discussed in this chapter and the reader is referred to Glowinski (1981) for a review of the application of push-pull techniques for studying the cat basal ganglia. More recently the push-pull technique has been applied to the measurement of endogenous amine release in the rat (Loullis, Hingten, Shea and Aprison 1980; Elghozi,

LeChan-Bui, Earnhardt, Meyer and Devynck 1981; Philips, Robson and Boulton 1982)—for review see Philippu (1984). The perfusion of the intact spinal cord has been described by Yaksh (1984). The two approaches discussed in this chapter are (*i*) intracranial dialysis and (*ii*) cerebrospinal fluid (CSF) sampling.

9.2.1   In vivo *intracranial dialysis*

This technique involves the implantation of a small dialysis tube into the brain area of interest to the investigator. The tube is then perfused with physiological saline and substances that are able to cross the dialysis membrane move along a diffusion gradient from the brain tissue into the perfusate. The perfusate is collected and the substances in it assayed (Ungerstedt, Herrera-Marschitz, Jungnelius, Stahle, Tossman and Zetter- ström 1982; Zetterström, Sharp, Marsden and Ungerstedt 1983; Hernandez, Paez and Hamlin 1983). The range of substances collected by the dialysis tube obviously depends on the molecular weight cut-off of the tubing (normally 5000 MW but tubes with pore sizes up to 50 000 MW are available). This approach can be applied not only to the collection and measurement of amines and their metabolites but also to amino acids (Tossman and Ungerstedt 1981) and providing the radioimmunoassays are adequately sensitive, peptides.

The original studies by Ungerstedt and his colleagues (see Ungerstedt *et al.* 1982) used a loop of dialysis tubing prepared from flexible cellulose tubing (Dow Co 0.25 mm o.d.) with a 5000 MW cut-off. Both ends of a 5 cm length of tubing are glued inside stainless steel cannulae (23 G) leaving approximately 5 mm exposed between the cannulae (Fig. 9.1*a*). This arrangement is then kept moist in physiological saline, folded into a loop and supported with a fine wire placed into the lumen of one half of the loop. The dialysis loop is then stereotaxically implanted into a brain region, for example the striatum. During perfusion one cannula is connected, via polythene tubing, to a microinfusion pump and perfused with physiological saline (147 mmol Na^+, 2.3 mmol Ca^{2+}, 4 mmol K^+, 155.6 mmol Cl^-, pH 6.0) at a rate of 1 or 2 $\mu l/min$. The perfusate is collected into an everted eppendorf tube placed on the end of the other cannula. To prevent oxida- tion of the amines in the collected perfusate the eppendorf tube contains 10 μl of perchloric acid (1 M) containing sodium metabisulphate. Normally 20 min samples are collected and directly assayed by LCEC for amines and their metabolites. The Dow tubing used in the original studies is no longer available but other tubing (e.g. Gambos) is equally effective. At Nottingham we have obtained our tubing from the Hospital Renal Dialysis Unit where the dialysis packs are discarded unused once the stated expiry date has passed. One such discarded pack provides adequate tubing for a life time's

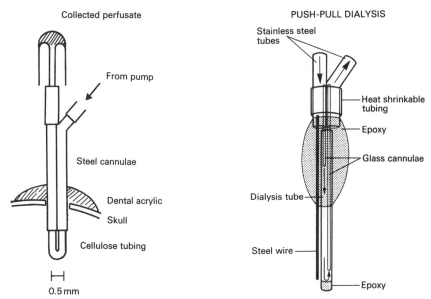

Fig. 9.1   Design of intracerebral dialysis systems. (*a*) Dialysis loop system based on that described by Ungerstedt *et al.* (1982). A slow perfusion pump is used to pump physiological saline into the stainless steel cannula, through the cellulose dialysis loop (MW cut off 5000), which is implanted in the brain region under investigation and out through the other cannula into the collecting tube. Materials enter the saline in the cellulose tubing from the brain. (*b*) Alternative dialysis system based on that described by Johnson and Justice (1983). The system is a closed push-pull system in which dialysis occurs within a closed dialysis bag.

work on intracerebral dialysis. The diameter of the dialysis loops described are about 0.5 mm and this obviously limits the brain areas that can be investigated. A modification is the push–pull dialysis cannula (Johnson and Justice 1983) in which the push–pull cannulae lie inside a dialysis tube sealed at the bottom with epoxy (Fig. 9.1*b*). This arrangement gives a system with a smaller diameter (0.25 mm) but a smaller surface area for exchange to occur. Not all microdialysis tubing is suitable for making loops (e.g. Amicon Vita-Fibre) in which case the push–pull arrangement or a modification of it, has obvious attractions.

The simplest suitable method for measuring the indoles in the perfusates is reverse-phase (Spherisorb, Ultrasphere 5 or 3 ODS 2) LCEC using 0.1 M acetate/citrate pH 4.6 buffer containing approximately 10 per cent methanol as mobile phase with the electrode potential set at $+0.7$ V. The precise methanol percentage and pH will depend on the source and condition of the chromatography column. This assay provides rapid separation of dihydroxyphenylacetic acid (DOPAC), serotonin, 5-HIAA, and homovanillic acid (HVA) (Fig. 9.2). If the potential is raised to $+0.9$ V,

222   Charles A. Marsden

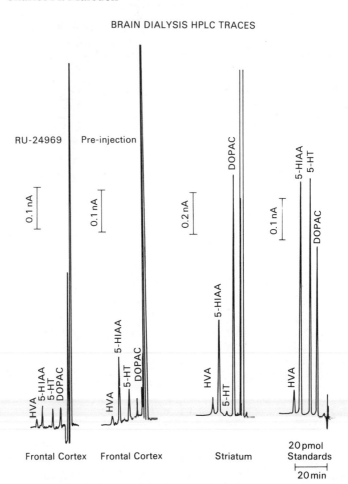

Fig. 9.2 Typical reverse phase LCEC recording of perfusates collected from the rat striatum and frontal cortex. Note the high DOPAC levels in the striatal sample compared to the frontal cortex but the similarity in the 5-HIAA levels in both regions. The final trace shows the effects of the 5-HT$_1$ receptor agonist RU 24969 on 5-HIAA in frontal cortex perfusates. The pre-injection frontal cortex sample was collected just before injection of RU 24969 (10 mg/kg) and the post-injection sample from the same rat is the sample collected 80 min after injection. Perfusate flow rate: 2 µl/min (20 min samples collected and directly injected onto the LCEC system (see text for details). (Data from Routledge and Marsden, unpublished)

tryptophan levels can also be assayed. An alternative is to use an ion pair reverse phase separation with a mobile phase containing 0.1 M NaH$_2$PO$_4$ buffer pH 3.6, 0.1 mM EDTA, 0.05 mM ion pair reagent (sodium octanyl sulphonic acid) and 9 per cent methanol. With this assay noradrenaline, adrenaline and dopamine will be measured as well as DOPAC, 5-HIAA and HVA but serotonin is very slowly eluted so if it is the primary interest the first assay is preferable.

Several studies have shown that recovery of both amines and their metabolites across the dialysis membrane is directly proportional to the concentration in the external medium (Zetterström *et al.* 1983; Johnson and Justice 1983; Sharp, Maidment, Brazell, Zetterström, Ungerstedt, Bennett and Marsden 1984*a*). With 5-HIAA this relationship has been shown over a concentration range of 5×10^{-5} to 10^{-7} M. The percentage recovery observed depends on the compound under study and the flow rate of the perfusate with recovery inversely related to flow rate. At high (4–6 µl/min) flow rates there is relatively little change in recovery while at lower flow rates (1–0.1 µl/min) there is a near inverse linear relationship between the two (Johnson and Justice 1983). In practical terms the optimum flow rate is that which allows adequate measurable material to be collected within the shortest sampling time. For amine measurements this is normally either 1 or 2 µl/min for 20 min which with dialysis loops gives recoveries of 8–15 per cent (Table 9.1) and adequate signal to noise ratio on the chromatograms (Fig. 9.2). With the push–pull dialysis system another factor that influences recovery is the distance between the ends of the inlet and outlet cannulae within the dialysis bag (Fig. 9.1 *b*). No detailed studies have compared the recovery of one amine alone with recovery of the same amine from a mixture of amines which is the situation *in vivo*. Preliminary studies in our laboratory suggest that serotonin recovery *in vitro* is lower when determined using a mixture of serotonin, 5-HIAA and DOPAC than from a solution of serotonin alone.

Table 9.1 *In vitro* recovery of DOPAC, 5-HIAA and serotonin through the dialysis tubing

	Perfusion rate 1 µl/min	2 µl/min
	Recovery (%)	
DOPAC	15	8
5-HIAA	15.75	9.5
Serotonin	14.50	8

Recovery was determined using loops described in Fig. 9.1(*a*) immersed in beakers containing 1×10^{-6} M of the relevant amine in physiological saline. The loops were perfused at either 1 or 2 µl/min using the same physiological saline (Marsden and Routledge 1984).

Most studies have concentrated on the measurement of extracellular dopamine, DOPAC and HVA in the striatum and nucleus accumbens and the effect of drugs on these levels (Ungerstedt *et al.* 1982; Zetterström *et al.* 1983; Zetterström and Ungerstedt 1984). The main finding regarding

Table 9.2 Dopamine, DOPAC, 5-HIAA and serotonin levels in dialysis per-fusates collected from the rat striatum and frontal cortex

	Striatum (pmol/50 μl)	Frontal cortex
Dopamine	3.9±0.4 (6)	Not measurable
DOPAC	91.9±6.6 (6)	1.2±0.04 (6)
Serotonin	1.2±0.1 (6)	1.7±0.1 (6)
5-HIAA	27.2±0.8 (6)	11.6±0.04 (6)

The values have been corrected for *in vitro* recovery through the dialysis tubes (see text for details and Table 9.1). (Data from Marsden and Routledge 1984, and unpublished data)

basal levels is the very high extracellular concentrations of the metabolites compared to dopamine with a difference of about 100:1 (Zetterström *et al.* 1983, Table 9.2). This marked difference is also apparent when extra-cellular levels of serotonin and 5-HIAA are compared in the frontal cortex and striatum of the rat (Marsden and Routledge 1984; Table 9.2). The results stress the efficiency with which extracellular amine transmitter levels are maintained by re-uptake while raising questions about the origins of the high levels of the metabolites in the extracellular space—are they a reflection of released transmitter or is the main component derived from intracellular metabolism not directly associated with release? These questions will be discussed in more detail later, together with appropriate data obtained with the other sampling techniques.

In summary, intracerebral dialysis provides a relatively simple method for sampling extracellular serotonin and 5-HIAA together with trypto-phan. The advantages include the ease with which the dialysis probes can be made and the positive identification of the substances collected because of the LCEC assay stage. Disadvantages are the relatively large probe size that limits the application of the method to the study of the bigger brain regions. Another problem is the long collection time required (usually 10–20 min) for each sample. This is particularly true when serotonin is being measured because of its low extracellular concentration combined with its low recovery rate (about 10 per cent) at a perfusion rate of 1 μl/min. When the main interest is the metabolite 5-HIAA this is not a limitation because of the vastly greater extracellular levels of 5-HIAA compared with sero-tonin (Table 9.2). A final limitation is the time that the dialysis loop remains usable *in vivo*. Several factors can alter the permeability of the membrane but glial cell growth is probably the major problem. Results show a constant baseline level of 5-HIAA over about 24 h, apart from circadian variations, but beyond 24 h the level starts to decline probably reflecting decreased recovery rather than a change in extracellular concentrations of 5-HIAA. The latter may also be a factor as glial cell growth around the dialysis loop may produce a compartment around the probe in which diffusion is restricted compared to diffusion in normal brain tissue. Similar considerations apply to all probes implanted chronically and while initial (24 h) measurements may reflect normal diffusion and extracellular levels, longer studies will be investigating changes within a compartment showing restricted diffusion characteristics (Cheng, Schenk, Huff and Adams 1979). The problem of declining baseline values needs careful consideration if experiments are planned aimed at investigating changes in serotonin turnover or release during physiological situations.

9.2.2.   *Continuous CSF sampling*

These techniques involve the chronic implantation of a catheter or cannulae into the ventricles so that small samples of CSF can be withdrawn at regular intervals and the levels of 5-HIAA and serotonin (if possible) assayed. This method has been used for studies in anaesthetized (Le Quan-Bui, Elghozi, Devynck and Meyer 1982) and freely moving rats (Danguir, Le Quan-Bui, Elghozi, Devynk and Nicolaidis 1982; Nielsen and Moore 1982; Sarna, Hutson, Tricklebank and Curzon 1983; Hutson, Sarna, Kantamaneni and Curzon 1984*a*) and cats (Degrell, Zenner, Kummer and Stock 1983) following earlier studies in humans (Garelis, Young, Lal and Sourkes 1974) primates (Baker and Ridley 1979) and dogs (Guldberg and Yates 1968). The new sensitive assays has allowed the extension of the technique to small rodents. The experimental approach is similar to brain dialysis in that steel cannulae (Degrell *et al.* 1983) or polyethylene catheters (Sarna *et al.* 1983) are implanted into the ventricles or cisterna magna (Fig. 9.3) and CSF samples are withdrawn, filtered and assayed by LCEC as described in the previous section on brain dialysis.

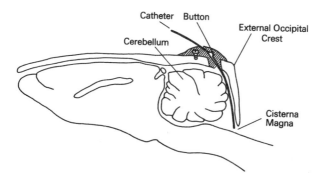

Fig. 9.3   Diagram showing the placement of a catheter into the rat cisterna magna for repeated sampling of CSF for amine metabolite measurements. (Redrawn with permission from Sarna *et al.* 1983)

In vivo measurements have confirmed earlier studies in humans (Garelis *et al.* 1974) and dogs (Guldberg and Yates 1968) showing high levels of the amine metabolites (DOPAC, 5-HIAA) but very low levels or, more usually, absence of the transmitters in the CSF. In the cat the level measured remained fairly constant over a sampling period of 150 days (Degrell *et al.* 1983), while in the rat 5-HIAA levels are stable over many hours (Fig. 9.4). Le Quan Bui *et al.* (1982) collected rat CSF samples in the anaesthetized rat at 15 min intervals and found basal 5-HIAA concentrations of about 2×10^{-6}M while serotonin was 3×10^{-8}M. Injection of an MAO inhibitor tranylcypromine increased serotonin (5×10^{-8}M) while decreasing 5-HIAA (0.5×10^{-6}M). When L-tryptophan was given after tranlycypromine there was a greater rise in serotonin than with MAO inhibitor alone (1.3×10^{-7}M).

Repeated CSF 5-HIAA measurements in rat have principally been used as an index of whole brain serotonin turnover using either probenecid induced increases or MAO inhibitor induced decreases in 5-HIAA (Sarna

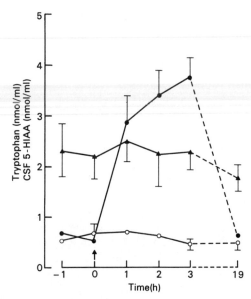

Fig. 9.4 CSF concentrations of 5-HIAA and tryptophan obtained by repeated sampling of CSF from the cirsterna magna of individual rats. The graph shows the mean values (n=4) of the effects of saline on CSF 5-HIAA (O) and probenecid (200 mg/kg i.p.) on 5-HIAA (●) and tryptophan (▲). Injections of saline or probenecid are indicated by the arrow. Note the marked rise in 5-HIAA after probenecid and the stability of the 5-HIAA levels in the saline controls. (Redrawn, with permission from Sarna *et al.* 1983)

et al. 1983; Hutson *et al.* 1984*a, b*). CSF 5-HIAA values rose linearly over 60 min post-probenecid administration and there was a reasonable correlation between serotonin turnover using brain or CSF 5-HIAA measurements after probenecid (Sarna *et al.* 1983; Hutson *et al.* 1984*a*) though the correlation was less impressive using DOPAC and HVA to determine dopamine turnover (Hutson *et al.* 1984*a*). The tryptophan hydroxylase inhibitor, *p*-chlorophenylalanine, also produced comparable decreases in serotonin turnover whether calculated from brain or CSF 5-HIAA changes (Sarna *et al.* 1983). Furthermore, after a tryptophan load (50 mg/kg) there is a similar clearance rate of tryptophan from CSF and striatal perfusates and similar maximum increases in 5-HIAA in striatal (+54 per cent) perfusates and CSF (+90 per cent) samples (Sarna, Hutson, Kantamaneni, Mootoo and Curzon 1984).

In the early studies using CSF amine metabolite measurements, both in animals and humans, only the non-conjugated metabolites were measured. It is now well-established that dopamine metabolites in the brain and CSF are found in both free and conjugated forms—the latter mainly as sulphates or glucuronides (Gordon, Markey, Sherman and Kopin 1976; Elchisak, Maas and Roth 1977; Dedek, Baumes, Tien-Duc, Gomeni and Korf 1979) and these studies have further indicated that probenecid blocks the transport of the free and conjugated acids from brain and CSF into blood. However, this does not seem to be a factor with either brain or CSF 5-HIAA measurements for while acid hydrolysis increases the level of DOPAC and HVA measured in CSF (Fig. 9.5) and brain dialysis samples (Routledge and Marsden, unpublished) it has not effect on 5-HIAA levels (Hutson *et al.* 1984*a*). The exact importance of conjugation with regard to brain and CSF serotonin and 5-HIAA needs to be clarified. We have evidence that while 5-HIAA levels in hypothalamic dialysis samples are unaffected by acid hydrolysis measurable serotonin increases. Interestingly, initial studies with the catecholamines have shown that drugs produce similar effects on both the free and sulphated amines in brain (Karoum, Chuang and Wyatt 1983) indicating measurement of either is equally informative. Similar studies need to be performed to establish whether the free and conjugated metabolites in brain dialysis and CSF samples show different changes in response to drug administration and also whether there are differences in turnover when it is determined using either free or sulphated metabolite measurements (Dedek *et al.* 1979).

Hutson and coworkers (1984*a*) have shown that in the rat there are considerable intra-individual variations in basal turnover values, measured using probenecid-induced accumulation of 5-HIAA. Values in a group of 7 rats ranged between about 0.6–2.0 nmols/ml/h (mean 1.24±0.56) and there was also considerable variation between successive values in individual rats although environmental variables (i.e. time of day) were

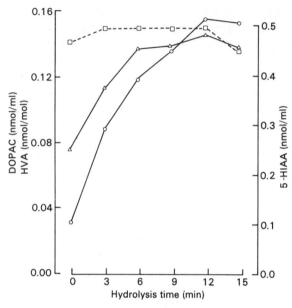

Fig. 9.5   The effect of acid hydrolysis for different times on rat CSF 5-HIAA (□), DOPAC (O) and HVA (△). Note the absence of effect on 5-HIAA levels but the large increase in the dopamine metabolites indicating the importance of sulphation in the latter case. (Data redrawn, with permission from Hutson *et al.* 1984*a, b*)

controlled. This variation may reflect toxicity caused by repeated pro benecid treatment rather than physiological variation and alternatives are obviously required to probenecid for such turnover measurements.

In summary, the serial sampling of rodent CSF is now a relatively simple method for obtaining repeated measurements of amine metabolites, though the best way of assessing repeated turnover values needs to be determined. The applications to which this approach has been put, i.e. monitoring of duirnal changes in catechol and indoleamine turnover and investigation of associations between pre-determined serotonin turnover and subsequent behaviour, will be discussed in the section of this chapter on applications. The disadvantage of the CSF sampling method is the inability to obtain data on regional variations in turnover or release in response to physiological or pharmacological manipulation.

9.3   *In vivo* voltammetry

The methods discussed so far for measuring endogenous transmitter release in brain depend on the collection of perfusate followed by the assay of the transmitters in it. A limitation of these techniques is the difficulty in providing information about minute-to-minute variations in release as samples can only be collected at 10–20 min intervals. One approach to this problem is the use of *in vivo* voltammetry. At present, however, the technique cannot be used to measure serotonin release but only serotonin metabolism as existing electrodes detect extracellular 5-HIAA not serotonin. The principles of the electrochemical method have previously been discussed in detail (Adams and Marsden 1982; Marsden, Brazell and Maidment 1984*a*) and so will only be dealt with briefly.

9.3.1   *Electrochemical principles*

The *in vivo* electrochemist has exploited a well-known problem for those working with catechol and indoleamines—their ease of oxidation (Fig. 9.6). The electrochemist has harnessed this reaction by carrying it out at the surface of a carbon (graphite) electrode which acts as the oxidizing agent. The transfer of oxygen to the hydroxy groups on the catechol (Fig. 9.7) results in a decrease in free energy and release of electrons. The electrochemist measures this electron release in the form of current, the amount of which is directly proportional to the number of molecules oxidized. The first practical application of electrochemistry in the neurosciences was the use of an electrochemical detector combined with high performance liquid chromatography for the very sensitive assay of amines (LCEC) (see Adams and Marsden 1982). The *in vivo* electrochemical technique involves the miniaturization of this detector and its implantation into the brain. The main problem with the *in vivo* technique is that, unlike LCEC, there is no

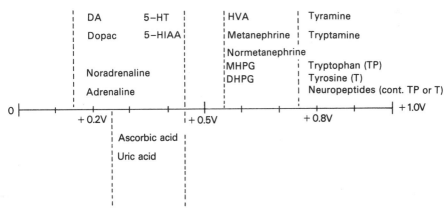

Fig. 9.6   Diagram showing the compounds commonly found in brain that are electroactive. At present *in vivo* voltammetry is only concerned with compounds that oxidize at potentials between 0.0 to +0.6 V.

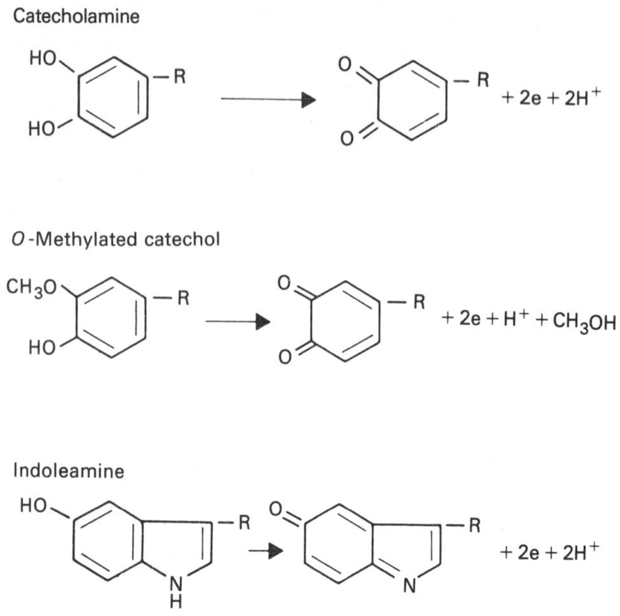

Catecholamine

O-Methylated catechol

Indoleamine

Fig. 9.7   The oxidation of catechol and indoleamines.

chromatographic separation of compounds with closely related oxidation potentials (Fig. 9.6).

Existing *in vivo* studies use a three electrode recording system with the potential needed to promote oxidation applied to the carbon or graphite working electrode. The potential is maintained with respect to a reference electrode by passing the required current through the working and auxiliary electrodes.

The most important factor in *in vivo* voltammetry is the working electrode which acts as an oxidizing agent by changing its electron energy state on application of suitable positive potential so that electrons are transferred from the oxidizable compound to the electrode with the concomitant oxidation of the compound. The ease at which compounds are oxidized depends on the presence of certain oxidizable groups (e.g. OH or NH in the case of indoleamines) and the potential at which oxidation occurs depends on how readily these groups are oxidized. Generally, compounds with similar chemical structures (i.e. dopamine and DOPAC or serotonin and 5-HIAA) oxidize at similar potentials (Fig. 9.6). This is the major limitation of *in vivo* electrochemistry and the emphasis at present is the production of electrodes with increased selectivity. Luckily there are relatively few compounds found in adequate amounts in the extracellular space which are electroactive, particularly within the potential range that can be applied to working electrodes implanted in the brain (about $-0.2-+0.7$ V). The lower limit is set by the reduction of oxygen and the higher limit by the oxidation of ions in the extracellular space—both of these events produce a dominant current which overshadows any other current generated.

All oxidative processes occur at the surface of the working electrode and there is no voltage or current flow across the brain tissue at the potentials used *in vivo* as the electrons generated are unable to move distances greater than a few molecular diameters. This means that the extracellular electrodes only detect *extracellular* compounds in its immediate vicinity (Cheng *et al.* 1979). From an electrochemical viewpoint the extracellular fluid pool should be considered as unstirred and so compounds only enter or leave it by diffusion along the concentration gradient. Thus, following the application of a potential to the working electrode all material that can be oxidized at its surface is immediately oxidized and this is followed by replenishment of the oxidizable material by diffusion until an equilibrium state is reached again. In practical terms this means that recordings must be made under conditions that allow the equilibrium state to be re-established between each recording—in this way a steady baseline current value is obtained. With chronic recordings made over days a further complication is the growth of glial and other cells which may impede the diffusion of the electroactive substances between the electrode surface and surrounding tissue. In this case recordings are being made from a pool of substances within a restricted compartment around the electrode rather than one showing free diffusion properties. Similar considerations apply to chronic measurements made using intracerebral dialysis or push–pull techniques.

The form of electrochemical signal monitored depends on two main factors. Firstly, the measurement technique employed and secondly, the nature of the working electrode.

9.3.2   *Measurement techniques*

These are described in detail elsewhere (see Marsden *et al.* 1984*a*; Gonon, Buda and Pujol 1984; Cespuglio, Faradji, Hahn and Jouvet 1984 ; Schenk and Adams 1984). There are basically two approaches. In the first case a square wave potential pulse is applied to the working electrode for a fixed time (i.e. $+0.5$ V for 1 s) and the total current generated is measured during the last $1/10$th of the fixed time period so as not to include the capacitance current in the measurement. This method is termed chronoamperometry and the important feature of it is that all compounds with oxidation potentials at or below the potential applied are oxidized giving a summed current response and so as such provides quantitative but not qualitative information. The advantage of chronoamperometry is the speed at which the measurements can be made.

The second approach is to apply a steadily increasing ramp potential, with a pre-determined range, to the working electrode. The oxidation of a compound is recorded as a peak of current as at its oxidation potential and beyond all the compound at the electrode surface is oxidized but not fully replenished by diffusion. Providing there is a reasonable difference between the oxidation potentials (i.e. 150 mV) of two compounds they will produce separate oxidation peaks so providing qualitative as well as quantitative information. Peak resolution can be improved by using modifications of the basic linear ramp technique. These modifications include differential pulse voltammetry (DPV) in which regular step potentials (duration about 30 ms, amplitude 50 mV, frequency 2/s) are superimposed on the steadily increasing ramp potential. Current generated is measured for a short period (20 ms) immediately before a pulse and immediately before the end of a pulse and the current difference plotted against the applied voltage. Most of the results discussed in this chapter have been obtained using DPV measurements. Another modification described by Gonon et al. (1984) is differential normal pulse voltammetry in which potential pulses of increasing amplitude are applied to the working electrode and at the end of each pulse the potential returns to the start potential (i.e. zero). Superimposed on these pulses is a short potential of fixed amplitude length which is applied towards the end of the main potential pulse. Current measurements are again made just before the start and just before the end of the superimposed pulse and the current difference plotted against the applied voltage.

9.3.3   Working electrode

Electrochemical signals recorded in vivo are in the nano-amp range so the working electrode needs to be made of material with very low residual current. This in effect has limited electrode construction to either powdered carbon mixed with a hydrophobic component (i.e. wax, silicon or paraffin oil) or pyrolytic carbon fibres. Once suitable materials have been found the major problem is to produce electrodes that are able to distinguish between electroactive compounds with similar oxidation potentials which are found in the extracellular fluid. The most important example is the problem caused by the very high levels of ascorbic acid in brain (Mefford, Oke and Adams 1981) which oxidizes at a very similar potential to dopamine and DOPAC at the surface of untreated carbon paste electrodes (Brazell and Marsden 1982a, b; O'Neill, Grunewald, Fillenz and Albery 1982a; Adams and Marsden 1982; Wightman and Dayton 1982). Another example is the similar oxidation potentials of the

CARBON FIBRE ELECTRODE

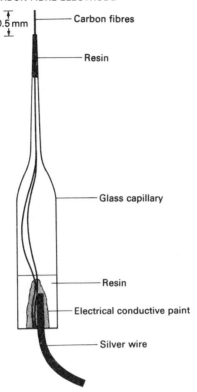

Fig. 9.8 The carbon fibre *in vivo* voltammetric electrode. For details of these electrodes see Sharp *et al.* 1984.

indoleamines and uric acid especially as it is now known that the latter is found in high concentrations in the brain (Zetterström *et al.* 1983). The next problem has been to produce electrodes sensitive to dopamine and serotonin but not DOPAC and 5-HIAA. The ascorbic acid problem has been largely overcome (Gonon, Buda, Cespuglio, Jouret and Pujol 1980, 1981*a*; Gonon *et al.* 1984) and there are now stearate modified paste electrodes that appear to be selective for dopamine to the exclusion of both ascorbic acid and DOPAC (Blaha and Lane 1984). The main problem with producing electrodes that are able to detect dopamine or serotonin rather than DOPAC or 5-HIAA is sensitivity as we now know that there is a great difference between the extracellular concentrations of the amines (54×10^{-8}M) and their metabolites (5×10^{-6}M). While it may be possible to detect extracellular striatal dopamine, where the concentrations are relatively high, the present electrodes will need to have improved sensitivity to

234　Charles A. Marsden

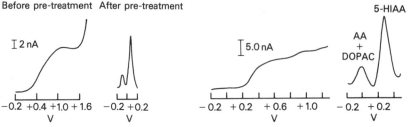

(a) ELECTRICAL PRE-TREATMENT

AA + DOPAC

Before pre-treatment　After pre-treatment

I 2 nA

−0.2 +0.4 +1.0 + 1.6　−0.2 +0.2
　　　　V　　　　　　　　V

(b) ELECTRICAL PRE-TREATMENT
AA + DOPAC + 5-HIAA
Before pre-treatment　　After pre-treatment

5-HIAA

I 5.0 nA

AA
+
DOPAC

− 0.2　+ 0.2　+ 0.6　+1.0　− 0.2　+ 0.2
　　　　V　　　　　　　　　　V

Fig. 9.9 Effects of electrical pre-treatment of the carbon fibre electrodes on the separation of (a) ascorbic acid and DOPAC and (b) ascorbic acid, DOPAC, and 5-HIAA *in vitro* using mixtures containing 100 μM of each and differential pulse voltammetry. Pre-treatment parameters (all performed with the electrodes in phosphate buffer pH 7.4). Note the absence of separation without electrical pre-treatment.

(a) 1. Triangular waveform 0−+3 V, 70 H_2 for 20 s; 2. +1.5 V DC for 5 s; 3. −0.9 V DC for 3 s; 4. +1.5 V DC for 5 s.

(b) 1. Triangular waveform 0−+3 V, 70 Hz for 20 s; 2. 0−+2 V, 70 Hz for 20 s; 3. 0−+1 V, 70 Hz for 20 s. (For details of electrical pre-treatment, see Gonon *et al.* 1984; Sharp *et al.* 1984)

be able to monitor basal dopamine or serotonin in other less intensely innervated regions.

Working electrodes are made from 2 or 3 pyrolytic carbon fibres (Le Carbone Lorraine 8 m OD ref AFT/F) supported in a pulled glass capillary (1.2 mm OD 0.69 mm ID) as described by Ponchon, Cespuglio, Gonon, Jouret and Pujol (1979) and Gonon, Fombarlet, Buda and Pujol (1981b). We have simplified the construction of these electrodes by making the electrical contact using conductive paint strengthened with polyester resin which is also used to seal the electrode tip. The carbon fibres protrude 0.5 mm beyond the end of the glass capillary. These working electrodes are used together with a miniature Ag/AgCl reference electrode and platinum auxillary electrode (Sharp *et al.* 1984a).

To obtain electrochemical separation between compounds it is essential to electrically pre-treat these electrodes prior to implantation. Slightly different pre-treatment conditions are used for electrodes that are to be used to monitor ascorbic acid (AA) and DOPAC from those for 5-HIAA measurements (Fig. 9.9). Full details of the pre-treatment conditions are given in Sharp *et al.* (1984a) (AA+DOPAC and 5-HIAA), Gonon, Cespuglio, Buda and Pujol 1983; Gonon *et al.* 1984 (AA+DOPAC) and Cespuglio *et al.* (1984) (5-HIAA). Basically the electrical pre-treatment

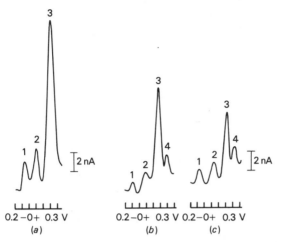

Fig. 9.10   Electrically pre-treated electrodes able to separate 4 peaks *in vivo*. Peak 1, Ascorbic acid; Peak 2, DOPAC; Peak 3, 5-HIAA+(uric acid); Peak 4, HVA. (*a*) Separation of ascorbic acid, DOPAC and 5-HIAA *in vitro*. (*b*) Same electrode implanted into the rat striatum. Note the 4 oxidation peaks. (*c*) The same electrode in the striatum following intrastriatal injection of uricase. Note the decrease in the height of Peak 3 and the improved separation between Peaks 2 and 3. (Data from Crespi, Sharp, Maidment and Marsden, in press)

involves the application to the electrode of a triangular potential wave-form (0.0–+3, 2 or 1.5 V) followed by the application of fixed potentials (i.e. −0.9 V then +1.5 V each for 3 s for AA and DOPAC separation). More recently we have shown that slight modification to the conditions used for the 5-HIAA electrodes allows the detection of AA, DOPAC, 5-HIAA and HVA with the single electrode (Fig. 9.10). Unfortunately, the clear separa-tion of the four compounds can only be maintained for up to six hours *in vivo*, while the original 5-HIAA electrode is stable for up to 48 h (Faradji, Cespuglio and Jouvet 1983). Prolonged recordings in rats have been described using a miniature micro-manipulator which allows electrodes to be replaced *in vivo* (Cespuglio *et al*. 1984). The fibre electrodes are implanted into specific brain regions using conventional techniques though care must be taken to avoid bleeding as blood contamination can cause falsely high 5-HIAA levels around the electrode surface. Care also needs to be taken to avoid damaging the electrode while penetrating the dura. For DPV recordings in anaesthetized preparations the electrodes are simply cemented in place and connected to the polarograph (e.g. Princeton, Brucker, Metrohm and Tacussel) and the output from this recorded on a flat-bed chart recorder or X-Y plotter. Chronoamperometric systems are not commercially available but are relatively simple to make (Lindsey, Kissort, Justice, Salamone and Neill 1980; Cheng, White and

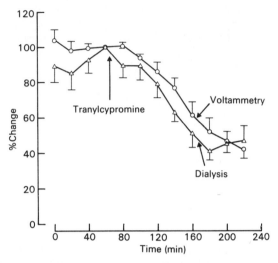

Fig. 9.11 Comparison of the effects of tranylcypromine (10 mg/kg) on the 5-HIAA in striatal dialysis samples and the height of Peak 3 (5-HIAA). Dialysis and voltammetry were performed simultaneously in the same animal. Results are the mean of 5 experiments ±S.E.M. Note the similarity in the decreases after the MAO inhibitor using the two methdds. (Data from Sharp *et al.* 1984*a*)

Adams 1980). With freely moving animals the only major difference is the need for a permanent head connector (e.g. Plastic Products, Roanoke, USA) and an electrically 'noise' free swivel system (e.g. 'metal brush' type).

Evidence that the carbon fibre electrodes pre-treated appropriately are able to monitor extracellular 5-HIAA changes has come from two main approaches. The first is pharmacological manipulation of the presumed 5-HIAA peak (Peak 3) (e.g. Cespuglio, Faradji, Ponchon, Buda, Riou, Gonon, Pujol and Jouvet 1981*a, b*; Crespi, Cespuglio and Jouvet 1982; Lamour, Rivot, Pointis and Ory-Lavollee 1983; Rivot, Ory-Lavollee and Chiang 1983*b*; Faradji *et al.* 1983; Cespuglio *et al.* 1984). These studies are summarized in Table 9.3 and while the results generally support the view that the indoleamine contributing to Peak 3 is 5-HIAA and not serotonin, there are some queries. The decrease in Peak 3 in the striatum after 5,7-DHT lesions, monoamine oxidase inhibition (Fig. 9.11) and 5-HTP decarboxylase inhibition is usually about 20 per cent less than the decrease in tissue 5-HIAA (Cespuglio *et al.* 1984). Conversely the increase in Peak 3 after 5-HTP (25 mg/kg) was considerably less than the increase in tissue 5-HIAA (Cespuglio *et al.* 1981*b*) (Table 9.3) while there was no change in Peak 3 following L-tryptophan (100 mg/kg) (Cespuglio *et al.* 1981*b*) although striatal 5-HIAA levels increase (Knott and Curzon 1974) and 5-HIAA increases in striatal perfusates (Sarna *et al.* 1984).

The differences between the change in Peak 3 and the biochemical data might suggest some other compound also contributes to the oxidation peak. Another explanation is that while the voltammetric electrodes measure extracellular 5-HIAA the tissue measurements include intra- and extracellular metabolites and that the drugs in Table 9.3 have a greater effect on intra- rather than extracellular metabolite levels. This latter view is supported by the finding that when intracerebral dialysis and voltammetry were performed simultaneously in the striatum of the same rat there was a very close correlation between the change in Peak 3 and the decrease in 5-HIAA in the perfusates after tranylcypromine (10 mg/kg) (Fig. 9.11) with maximal decreases of 58 per cent and 54 per cent respectively (Sharp

Table 9.3   Changes in Peak 3 recorded with carbon fibre electrodes in the striatum following lesions or drug administration

	Voltammetry* (%)	Tissue 5-HIAA† (%)
Treatments that increase Peak 3		
Stimulation of the raphe nuc (30 min)	+15	+40‡
L-Tryptophan (100 mg/kg)	+10	+32**
5-HTP (25 mg/kg)	+55	+230
	+97	
Reserpine (10 mg/kg)	+20	+143
Probenecid (200 mg/kg)	+20	+47
Uric acid (10 μg intra striatum)	+380	nd
Treatments that decrease Peak 3		
Medial forebrain bundle lesion	−65	−85
5,7-DHT lesion	−64	−83
p-Chlorophenylalamine (400 mg/kg)	−72	−94
Decarboxylase inhibitors		
(*a*) NSD 1015 (50 mg/kg)	−45	−52
(*b*) RO4.4602 (800 mg/kg)	−50	−35
MAO inhibitor		
(*a*) Clorgyline (10 mg/kg)	−40	−55
(*b*) Tranylcypromine (10 mg/kg)	−58	−82
Uricase (0.2 units)	−32	nd
Uricase (0.2 units)+pargyline (75 mg/kg)	−100	nd

Data from: * Cespuglio *et al.* (1981*b*, 1984), Crespi *et al.* (1984), Sharp *et al.* (1984*a*); †Biochemical assays of tissue levels (HPLC+ECD or fluorimetry); ‡Marsden and Curzon (1976); **Knott and Curzon (1974); nd, not done.
Comparable changes in Peak 3 have been observed in the suprachiasmatic nucleus (Faradji *et al.* 1983; Martin and Marsden 1984), dorsal raphe nucleus (Crespi *et al.* 1982; Echizen and Freed 1983 (carbon paste electrodes)), dorsal horn of the spinal cord (Rivot *et al.* 1983) and (Lamour *et al.* 1983) frontal cortex (Brazell and Marsden, unpublished). Kennett and Joseph (1982) using carbon paste electrodes with linear ramp measurements in the hippocampus found a decrease with PCPA, increases with L-tryptophan, *p*-chloroamphetamine, fluoxetine and probenecid but no change with MAO inhibitors (pargyline or nialamide).

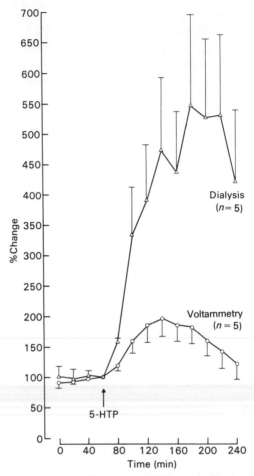

Fig. 9.12   Comparison of the effects of 5-HTP (25 mg/kg) on the 5-HIAA in striatal dialysis samples and the height of Peak 3 (5-HIAA). Dialysis and voltammetry were performed simultaneously in the same animal. Results are the mean of 5 experiments ±S.E.M. Note the rise in both the dialysis 5-HIAA and Peak 3 but the marked difference in the maximal response.

et al. 1984*a*). In contrast, however, are the results obtained following 5-HTP (25 mg/kg) administration. In this case although both Peak 3 and 5-HIAA in the perfusates increased there was a large difference in the maximal response (+97 per cent in Peak 3 and +447 per cent 5-HIAA) (Fig. 9.12) (Sharp *et al.* 1984*a*).

The failure of Peak 3 to keep pace with the increased 5-HIAA levels may in part relate to the sensitivity of the electrodes to such high concentrations

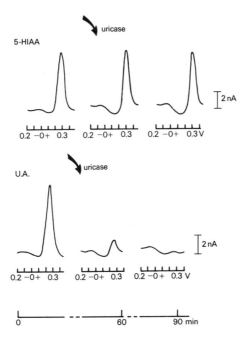

Fig. 9.13   The *in vitro* oxidation of 5-HIAA and uric acid (100 μM) and the effect of uricase (0.2 units) on this oxidation using carbon fibre electrodes. Note the similarity of the oxidation potential of 5-HIAA and uric acid and the abolition of the uric acid oxidation following addition of uricase but the absence of effect of this enzyme on 5-HIAA oxidation. (Redrawn, with permission from Crespi *et al.* 1984)

of 5-HIAA as the electrodes only show a linear response over a concentration range of 5–100 μM and above this the response flattens. Alternatively, the pool of 5-HIAA sampled by dialysis may differ from that sampled by voltammetry with damaged tissue pools making a significant contribution to the former. However, there still remains the possibility that some other compound contributes to Peak 3 and the main candidate is uric acid which is found in high concentrations in the brain extracellular space (Zetterström *et al.* 1983). Recently it has been shown (Crespi, Sharp, Maidment and Marsden 1984) that uric acid and 5-HIAA have similar oxidation potentials and that uricase, which converts uric acid to the electroactively inert allantoin, abolishes the uric acid oxidation peak *in vitro* (Fig. 9.13). Intrastriatal injection of uric acid (10 μg) increases the height of the striatal Peak 3 (Fig. 9.14, Table 9.2) while injection of uricase decreased Peak 3 by about 30 per cent and when this is followed by injection of the MAO inhibitor pargyline (75 mg/kg) the peak is abolished within the subsequent 3 h (Fig. 9.14). These results strongly suggest that uric acid does contribute about 30 per cent to Peak 3 recorded with electrochemically pre-treated fibre electrodes and so this needs to be considered when interpreting changes in this peak. Interestingly, the separation of Peaks 2 (DOPAC) and 3 in the electrodes pre-treated to record AA, DOPAC, 5-HIAA and HVA is improved following intracerebral injection of uricase (Fig. 9.10) further indicating the contamination of Peak 3 with uric acid. Data is now required on drug-induced effects on brain uric acid levels, as these might explain some of the differences between changes in Peak 3 and 5-HIAA levels, as well as information on the origins of brain uric acid (i.e. is it produced within the brain or peripherally?). Another suggested contaminant of the 5-HIAA Peak is glutathione (O'Neill *et al.* 1982*a*) although this study used carbon paste electrodes which have different characteristics from fibres.

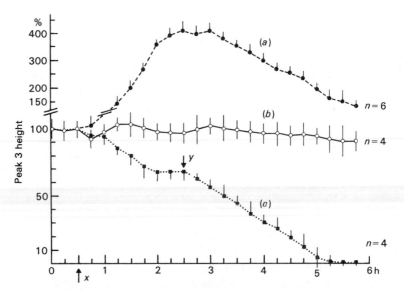

Fig. 9.14   Graph showing the effect of (*a*) intrastriatal injection of uric acid (10 µg) on the height of the striatal Peak 3 (injection at upward ↑ labelled *x*). (*b*) Intrastriatal injection of vehicle. (*c*) Intrastriatal injection of uricase (0.2 units) followed by pargyline (75 mg/kg) (injected at downward ↓ labelled *y*). Note the increase in Peak 3 following uric acid and the 30% decrease after uricase with the total loss of the peak when uricase was following by pargyline. (Redrawn, with permission from Crespi *et al.* 1984)

The original carbon paste electrodes suffered from an inability to separate ascorbate from catechol oxidation or more particularly dopamine from DOPAC oxidation (for details of the early studies see Adams and Marsden 1982). Recent work has demonstrated two ways in which paste electrodes may be converted into ion selective electrodes able to detect dopamine in the striatum but not either of the acids ascorbate or DOPAC. The first type is the stearate modified electrode (Yamamoto, Lane and Freed 1982; Keller, Stricker and Zigmond 1983; Blaha and Lane 1984) and the second type is the nafion coated (a polysulphonated derivative of teflon) electrode (Gerhardt, Oke, Nagy and Adams 1984). With the first type the inclusion of stearate in the paste electrostatically inhibits access of anions while the nafion is highly permeable to cations (i.e. dopamine at physiological pH) but impermeable to anions (i.e. ascorbic acid and DOPAC). These ideas have yet to be adopted for making electrodes that are sensitive to serotonin but insensitive to 5-HIAA. There are, however, certain problems and limitations to this approach. Adsorbtion of the catecholamine on to the electrode surface may be a greater problem with these selective electrodes so that the electrode sensitivity and diffusion between the electrode and surrounding tissue will alter during *in vivo* recordings. More problematical is the sensitivity required for electrodes able to measure either dopamine or serotonin in regions other than those with extremely rich innervation of the transmitter of interest (i.e. dopamine in the striatum). More attention is needed to improve the sensitivity of the *in vivo* voltammetric electrodes.

In summary, *in vivo* voltammetry at present offers the possibility of making rapid and repeated measurements of 5-HIAA in various brain regions. However, there remains the problem of a contribution (30 per cent) from brain uric acid to the '5-HIAA' peak recorded. In the future there may be serotonin selective electrodes but practical application of such electrodes will have to be accompanied by improved electrode sensitivity as basal extracellular serotonin concentrations are below the sensitivity of existing carbon paste or fibre electrodes.

9.4   Applications of *in vivo* measurement techniques

9.4.1   *Anaesthetized v. unanaesthetised preparations*

It is now clear that there are differences in drug responses both in terms of amine release, turnover and neuronal firing between the anaesthetized and freely moving animal. For example, chloral hydrate only produces a small decrease in the firing of cat dorsal raphe neurones but prevents their excitation by visual or auditory stimuli while augmenting their inhibition by an α_1-receptor antagonist (Heym, Steinfels and Jacobs 1984). With the marked improvement in the techniques available it would seem appropriate that most *in vivo* measurements of either metabolism or release were performed with freely moving animals. The exceptions might include strictly pharmacological studies such as identification of receptor sub-types or micromapping studies. Most of the applications discussed will centre on measurements in the unanaesthetized animal.

9.4.2   Micro-mapping of serotonin systems in brain regions

The only one of the techniques suitable for detailed mapping of the serotonin innervation of specific brain areas is *in vivo* voltammetry using the eight carbon fibre electrodes. Crespi *et al.* (1982) have used the technique to map the 5-HIAA distribution within the dorsal, medial, pontis and magnus raphe nuclei demonstrating that the maximum 5-HIAA peak coincides with the largest concentration of serotonin cell bodies. Lamour and coworkers (1983) used carbon fibres to map the 5-HIAA signal in the somatosensory cortex and found a heterogenous distribution with the maximum 5-HIAA peak in the superficial cortex. Stimulation of the dorsal raphe increased the 5-HIAA peak but this increase was absent in *p*-CPA treated rats (Rivot *et al.* 1983*a*). These results indicate a serotonergic pathway from the dorsal raphe to the superficial layer of the somatosensory cortex, the function of which could be to modulate the activity of superficial cortical cells or neurones in deeper layers with dendrites reaching the superficial area.

9.4.3   Transmitter interactions

There are numerous *in vitro* release studies investigating the effects of amines and other neurotransmitter substances on serotonin release (e.g. Reubi, Emson, Jessell and Iversen 1978) and *in vivo* push–pull studies in the encephalé isolé cat measuring release of [³H]-serotonin have demonstrated reduced release in the substantia nigra and striatum following intranigral administration of dopamine or L-glutamic acid (Héry, Soubrié, Bourgoin, Motastruc, Artaud and Glowinski 1980; Reisine *et al.* 1982) and increased release in response to Substance P (Reisine *et al.* 1982). With the present methods for measuring endogenous release or metabolism in the freely moving animal, it should in future be possible to monitor changes in serotonin release, concurrently with other amine transmitters, in response to administered transmitters and drugs. Relatively few studies have so far been performed along these lines especially with regard to serotonin release. Thyrotrophin releasing hormone (TRH) and its stable analogues increase endogenous dopamine release *in vitro* in the accumbens but not the striatum (Sharp, Bennett and Marsden 1982). With *in vivo* voltammetry CG3509, a stable analogue of TRH, also selectively increases the DOPAC oxidation peak in the nuc accumbens but not in the striatum without altering the 5-HIAA peak (Sharp, Brazell, Bennett and Marsden 1984*b*).

9.4.4   *Pharmacological identification* in vivo *of serotonin receptor responses*

While there have been numerous studies investigating the effects of peripherally administered drugs on catecholamine and indoleamine release and metabolism to validate the methods as already described in previous sections (i.e. Table 9.2), fewer studies have used the techniques to identify responses associated with particular amine receptors (Maidment and Marsden 1983). The distinction between serotonin receptor sub-types (see Chapter 4) is mainly based on ligand binding studies although increasing effort is being made to correlate binding with functional data such as behaviour or serotonin release. Recently, Baumann and Waldmeier (1984) have compared changes in the 5-HIAA voltammetric peak with results using release of [^{3}H]-serotonin from electrically stimulated cortical slices *in vitro* (Baumann and Waldmeier 1981) in response to serotonin agonists and antagonists. In general, there was a good correlation between the voltammetric and *in vitro* data with the serotonin receptor antagonists increasing 5-HIAA and increasing [^{3}H]-serotonin release while the agonists decreased 5-HIAA and release. These results indicate that *in vivo* voltammetry could provide a relatively simple means of detecting the action of drugs in specific brain regions that alter serotonin autoreceptor activity.

We have shown that the 5-HT$_1$ receptor agonist 5-methoxy-3(1,2,3,6-tetrahydropyridine-4-yl)H-indole (RU 24969) decreases the 5-HIAA peak in the frontal cortex (Fig. 9.15) and the suprachiasmatic nucleus. The decrease in the front cortex correlates well with changes in 5-HIAA in dialysis perfusates (Fig. 9.15) and decreased release of [^{3}H]-serotonin from cortical slices (Routledge and Marsden 1984; Marsden, Maidment, Brazell and Sharp 1984*b*; Marsden *et al.* in preparation), again suggesting that measurement of extracellular 5-HIAA is an index of serotonin neuronal function.

9.4.5   *Monitoring changes in serotonin release and metabolism in response to physiological and behavioural stimuli*

Both CSF metabolite measurements and *in vivo* voltammetry have been used to monitor circadian changes in serotonin turnover. The turnover of both dopamine and serotonin, measured by sampling cisternal CSF is significantly greater during the dark (active period) than during the light period, although the basal metabolite (5-HIAA and DOPAC and HVA) levels were not significantly different (Hutson, Sarna and Curzon 1984*b*). Measurements of the 5-HIAA peak over 24 h periods have also been made

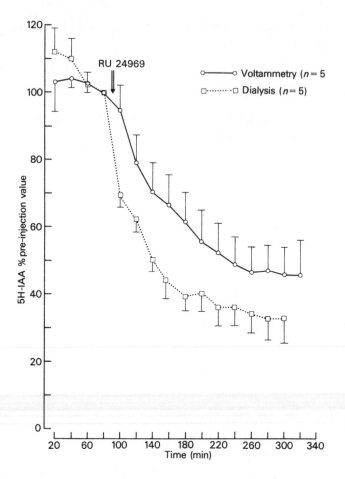

Fig. 9.15 Effect of the 5-HT$_1$ receptor agonist RU 24969 on the 5-HIAA oxidation peak and 5-HIAA in dialysis perfusates from the rat frontal cortex.

in the suprachiasmatic nucleus, the nucleus considered to regulate circadian variations, by Faradji *et al.* (1983) who found that not only was there a circadian variation with low 5-HIAA during the light period, in agreement with previous tissues studies (Héry, Rouer and Glowinski 1972), but also spontaneous changes in the 5-HIAA peak that corresponded with periods of activity indicating higher serotonin turnover and release in the active rat (Puizillot, Gaudin-Chazal, Daszuta, Seyfritz and Ternaux 1979). Other *in vivo* voltammetric studies in rat have indicated increased dopamine turnover and extracellular ascorbic acid during the dark phase (O'Neill, Fillenz and Albery 1982*b*), again suggesting a relationship between increased dopaminergic function and motor activity (Iversen 1977).

The development of the *in vivo* techniques to a state where continuous recordings in the freely moving animal can be made over long periods will markedly alter the possibility of making correlations between behaviour and endogenous serotonin release. At present only a few studies using the techniques described along these lines have been attempted. Immobilization has been shown to increase an 'indole' oxidation peak in the rat (Kennett and Joseph 1982) which the authors related to increased brain tryptophan. These results however, need to be repeated as the original observations were made with unmodified carbon fibre electrodes. Other voltammetric studies have indicated increases in dopaminergic activity in the striatum following various environmental stimuli (Keller *et al.* 1983).

9.5   Conclusions

There are now several methods for measuring whole and regional brain serotonin turnover and release. Considerable effort has been put into establishing these methods and validating the measurements made. They have been used for only a few studies either investigating the physiological role or the detailed regional pharmacology of the serotonin neurones in the brain. It is these areas that should be the most rewarding areas for future work using the *in vivo* techniques especially combined with measurement of unit activity in the freely moving animal (e.g. Heym, Steinfels and Jacobs 1982). The main technical advantage will come with the development of dialysis or electrode systems that give stable baseline readings over prolonged periods. Such systems will allow the investigation of the chronic effects of drugs on serotonin release combined with behavioural measurements.

Acknowledgements

Studies from the author's laboratory were supported financially by the Wellcome Trust, Medical Research Council and ICI plc. I should like to thank N. T. Maidment, M. P. Brazell, A. Nisbet, F. Crespi and Carol Routledge for providing the unpublished data contained in this chapter.

References

Adams, R. N. and Marsden, C. A. (1982). Electrochemical detection methods for monoamine measurements *in vitro* and *in vivo*. In *Handbook in psychopharmacology* (eds. L. L. Iversen, S. D. Iversen and S. H. Snyder) Vol. 15, p. 1. Academic Press, New York.

Baker, H. F. and Ridley, R. M. (1979). Increased HVA levels in primate ventricular CSF following amphetamine administration. *Brain Res.* **167**, 206.

Baumann, P. A. and Waldmeier, P. C. (1981). Further evidence for negative feedback control of serotonin release in the central nervous system. *Naunyn-Schmiedeberg's Arch. Pharmac.* **317**, 36.

—— —— (1984). Negative feedback control of serotonin release *in vivo:* Comparison of 5-hydroxyindoleacetic acid levels measured by voltammetry in conscious rats and by biochemical techniques. *Neuroscience* **11**, 195.

Blaha, C. D. and Lane, R. F. (1984). Direct *in vivo* electrochemical monitoring of dopamine release in response to neuroleptic drugs. *Eur. J. Pharmac.* **98**, 113.

Brazell, M. P. and Marsden, C. A. (1982*a*). Differential pulse voltammetry in the anaesthetized rat: identification of ascorbic acid, catechol and indoleamine oxidation peaks in the striatum and frontal cortex. *Br. J. Pharmac.* **75**, 539.

—— —— (1982*b*). Intracerebral injection of ascorbate oxidate—effect on *in vivo* electrochemical recordings. *Brain Res.* **249**, 167.

Cespuglio, R., Faradji, H., Hahn, Z. and Jouvet, M. (1984). Voltammetric detection of brain 5-hydroxyindoleamines by means of electrochemically treated carbon fibre electrodes. Chronic recordings for up to one month with moveable cerebral electrodes in the sleeping or waking rat. In *Measurement of neurotransmitter release* in vivo (ed. C. A. Marsden). John Wiley, Chichester. (In press).

—— —— Ponchon, J. L., Buda, M., Riou, F., Gonon, F., Pujol, J.-F. and Jouvet, M. (1981*a*). Differential pulse voltammetry in brain tissue: I. Detection of 5-hydroxyindoles in the rat striatum. *Brain Res.* **223**, 287.

—— —— —— —— —— —— —— —— (1981*b*). Differential pulse voltammetry in brain tissue: II. Detection of 5-hydroxyindoleacetic acid in the rat striatum. *Brain Res.* **223**, 299.

Cheng, H-Y., Schenk, J., Huff, R. and Adams, R. N. (1979). *In vivo* electrochemistry: behaviour of micro-electrodes in brain tissue. *J. Electroanalyt. Chem.* **100**, 23.

—— White, W. and Adams, R. N. (1980). Microprocessor controlled apparatus for *in vivo* electrochemical measurement. *Anal. Chem.* **54**, 1384.

Crespi, F., Cespuglio, R. and Jouvet, M. (1982). Differential pulse voltammetry in brain tissue. III. Mapping of the rat serotoninergic raphe nuclei by electrochemical detection of 5HIAA. *Brain Res.* **270**, 45.

—— Sharp, T., Maidment, N. and Marsden, C. A. (1984). Differential pulse voltammetry *in vivo*—evidence that uric acid contributes to the indole oxidation peak. *Neurosci. Letters* **43**, 203.

——— ——— ——— ——— (1984). Differential pulse voltammetry: simultaneous *in vivo* measurement of ascorbic acid, catechols and 5-hydroxyindoles in the rat striatum. *Brain Res.* (In press).

Conti, J. C., Strope, E., Adams, R. N. and Marsden, C. A. (1978). Voltammetry in brain tissue: chronic recording of stimulated dopamine and 5-hydroxytryptamine release. *Life Sci.* **23**, 2705.

Danguir, J., Le Quan-Bui, K. H., Elghozi, J. L., Devynck, M. A. and Nicolaidis, S. (1982). LCEC monitoring of 5-hydroxyindolic compounds in the cerebrospinal fluid of the rat related to sleep and feeding. *Brain Res. Bull.* **8**, 293.

Dedek, J., Baumes, R., Tien-Duc, N., Gomeni, R. and Korf, J. (1979). Turnover of free and conjugated (sulphonyloxy) dihydroxyphenylacetic acid and homovanillic acid in rat striatum. *J. Neurochem.* **33**, 687.

Degrell, I., Zenner, K., Kummer, P. and Stock, G. (1983). Monoamine metabolites in the CSF of conscious unrestrained cats. *Brain Res.* **277**, 283.

Elchisak, M. A., Maas, J. W. and Roth, R. H. (1977). Dihydroxyphenylacetic acid conjugate: natural occurrence and demonstration of probenecid-induced accumulation in rat striatum, olfactory tubercles and frontal cortex. *Eur. J. Pharmac.* **41**, 369.

Elghozi, J. L., Le Quan-Gui, K. H., Earnhardt, J. T., Meyer, P. and Devynck, M. A. (1981). *In vivo* dopamine release from the anterior hypothalamus of the rat. *Eur. J. Pharmac.* **73**, 199.

Faradji, H., Cespuglio, R. and Jouvet, M. (1983). Voltammetric measurements of 5-hydroxyindole compounds in the suprachiasmatic nuclei: Circadian fluctuations. *Brain Res.* **279**, 111.

Gaddum, J. H. (1961). Push–pull cannulae. *J. Physiol. (Lond.)* **155**, 1P.

Garelis, E., Young, S. N., Lal, S. and Sourkes, R. L. (1974). Monoamine metabolites in lumbar CSF: The question of their origin in relation to clinical studies. *Brain Res.* **79**, 1.

Gerhardt, G. A., Oke, A. F., Nagy, G. and Adams, R. N. (1984). Nafion-coated electrodes with high selectivity for CNS electrochemistry. *Brain Res.* **290**, 390.

Glowinski, J. (1981). *In vivo* release of transmitters in the cat basal ganglia. *Fed. Proc.* **40**, 135.

Gonon, F., Buda, M. and Pujol, J-F. (1984). Treated carbon fibre electrodes for measuring catechols and ascorbic acid. In *Measurement of neurotransmitter release* in vivo. (ed. C. A. Marsden) p. 553. John Wiley, Chichester.

——— Cespuglio, R., Buda, M. and Pujol, J-F. (1983). *In vivo* electrochemical detection of monoamine derivatives. In *Methods in biogenic amine research.* (eds. H. Parvez, S. Parvez and I. Nagatsu). Elsevier, Amsterdam.

—— Fombarlet, C. M., Buda, M. J. and Pujol, J-F. (1981b). Electrochemical treatment of pyrolytic carbon fibre electrodes. *Anal. Chem.* **53**, 1386.

—— Buda, M., Cespuglio, R., Jouvet, M. and Pujol, J-F. (1980). *In vivo* electrochemical detection of catechols in the neostriatum of anaesthetised rats: dopamine or DOPAC? *Nature (Lond.)* **286**, 902.

—— —— —— —— —— (1981a). Voltammetry in the striatum of chronic freely moving rats: detection of catechols and ascorbic acid. *Brain Res.* **223**, 69.

Gordon, E. K., Markey, S. P., Sherman, R. L. and Kopin, I. J. (1976). Conjugated 3,4-dihydroxyphenylacetic acid (DOPAC) in human and monkey cerebro-spinal fluid and rat brain and the effects of probenecid treatment. *Life Sci.* **18**, 1285.

Guldberg, H. C. and Yates, C. M. (1968). Some studies of the effect of chlorpromazine, reserpine and dihydroxyphenylacetic acid and 5-hydroxy-3-ylacetic acid in ventricular cerebrospinal of the dog using the technique of serial smmpling of the cerebrospinal fluid. *Br. J. Pharmac.* **33**, 457.

Hernandez, L., Paez, X. and Hamlin, C. (1983). Neurotransmitter extraction by local intracerebral dialysis in anaesthetized rats. *Pharmac. Biochem. Behav.* **18**, 159.

Héry, F., Rouer, E. and Glowinski, J. (1972). Daily variations in serotonin metabolism in the rat brain. *Brain Res.* **43**, 445.

—— Simonnet, G., Bourgoin, S., Soubrié, P., Artaud, F., Hamon, M. and Glowinski, J. (1979). Effect of nerve acvitiy on the *in vivo* release of ^{3}H-serotonin continuously formed from L-^{3}H-tryptophan in the caudate nucleus of the cat. *Brain Res.* **169**, 317.

—— Soubrié, P., Bourgoin, S., Motastruc, J. L., Artuaud, F. and Glowinski, J. (1980). Dopamine released from dendrites in the substantia nigra controls and nigral striatal release of serotonin. *Brain Res.* **193**, 143.

Heym, J., Steinfels, G. F. and Jacobs, B. L. (1982). Activity of serotonin-containing neurones in the nucleus raphe pallidus of freely moving cats. *Brain Res.* **251**, 259.

—— —— —— (1984). Chloral hydrate anaesthesia alters the responsiveness of central serotonergic neurones in the cat. *Brain Res.* **291**, 63.

Holloway, J. A. (1975). Norepinephrine and serotonin: specificity of release with rewarding electrical stimulation of the brain. *Psychopharmacology* **42**, 127.

Holman, R. B. and Vogt, M. (1972). Release of 5-hydroxytryptamine from caudate nucleus and septum. *J. Physiol. (Lond.)* **223**, 243.

Hutson, P. H., Sarna, G. S. and Curzon, G. (1984b). Determination of daily variation of brain 5-hydroxytryptamine and dopamine turnovers and of the clearance of their acidic metabolites in conscious rats by repeated sampling of cerebrospinal fluid. *J. Neurochem.* (In press).

—— —— Kantamaneni, B. D. and Curzon, G. (1984*a*). Concurrent determination of brain dopamine and 5-hydroxytryptamine turnovers in individual freely moving rats using repeated sampling of cerebrospinal fluid. *J. Neurochem.* **43**, 151.

Iversen, S. D. (1977). Brain dopamine systems and behaviour. In *Handbook of psychopharmacology* (eds. L. L. Iversen, S. D. Iversen and S. H. Snyder) Vol. 8, p. 333. Plenum Press, New York.

Johnson, R. D. and Justice, J. B. (1983). Model studies for brain dialysis. *Brain Res. Bull.* **10**, 567.

Joseph, M. H. and Kennett, G. A. (1983). Stress-induced release of 5HT in the hippocampus and its dependence on increased tryptophan availability. An *in vivo* electrochemical study. *Brain Res.* **270**, 251.

Karoum, F., Chuang, L-W. and Wyatt, R. J. (1983). Biochemical and pharmacological characteristics of conjugated catecholaemines in the rat brain. *J. Neurochem.* **40**, 1735.

Keller, R. W., Stricker, E. M. and Zigmond, M. J. (1983). Environmental stimuli but not homeostatic challenges produce apparent increases in dopaminergic activity in the striatum: An analysis by *in vivo* voltammetry. *Brain Res.* **279**, 159.

Kennett, G. A. and Joseph, M. H. (1982). Does *in vivo* voltammetry in the hippocampus measure 5HT release? *Brain Res.* **236**, 305.

Knott, P. J. and Curzon, G. (1974). Effect of increased rat brain trytophan on 5-hydroxytryptamine and 5-hydroxyindolylacetic acid in the hypothalamus and other brain regions. *J. Neurochem.* **22**, 1065.

Lamour, Y., Rivot, J. P., Pointis, D. and Ory-Lavollee, L. (1983). Laminar distribution of serotonergic innervation in rat somatosensory cortex, as determined by *in vivo* electrochemical detection. *Brain Res.* **259**, 163.

Le Quan-Bui, K. H., Elghozi, J. L., Devynck, M. A. and Meyer, P. (1982). Rapid liquid chromatographic determination of 5-hydroxyindoles and dihydroxyphenylacetic acid in cerebrospinal fluid of the rat. *Eur. J. Pharmac.* **81**, 315.

Lindsey, W. S., Kissort, B. L., Justice, J. B., Salamone, J. D. and Neill, D. B. (1980). An automated electrochemical method for *in vivo* monitoring of catecholamine release. *J. Neurosci. Methods* **2**, 373.

Loullis, C. C., Hingten, J. N., Shea, P. A. and Aprison, M. H. (1980). *In vivo* determination of endogenous biogenic amines in rat brain using HPLC and push–pull cannula. *Pharmac. Biochem. Behav.* **12**, 959.

Maidment, N. T. and Marsden, C. A. (1983). *In vivo* changes in dopamine metabolism in the rat nucleus accumbens after infusion of haloperidol into the ventral tegmental area. *Prog. Neuro.-Psychopharmac. Biol. Psychiat.* **S1**, 214.

Marsden, C. A. (1984) (ed.). *Measurement of neurotransmitter release* in vivo. John Wiley, Chichester.

—— and Routledge, C. (1984). *In vivo* measurements of DOPAC, 5HIAA and 5HT in specific brain regions by intracerebral dialysis. *Br. J. Pharmac.* (In press).

—— Brazell, M. P. and Maidment, N. T. (1984*a*). An introduction to *in vivo* electrochemistry. In *Measurement of neurotransmitter release* in vivo (ed. C. A. Marsden) p. 127. John Wiley, Chichester.

—— Maidment, N. T., Brazell, M. P. and Sharp, T. (1984*b*). Application of *in vivo* voltammetry to the development of new drugs. *Clin. Neuropharmac.* **7**, Supp. 1, S197.

—— Conti, J., Strope, E., Curzon, G. and Adams, R. N. (1979). Monitoring 5-hydroxytryptamine release in the brain of the freely moving unanaesthetised rat using *in vivo* voltammetry. *Brain Res.* **171**, 85.

Martin, K. F. and Marsden, C. A. (1984). *In vivo* diurnal variations of 5-HT release in hypothalamic nuclei. In *Circadian rhythms in the CNS* (eds. P. H. Redfern, J. A. Davis, I. C. Campbell and K. F. Martin). Macmillan Press, Basingstoke. (In press).

Mefford, I. N. (1981). Application of high performance liquid chromatography with electrochemical detection to neurochemical analysis. Measurement of catecholamines, serotonin and metabolites in rat brain. *J. Neurosci. Methods* **3**, 207.

—— Oke, A. F. and Adams, R. N. (1981). Regional distribution of ascorbate in human brain. *Brain Res.* **212**, 223.

Myers, R. D. and Beleslin, D. R. (1971*a*). Changes in serotonin release in hypothalamus during cooling and warming in the monkey. *Am. J. Physiol.* **220**, 1746.

Myers, R. D. and Beleslin, D. R. (1971*b*). The spontaneous release of 5-hydroxytryptamine and acetylcholine within the diencephalon of the unanaesthetised rhesus monkey. *Exp. Brain Res.* **11**, 539.

Nielsen, J. A. and Moore, K. E. (1982). Measurements of metabolites of dopamine and 5-hydroxytryptamine in cerebroventricular perfusates of unanaesthetised freely moving rats: selective effects of drugs. *Pharmac. Biochem. Behav.* **16**, 131.

O'Neill, R. D., Fillenz, M. and Albery, W. J. (1982*b*). Circadian changes in homovanillic acid and ascorbate levels in the rat striatum using microprocessor controlled voltammetry. *Neurosci. Letters* **34**, 189.

—— Grunewald, R. A., Fillenz, M. and Albery, W. J. (1982*a*). Linear sweep voltammetry with carbon paste electrodes in the rat striatum. *Neuroscience* **7**, 1945.

Philippu, A. (1984). Use of push–pull cannulae to determine the release of endogenous neurotransmitters in distinct brain areas of anaesthetised and freely moving animals. In *Measurement of neurotransmitter release* in vivo (ed. C. A. Marsden) p. 3. John Wiley, Chichester.

Philips, S. R., Robson, A. M. and Boulton, A. A. (1982). Unstimulated and amphetamine-stimulated release of endogenous noradrenaline and dopamine from rat brain *in vivo. J. Neurochem.* **38**, 1106.

Ponchon, J. L., Cespuglio, R., Gonon, F., Jouvet, M. and Pujol, J. F. (1979). Normal pulse polarography with carbon fibre electrodes for *in vitro* and *in vivo* determinations of catecholamines. *Anal. Chem.* **51**, 1483.

Puizillot, J. J., Gaudin-Chazal, G., Daszuta, A., Seyfritz, N. and Ternaux, J. P. (1979). Release of endogenous serotonin from encephalé isolé cats. II. Correlations with raphe neuronal activity and sleep and wakefulness. *J. Physiol. (Paris)* **75**, 531.

Reisine, T., Soubrié, P., Artaud, F. and Glowinski, J. (1982). Application of L-glutamic acid and substance P to the substantia nigra modulates *in vivo* [^{3}H]-serotonin release in the basal ganglia of the cat. *Brain Res.* **236**, 317.

Reubi, J. C., Emson, P. C., Jessell, T. M. and Iversen, L. L. (1978). Effect of GABA, dopamine and Substance P on the release of newly synthetized [^{3}H]-5-hydroxytryptamine fom rat substantia nigra *in vitro. Naunyn-Schmiedeberg's Arch. Pharmac.* **304**, 271.

Rivot, J. P., Lamour, Y., Ory-Lavollee, L. and Pointis, D. (1983*a*). *In vivo* electrochemical detection of 5-hydroxyindoles in rat somato-sensory cortex: Effect of the stimulation of the serotonergic pathways in normal and PCPA pretreated animals. *Brain Res.* **275**, 164.

—— Ory-Lavollee, L. and Chiang, C. Y. (1983*b*). Differential pulse voltammetry in the dorsal horn of the spinal cord of the anaesthetized rat: Are the voltammograms related to 5HT and/or to 5HIAA? *Brain Res.* **275**, 311.

Routledge, C. and Marsden, C. A. (1984). *In vivo* measurement of 5HT and 5HIAA by intracerebral dialysis. Effects of $5HT_1$ receptor agonists RU24969 and 8-hydroxy-2-(di-*N*-propylamino)tetralin. *9th C.I.N.P. Congress,* 1984.

Sarna, G. S., Hutson, P. H., Trickelbank, M. D. and Curzon, G. (1983). Determination of brain 5-hydroxytryptamine turnover in freely moving rats using repeated sampling of cerebrospinal fluid. *J. Neurochem.* **40**, 383.

—— —— Kantamaneni, B. D., Mootoo, S. and Curzon, G. (1984). Striatal dialysate and cisternal CSF as indices of changes in rat brain indole metabolism after a tryptophan load. *4th Eur. Winter Conf. on Brain Res., Coucheval,* March 1984.

Schenk, J. O. and Adams, R. N. (1984). Chronoamperometric measurement in the central nervous system. In *Measurement of neurotransmitter release* in vivo. (ed. C. A. Marsden). p. 193. John Wiley, Chichester.

Sharp, T., Bennett, G. W. and Marsden, C. A. (1982). Thyrotrophin-releasing hormone analogues increase dopamine release from slices of rat brain. *J. Neurochem.* **39**, 1763.

——— Brazell, M. P., Bennett, G. W. and Marsden, C. A. (1984*b*). The TRH analogue CG3509 increases *in vivo* catechol/ascorbate oxidation in the rat n. accumbens but not the striatum. *Neuropharmacology* **23**, 617.

——— Maidment, N. T., Brazell, M. P., Zetterström, T., Ungerstedt, U., Bennett, G. W. and Marsden, C. A. (1984*a*). Changes in monoamine metabolites measured by simultaneous *in vivo* differential pulse voltammetry and intracerebral dialysis. *Neuroscience* **12**, 1213.

Ternaux, J. P., Boireau, A., Bourgoin, S., Hamon, M., Héry, F. and Glowinski, J. (1976). *In vivo* release of 5HT in the lateral ventricle of the rat: Effects of 5-hydroxytryptophan and tryptophan. *Brain Res.* **10**, 533.

——— Héry, F., Hamon, M., Bourgoin, S. and Glowinski, J. (1977). 5HT release from ependymal surface of the caudate nucleus in encephale isole cats. *Brain Res.* **132**, 575.

Tossman, U. and Ungerstedt, U. (1981). Neuroleptic action on putative aminoacid neurotransmitters in the brain studied with a new technique of brain dialysis. *Neurosci. Letters* **S7**, S479.

Ungerstedt, U., Herrera-Marschitz, M., Jungnelius, U., Stahle, L., Tossman, U. and Zetterström, T. (1982). Dopamine synaptic mechanisms reflected in studies combining behavioural recordings and brain dialysis. *Adv. in the Biosciences* **37**, 219.

Wightman, R. M. and Dayton, M. A. (1982). Voltammetric techniques for the analysis of biogenic amines. In *Analysis of biogenic amines* (eds. R. T. Coutts and C. G. Baker) Vol. 4A. Elsevier, New York.

Yaksh, T. (1984). Spinal superfusion in the rat and cat. In *Measurement of neurotransmitter release* in vivo (ed. C. A. Marsden) p. 107. John Wiley, Chichester.

Yamamoto, B. K., Lane, R. F. and Freed, C. R. (1982). Normal rats trained to circle show asymmetric caudate dopamine release. *Life Sci.* **30**, 2155.

Zetterström, T. and Ungerstedt, U. (1984). Effects of apomorphine on the *in vivo* release of dopamine and its metabolites studied by brain dialysis. *Eur. J. Pharmac.* **97**, 29.

——— Sharp, T., Marsden, C. A. and Ungerstedt, U. (1983). *In vivo* measurement of dopamine and its metabolites by intracerebral dialysis: changes after *d*-amphetamine. *J. Neurochem.* **41**, 1769.

10

Involvement of serotonin in the action of hallucinogenic agents

RICHARD A. GLENNON

10.1 Introduction

Over twenty years ago, Woolley (1962) wrote that '... analogs of serotonin which cause hallucinations ... in normal man exhibit serotonin-like actions on several tissues including some brain cells.' This statement was based on the results of earlier studies by his group, and by others, on hallucinogenic and related agents such as lysergic acid diethylamide (LSD), and derivatives of N,N-dimethyltryptamine and β-carboline. A link had been forged between the neurotransmitter serotonin and the mechanism of action of certain hallucinogenic agents; yet, today, the nature of this relationship remains relatively obscure and somewhat elusive. Nevertheless, during the ensuing period of time, this relationship has gained considerable support.

The study of hallucinogenic agents is plagued by several problems, not the least being an exact classification of agents that produce this effect. The work to be discussed in this chapter will be concerned only with classical hallucinogens, i.e. agents that possess an indolealkylamine (derivatives of tryptamine, β-carboline and lysergic acid) or a simple phenalkylamine (derivatives of phenethylamine and phenylisopropylamine) backbone (Fig. 10.1). The pharmacology and structure-activity relationships of these agents have been recently reviewed (Shulgin 1982; Nichols and Glennon 1984).

Another major problem encountered in this field of study is the paucity of reliable human hallucinogenic data; this is further compounded by the subjective nature of the response being measured, which does not necessarily lend itself to a distinct endpoint, and by the general lack of clinical dose-effect studies. As a result, human potency data are relatively scarce, and what is available may be difficult to interpret. On the other hand, results of a variety of animal studies are available and, in general, such studies have used sufficient numbers of subjects, and have employed a wide

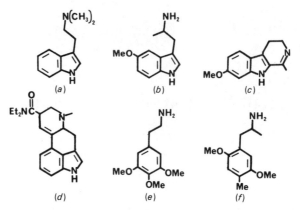

Fig. 10.1   Examples of indolealkylamine and phenalkylamine hallucinogens of the tryptamine (*a*, DMT, *b*, 5-OMe α-MeT), β-carboline (*c*, harmaline), lysergic acid (*d*, LSD), phenethylamine (*e*, mescaline) and phenylisopropylamine (*f*, DOM) subtypes.

enough range of doses such that statistical analysis of the data are possible. The problem here, of course, is that because of the subjective nature of the hallucinogenic response, it is difficult to determine, with any certainty, to what effect the animals are responding.

There are two basic approaches to studying hallucinogenic agents. The first approach is to examine one or two individual agents *in toto,* while the second is to study a large series of agents in relatively less detail. As a consequence, much is known about certain agents, such as LSD and mescaline, while comparatively less is known about others. The second approach, while suffering from several disadvantages including the lack of availability of large series of compounds, does allow for intra- and inter-class comparisons and formulation of structure-activity relationships. This approach also serves to identify interesting novel agents for subsequent more detailed examination.

The effect and potencies of hallucinogenic agents in man have been reviewed (Shulgin 1978, 1982; Glennon 1983; Nichols and Glennon, 1984). However, because studies employing animals are more readily conducted and a greater population of subjects can be employed, and because they allow a greater freedom with respect to pharmacological and neuro-chemical manipulation, such studies provide valuable information with respect to the mechanism of action of hallucinogenic agents. Hallucinogens produce a variety of effects in animals; in addition to producing certain gross behavioural effects, hallucinogenic agents have been shown to, for example, produce hyperthermia in rodents and rabbits, elicit a 'serotonin syndrome' in rodents, produce a flexor/stepping reflex in chronic spinal

dogs, produce head twitch in mice, affect fixed ratio responding ('hallucinogenic pause') by rats, produce limb flick (and other) behaviour in cats,
and serve as discriminative stimuli in stimulus control of behaviour in a
variety of animal species. It has also been demonstrated that each of the
above eight hallucinogen-induced behavioural changes can be attenuated
by co-administration or pre-treatment of the animals with serotonin antagonists (for example, see Wallach, Friedman and Gershon 1972; Huang
and Ho 1975; Martin and Sloan 1977; Martin, Vaupel, Nozaki and Bright
1978; Commissaris, Semeyn, Moore and Rech 1980; Green 1981; Jacobs
1982; Glennon, Rosecrans and Young 1983c). It is not being suggested that
any of these behavioural changes constitutes an ideal model of hallucinogenic activity, nor is it being denied that in several instances the effects
produced by a particular agent may have been antagonized by other neurotransmitter antagonists (although this is not normally found to be the case).
The point being made is that serotonin has certainly been implicated as
playing some role, direct or indirect, in the mechanism of action of these
agents. Hallucinogens do appear to affect other neurotransmitters, including, for example, dopamine, noradrenaline and histamine (see reviews by
Ehrenpreis and Teller 1972; Martin and Sloan 1977; Jacobs 1982), but
these effects may not be significant with respect to hallucinogenic activity,
particularly where some of these effects are observed only at very high
doses. Current evidence does not support the notion that these other
neurotransmitters are consistently involved in the mechanism of action of
hallucinogenic agents. Jacobs (1982) has reviewed the available data and
has concluded that although various hallucinogens interact with neurotransmitters other than serotonin, they do not do so universally.
Nevertheless, additional research addressing this problem is necessary
and warranted.

A comprehensive review of all the data implicating serotonin as playing a
role in the mechanism of action of hallucinogenic agents would be a for-
midable task and is far beyond the scope of this chapter. The interested
reader is directed to overviews presented by Downing (1964), Brawley and
Duffield (1972), Brimblecombe and Pinder (1975), Sankar (1975), Martin
and Sloan (1977), Freedman and Halaris (1978), Glennon and Rosecrans
(1981), Jacobs (1982), and Glennon (1983). These articles describe the *in
vitro* receptor binding properties, as well as the neurochemical, pharmacological and behavioural effects of hallucinogenic agents in light of a
possible serotonergic mechanism.

The following discussion will outline our approach to studying the serotonergic aspects of the classical hallucinogens. The studies can be divided
into three basic categories: (*i*) *in vitro* receptor affinity studies, (*ii*) drug
discrimination studies, and (*iii*) ligand-receptor binding studies in brain.
The receptor affinity studies were an attempt to measure the serotonin

receptor affinities of classical hallucinogens, to formulate structure-activity relationships, and to determine whether or not a relationship exists between activity/potency and receptor affinity. The drug discrimination studies were directed toward determining whether there was a common effect shared by phenalkylamine and indolealkylamine hallucinogens; agents not producing a similar effect are most likely not acting via a similar mechanism. Tests with various serotonin agonists and antagonists were also employed to help elucidate possible mechanisms of action. Finally, we are currently investigating central serotonin binding sites in the same manner that we studied serotonin receptors. Our progress in each of these areas, along with the results of other contemporary investigators working in this field, are now described.

10.2   Receptor affinity studies

Our work on the mechanism of action of hallucinogenic agents began with the question of a possible relationship between human hallucinogenic potency and serotonin receptor affinity. Our thinking was shaped by the early studies of Vane (1959), Barlow and Khan (1959), and Offermeier and Ariens (1966*b*), who, using *in vitro* serotonin receptor preparations, demonstrated the influence of substituent groups on the serotonergic activity of various indolealkylamines. Based on the results of a study by Offermeier and Ariens (1966*a*), the isolated rat fundus preparation was chosen by us because of its apparent advantages over other *in vitro* preparations. Winter and Gessner (1968) had determined that the fundus preparation possesses PRT (phenoxybenzamine-resistant tryptamine) as well as serotonin receptors, and that serotonin interacts exclusively with the latter while certain other agents interact with both. Thus, the affinity of a series of agents was determined by examining their ability to compete with serotonin for its receptors rather than studying their direct agonistic effects. Briefly, a series of dose-response curves was obtained for serotonin in the absence and in the presence of increasing concentrations of test compound. Parallel shifts of the dose-response curves suggest that serotonin and the test compound are acting on the same receptor. ED_{50} values for serotonin were determined for each of these curves and the method of Arunlakshana and Schild (1959) was used to calculate pA_2 values (i.e. $-\log K_B$) or 'apparent affinities'. (Whereas we have usually reported our affinity data as pA_2 values, K_B will be used herein in order to facilitate comparisons.)

Most of the indolealkylamine analogues were found to interact with the serotonin receptors in a competitive manner as determined either by the slope of their Schild plots or by double-reciprocal plots. In this way, it was possible to formulate structure-activity relationships for a series of

Table 10.1 Rat fundus serotonin receptor affinities for some representative indolealkylamines

Agent	R	R'	K_B (nM)*
Tryptamine	H	H	525
DMT	H	Me	1000
DET	H	Et	1600
Psilocin	4-OH	Me	150
4-OMe DMT	4-OMe	Me	700
4-TMT	4-Me	Me	140
	4-NH$_2$	Me	525
Bufotenin	5-OH	Me	40
5-OMe DMT	5-OMe	Me	80
5-OMe DET	5-OMe	Et	100
5-OMe DPT	5-OMe	nPr	300
5-TMT	5-Me	Me	300
	5-NH$_2$	H	40
	5-NH$_2$	Me	80
6-OMe DMT	6-OMe	Me	1700
6-TMT	6-Me	Me	—†
7-OH DMT	7-OH	Me	13 000
7-OMe DMT	7-OMe	Me	4700
7-TMT	7-Me	Me	500
5-OMe 7-TMT	5-OMe, 7-Me	Me	250
	5,7-(OMe)$_2$	Me	3200
7-Br DMT	7-Br	Me	300
Harmine			3400
Harmaline			1500
6-OMe Harmalan			700

*Determined from $pA_2 = -\log K_B$; values are approximate. †Non-competitive interaction as determined by slope of Schild plot.

indolealkylamines; some representative 'affinity' data are presented in Table 10.1. In general, as the size of the di-alkyl substituent on the terminal amine was increased, affinity decreased in the order H>Me>Et>Pr, with the di-methyl derivatives usually possessing half the affinity of their corresponding primary amine. Hydroxy or amino substitution at the 4- or 5-position enhanced affinity while a hydroxy group at the 7-position decreased affinity; methylation of the 4-OH or 5-OH group in the DMT series, to afford 4-OMe and 5-OMe DMT, respectively, decreased affinity while methylation of the 7-hydroxyl group increased affinity, such that with respect to methoxy substituents, 5-OMe>4-OMe>6-OMe>7-OMe.

Methylation of the 4-, 5- or 7-position of DMT enhanced affinity in the order 4-Me>5-Me>7-Me, while methylation of DMT at the 6-position resulted in a compound that did not interact with the serotonin receptors in a competitive manner. This same effect was observed when a 6-methyl group was introduced into 5-OMe DMT to afford 5-OMe-6-TMT.

Human potency data are unavailable for most of these agents, and where such data are available, they were not necessarily obtained using the same route of administration; this makes it difficult to compare receptor affinity with human potency. Nevertheless, there is an apparent relationship between serotonin receptor affinity and hallucinogenic potency for those few agents known to be active in man including 5-OMe DMT, psilocin, DMT and DET. It might be noted that differences in affinity are relatively small, such as the difference between the affinity of DMT and 5-OMe DMT, but it might also be noted that differences in human potency are also small, such as the 10-fold difference between these same two agents in man.

Demethylation of the terminal amine of, for example, DMT and 5-OMe DMT, to produce tryptamine and 5-methoxytryptamine, result in a doubling of serotonin receptor affinity, and yet these agents are inactive as hallucinogenic agents. O-Demethylation of 5-OMe DMT, to produce bufotenin, also results in decreased activity in man. This may be a consequence either of the inability of these agents to penetrate the blood-brain barrier or the result of rapid metabolic degradation by, for example, monoamine oxidase, or both. This has been discussed in greater detail elsewhere (Glennon and Rosecrans 1981; Nichols and Glennon 1984). Thus, not all compounds that possess a reasonable serotonin receptor affinity are hallucinogenic in man, although those agents of the series that are hallucinogenic, particularly those that are amongst the more potent, do possess high serotonin receptor affinities.

If demethylation of DMT abolishes activity for one (or both) of the above reasons, it might be possible to overcome this problem by α-methylation to afford α-methyltryptamine (α-MeT). Table 10.2 shows that α-methylation of tryptamine has little effect on affinity when the racemate is examined. Furthermore, α-methylation introduces a chiral centre; resolution of the racemic mixture and evaluation of the individual optical isomers reveals that the serotonin receptors of the fundus preparation display stereoselectivity (Glennon 1979). Both isomers interact with the serotonin receptors but the S(+)-isomer possesses ten times the affinity of its enantiomer. Using the same rationale, we examined the affinity of racemic 5-OMe α-MeT and found this agent to possess a very high affinity. 5-Methoxy-α-methyltryptamine was independently determined to be amongst the most potent of the tryptamine hallucinogens in man (Kantor, Dudlettes and Shulgin 1980).

Using a drug discrimination paradigm (which is described below) with

Table 10.2   Effect of α-alkyl groups on the serotonin receptor affinity of some indolealkylamines

Agent	R	R'	K_B (nM)*
Tryptamine	H	H	525
($\pm$)-α-MeT	H	Me	550
S(+)-α-MeT	H	Me	350
R(−)-α-MeT	H	Me	3200
($\pm$)-α-EtT	H	Et	5900
5-OMe Tryptamine	OMe	H	30
($\pm$)-OMe α-MeT	OMe	Me	20

*Determined from pA_2=−log K_B; values are approximate.

1-(2,5-dimethoxy-4-methylphenyl)-2-aminopropane (DOM) as the training drug, ED_{50} values for stimulus generalization were determined for a small series of tryptamine analogues. Figure 10.2 reveals that there is a direct relationship between serotonin receptor affinity and potency within this series. Those agents that were the most potent possess the greatest affinity; as potency decreases, so does serotonin receptor affinity.

The β-carbolines are another class of hallucinogenic indolealkylamines (Schultes 1982); these agents have received relatively little attention over the years, and none has been demonstrated to be much more active in man than DMT. While certain of these agents are known to be inhibitors of monoamine oxidase (Buckholtz and Boggan 1977), Naranjo (1967) has demonstrated that derivatives such as harmaline and 6-methoxyharmalan are hallucinogenic in man. Harmine, harmaline and 6-methoxyharmalan (Table 10.1) were also evaluated in the drug discrimination paradigm; harmaline and 6-methoxyharmalan were found to produce effects similar to that of the training drug (i.e. DOM), while DOM-stimulus generalization did not occur with harmine (Fig. 10.2).

Another class of hallucinogenic agents, for which there are available somewhat more human data, are the phenalkylamines. Vane (1960) is credited for dispelling the notion that phenalkylamines, including amphetamine, interact solely via an adrenergic mechanism and established that such compounds could produce an agonistic effect by interaction with the serotonin receptors of an isolated tissue preparation. These findings have since been confirmed and extended by other investigators (e.g. Innes 1963; Cheng and Long 1973; Cheng, Long, Nichols and Barfknecht 1974; Dyer,

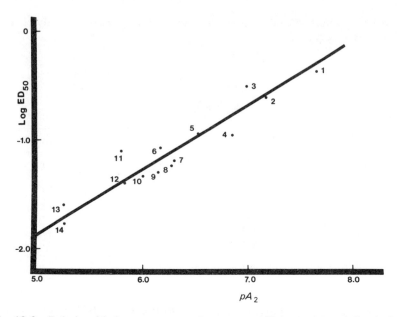

Fig. 10.2   Relationship between serotonin receptor affinity (pA_2) and discrimination-derived ED_{50} values (μmol/kg) for several indolealkylamines; mescaline is included for comparative purposes. 5-OMe α-MeT (1), 5-OMe DMT (2), 5-OMe DET (3), 4-TMT (4), S(+)-α-MeT (5), 4-OMe DMT (6), (±)-α-MeT (7), 7-TMT (8), 6-OMe harmalan (9), DMT (10), DET (11), harmaline (12), (±)-α-EtT (13), mescaline (14).

Nichols, Rusterholz and Barfknecht 1973; Nair 1974), who have found that substituents on the aromatic nucleus of amphetamine greatly enhance serotonergic activity. One of the more interesting studies is that reported by Dyer and coworkers (1973) who examined the contractile effect of a series of phenalkylamine isomers on sheep umbilical vasculature; an effect which is presumably mediated via an agonistic interaction with serotonin receptors. The interaction of these agents with the receptor was stereoselective in that the R(−)-isomers were more potent than their enantiomers (Table 10.3); their agonistic effects were antagonized by the antagonist 2-bromo LSD (BOL). Furthermore, there was a degree of correspondence between the concentrations of compound necessary to produce their agonistic effect and the human hallucinogenic potencies of these agents. Dyer and Gant (1973) had previously reported that LSD, bufotenin and psilocin produce similar agonistic effects in this preparation, and, subsequently, other phenalkylamine analogues have also been examined (Shulgin and Dyer 1975; Nichols, Shulgin and Dyer 1977).

Huang and Ho (1974) were the first to report that the phenalkylamine hallucinogen DOM causes contraction of the isolated rat fundus strip and

Table 10.3 Agonistic potency of several phenylisopropylamines on sheep umbilical vasculature serotonin receptor

Agent	ED_{30} ($\times 10^9$M)	S/R*
R(−)-DOB	2.4	7.5
S(+)-DOB	18.0	
R(−)-DOET	4.8	4.5
S(+)-DOET	21.5	
R(−)-DOM	24.0	5.8
S(+)-DOM	139.7	
R(−)-2,5-DMA	800.0	
Serotonin	205.0	

Data from: Dyer *et al.* (1973). *Enantiomeric potency ratio.

Standridge, Howell, Gylys, Partyka and Shulgin (1976) found that R(−)-DOM was twice as potent as S(+)-DOM in this respect. Glennon, Liebowitz and Mack (1978), using this same preparation, investigated the serotonin receptor affinities of a small series of phenalkylamines and reported that DOM and DOB possessed affinities similar to, or greater than, that of 5-OMe DMT; they commented that those indolealkylamine and phenalkylamine hallucinogens that were most potent in man possessed high serotonin affinities. They subsequently investigated the affinities of a larger series of phenalkylamines and some of these data are reproduced in Table 10.4. None of the monomethoxy phenylisopropylamines displayed a high affinity for the serotonin receptors. It is also quite clear that of the dimethoxy (DMA) derivatives, there appears to be a special significance attached to the 2,5-dimethoxy substitution pattern of 2,5-DMA. Within the trimethoxy (TMA) series, again, 2,5-dimethoxy substitution appears to be important. However, even with this pattern, the presence of a 3-position substituent reduces affinity (i.e. compare 2,5-DMA and 2,4,5-TMA with 2,3,5-TMA); this efffect is consistent and will be seen again. Further substitution at the 4-position of 2,5-DMA generally serves to enhance affinity; with respect to 4-alkyl substituents, affinity follows the following order: Pr=Et>Me>Bu>Am, although the differences in affinity are quite small. Halogenation at the 4-position of 2,5-DMA results in compounds with the greatest affinity, e.g. DOB and DOI. Again, transposition of the 4-methyl group of DOM or the 4-bromo group of DOB, to afford 3-Me 2,5-DMA and 3-Br 2,5-DMA, respectively, results in a dramatic decrease in affinity. In general, mono- and di-N-methylation of the terminal amine halves affinity, and removal of the α-methyl group has no effect on affinity when compared to the affinity of the corresponding racemate. The interaction is usually stereoselective in that the affinities of the R(−)-isomers is greater than that of their racemates or S(+)-enantiomers. Further, there

Table 10.4 Rat fundus serotonin receptor affinities for some representative phenalkylamines

Agent	R_2	R_3	R_4	R_5	R_6	R′	K_B (nM)*
Amphetamine (PIA)	H	H	H	H	H	Me	5400
Phenethylamine(PEA)	H	H	H	H	H	H	5500
OMA	OMe	H	H	H	H	Me	2900
2-OMe PEA	OMe	H	H	H	H	H	3000
MMA	H	OMe	H	H	H	Me	1180
3-OMe PEA	H	OMe	H	H	H	H	1280
PMA	H	H	OMe	H	H	Me	7000
R(−)-PMA	H	H	OMe	H	H	Me	4200
S(+)-PMA	H	H	OMe	H	H	Me	7000
4-OMe PEA	H	H	OMe	H	H	H	7900
4-OH PIA	H	H	OH	H	H	Me	9000
4-Me PIA	H	H	Me	H	H	Me	3000
2,3-DMA	OMe	OMe	H	H	H	Me	2900
2,4-DMA	OMe	H	OMe	H	H	Me	2500
2,5-DMA	OMe	H	H	OMe	H	Me	150
S(+)-2,5-DMA	OMe	H	H	OMe	H	Me	2200
2,5-di OMe PEA	OMe	H	H	OMe	H	H	140
2,6-DMA	OMe	H	H	H	OMe	Me	8100
3,4-DMA	H	OMe	OMe	H	H	Me	3500
3,5-DMA	H	OMe	H	OMe	H	Me	2800
2,3,4-TMA	OMe	OMe	OMe	H	H	Me	8500
2,3,5-TMA	OMe	OMe	H	OMe	H	Me	4200
2,4,5-TMA	OMe	H	OMe	OMe	H	Me	150
2,4,6-TMA	OMe	H	OMe	H	OMe	Me	525
	OH	H	H	OMe	H	Me	100
	H	H	OH	OMe	H	Me	6500
	OMe	H	OMe	SMe	H	Me	1700
3-Me 2,5-DMA	OMe	Me	H	OMe	H	Me	4700
3,4,5-TMA	H	OMe	OMe	OMe	H	Me	2500
Mescaline	H	OMe	OMe	OMe	H	H	2250
DOM	OMe	H	Me	OMe	H	Me	75
R(−)-DOM†	OMe	H	Me	OMe	H	Me	70
S(+)-DOM†	OMe	H	Me	OMe	H	Me	400
DOET	OMe	H	Et	OMe	H	Me	65
DOPR	OMe	H	nPr	OMe	H	Me	65
DOBU	OMe	H	nBu	OMe	H	Me	90
DOAM	OMe	H	nAm	OMe	H	Me	100
4-OEt 2,5-DMA	OMe	H	OEt	OMe	H	Me	170
DOF	OMe	H	F	OMe	H	Me	65
DOB	OMe	H	Br	OMe	H	Me	45
S(+)-DOB	OMe	H	Br	OMe	H	Me	120
3 Br 2,5 DMA	OMe	Br	H	OMe	H	Me	5400
DOI	OMe	H	I	OMe	H	Me	30
R(−)-DOI	OMe	H	I	OMe	H	Me	20

*Determined from $pA_2 = -\log K_B$; values are approximate. †Enantiomeric potency ratio: S/R=5.7.

appears to be a correspondence between affinity and human hallucinogenic potency in man; those agents that are most active possess relatively
high serotonin receptor affinities (Glennon, Liebowitz and Anderson
1980; Glennon, Young, Bennington and Morin 1982*b*).

If indolealkylamines and phenylisopropylamines both interact with
serotonin receptors such that there exists a common aromatic site and a
common amine site, it might be anticipated that this type of interaction
would accommodate the α-methyl groups of both series in a similar
manner. That is, if the isomer of α-methyltryptamine with the greater
affinity (i.e. *S*-isomer) interacts as shown in Fig. 10.3. (i.e. with the methyl
group behind the plane of the molecule), then the phenylisopropylamines
should interact in a similar manner (shown for DOI in Fig. 10.3). This
would correspond to the R(−)-isomer of DOI. As already discussed, the
R(−)-isomers of the hallucinogenic phenylisopropylamines have been
found to constitute the eutomeric series; this has been discussed in greater
detail (Glennon 1983) and is in agreement with the findings of Dyer *et al.*
1973) that R(−)-isomers of certain 2,5-dimethoxy-substituted phenylisopropylamines are more potent serotonin agonists than their corresponding
S(+)-isomers. There are additional similarities between the interaction of
phenalkylamines and indolealkylamines with fundas serotonin receptors.

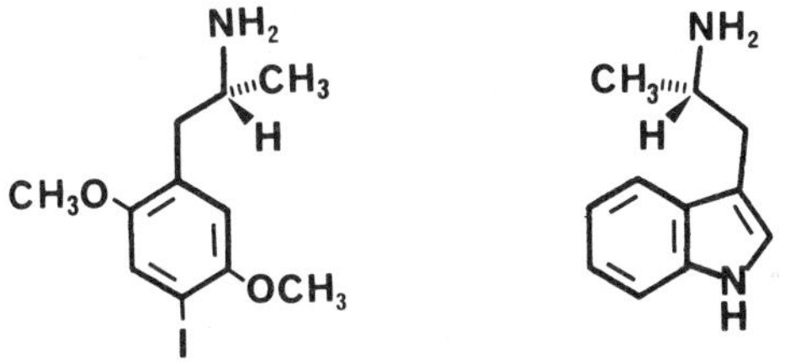

Fig. 10.3   Structures of R(−)-DOI and S(+)-α-methyltryptamine.

For example, if compounds *A* and *B* (Fig. 10.4) interact such that sites a.–c.
are common, it should be possible to demonstrate similar trends upon
substitution at these positions. Methoxylation at the a.-position enhances
affinity in both series, and de-methylation of this methoxy group, to afford
the a.-OH derivatives, further enhances affinity. Moving the a.-OMe group
to the b.-position decreases affinity by about 20-fold, while the c.-OMe
derivatives possess an even slightly lower affinity. De-methylation of the c.-
OMe group results in a further decrease in affinity. In both series, the effect
of an α-methyl group is similar, except for stereochemistry (as already
discussed), and, extention of an α-methyl to an α-ethyl group reduces
affinity. *N*-Mono- or *N,N*-dimethylation in the *A*-series halves affinity,
while de-methylation of the terminal amine in the *B*-series doubles affinity.

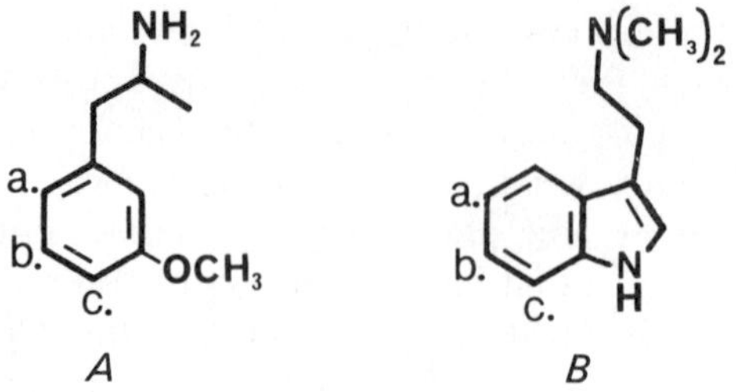

Fig. 10.4   A comparison of the structures of a phenylisopropylamine (*A*) and an indolealkylamine (*B*).

Thus, there is reason to believe that at least some phenalkylamines and indolealkylamines can interact with serotonin receptors in a similar manner. Most of these parallel changes involved a single substituent; however, when multiple substituents are involved, as for example, in comparing a.,c.-dimethoxy derivatives, of *A* and *B* (Fig. 10.4), the effects on affinity are not always parallel. This may be the result of different steric interactions arising from substituent adjacency, or, different modes of binding may be involved.

Not all agents with high serotonin receptor affinities are necessarily hallucinogenic in man, however, the overall results of all the above-mentioned studies implicate a role for serotonin in the mechanism of action of the classical hallucinogens. On the other hand, these serotonin receptors are 'peripheral serotonin sites' and their relationship to central receptors has yet to be determined. With the discovery of multiple central serotonin binding sites, our studies with rat fundus receptors have been temporarily abandoned until a relationship (if one exists) can be firmly established with the binding characteristics of some central site.

10.3   Drug discrimination studies

Drug discrimination is, in essence, an *in vivo* drug 'detection' procedure. In the drug discrimination paradigm, animals are trained to emit a particular response under one set of conditions (e.g. a drug state) and a different response under a different set of conditions (e.g. a non-drug state). For example, using standard two-lever operant chambers, animals can be trained to associate a particular drug state with one of the two levers, and the non-drug state with the other lever such that the animals will respond on one lever when given drug, and on the other lever when given saline. The drug discrimination paradigm has proven to be quite useful for the study of various centrally-active agents; the stimulus (interoceptive cue) produced by a drug is quite specific, and the animals are very sensitive to minute alterations in dose. Recent reviews on drug-induced discrimination with particular emphasis on hallucinogenic agents include those by Glennon, Rosecrans and Young 1982*a*; Glennon *et al.* 1983*c* and Appel, White and Holohean (1982); for a more general overview, see Colpaert and Slangen (1982). Once animals have been trained to discriminate the stimulus effects of a drug (i.e. the training drug) from those saline, several different approaches can be taken to study mechanism of action. Two of these approaches will be discussed, (*a*) stimulus generalization studies, and (*b*) stimulus antagonism studies.

Drugs that produce similar subjective effects in man often generalize (transfer) to one another in tests of discriminative control of behaviour. The rationale behind this approach is that animals trained to recognize or discriminate a dose of a particular training drug will exhibit stimulus generalization to other agents (challenge drugs) that produce a similar effect. It is likely that animals respond to the effects of a challenge drug, in tests of stimulus generalization, to the extent that the effects of that drug contains pharmacological components which overlap with those produced by the training dose of the training drug. Thus, it has been proposed that the possibility of drug-appropriate responding occurring with a challenge drug is a function of the proportion of pharmacological effects in the test set, which because they formed part of the set of pharmacological effects present during training, are associated with reinforcement during discrimination training (Glennon *et al.* 1983*c*). In this way, agents can be compared for similarity of effect; the results of such studies provide both qualitative and quantitative information. It should be emphasized, however, that simply because two agents produce similar effects (i.e. result in stimulus generalization) does not necessarily imply commonality of mechanism of action.

In stimulus antagonism studies, animals, trained to discriminate a training drug from saline, can be pre-treated with various agents in an attempt to antagonize the effect of the training drug. From a theoretical standpoint, the effect produced by a training drug will only be blocked by antagonists that somehow interfere with its mechanism of action. This is a particularly effective strategy that is now becoming widely employed to study mechanisms of action (Colpaert and Slangen 1982).

Hirschhorn and Winter (1971) were the first to demonstrate that the hallucinogens LSD and mescaline would serve as discriminative stimuli in animals. Over the next decade, it was determined that the stimulus effects of these two agents could be attenuated by pre-treatment of the animals with any one of a number of serotonin antagonists including cinanserin, BOL, methiothepin, cyproheptadine and pizotifen (Browne and Ho 1975; Winter 1975, 1978; Kuhn, White and Appel 1978; Silverman and Ho 1980; Minnema 1981; Colpaert, Niemegeers and Janssen 1982). Appel *et al.* (1982) have commented that apart from central serotonin antagonists,

266   Richard A. Glennon

no other type of neurotransmitter antagonist (e.g. those for dopamine, noradrenaline, opiate), serotonin uptake inhibitor, peripheral serotonin antagonist (e.g. xylamidine), or indeed any other compound, has been shown to block the discriminative stimulus effects of LSD. On the other hand, Colpaert *et al.* (1979, 1982) have reported that at sufficiently high doses, most serotonin antagonists can mimic, to varying degrees, the effect of LSD, suggesting that some antagonists may possess complex, mixed-agonist antagonist properties.

Silverman and Ho (1980) demonstrated that racemic DOM serves as a discriminative stimulus in rats, and that the DOM-stimulus can be anta-gonized by the serotonin antagonists methysergide and cinanserin, but not by the dopamine antagonist haloperidol. We have shown that pizotifen is also an effective antagonist of the DOM-stimulus (Young, Glennon and Rosecrans 1980). Using rats trained to discriminate 1.0 mg/kg (i.p.) of DOM from saline, we have conducted stimulus generalization studies on well over 100 indolealkylamine and phenalkylamine analogues in order to determine commonality of effect and to develop *in vivo* structure-activity relationships (Glennon *et al.* 1983*c*). Although DOM is a phenylisopro-pylamine hallucinogen, the DOM-stimulus generalizes to various indolealkylamines as well as to phenalkylamines; this suggests that these agents, where generalization occurs, are capable of producing similar behavioural (discriminative) effects in the animals.

With respect to the indolealkylamines, the general structure-activity relationships appear to echo those derived from the receptor affinity studies. For example, 5-OMe DMT is more potent than 4-OMe which is more potent than either DMT or 6-OMe DMT. The α-methyl derivatives are approximately twice as potent as their *N,N*-dimethyl counterparts (e.g. comparing DMT with α-MeT or 5-OMe DMT with 5-OMe α-MeT) and, the (+)-isomers of the α-methyl derivatives are more potent than their racemates and/or (−)-enantiomers (Glennon, Jacyno and Young 1983*a*; Glennon, Young and Jacyno 1983*d*). The one obvious difference is that while *N,N*-diethyl derivatives usually possess a somewhat lower serotonin receptor affinity than their corresponding N,N-dimethyl derivatives, it is the diethyl derivatives that are more potent in the drug discrimination paradigm; this, however, might reflect differences in *in vivo* distribution or metabolism or both, and the differences are minimized when activity is compared on a molar basis. The relationship between serotonin receptor affinity and discrimination-derived potencies (ED_{50} values) is shown in Fig. 10.2 for several indolealkylamines. LSD and two β-carbolines, harmaline and 6-methoxyharmalan, also produce DOM-like effects.

The DOM-stimulus generalizes to a large number of phenalkylamines; again, the 2,5-dimethoxy substitution pattern appears to be important for stimulus generalization, but several 2,4-dimethoxy analogues are also quite potent (Glennon and Young 1982). With respect to the phenylisopropyl-amines. *N*-methylation reduces activity; in this series, removal of the α-methyl group, which would result in a primary amine that would be unprotected to oxidative deamination, also decreases activity. Substitution at the 4-position of 2,5-DMA by small alkyl groups (e.g. DOET, DOPR) or by a bromo or iodo group, (e.g. DOB, DOI) results in the most potent agents, and the (−)-isomers are more potent than their racemates or (+)-enantiomers. A relationship between potencies and serotonin receptor affinities within the 2,5-DMA series has been reported (Glennon *et al.* 1982*b*). One question that might be asked at this point is whether there is a relationship between discrimination-derived potency and human hallu-cinogenic potency. We have now shown that such a relationship exists for approximately 30 indolealkylamine and phenalkylamine derivatives for which human potency data are available (Glennon *et al.* 1982*b*).

The indolealkylamine hallucinogen 5-OMe DMT also serves as a discri-minative stimulus in rats; the effects of 5-OMe DMT can be attenuated by pre-treatment of the animals with the serotonin antagonist pizotifen (Glennon, Rosecrans, Young and Gaines 1979). Furthermore, stimulus generalization occurs between 5-OMe DMT, LSD and DOM regardless of which of the three is used as the training drug, and, the stimulus effects of all three can be attenuated by pre-treatment of the animals with serotonin antagonists. If these compounds produce their behavioural effects via an agonistic interaction at serotonin sites, it might be possible to demonstrate stimulus generalization, in 5-OMe DMT-, LSD- or DOM-trained rats, to a serotonin agonist. LSD-stimulus generalization occurs with the purported agonists MK-212 (Fig. 10.5) and quipazine (Colpaert *et al.* 1979; White

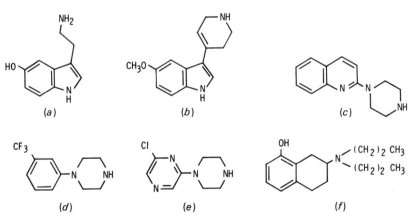

Fig. 10.5   Agents with activity as serotonin agonists: (*a*) serotonin, (*b*) RU 24969, (*c*) quipazine, (*d*) TFMPP, (*e*) MK-212, (*f*) 8-OH DPAT.

and Appel 1982). Quipazine produces partial generalization (71 per cent appropriate responding) when administered to 5-OMe DMT-trained animals, and complete generalization in DOM-trained animals (Glennon *et al.* 1982*a*). In fact, quipazine itself has been used as a training drug and quipazine-stimulus generalization occurs to the hallucinogens LSD, psilocybin, mescaline and 5-OMe DMT (Winter 1979; White, Appel and Kuhn 1979; Glennon *et al.* 1982*a*). On the other hand, administration of the serotonin agonists TFMPP and RU 24969 (Fig. 10.5) to DOM-trained animals results in saline-like responding at low doses and disruption of behaviour at higher doses (Glennon, Young and Rosecrans 1983*e*); that is, the effects of DOM appear to be similar to those of quipazine, but dissimilar to those of TFMPP and RU 24969. Other pharmacological dissimilarities have also been recently noted for quipazine v. these latter agonists (e.g. Green 1981; Ahlenius, Larsson, Svensson *et al.* 1981; Lucki and Frazer 1982; Cohen and Fuller 1983; Gardner and Guy 1983; Green, Guy and Gardner 1984) and it has been suggested that these agonists might act on different sub-populations of serotonin receptors, a suggestion supported both by ligand-receptor binding studies (Chapter 4) and other behavioural investigations (Chapter 12).

   The last set of discrimination studies to be discussed are those involving new serotonin antagonists. Leysen, Niemegeers, van Nueten and Laduron (1982) found that ketanserin and pirenperone were capable of blocking mescaline-induced head twitches in mice; pirenperone also blocks 5-HTP-induced head twitches in mice (Green, O'Shaughnessy, Hammond, Schachter and Grahame-Smith 1983) and quipazine-induced head twitches in rats (Colpaert and Janssen 1983). Colpaert *et al.* (1982) have also found that pirenperone effectively antagonizes the stimulus effects of LSD in LSD-trained animals, and that pirenperone is not only very potent, but that is lacks the partial agonistic (i.e. generalization) effect observed with many other serotonin antagonists. Both ketanserin and pirenperone (Fig. 10.6) are effective in attenuating the DOM-stimulus; the ED_{50} value for pirenperone is 10 µg/kg at a DOM training dose of 1 mg/kg (Glennon *et al.* 1983*e*). In this same study, it was demonstrated that DOM-stimulus generalization to LSD, 5-OMe DMT, mescaline and quipazine were all antagonized by pirenperone and that the slopes of the dose-response curves for pirenperone antagonism were parallel (Fig. 10.7).

   The general conclusion of these studies is that serotonin antagonists can block the discriminative stimulus effects of various classical hallucinogens. Based on the most recent findings, it appears that these hallucinogens may act as agonists at certain subpopulations of serotonin receptors to produce their stimulus effects. This will be discussed in more detail in the following section.

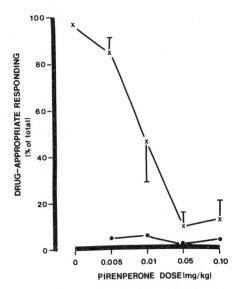

Fig. 10.6   Effect of pirenperone pre-treatment on the drug-appropriate responding produced by either 1.0 mg/kg of DOM (×) or saline (●) in rats trained to
discriminate 1.0 mg/kg of DOM from saline. (From Glennon *et al.* 1983*e*)

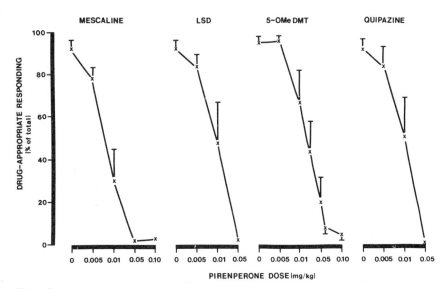

Fig. 10.7   Antagonism of DOM-stimulus generalization to mescaline (25 mg/kg),
LSD (0.1 mg/kg), 5-OMe DMT (3.0 mg/kg) and quipazine (3.0 mg/kg) by doses of
the 5-HT$_2$ antagonist pirenperone. (From Glennon *et al.* 1983*e*)

10.4   Ligand-receptor binding studies in brain

There has been from the very beginning an intimate connection between hallucinogenic agents, notably LSD, and central serotonin binding (receptor?) site studies. [³H]-LSD was found to bind saturably, reversibly and stereospecifically to brain membrane preparations at a site initially proposed to be a central serotonin receptor. In other studies, it was found that LSD was a potent displacer of [³H]-serotonin binding; both tritiated radioligands could by displaced by certain indolealkylamine and phenalkylamine hallucinogens (e.g. Farrow and van Vunakis 1973; Bennett and Snyder 1975; Bennett and Snyder 1976; G. M. B. Fillion, Rousells, M. P. Fillion *et al.* 1978; Green, Johnson, Weinstein, Kang and Chou 1978; Whitaker and Seeman 1978). Further ligand-binding studies led Peroutka and Snyder (1979) to suggest the existence of two distinct types of serotonin binding sites: (*a*) 5-HT$_1$ sites, those defined by high-affinity binding of [³H]-serotonin and (*b*) 5-HT$_2$ sites, those defined by [³H]-spiperone binding in rat frontal cortex. It has since been shown that spiperone can distinguish between two sub-populations of 5-HT$_1$ sites, 5-HT$_{1A}$ sites, for which it displays a relatively high affinity, and 5-HT$_{1B}$ sites (Pegido, Yamamura and Nelson 1981; Nelson, Pedigo and Yamamura 1981; Nelson 1982), and the suggestion has been made that [³H]-ketanserin may be a more suitable ligand than [³H]-spiperone for studying 5-HT$_2$ sites (Leysen *et al.* 1982; Leysen and Tollenaere 1982; Janssen 1983; and see Section 4.4).

In collaboration with Dr David Nelson at the University of Arizona we have examined the ability of various indolealkylamines to bind at 5-HT$_1$ sites in order to determine if a relationship exists between binding and behavioural potency. Some of the binding data are shown in Table 10.5. At first glance, there seems to be some similarity between these results and both behavioural potency data and, for that matter, serotonin receptor affinity (i.e, Table 10.4). However, this issue is complicated by the heterogeneity of 5-HT$_1$ binding sites; it appears that certain of these agents act differently at, and can discriminate between, 5-HT$_{1A}$ and 5-HT$_{1B}$ sites (Smit, Glennon, Yamamura and Nelson 1981; Nelson, Weck and Taylor 1983). As mentioned earlier, 5-OMe DMT and DOM generalize to one another in tests of discriminative control of behaviour, regardless of which is used as the training drug, and that R(−)-DOM is nearly an order of magnitude more potent than its enantiomer. However, DOM displays a relatively low affinity for 5-HT$_1$ sites (IC$_{50}$=>1000 nM) and binding is not stereoselective when R(−)-DOM is compared with S(+)-DOM (Mason and Glennon, unpublished data). DeJong, Huggins, Fournier and Makriyannis (1982) also report that racemic DOM is a weak competitor for serotonin binding sites (IC$_{50}$=13 000 nM); however, these investigators

Table 10.5   Inhibition of [^{3}H]-serotonin binding in rat cortical membranes

Agent	R'	R"	R_4	R_5	R_6	IC_{50} (nM)
Tryptamine	H	H	H	H	H	176
Serotonin	H	H	H	OH	H	4
α-MeT	Me	H	H	H	H	5700
α-EtT	Et	H	H	H	H	9500
5-OMe α-MeT	Me	H	H	OMe	H	324
DMT	H	Me	H	H	H	137
4-OMe DMT	H	Me	OMe	H	H	224
5-OMe DMT	H	Me	H	OMe	H	71
6-OMe DMT	H	Me	H	H	OMe	631
DET	H	Et	H	H	H	176
5-OMe DET	H	Et	H	OMe	H	113
5-OMe DPT	H	Pr	H	OMe	H	180
DIPT	H	iPr	H	H	H	961

Data from: Smit *et al.* (1981), Nelson *et al.* (1983), Glennon *et al.* (1983*b*), Nelson and Glennon (unpublished data).

examined a dozen related phenalkylamines and found that their affinities, albeit low, did correlate with their rank-order of human hallucinogenic potencies.

Ketanserin and pirenperone display high affinity for 5-HT$_2$ sites and low affinity for 5-HT$_1$ sites. Because these agents effectively antagonize the discriminative stimulus effects of hallucinogens such as DOM and LSD, the mechanism(s) of action of the latter compounds might involve 5-HT$_2$ sites. This may be an indirect effect, or it may be a direct effect involving the interaction of these agents at 5-HT$_2$ sites. Together with Dr Milt Titler of the University of Toronto (currently at Albany Medical College) we are currently investigating the latter possibility by examining the ability of a series of phenalkylamines to displace [^{3}H]-ketanserin binding to rat frontal cortex homogenates (Battaglia, Shannon, Glennon and Titler 1983; Battaglia, Glennon and Titler 1983). Some of these results are shown in Table 10.6; comparative 5-HT$_1$ binding data are also presented. Once again, the 2,5-dimethoxy-substitution pattern appears to be an important factor; further substitution at the 4-position of 2,5-DMA by an alkyl (e.g. DOM) or halo (e.g. DOB, DOI) group enhances binding. Additional studies are necessary before any conclusions can be drawn, but the results obtained thus far are consistent with a possible 5-HT$_2$ mechanism.

Table 10.6   Affinities of selected phenylisopropylamines for 5-HT$_1$ and 5-HT$_2$ binding sites

	R	R$_2$	X	R$_5$	5-HT$_1$	5-HT$_2$
					— K_i (nM)	
R(−)-DOI	H	OMe	I	OMe	2290	10
(±)-DOI	H	OMe	I	OMe	2240	19
(±)-DOB	H	OMe	Br	OMe	3340	63
(±)-DOM	H	OMe	Me	OMe	2880	100
(±)-N-Me DOM	Me	OMe	Me	OMe	3865	410
(±)-2,4,5-TMA	H	OMe	OMe	OMe	>30 000	1645
(±)-MEM	H	OMe	OEt	OMe	>30 000	2220
(±)-2,5-DMA	H	OMe	H	OMe	1300	5200
(±)-PMA	H	H	OMe	H	>60 000	33 600
(±)-3,4-DMA	H	H	OMe	OMe	>60 000	43 300

Data from: Battaglia *et al.* (1983), Shannon *et al.* (1984).

As mentioned earlier, the DOM-stimulus was found to generalize to quipazine, but not to TFMPP or RU 24969. According to Leysen and Tollenaere (1982), quipazine displays some selectivity for 5-HT$_2$ sites, while current evidence suggests that the other two serotonin agonists may exert more of a 5-HT$_1$-related effect (see Section 12.5). In a recent series of studies, for example, 8-hydroxy-2-(di-*n*-propylamino)tetralin (8-OH DPAT) has been identified as a new serotonin agonist; this agent displays a high affinity for 5-HT$_1$ sites and is only weakly active at 5-HT$_2$ sites (Middlemiss and Fozard 1983). The pharmacological effects of this agent are more similar to those of RU 24969 than to those of quipazine. Tritiated 8-OH DPAT labels what are presumably hippocampal 5-HT$_1$ sites and RU 24969 is several-hundred times more potent than quipazine in displacing this ligand (Gozlan, El Mestikawy, Pichat, Glowinski and Hamon 1983, and see Section 12.6). Interestingly, administration of doses of 8-OH DPAT to animals trained to discriminate DOM from saline does not result in stimulus generalization (Glennon, Hauck and Young, unpublished results).

Taken together, these studies suggest that some hallucinogens, e.g. DOM and related agents, might produce their *in vivo* effects via a direct agonistic interaction at 5-HT$_2$ sites. If this is the case, then, to date, DOI, DOB and DOM are amongst the most potent known 5-HT$_2$ agonists.

10.5   Conclusions and speculations

The evidence accumulated to date strongly favours a role for serotonin in the mechanism of action of certain hallucinogenic agents. On the basis of various animal studies, it has been determined that serotonin antagonists can block the behavioural effects produced by many of the classical hallucinogens; furthermore, most of the effective antagonists, though not necessarily specific for serotonin, do display selectivity for 5-HT_2 v. 5-HT_1 binding sites (e.g. see Chapter 4 and Leysen and Tollenaere 1982). Receptor affinity studies reveal that those agents that are known to be hallucinogenic in man, and/or are behaviourally potent in animals, possess a higher serotonin receptor affinity than those agents which are less active or inactive. However, not all agents with high receptor affinities are necessarily active (e.g. bufotenin), and the exact relationship between these serotonin receptors and central serotonin binding sites remains unclear.

Drug discrimination studies, using racemic DOM as the training drug, have offered a method of classifying those agents that produce similar stimulus effects in animals. Various phenalkylamines and indolealkylamines are capable of producing stimulus effects similar to those of DOM. There also seems to be a significant correlation between discrimination-derived ED_{50} values and human hallucinogenic potencies, where such data are available. Furthermore, the stimulus effects of LSD and DOM, and DOM-stimulus generalization to the hallucinogens LSD, mescaline and 5-OMe DMT, can be effectively attenuated by pre-treatment of the animals with low doses of the 5-HT_2 antagonist pirenperone.

Certain potent hallucinogenic agents including LSD, DOM, DOB and DOI display a high affinity for central 5-HT_2 binding sites. It is still too early to conclude that 5-HT_2 sites might constitute a 'hallucinogen receptor', because (a) LSD, for example, interacts at both 5-HT_1 and 5-HT_2 sites (as well as at other neurotransmitter sites) (b) too few agents have been investigated, and (c) the binding properties of most of these hallucinogens for other neurotransmitter sites has not yet been fully investigated. Nevertheless, current data support the notion that there may be involvement of a 5-HT_2 mechanism in the effects produced by hallucinogens. The hypothesis developed earlier in our laboratories that, in drug discrimination studies, the stimulus cue to which the animals might be responding when administered these agents is a serotonin agonist effect (Rosecrans and Glennon 1979), may now be in need of modification. At least when DOM is employed as the training drug, the stimulus cue may be a measure of agonism at 5-HT_2 sites (or a certain sub-population of such sites). However, the exact role of 5-HT_1 sites still awaits elucidation, particularly where certain indolealkylamines possess a high affinity for these sites.

Because of the close structural analogy between serotonin and the indolealkylamines, it is not surprising that the mechanism of action of indolealkylamine hallucinogens might involve a serotonergic mechanism. The structures of the phenalkylamine hallucinogens, however, more closely resemble those of noradrenaline or dopamine than that of serotonin; this may be one reason why there has been speculation of catecholaminergic involvement in the mechanism of action of these agents. The results presented herein strongly support a serotonergic mechanism for both classes of agents, and even though catecholaminergic involvement can not yet be ruled out for the phenalkylamine hallucinogens, any theory concerning the mechanism of action of these agents must accommodate the rapidly growing body of data implicating serotonergic invovlement.

Finally, there is the intriguing possibility, if indeed hallucinogens such as DOM work via a 5-HT_2 agonist interaction, that appropriate structural modifications could give rise to 5-HT_2 agonists that are not hallucinogenic. In other words 5-HT_2 agonist action might simply be a predisposing factor for such activity; hallucinogens might be 5-HT_2 agonists but will all 5-HT_2 agonists be hallucinogens?

Acknowledgements

I would like to acknowledge the US Public Health Service grant DA-01642 for supporting much of the work from our laboratory, and to express my appreciation to the graduate students, post-doctoral fellows, collaborators and technicians who have contributed to this endeavour.

References

Ahlenius, S., Larsson, K., Svensson, L., Hjorth, S., Carlsson, A., Lindberg, P., Wikstrom, H., Sanchez, D., Arvidsson, L-E., Hacksell, U. and Nilsson, J. L. G. (1981). Effects of a new type of 5-HT receptor agonist on male rat sexual behavior. *Pharmac. Biochem. Behav.* **15,** 785.

Appel, J. B., White, F. J. and Holohean, A. M. (1982). Analyzing mechanism(s) of hallucinogenic drug action with drug discrimination procedures. *Neurosci. Biobehav. Rev.* **6,** 529.

Arunlakshana, O. and Schild, H. O. (1959). Some quantitative uses of drug antagonists. *Br. J. Pharmac.* **14,** 48.

Barlow, R. B. and Khan, I. (1959). Actions of some analogues of 5-hydroxytryptamine on the isolated rat uterus and rat fundus strip preparations. *Br. J. Pharmac.* **14,** 265.

Battaglia, G., Shannon, M., Glennon, R. A. and Titler, M. (1983). Hallucinogenic drug interactions with S_1 and S_2 cortical serotonin receptors. *Soc. Neurosci. Abstr.* **9,** 1157.

Bennett, J. P. and Snyder, S. H. (1975). Stereospecific binding of *d*-lysergic acid diethylamide (LSD) to brain membranes: Relationship to serotonin receptors. *Brain Res.* **94,** 523.

—— —— (1976). Serotonin and lysergic acid diethylamide binding in rat brain membranes: Relationship to postsynaptic serotonin receptors. *Mol. Pharmac.* **12,** 373.

Brawley, P. and Duffield, J. C. (1972). The pharmacology of hallucinogens. *Pharmac. Rev.* **24,** 31.

Brimblecombe, R. W. and Pinder, R. M. (1975). *Hallucinogenic agents.* Wright-Scientechnica, Bristol.

Browne, R. G. and Ho, B. T. (1975). Role of serotonin in the discriminative stimulus properties of mescaline. *Pharmac. Biochem. Behav.* **3,** 429.

Buckholtz, N. S. and Boggan, W. O. (1977). Monoamine oxidase inhibition in brain and liver produced by β-carbolines: Structure-activity relationships and substrate specificity. *Biochem. Pharmac.* **26,** 1991.

Cheng, H. C. and Long, J. P. (1973). Effects of *d*- and *l*-amphetamine on 5-hydroxytryptamine receptors. *Archs. Int. Pharmacodyn.* **204,** 124.

—— Long, J. P., Nichols, D. E. and Barfknecht, C. F. (1974). Effects of psychotomimetics on vascular strips; studies of methoxylated amphetamines and optical isomers of 2,5-dimethoxy-4-methylamphetamine and 2,5-dimethoxy-4-bromoamphetamine. *J. Pharmac. Exp. Ther.* **188,** 114.

Cohen, M. L. and Fuller, R. W. (1983). Antagonism of vascular serotonin receptors by *m*-chlorophenylpiperazine and *m*-trifluoromethylphenylpiperazine. *Life Sci.* **32,** 711.

Colpaert, F. C. and Janssen, P. A. J. (1983). The head-twitch response to the intraperitoneal injection of 5-hydroxytryptophan in the rat: Antagonist effects of purported 5-hydroxytryptamine antagonists and of pirenperone, an LSD antagonist. *Neuropharmacology* **22,** 993.

—— and Slangen, J. L. (1982). *Drug discrimination: Applications in CNS pharmacology.* Elsevier, Amsterdam.

—— Niemegeers, C. J. E. and Janssen, P. A. J. (1982). A drug discrimination analysis of lysergic acid diethylamide (LSD): In vivo agonist and antagonist effects of purported 5-hydroxytryptamine antagonists and of pirenperone, an LSD antagonist. *Neuropharmacology* **22,** 993.

—— —— —— (1979). In vivo evidence of partial agonist activity exerted by purported 5-hydroxytryptamine antagonists. *Eur. J. Pharmac.* **58,** 509.

Commissaris, R. L., Semeyn, D. R., Moore, K. E. and Rech, R. H. (1980). The effects of 2,5-dimethoxy-4-methylamphetamine (DOM) on operant behavior: Interactions with other neuroactive agents. *Commun. Psychopharmac.* **4,** 393.

DeJong, A. P., Huggins, F., Fournier, D. and Makriyannis, A. (1982). Inhibition of ³H-5-HT binding to rat brain membranes by psychotomimetic amphetamines. *Eur. J. Pharmac.* **83,** 305.

Downing, D. F. (1964). Psychotomimetic compounds. In *Psychopharmacological agents* (ed. M. Gordan) Vol. 1, p. 555. Academic Press, New York.

Dyer, D. C. and Gant, D. W. (1973). Vasoconstriction produced by hallucinogens on isolated sheep and human umbilical vasculature. *J. Pharmac. Exp. Ther.* **184**, 366.

—— Nichols, D. E., Rusterholz, D. B. and Barfknecht, C. F. (1973). Comparative effects of stereoisomers of psychotomimetic phenylisopropylamines. *Life Sci.* **13**, 885.

Ehrenpreis, S. and Teller, D. N. (1972). Interaction of drugs of dependence with receptors. In *Chemical and biological aspects of drugs of dependence* (eds. S. J. Mule and H. Briss) p. 177. CRC Press, Cleveland.

Farrow, J. T. and van Vunakis, H. (1973). Characteristics of *d*-lysergic acid diethylamide binding to subcellular fractions derived from rat brain. *Biochem. Pharmac.* **22**, 1103.

Fillion, G. M. B., Rouselle, J-C., Fillion, M-P., Beaudoin, D. M., Goiny, G. M., Deniau, J-M. and Jacob, J. J. (1978). High-affinity binding of ^{3}H-5-hydroxytryptamine to brain synaptosomal membranes: Comparison with ^{3}H-lysergic acid diethylamide binding. *Mol. Pharmac.* **14**, 50.

Freedman, D. X. and Halaris, A. E. (1978). Monoamines and the biochemical mode of action of LSD at synapses. In *Psychopharmacology: A generation of progress.* (eds. M. A. Lipton, A. DiMascio and K. F. Killam) p. 347. Raven Press, New York.

Gardner, C. R. and Guy, A. P. (1983). Behavioural effects of RU-24969, a 5-HT$_1$ receptor agonist, in the mouse. *Br. J. Pharmac.* **78**, 96 P.

Glennon, R. A. (1979). The effect of chirality on serotonin receptor affinity. *Life Sci.* **24**, 1487.

—— (1983). Hallucinogenic Phenylisopropylamines: Stereochemical Aspects. In *Stereoisomers: drugs in psychopharmacology* (ed. D. F. Smith). CRC Press, Boca Raton.

—— and Rosecrans, J. A. (1981). Speculations on the mechanism of action of hallucinogenic indolealkylamines. *Neurosci. Biobehav. Rev.* **5**, 197.

—— and Young, R. (1982). Comparison of behavioral properties of di- and tri-methoxyphenylisopropylamines. *Pharmac. Biochem. Behav.* **17**, 603.

—— Jacyno, J. M. and Young, R. (1983a). A comparison of the behavioral properties of (±)-, (+)- and (−)-5-methoxy-α-methyltryptamine. *Biol. Psychiat.* **18**, 493.

—— Liebowitz, S. M. and Anderson, G. M. (1980). Serotonin receptor affinities of psychoactive phenalkylamine analogs. *J. Med. Chem.* **23**, 294.

—— —— and Mack, E. C. (1978). Serotonin receptor affinities of several hallucinogenic phenalkylamine and *N,N*-dimethyltryptamine analogues. *J. Med. Chem.* **21**, 822.

—— Rosecrans, J. A. and Young, R. (1982a). The use of the drug discrimination paradigm for studying hallucinogenic agents. In *Drug discrimination: Applications in CNS pharmacology* (eds. F. C. Colpaert and J. L. Slangen) p. 69. Elsevier, Amsterdam.

—— —— —— (1983c). Drug-induced discrimination: A description of the paradigm and a review of its specific application to the study of hallucinogenic agents. *Med. Res. Rev.* **3**, 289.

—— Young, R. and Jacyno, J. M. (1983d). Indolealkylamine and phenalkylamine hallucinogens: Effect of α-methyl and N-methyl substituents on behavioral activity. *Biochem. Pharmac.* **32**, 1267.

—— —— and Rosecrans, R. A. (1983e). Antagonism of the effects of the hallucinogen DOM and the purported 5-HT agonist quipazine by 5-HT$_2$ antagonists. *Eur. J. Pharmac.* **91**, 189.

—— Jacyno, J. M., Young, R., McKenney, J. D. and Nelson, D. L. (1983b). Synthesis and evaluation of a novel series of N,N-dimethylisotryptamines. *J. Med. Chem.* (In press).

—— Rosecrans, J. A., Young, R. and Gaines, J. (1979). Hallucinogens as discriminative stimuli: Generalization of DOM to a 5-methoxy-N,N-dimethyltryptamine stimulus. *Life Sci.* **24**, 993.

—— Young, R., Bennington, F. and Morin, R. D. (1982b). Behavioral and serotonin receptor properties of 4-substituted derivatives of the hallucinogen 1-(2,5-dimethoxyphenyl)-2-aminopropane. *J. Med. Chem.* **25**, 1982.

Green, A. R. (1981). Pharmacological studies on serotonin-mediated behavior. *J. Physiol. (Paris)* **77**, 437.

—— O'Shaughnessy, K., Hammond, M., Schachter, M. and Grahame-Smith, D. G. (1983). Inhibition of 5-hydroxytryptamine-mediated behavior by the putative 5-HT$_2$ antagonist pirenperone. *Neuropharmacology.* **22**, 573.

—— Guy, A. P. and Gardner, C. R. (1984). The behavioural effects of RU 24969, a suggested 5-HT receptor agonist in rodents and the effect on the behaviour of treatment with antidepressants. *Neuropharmacology* **23**, 655.

Green, J. P., Johnson, C. L., Weinstein, H., Kang, S. and Chou, D. (1978). Molecular determinants for interaction with the LSD receptor: Biological studies and quantum chemical analysis. In *The psychopharmacology of hallucinogens* (eds. R. C. Stillman and R. E. Willette) p. 28. Pergamon Press, New York.

Gozlan, H., El Mestikawy, S., Pichat, L., Glowinski, J. and Hamon, M. (1983). Identification of presynaptic serotonin autoreceptors using a new ligand: ^{3}H-PAT. *Nature* **305**, 140.

Hirschhorn, I. and Winter, J. C. (1971). Mescaline and lysergic acid diethylamide (LSD) as discriminative stimuli. *Psychopharmacologia,* **22**, 64.

Huang, J. and Ho, B. T. (1974). Effect of 2,5-dimethoxy-4-methylamphetamine on heart and smooth muscle contraction. *J. Pharm. Pharmac.* **26,** 69.

—— —— (1975). Some pharmacological actions of 2,5-dimethoxy-4-ethylamphetamine (DOET) in rats and mice. *J. Pharm. Pharmac.* **27,** 18.

Innes, I. R. (1963). Action of dexamphetamine on 5-hydroxytrypamine receptors. *Br. J. Pharmac.* **21,** 427.

Jacobs, B. L. (1982). Mechanism of action of hallucinogenic drugs: Focus upon postsynaptic serotonergic receptors. In *Psychopharmacology 1. Part 1: Preclinical psychopharmacology* (eds. D. G. Grahame-Smith and D. J. Cowen). Excerpta Media, Amsterdam.

Janssen, P. A. J. (1983). 5-HT$_2$ Receptor blockade to study serotonin-induced pathology. *Trends Pharmac Sci.* **4,** 198.

Kantor, R. E., Dudlettes, S. D. and Shulgin, A. T. (1980). 5-Methoxy-α-methyltryptamine (α-*O*-dimethylserotonin): A hallucinogenic homolog of serotonin. *Biol. Psychiat.* **15,** 349.

Kuhn, D. M., White, F. J. and Appel, J. B. (1978). The discriminative stimulus properties of LSD: Mechanism of action. *Neuropharmacology* **17,** 257.

Leysen, J. E. and Tollenaere, J. P. (1982). Biochemical models for serotonin receptors. *Ann. Rep. Med. Chem.* **17,** 1.

—— Niemegeers, C. J. E., van Nueten, J. M. and Laduron, P. M. (1982). ^{3}H-Ketanserin (R-41, 468), a selective ^{3}H-ligand for serotonin-2 receptor binding sites. *Mol. Pharmac.* **21,** 301.

Lucki, I. and Frazer, A. (1982). Behavioral effects of indole- and piperazine-type serotonin receptor agonists. *Abstr. Soc. Neurosci.* **8,** 101.

Martin, W. R. and Sloan, J. W. (1977). Pharmacology and classification of LSD-like hallucinogens. In *Drug addiction* (ed. W. R. Martin) Vol. II, p. 305. Springer-Verlag, Berlin.

—— Vaupel, D. B., Nozaki, M. and Bright, L. D. (1978). The identification of LSD-like hallucinogens using the chronic spinal dog. *Drug Alcohol Dependence* **3,** 113.

Middlemiss, D. N. and Fozard, J. R. (1983). 8-Hydroxy-2-(di-*n*-propylamino) tetralin discriminates between subtypes of 5-HT$_1$ recognition sites. *Eur. J. Pharmac.* **90,** 151.

Minnema, D. J. (1981). *An examination of the central nervous system mechanisms underlying discriminative stimulus properties of lysergic acid diethylamide.* Doctoral Thesis, Medical College of Virginia, Virginia Commonwealth University, Richmond, Virginia.

Nair, X. (1974). Contractile responses of guinea pig umbilical arteries to various hallucinogenic agents. *Res. Commun. Chem. Path. Pharmac.* **9,** 535.

Naranjo, C. (1967). Psychotropic properties of the harmala alkaloids. In

Ethnopharmacological Search for Psychoactive Drugs (eds. D. H. Efron, B. Holmstedt and N. S. Kline) p. 385. U.S. Government Printing Office, Washington, D.C.

Nelson, D. L. (1982). Central serotonergic receptors: Evidence for heterogeneity and characterization by ligand binding. *Neurosci. Biobehav. Rev.* **6**, 499.

—— Pedigo, N. W. and Yamamura, H. I. (1981). Multiple ^{3}H-5-hydroxytryptamine binding sites in rat brain. *J. Physiol. (Paris)* **77**, 369.

—— Weck, B. and Taylor, W. (1983). Discrimination of heterogeneous serotonin$_1$ receptors by indolealkylamines. In *CNS receptors—From molecular pharmacology to behavior* (eds. F. V. DeFeudis and P. Mandel). Raven Press, New York. (In press).

Nichols, D. E. and Glennon, R. A. (1984). Medicinal chemistry and structure-activity relationships of hallucinogenic agents. In *Hallucinogens: neurochemical, behavioral and clinical perspectives* (ed. B. L. Jacobs), p. 95. Raven Press, New York.

—— Shulgin, A. T. and Dyer, D. C. (1977). Directional lipophilic character in a series of psychotominetic phenethylamine derivatives. *Life Sci.* **21**, 569.

Offermeier, J. and Ariens, E. J. (1966a). Serotonin I. Receptors involved in its action. *Archs. Int. Pharmacodyn.* **164**, 192.

—— —— (1966b). Serotonin II. Structural variation and action. *Archs. Int. Pharmacodyn.* **164**, 216.

Pedigo, N. W., Yamamura, H. I. and Nelson, D. L. (1981). Discrimination of multiple ^{3}H-5-hydroxytryptamine binding sites by the neuroleptic spiperone in rat brain. *J. Neurochem.* **36**, 220.

Peroutka, S. J. and Snyder, S. H. (1979). Multiple serotonin receptors: Differential binding of ^{3}H-5-hydroxytryptamine, ^{3}H-lysergic acid diethylamide and ^{3}H-spiroperidol. *Mol. Pharmac.* **16**, 687.

Rosecrans, J. A. and Glennon, R. A. (1979). Drug-induced cues in studying mechanisms of drug action. *Neuropharmacology* **18**, 981.

Sankar, D. V. S. (1975). *LSD: A total study.* PJD Publications, Westbury, New York.

Schultes, R. E. (1982). The β-carboline hallucinogens of South America. *J. Psych. Drugs* **14**, 205.

Shannon, M., Battaglia, G., Glennon, R. A. and Titler, M. (1984). S$_2$ and S$_1$ serotonin receptor binding properties of derivatives of the hallucinogen 1-(2,5-dimethoxyphenyl)-2-aminopropane. *Eur. J. Pharmac.* **102**, 23.

Shulgin, A. T. (1978). Psychotomimetic drugs: structure activity relationships. In *Handbook of psychopharmacology* (eds. L. L. Iversen, S. D. Iversen and S. H. Snyder) Vol. 11, p. 243. Pleunum Press, New York.

—— (1982). Chemistry of psychotomimetics. In *Psychotropic agents Part III* (eds. F. Hoffmeister and G. Stille) p. 3. Springer-Verlag, Berlin.

—— and Dyer, D. C. (1975). Potential psychotomimetics. 4-Alkyl-2,5-dimethoxyphenylisopropylamines. *J. Med. Chem.* **18,** 1201.

Silverman, P. B. and Ho, B. T. (1980). The discriminative stimulus properties of 2,5-dimethoxy-4-methylamphetamine (DOM): Differentiation from amphetamine. *Psychopharmacology* **68,** 209.

Smit, M. H., Glennon, R. A., Yamamura, H. I. and Nelson, D. L. (1981). Multiple serotonin receptors discriminated by analogs of 5-methoxytryptamine. *Soc. Neurosci. Abstr.* **7,** 7.

Standridge, R. T., Howell, H. G., Gylys, J. A., Partyka, R. A. and Shulgin, A. T. (1976). Phenalkylamines with potential psychotherapeutic activity. 2-Amino-1-(2,5-dimethoxy-4-methylphenyl)-butanes. *J. Med. Chem.* **19,** 1400.

Vane, J. R. (1959). The relative activities of some tryptamine analogues on the isolated rat stomach strip preparation. *Br. J. Pharmac.* **14,** 87.

—— (1960). The actions of sympathomimetic amines on tryptamine receptors. In *Adrenergic Mechanisms* (eds. J. R. Vane, G. E. W. Wolstenholme and M. O'Connor) p. 356. Little, Brown Publishers, Boston.

Wallach, M. B., Friedman, E. and Gershon, S. (1972). 2,5-Dimethoxy-4-methylamphetamine (DOM), A neuropharmacological examination. *J. Pharmac. Exp. Ther.* **182,** 145.

White, F. J. and Appel, J. B. (1982). Training dose as a factor in LSD-saline discrimination. *Psychopharmacology* **76,** 20.

—— —— and Kuhn, D. M. (1979). Discriminative stimulus properties of quipazine: Direct serotonergic mediation. *Neuropharmacology* **18,** 143.

Winter, J. C. (1975). Blockade of the stimulus properties of mescaline by a serotonin antagonist. *Archs. Int. Pharmacodyn.* **214,** 250.

—— (1978). Stimulus properties of phenethylamine hallucinogens and lysergic acid diethylamide: The role of 5-hydroxytryptamine. *J. Pharmac. Exp. Ther.* **204,** 416.

—— (1979). Quipazine-induced stimulus control in the rat. *Psychopharmacology* **60,** 265.

—— and Gessner, P. K. (1968). Phenoxybenzamine antagonism of tryptamines, their indene isosteres and 5-hydroxytryptamine in the rat stomach fundus preparation. *J. Pharmac. Exp. Ther.* **162,** 286.

Whitaker, P. M. and Seeman, P. (1978). High-affinity ^{3}H-serotonin binding to caudate: Inhibition by hallucinogenic agents and serotonergic drugs. *Psychopharmacology* **59,** 1.

Woolley, D. W. (1962). *The biochemical bases of psychoses,* or *The serotonin hypothesis about mental disease,* p. 181. John Wiley, New York.

Young, R., Glennon, R. A. and Rosecrans, J. A. (1980). Discriminative stimulus properties of the hallucinogenic agent DOM. *Commun. Psychopharmac.* **4,** 501.

11

Pharmacological studies of the role of serotonin in animal models of anxiety

COLIN R. GARDNER

11.1 Introduction

Early studies of the pharmacology of models of anxiety led to the suggestion that central serotonin systems mediate aversive responses (Cook and Sepinwall 1973; Stein, Wise and Belluzi 1977b) and that the anxiolytic effects of drugs such as benzodiazepines resulted from a reduction in the activity of these serotonin systems (Stein, Wise and Berger 1973). Recent pharmacological studies have yielded results inconsistent with these hypotheses (Schenberg and Graeff 1978; Shephard, Buxton and Broadhurst 1982) indicating the need for a critical reappraisal of the role or serotonin in models of anxiety and in the actions of anxiolytics.

It is not the purpose of this Chapter to consider the validity of the different models of anxiety. This could form the basis of another substantial review. The serotonin-related pharmacology has in general been performed using well-established animal models of anxiety and it has therefore been possible to concentrate on these models with the inclusion of others of particular interest.

11.2 Effect of lesions of serotonin pathways on animal models of anxiety

11.2.1 *Punished behaviour*

Depletion of serotonin with the tryptophan hydroxylase inhibitor (PCPA) has shown increases in food motivated, punished behaviour. The first multischedule model of anxiety routinely used was developed by Geller and Seifter (1960) and this 'conflict' method (Geller and Blum 1970) and several similar paradigms (Robichaud and Sledge 1969; Stein *et al.* 1973; Cook and Sepinwall 1975a; Tye, Iversen and Green 1979; Shephard *et al.*

1982; S. E. Green, Hodges and Summerfield 1983) have demonstrated release of punished behaviour with PCPA. PCPA depletes brain catecholamines to some extent in addition to serotonin. However, the effect of PCPA on punished behaviour probably involves depletion of serotonin as it can be reversed by the serotonin precursor 5-HTP (Robichaud and Sledge 1969; Geller and Blum 1970; Stein *et al.* 1973; Tye *et al.* 1979; S. E. Green *et al.* 1983) but not by the catecholamine precursor L-Dopa (Tye *et al.* 1979). However, the release of punished behaviour by PCPA has not invariably been observed despite its effectiveness in reducing brain serotonin (Blakely and Parker 1973; Cook and Sepinwall 1975*a*). Cook and Sepinwall (1975*b*) observed increased punished responding after PCPA methyl ester HCl but no such effect after PCPA free salt, despite similar degrees of serotonin depletion. Such a difference cannot be easily explained as being due to the relatively low shock levels employed (Cook and Sepinwall 1975*a*).

The effects of degenerative lesions of serotonin pathways in the brain provided further support for the involvement of serotonin in mediating the effects of punishment in this type of anxiety model. The serotonin neurotoxins 5,6-DHT and 5,7-DHT also release punished responding after intraventricular injection (Stein, Wise and Belluzzi 1975; Tye, Everitt and Iverson 1977) and after injection into the ascending serotonin pathways at the level of the medial forebain bundle (Tye *et al.* 1977). However, no increase in punished behaviour was observed by Thiebot, Hamon and Soubrié (1982) three weeks after injection of 5,7-DHT into the dorsal raphe nucleus although it remains possible that there would have been an effect at earlier times (see Sepinwall 1983).

Another conflict model of anxiety which is commonly used is based on the suppression of drinking in thirsty rats by punishment (foot shock or mouth shock). Lesions of central serotonin pathways have not yielded consistent effects in such models. Electrolytic lesions of the median raphe nucleus released punished licking, with a greater effect in a conditioned emotional response paradigm than in response-contingent punished drinking (Wirtshafter and Asin 1981). PCPA did induce a statistically significant increase in punished drinking, but the effect was of extremely low magnitude in comparison to those of reference anxiolytic agents in the same model (Petersen and Lassen 1981). In contrast, 5,6-DHT (but not 6-hydroxydopamine) injected intraventricularly increased punished drinking to a similar extent as diazepam (Lippa, Nash and Greenblatt 1979*a*). However, Commissaris, Lyness and Rech (1981) observed no increase in punished drinking from 4–5 days after 5,7-DHT in a model in which rats were tested repeatedly.

Foot shock suppression of ongoing locomotor activity in mice is reversed by anxiolytic agents, but PCPA was ineffective. However, the degree of serotonin depletion was not determined (Aron, Simon, Larousse and Boissier 1971).

However, not all authors have considered the effects of serotonin depletion to be a reduction in aversiveness. A lengthening of the post-reinforcement pause by PCPA in a fluid rewarded FR20 schedule was interpreted as being due to an increase in the aversiveness of environmental stimuli (Campbell, Brown and Seiden 1971). Milk drinking, grooming, petting and suckling in cats are associated with EEG synchronization. This group of responses have been generically called 'relaxation behaviour'. The EEG synchronization due to milk drinking is reduced by PCPA and enhanced by 5-HTP (Cervantes, Ruelas and Beyer 1983). Lesions of dorsal and median raphe nuclei result in increased reactivity to the experimenter and enhanced shock-induced fighting which were thought to represent increased defensive behaviour (Albert and Walsh 1982). Lesions of ascending serotonin pathways increase locomotor activity (Carey 1976; Geyer, Puerto, Menkes, Segal and Mandell 1976; Segal 1976) and increase responsiveness to acoustic startle (Davis 1980). Increased sensitivity to electric shock has been observed after PCPA (Tenen 1967; Harvey and Lints 1971). It is difficult to reconcile release of punished behaviour with increased sensitivity to the punishment. The effect of serotonin depletion in the conflict situation could be a complex function of changes in sensitivity to both aversive and positively reinforcing stimuli.

However, the most generally accepted view is that there exists in the brain a serotonin-mediated 'punishment' system (Stein *et al.* 1973, 1977*b*; Crow and Deakin 1981) which, when serotonin is depleted results in release of punishment-suppressed behaviour, but it should be noted that such releases of punished behaviour have not been consistently observed in all studies and this may depend upon the method employed as well as other factors.

Some possible alternative interpretations for the effects of serotonin depletion on punished behaviour could be considered (Iversen 1980). Many of these punished behaviours are motivated by appetitive rewards (food or drink). Depletion of central serotonin pathways leads to hyperphagia (Blundell 1977; Hoebel 1977; Blundell and Latham 1979*b*) although this can be complicated by peripheral serotonin depletion with PCPA (Hoebel 1977). Similarly, depletion of central serotonin can result in polydipsia (Barofsky, Grier and Pradhan 1980) and such an effect can give rise to increased punished responding in drinking conflict (Patel and Malick 1980; Cooper 1983; Leander 1983). Thus it is possible that serotonin depletion may enhance appetitive motivation rather than decrease the response to aversion in these conflict procedures. Increased sensitivity to environmental stimuli might constitute a stress leading to the phenomenon of stress-induced eating (Morley, Levine and Rowland 1983). Such a phenomenon might prevail despite punishment of the food-motivated behaviour, leading to a measured increase in punished responding. Similarly, increased responsiveness to environmental stimuli [possibly by inhibiting serotonin-mediated 'stress analgesia' (Terman, Lewis and Liebeskind 1983)] might lead to an enhancement of opioid-mediated stress analgesia (Kelly 1983) which does not involve serotonin and is activated by different qualities of stressful stimuli. Such a change might result in an increase in punished responding whilst in the conflict apparatus. In fact, naloxone has been shown to enhance the effects of foot shock punishment (Young 1980) and taste aversion (Pilcher, Stolerman and D'Mello 1978) and to reverse the effects of benzodiazepines in such tests (Billingsley and Kubena 1978; Duka, Cumin, Haefely and Herz 1981). This leaves it possible that the opioid mechanism is operative in these circumstances. Whilst this suggestion is speculative, it may be worth further investigation.

11.2.2   Other models of anxiety

Attempts have been made to develop models of anxiety which employ more biologically relevant stimuli (than electric shock) to induce a state of anxiety. Recently a simple model has been developed which does not involve foot shock punishment and is sensitive to anxiolytic benzodiazpines and non-sedative anxiolytic candidates (see Table 11.1). This model is based on the emission of ultrasonic vocalizations induced by separation and physical manipulation of 10–13-day-old rat pups (Gardner 1984). These vocalizations are inhibited by anxiolytic agents at doses having no depressant effect on overt behaviour and postnatal PCPA pre-treatment induces a similar response. Postnatal pre-treatment with 6-hydroxydopamine did not inhibit ultrasound production (Hard, Engel and Musi 1982). Similarly a freezing response (which is generally considered to reflect fear or anxiety) can be induced in 25-day-old pups by audiogenic stimulation and this is also inhibited by neonatal treatment with 5,7-DHT (Hard, Ahlenius and Engel 1983).

   File and Hyde (1978) proposed that uncertainty as to source, nature and timing of events is a potent cause of anxiety. Using this principle they observed a reduction of social interaction between pairs of male rats placed in a novel, well-lit environment. This model has subsequently been shown to be sensitive to the anxiolytic effects of benzodiazepines, which increase the suppressed social interaction (File 1980). In this model, both PCPA pre-treatment (File and Hyde 1977) and 5,7-dihydroxytryptamine lesions of the dorsal raphe (File, Hyde and MacLeod 1979) and the lateral septum, one of the dorsal raphe projection areas (Clarke and File 1981), produced a behavioural profile similar to that seen in rats treated chronically with benzodiazepines. In contrast, 6-hydroxydopamine lesions in the lateral

Table 11.1 Effect of metergoline and some anxiolytic drugs on stress induced ultrasounds in rat pups

	Percentage inhibition of ultrasounds (dose mg/kg i.p.)								
	0.2	0.5	1	2	2.5	4	5	10	20
Metergoline	13.9		31.0*		60.5*		72.6*	97.5*	
Chlordiazepoxide	30.9	46.7*	43.4*		73.2*			74.3*	
Diazepam	50.9*	71.4*	83.0*	72.9*		61.1*			
Premazepam				72.3*			88.3*	80.3*	90.3*
CL 218872		36.2	73.5*	17.2			39.5*	59.4*	69.0*

The number of ultrasounds (42 ± 5 kHz) induced by 30 s suspension by the tail were recorded with a QMC Bat Detector and the mean number from the drug treated pups ($n \geqq 5$) was compared with that of vehicle treated littermate controls. Drugs were administered 30 min prior to testing. *p <0.05 (Student's t test). Metergoline induced a response similar to the non-sedative anxiolytics premazepam and CL 218872 in reducing ultrasounds without affecting overt behaviour.

septum did not show the same effect. However, not all such serotonin pathway lesions have this effect on social behaviour. 5,7-Dihydroxytryptamine injections into the medium raphe or both the dorsal and median raphe nuclei or into the amygdala decreased social interaction (File and Deakin 1980; File, James and MacLeod 1981) which may result from a general decrease in activity following the lesions. These observations highlight the possibility of different influences of serotonin pathways on specific behaviours, even of opposing influences of different ascending serotonin pathways on some behaviours.

A similar hypothesis was proposed by Schenberg and Graeff (1978) on the basis of studies of escape behaviour after aversive electrical stimulation of the dorsal periaqueductal grey (PAG) region of the brain. In addition to the serotonin system involved in conflict behaviour, a second serotonin system involved in PAG stimulated responses has been proposed on which anxiolytic drugs and the reduction of activity of serotonin pathways have opposite effects. PCPA enhances the fear-like nature of the response to PAG stimulation in the rat (Kiser and Lebovitz 1975; Moriyama, Ichimaru and Gomita 1982). However, PCPA did not affect a similar response observed in cats (Wada and Matsuda 1971).

11.2.3   *General considerations*

In summary, the proposed anti-punishment (equated to anti-anxiety) effect of serotonin depletion, although not consistently observed, is supported by the effects of serotonin depletion in stress-induced ultrasounds and dorsal raphe lesions in the social interaction model of anxiety. Such models avoid the possibility of 'false positive' effects due to appetitive effects or changes in sensitivity to foot shock stimuli. However, the enhancement by PCPA of fear-like behaviour after PAG stimulations suggests heterogeneity in the involvement of separate serotonin pathways in different models of anxiety. The role of different serotonin pathways in the PAG stimulation response remains to be investigated. PCPA does not seem to have anxiolytic actions in man (see File 1981) and induced irritability and mild, psychological changes in patients with carcinoid tumours (Cremata and Koe 1966; Engelman, Lovenberg and Sjoerdsma 1967). This questions the clinical relevance of its activity in some models of anxiety.

11.3   Stimulation of central serotonin systems

11.3.1   *Electrical and direct chemical stimulation of raphe nuclei*

Increased dorsal raphe neurone firing has been correlated with fearful interactions of tree shrews (Walletschek and Raab 1982) but, there is little support for the involvement of the dorsal raphe nucleus in induction of anxiety responses from experiments involving electrical stimulation of the nucleus. Jacobs, Asher and Dement (1973) observed mild arousal and orientating responses in cats with 10 s trains of square wave stimuli at 10 Hz but little behavioural effect at 3 Hz. This contrasts with several earlier reports of induction of sedation, electroencephalogram synchronization and sleep induced by low frequency stimulation (Kostowski, Giacolone, Garattini and Valzelli 1969; Gumulka, Samanin, Valzelli and Consolo 1971; Kostowski 1971). Indeed dorsal raphe stimulation in rats with 100 ms pulse trains at 2 Hz markedly inhibited PAG stimulation fear-like behaviour (Kiser, Brown, Sanghera and German 1980). By itself the raphe stimulation had no overt behavioural effects. Dorsal raphe stimulation has resulted in reduced food intake (Siegel and Brownstein 1975) but in none of these studies were there signs of aversive responses.

Contrary to this, stimulation of the median raphe nucleus with a 60 Hz sine wave induced behavioural inhibition (lever pressing) accompanied by defecation, crouching, micturition, piloerection and teeth chattering. This behavioural syndrome was likened to that seen from rats in stressful situations (Graeff and Silveiera Filho 1978).

Chemical stimulation of the dorsal raphe nucleus in rats by application of crystalline carbachol suppressed both punished and unpunished lever pressing for food in three rats tested. Implantation sites outside the raphe nucleus had little or no effect (Stein *et al.* 1977*b*). Oxazepam reversed this suppression in all of these rats.

Gray (1976) has argued that a serotonin-mediated behavioural inhibitory system is activated by a range of aversive stimuli such as conditioned warning stimuli, sight of predators or dominant con-specific, novelty and conditioning stimuli associated with non-reward. In man, it was proposed that the activation of the behavioural inhibition system would be associated with experienced fear, anxiety or disappointment, depending on the environmental circumstances (Gray 1976). The above evidence from experiments involving raphe stimulation supports the role of ascending serotonin pathways in suppression of ongoing behaviour but lends little support to the proposed induction of an anxiety experience. Anxiety has been associated with the inhibition of inappropriate ongoing activity but also with induction of other 'anxious' behaviours (see Graeff and Silveira Filho 1978; Gardner, Piper and Kidd 1981) and displacement behaviours, and there is little evidence that activation of serotonin systems leads to such behaviours. Indeed, suppression of food-motivated lever pressing by 5-HTP has been suggested by a model of depression (Nagayama, Hingtgen and Aprison 1981).

11.3.2   *Central administration of serotonin*

Intraventricular serotonin decreases several behaviours such as self-stimulation (Wise, Berger and Stein 1973), locomotor and exploratory activity (Herman 1975) and startle responses (Geyer, Warbritton, Menkes, Zook and Mandrell 1975). Injection of serotonin bilaterally into the nucleus accumbens, a terminal area for serotonin fibres ascending from the median raphe nucleus, markedly reduced amphetamine-induced loco-motor hyperactivity (Pycock, Horton and Carter 1978). However, in the conflict procedure a complex response was observed (Stein, Belluzzi and Wise 1977*a*). An initial intense but transient behavioural depression was followed by facilitation which included release of punished behaviour. This gave way to long-lasting behavioural suppression. Both unpunished and punished behaviour was suppressed, including that released by oxazepam. Intracarotid serotonin has been observed to induce first arousal and then longer lasting hypersynchrony of the electroencephalogram (Koella and Czicman 1966). This suggests that a unified theory of the actions of sero-tonin on behaviour may be a simplification. Indeed, on the basis of lesion studies, it has been proposed that activation of ascending serotonin path-ways would lead to increased locomotor activity in an open field (Hole, Fuxe and Jonsson 1976).

11.3.3   *Interpretation problems with serotonin stimulation*

The actions of drugs on central serotonin function are complicated by the existence of many possible sites of action. There are at least two function-ally separate postsynaptic serotonin receptors. They have been designated 5-HT$_1$ (those labelled by [^{3}H]-serotonin) and 5-HT$_2$ (those labelled by [^{3}H]-spiperone) (Peroutka and Snyder 1981; Blackshear, Steranka and Sanders-Bush 1981; Ennis and Cox 1982; Nelson 1982; Quik and Azmitia 1983, and see Chapter 4). In addition, there are receptors on serotonin nerve terminals, activation of which inhibits serotonin release (Cerrito and Raiteri 1979; Cox and Ennis, 1982; Martin and Sanders-Bush 1982; Mounney, Brady, Carroll, Fisher and Middlemiss 1982; Gothert and Schlicker 1983 and see Chapter 2). This type will be referred to as pre-synaptic receptors in distinction from receptors which exist on the serotonin cell bodies which are referred to as autoreceptors (de Montigny and Aghajanian 1977; Aghajanian 1978) which may be innervated by collaterals from these serotonin cells or by axons from serotonin cells in other nuclei (Wang and Aghajanian 1978; Mosko, Haubrich and Jacobs 1977). Furthermore, there are serotonin receptors on the nerve terminals of neurones which release other neurotransmitters (Ennis, Kemp and Cox 1981) which may or may not be innervated (Fig. 11.1). Even assuming that

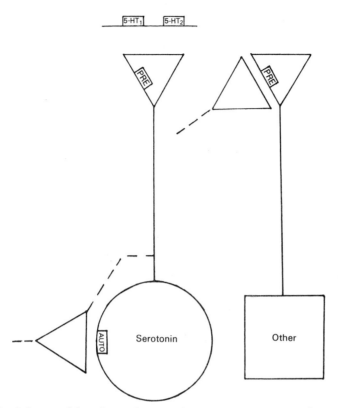

Fig. 11.1 Scheme of locations of serotonin receptors on serotonin and other neurones. There are two main divisions of postsynaptic receptors (5-HT$_1$ and 5-HT$_2$) as well as presynaptic receptors on nerve terminals (PRE) of both serotonin and other neurones. In addition there are serotonin autoreceptors on serotonin cell bodies (AUTO). Broken lines represent neuronal circuitry not certainly established.

the effects of drugs on behaviour are not mediated via effects on serotonin receptors in the periphery (Wallis 1981) or other pharmacological actions on serotonin or other neurotransmitter systems their effects still depend on their relative potencies and agonist/antagonist nature at the different sites as well as the relative functional significance of each site.

Most agonists have effects on both 5-HT$_1$ and 5-HT$_2$ receptors and, despite inhibition of serotonin cell activity and serotonin release via activation of autoreceptors and presynaptic receptors, they still activate the postsynaptic receptors of most or all serotonin pathways and this may result in most or all their behavioural effects. Quipazine is particularly interesting in that it is an agonist at postsynaptic 5-HT$_1$ and 5-HT$_2$ receptors (Green, Youdim and Grahame-Smith 1976; Malick, Doren and Barnett 1977; Jacoby, Howd, Levin and Wurtman 1975; Fuller, Mason and

Molloy 1980) and at autoreceptors (Blier and de Montigny 1983) but an antagonist at presynaptic receptors (Martin and Sanders-Bush 1982*b*), as well as some peripheral receptors (Lansdown, Nash, Preston, Wallis and Williams 1980). This different profile of activity on the receptors may not result in any different functional activation of serotonin systems as there would be little release of serotonin, due to reduced cell firing following autoreceptor activation, for the blockade of presynaptic receptors to potentiate. The postsynaptic receptor activation would remain the predominant effect.

The effects of serotonin agonists on behaviour are further complicated by the motor effects they evoke, particularly in rodents. The dominant overt effects of agonists are a 'myoclonic' syndrome including head-twitching in mice (see Section 12.3) and a syndrome of reciprocal forepaw treading, hindlimb abduction, lateral head weaving, Straub tail and 'wet dog shakes' in rats (see Section 12.2). All these effects may be mediated via descending serotonin pathways (Jacobs and Klemfuss 1975; Gerson and Baldessarini 1980). Study of this syndrome would therefore be inappropriate for investigation of the ascending serotonin pathways and might influence the responses of animals in models of anxiety. However, lesion of ascending serotonin pathways led to enhancement of the syndrome, which was interpreted as super-sensitivity of serotonin receptors (Hole *et al.* 1976). This might be explained by the presence of multibranched serotonin neurones with both ascending and descending fibres (Lovick and Robinson 1983). Nevertheless, disruption of other behaviours by this syndrome makes interpretation of responses in models of anxiety difficult.

11.3.4   *Direct and indirect serotoninomimetics*

5-HTP suppresses food reward behaviour in the pigeon (Aprison and Ferster 1961) and the rat (Nagayama *et al.* 1981). Furthermore, the long-lasting serotonin agonist, α-methyltryptamine, suppresses both punished and unpunished food-reward behaviour in the pigeon (Graeff and Schoenfeld 1970) and in the rat (Winter 1972; Stein *et al.* 1973). Another agonist, *N,N*-dimethyltryptamine, also decreases both punished and unpunished responses (Winter 1972) although 5-MeODMT has recently been shown to decrease unpunished but not punished responding for sweetened condensed milk (Shephard *et al.* 1982). The non-tryptamine agonist, quipazine, decreased both punished and unpunished drinking of a liquid diet (Commissaris and Rech 1982). This effect could be reversed by the serotonin antagonist, metergoline. As increased anxiety would lead to depression of all activity it is not possible to determine whether these effects are general depression of motor systems or anxiety-induced behavioural inhibition. As with serotonin depletion, the effect of direct or indirect serotonin agonists on hunger and thirst must be considered. These agents decreased both food and water intake (Blundell and Latham 1979*a, b*; Clineschmidt 1979; Samanin, Mennini, Ferraris, Bendotti, Borsini and Garattini 1979; Soulairac and Soulairac 1970) and such actions could contribute to the decreases in food or liquid motivated behaviour, whether punished or unpunished. It is possible that recently observed increases in conditioned avoidance in a shuttle box by *m*-chlorophenylpiperazine (Ventulani, Sandone, Bednarczk and Hano 1982) may represent increased reactivity to potential punishment.

Aversive effects due to electrical brain stimulation, however, are reduced by serotonin agonists. Stimulation of cerebral structures, usually the dorsal periaqueductal grey and lateral hypothalamus induces escape reactions which are reduced by administration of 5-HTP (Patkina and Lapin 1976; Kiser, German and Lebovitz 1978; Moriyama *et al.* 1982), 5-MeODMT (Sinden and Atrens 1978; Cazala and Garrigues 1983) or local perfusion of serotonin (Leroux and Myers 1975). It is not clear if the serotonin-mediated decrease of behaviour is a general effect on all behaviours including anxiety responses or whether there are opposed serotonin influences on anxiety which are uncovered by the different methodologies (Schenberg and Graeff 1978).

11.3.5   *Clinical activity of serotoninomimetics*

There is little clinical evidence supporting an anxiety inducing effect following enhancement of central serotonin systems. An interesting syndrome is induced by excessive atmospheric cations. The syndrome includes anxious feelings, is associated with high serum serotonin and has been called a 'serotonin irritation syndrome' (Giannini 1979). Clinical studies with serotonin precursors, L-tryptophan and 5-HTP, however, have not shown induction of anxiety (Trimble, Chadwick, Reynolds and Marsden 1975; Puhringer, Wirz-Justice and Lancranjan 1976; Cole, Hartmann and Brigham 1980). In fact, an investigation of the anxiolytic properties of L-tryptophan has been suggested (Wilbur and Kulik 1981). L-Tryptophan has been claimed to be useful in obsessive-compulsive neuroses (Yaryura-Tobias and Bhagavan 1977). Early clinical studies in depression suggested a sub-type of patients showing mainly agitation and anxiety which responded to L-tryptophan (Fujiwara and Otsuki 1974). Furthermore, clinical experience with fenfluramine, a serotonin releaser used as an anorectic (Pinder, Brogden, Sawyer, Speight and Avery 1975), and the serotonin agonists quipazine and MK-212 (Parati, Zanardi, Cocchi, Caraceni and Muller 1980) has shown no consistent anxiogenic effects.

On the other hand serotonin re-uptake inhibitors, which might be expected to enhance serotonin transmission acutely, have been shown to be effective in treating phobic anxiety states (zimelidine: Evans, Best, Moore and Cox 1980; Koczkas, Holmberg and Wedin 1981; Evans and Moore 1981) and obsessive compulsive disorders, phobic disorders and panic attacks (chlorimipramine: Ananth, Solyom and Bryntwick 1979; Gloger, Grunhaus, Birmacher and Troudart 1981).

Thus, although there is little direct evidence, clinical experience lends more support to enhancement of serotonin systems resulting in an anxiolytic effect rather than induction of anxiety. However, this does not exclude the possibility that pharmacologically distinct sub-types exist within the entire spectrum of anxiety symptoms. For instance, Wilbur and Kulik (1981) speculated that conflict models may be predictive of anticipation anxiety but not of panic anxiety, atypical depression or other syndromes with morbid anxiety.

11.3.6   *Functionally distinct serotonin pathways*

The demonstration of separate serotonin binding sites (Peroutka and Snyder 1981) has led to attempt to demonstrate different functions for the different sites. Electrophysiological studies have identified an inhibitory response of serotonin which may be mediated via $5\text{-}HT_1$ receptors and an excitatory response mediated via $5\text{-}HT_2$ receptors (Jones 1982; Bradshaw, Stoker and Szabadi 1983).

The discovery of a serotonin agonist, selective for $5\text{-}HT_1$ sites has enabled the behavioural investigation of the functional role of these sites and such studies are described in detail in Section 12.5. At this point I shall merely point out that RU 24969 is a potent displacer of $[^3H]$-serotonin binding (Hunt, Nedelec, Euvrard and Boissier 1981; Hunt and Oberlander 1981) but induces only part of the behavioural syndrome describe for other serotonin agonists; hyperlocomotion and hyper-reactivity (Gardner and Guy 1983; Green, Guy and Gardner 1984). It is doubtful whether this type of hyper-reactivity which is induced by most serotonin agonists (Jacobs 1976; Deakin and Green 1978; Green, O'Shaugnessy, Hammond, Schachter and Grahame-Smith 1983 b; Segal and Weinstock 1983), can be associated with anxiety. In a licking conflict procedure (Gardner *et al.* 1981) RU 24969 induced decreases in punished responding (Fig. 11.2). However, it is not possible to ascribe this effect to an enhancement of responses to punishment in view of hyperlocomotion induced at the effective doses which may have disrupted all other behaviour. However, the functional significance of $5\text{-}HT_1$ sites is still not clear when the activity of 8-hydroxy-2-(di-*n*-propylamino)-tetralin (8-OH-DPAT) is considered. This substance is selective for a high affinity $5\text{-}HT_{1A}$ sub-site, being in the order of 1000 times less potent on the low affinity $5\text{-}HT_{1B}$ sub-site. Its

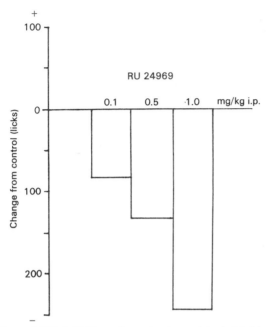

Fig. 11.2 Effect of RU 24969 on foot shock-punished drinking in thirsty rats (Gardner *et al.* 1981). Results are expressed as the difference in the mean number of licks per 5 minutes of groups of 8 drug-treated rats and that of a within-experiment control group. Having previously experienced the shock in a non-drug situation 4/8 control rats did not respond ($<$40 licks) on the test day. The proportion was similar for rats treated with 0.1 and 0.5 mg/kg RU 24969 (3/8, 4/8) but all the rats treated with 1 mg/kg RU 24969 did not respond.

potency on 5-HT$_2$ sites is similar to that on 5-HT$_{1B}$ sites (Middlemiss and Fozard 1983). This substance induces the full syndrome of serotonin agonist effects (Hjorth, Carlsson, Lindberg *et al.* 1982 and see Section 12.6). Further studies are clearly required to determine the mechanism underlying the selective behavioural syndrome induced by RU 24969.

Several lines of evidence previously discussed have led to the suggestion that there are different and functionally opposed serotonin systems in the brain (Schenberg and Graeff 1978; File *et al.* 1981). Such a functional antagonism was recently suggested to underlie serotonin-related 'wet dog shake' behaviour in rats (Harrison-Read 1984). Some of the serotonin agonist-induced behaviours are mediated via subsequent effects on central dopamine pathways (Green and Grahame-Smith 1974). Hyperlocomotion and hyper-reactivity are probably mediated via a dopamine system (Deakin and Dashwood 1981). This remains true when these behaviours are selectively induced with RU 24969 (Oberlander and Boissier 1981; Oberlander 1983; Green *et al.* 1984). Serotonin antagonists such as metergoline block agonist-induced forepaw treading, head weaving and hindlimb abduction in rats but potentiate the hyperlocomotion (Crow and Deakin 1977; Green *et al.* 1984). This has led to the suggestion that hyperlocomotion is induced by activation of one serotonin system, possibly involving 5-HT$_1$ sites and mediated via a dopamine system, and that activation of another serotonin system involving 5-HT$_2$ sites leads to a functionally opposite effect at the level of the dopamine system (Green *et al.* 1984 and Section 12.5).

In such functionally opposed serotonin systems do exist then they may influence brain regions involved in the generation of anxiety (Schenberg and Graeff 1978; File *et al.* 1981). Thus, the effects of modulation of serotonin systems on models of anxiety would depend on the brain regions modulated and which serotonin pathways were most involved in the particular model studied.

11.4   Injection of chemicals into the raphe nuclei

An alternative means of exciting or inhibiting central serotonin pathways is to inject chemicals into the raphe nuclei to affect the serotonin cell bodies directly. The suppression of behaviour by application of carbachol into the dorsal raphe nucleus, presumably to excite serotonin neurones, has already been mentioned (Stein *et al.* 1977*a*).

11.4.1   *Serotonin*

Serotonin itself will modulate the activity of neurones in the raphe nuclei. A body of evidence from studies in anaesthetised rats suggests that serotonin and serotonin agonists inhibit neurones in the dorsal raphe nucleus when they are applied by iontophoresis (de Montigny and Aghajanian 1977; Aghajanian 1978; Blier and de Montigny 1983). Furthermore, there is probably serotonin-mediated collateral inhibition in this nucleus (Wang and Aghajanian 1978). Intravenous injection of 5-MeODMT and LSD in cats also strongly inhibits dorsal raphe neurones although neurones in the raphe pallidus were relatively insensitive (Jacobs, Heym and Rasmussen 1983). Neurones in the raphe magnus of the cat were excited by serotonin (Llewelyn, Azami and Roberts 1983), an observation similar to that in the raphe pontis and raphe medianus, where excitation was the predominant response (Couch 1970; 1976). It is not certain however, that all of these responses are from serotonergic neurones.

Serotonin has been injected into the dorsal raphe nucleus of rats performing a conflict procedure. On the basis of the above evidence the neurones in this nucleus would be inhibited by serotonin. A release of punished behaviour was observed (Soubrié, Thiebot, Jobert and Hamon 1981; Thiebot *et al.* 1982).

11.4.2   *GABA and benzodiazepines*

In a similar way γ-aminobutyric acid (GABA) and the benzodiazepines diazepam and chlordiazepoxide, both of which might be expected to inhibit neuronal activity (Gallager 1978), released punished behaviour. Serotonin and chlordiazepoxide were synergistic. An inhibitory effect of chlordiazepoxide applied to the dorsal raphe nucleus was indicated by reduced release of serotonin in terminal areas of serotonin axons (Soubrié, Blas, Ferron and Glowinski 1983). The effect of chlordiazepoxide was probably due to an action on serotonin neurones as it was prevented by intra-raphe 5,7-DHT pre-treatment (Soubrié *et al.* 1981; Thiebot, Jobert and Soubrié 1980; Thiebot *et al.* 1982). One inconsistency arose from the injection of the potent GABA receptor agonist, muscimol. Intra-raphe muscimol has been shown to increase food intake and locomotor activity (Przewlocka, Stala and Scheel-Kruger 1979; Sainati and Lorens 1982), but it did not release punished behaviour (Thiebot *et al.* 1980). It would be of interest to investigate a range of inhibitors of neuronal activity, in order to further studies of the effect of inhibition of serotonin-releasing cells on anxiety behaviours.

These data in general confirm that inhibition of serotonin neurones in the dorsal raphe nucleus leads to a release of punished behaviour. However, as discussed for the effect of lesions of serotonin neurones, it is not certain whether this response represents anxiolytic activity or a consequence of manipulation of a serotonin system involved in other physiological functions.

11.4.3   α *-adrenoreceptor antagonists*

One group of substances which inhibit dorsal raphe neurones is the α-adrenoreceptor antagonists. Whether administered onto the neurones directly or intravenously, a range of α_1-antagonists markedly inhibited the neuronal activity (Baraban and Aghajanian 1980*a, b*). These observations formed the basis of the hypothesis that noradrenergic innervation of the dorsal raphe plays an essential role in maintaining the tonic activity of serotonin-containing cells there. Noradrenergic nerve terminals have been identified on the serotonin-containing cells in the dorsal raphe nucleus (Baraban and Aghajanian 1981). These terminals do not appear to be of cells in the locus coeruleus or the A_1 and A_2 cell body groups (Anderson, Pasquier, Forbes and Morgane 1977; Massari, Tizabi and Jacobowitz 1979) although the latter study indicated a pathway from A_1 and A_2 to the median raphe nucleus. Thus the anatomical location of the cell bodies innervating the dorsal raphe nucleus remains to be determined.

However, administration of α_1-adrenoceptor antagonists leads to inhibition of dorsal raphe neurones. When administered intraperitoneally prazosin, phentolamine and piperoxan all increased punished drinking in a licking conflict (Gardner and Piper 1982). Although it would be tempting to associate this effect with inhibition of serotonin neurone activity, α_1-adrenoceptor antagonists induce a range of pharmacological effects including sedation. Parallel measurements of unpunished drinking indicated that increased thirst may be responsible for the effect. Furthermore, a similar increase in punished drinking induced by the selective α_2-adrenoceptor antagonist, yohimbine, which has been observed to induce anxiety (Charney, Heninger and Redmond 1983), may dissociate this effect from inhibition of serotonin neurones and from an anxiolytic drug action.

11.5   Serotonin antagonists

11.5.1   *Classical antagonists*

Initial studies with serotonin antagonists in food-motivated conflict procedures suggested that these agents produced release of punished behaviour of similar magnitude to that produced by benzodiazepines. This was directly claimed by Graeff and Schoenfeld (1970) for methysergide and bromoLSD (BOL148) in a key pecking study in pigeons. Unpunished behaviour was not decreased at doses showing peak effects. Both punished and unpunished behaviour was decreased at higher doses of both drugs. Recently this work has been extended to include the more specific serotonin antagonist, metergoline (Leone, de Aguiar and Graeff 1983). Once again, a clear increase in punished responding of the pigeons was observed with no consistent effect on unpunished responding.

However, studies in rat models have not yielded such clearcut results. Winter (1972) observed an increase in punished responding for food with methysergide at doses in a reasonable range for blockade of serotonin receptors (3 and 6 mg/kg i.p.) but a similarly relevant dose range of cinanserin (3–25 mg/kg i.p.) was without effect. A large dose of cinanserin (60 mg/kg i.p.) was subsequently reported to increase food-motivated punished responding (Geller, Hartmann and Croy 1974). A similar release of punished behaviour has not been observed at similarly large doses of cinanserin (Kilts, Commissaris and Rech 1981). Methysergide was also inactive in this procedure. Cinanserin has, however, been observed to increase punished responding at 15 mg/kg i.p. although 7.5 and 30 mg/kg were without effect and 60 mg/kg i.p. decreased both punished and unpunished behaviour (Cook and Sepinwall 1975*b*). Methysergide has also been observed to evoke small increases in punished responding (Cook and

Sepinwall 1975b; Stein *et al.* 1973) and the range of antagonists tested has been increased to include cyproheptadine (Sepinwall and Cook 1980) which also induces increases in punished responding which are small relative to the effects of benzodiazepines but were significant over a wide dose range (5–40 mg/kg p.o.). The reported antagonism of this action of cyproheptadine by the benzodiazepine antagonist Ro 15-1788 (Barrett and Brady 1983) calls the pharmacological specificity of one of these agents into question. Should this be a pharmacological rather than a physiological antagonism then one possible explanation is that cyproheptadine induces its weak anticonflict action by some action on benzodiazepine mechanisms. Unless such an effect was subsequent to serotonin receptor blockade, the specificity of cyproheptadine as a serotonin antagonist is further questioned. For example, cyproheptadine interacts with a GABA-linked chloride ionophore with a similar potency to pentobarbitone (Squires, Casida, Richardson and Saederup 1983) and this could explain its activity in some models of anxiety. In a food-motivated conflict based on the method of Cook and Sepinwall (1975a) the more specific serotonin antagonist metergoline showed no significant effect on punished lever pressing at doses which produce a substantial block of serotonin-mediated behaviours (Fig. 11.3).

Inconsistencies in the activities of serotonin antagonists in conflict procedures become more apparent when studies using licking conflict are considered. Early studies observed an increase in punished lever pressing for water in rats with methysergide and cyproheptadine (Graeff 1974), although cyproheptadine increases unpunished responding in a similar procedure (Graeff 1976). Subsequently, a range of serotonin antagonists have been without effect in increasing punished drinking. Methysergide and ciinanserin (Kilts *et al.* 1981; Petersen and Lassen 1981), metergoline (Commissaris and Rech 1982) and cyproheptadine (Petersen and Lassen 1981) have all been without effect, with the exception of a significant increase with the highest dose (56 mg/kg i.p.) of cinanserin (Kilts *et al.* 1981). Using the method of Gardner *et al.* (1981) the inactivity of methysergide and cinanserin was confirmed and pizotifen was also found to be without effect in increasing punished drinking (Table 11.2). Metergoline evoked a small but statistically significant increase in punished licking at 20 and 40 mg/kg i.p.

In anxiety models which do not involve foot shock punishment different results have been obtained in different tests. Stress-induced ultrasounds in rat pups are reduced by metergoline at doses which have no overt effect on the animals (Table 11.1), further suggesting that serotonin is involved in this model of anxiety. However, metergoline and methysergide only decreased behaviour in the social interaction model (File 1981). Using a modification of this social interaction model of anxiety, which is less sensitive to the behaviourally depressant actions of benzodiazepines (Gardner and Guy

298   Colin R. Gardner

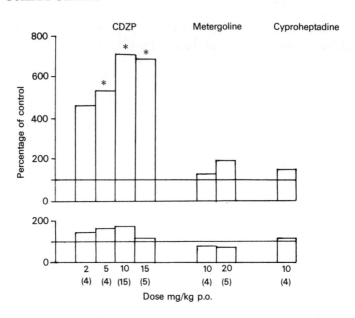

Fig. 11.3   Effects of serotonin antagonists and chlordiazepoxide (CDZP) on alter-
nating 4 min FI 30 s unpunished responding (lower histograms) and 3 min FR5 foot
shock-punished lever pressing for food pellets (upper histograms). Responding after
administration of a drug (30 min prior to the start of the 35 min test) is expressed as a
percentage of the median lever presses in the 10 sessions from two previous, stable,
vehicle-treated response days. *p <0.05, Mann Whitney U test performed on the
absolute data. CDZP was studied in the same rats, being tested on established stable
baselines either before and/or after the testing of the serotonin antagonists. The
number of observations is shown below in parentheses.

1984), metergoline was without effect (Fig. 11.4) even at a dose in excess of
that required to fully block serotonin-mediated behaviours (20 mg/kg p.o.).
Thus, in the tests so far discussed, serotonin antagonists are not consistently active.

The situation is complicated further when the fear-like syndrome
induced by electrical brain stimulation is considered. Lever pressing suppressed by stimulation of the dorsal periaqueductal grey region was not
released by methysergide or cyproheptadine although chlordiazepoxide
induced dose-dependent release of punished responding (Morato de
Carvalho, de Aguiar and Graeff 1981). However, escape latencies from
dorsal periaqueductal grey stimulation were decreased by cyproheptadine
and methysergide (Schenberg and Graeff 1978; Moriyama et al. 1982;
Clarke and File 1982) although metergoline was ineffective in a similar
model (Bovier, Broekkamp and Lloyd 1981). The enhancement of fear-like

Table 11.2 Effect of serotonin antagonists and some antidepressants on licking conflict

	Increase in punished licks (dose mg/kg i.p.)									
	1	2	4	5	8	10	12	20	40	60
Diazepam	146	599*	636*		669*		631*			
Metergoline	93	38		7		91		118	149	
Metergoline				105		74		217*	257*	(1 h pre-
Methysergide		32		86		57		−7		treatment)
Cinanserin				118		−31		78	20	134
Pizotifen				90		−7		60	−18	
Amitryptyline				50		0		62	221	
Trazodone				−21		164		93	−6	
Mianserin				69		38		128	244	

The number of licks in a 5 min period were counted with a capacitance sensor. Foot shock punishment was delivered every 100th lick (Gardner *et al.* 1981). The difference between the mean number of licks in a vehicle treated control group ($n \geqq 8$) and the mean in a drug treated group is shown above. *p <0.05 (Student's t test) calculated from the absolute data. All drugs were administered 30 min prior to testing except where otherwise stated.

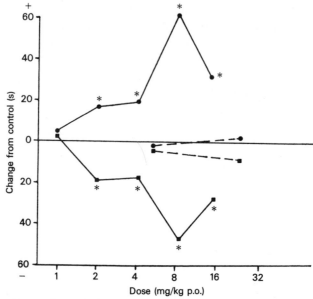

Fig. 11.4 Comparison of the effect of metergoline (– – – –) and chlordiazepoxide (————) on the time in s spent in active social interaction (●) and agression (■) during a 5 min observation of pairs of male rats, using the method of Gardner and Guy (1984). Results are expressed as the difference of the mean times for groups of 7 rats treated with drug and the time of a within-experiment control group. *p <0.05, Student's t test using the absolute data.

behaviour with serotonin antagonists is similar to the effects of serotonin depletion.

11.5.2   β-Adrenoreceptor antagonists

Established β-adrenoreceptor antagonists have recently been shown to interact with serotonin receptors, both the 5-HT$_1$, and 5-HT$_2$ receptor sub-types (Middlemiss, Blakeborough and Leather 1977; Green, Johnson and Nimgaonkar 1983*a*; Nahorski and Willcocks 1983). Although no specific block of serotonin responses on brain stem neurones was observed with propranolol and pindolol (Bradley and Gladman 1981) a range of β-adrenoreceptor antagonists block some serotonin-mediated behaviours (Green and Grahame-Smith 1976; Weinstock, Weiss and Gitter 1977; Segal and Weinstock 1982). Propranolol and other β-antagonists are thera-peutically useful in some anxiety states such as anxiety due to dentistry and stage fright (Kelly 1980; C. O. Brantigan, T. A. Brantigan and Joseph 1982), but this effect is associated with marked inhibition of autonomic signs, particularly increased heart rate (Taeber, Appel, Badian *et al.* 1980) and may result from disruption of positive feedback of these symptoms to the anxiety state (Lui, Debus and Janke 1978). This effect is peripheral and independent of serotonin receptor blockade as evidenced by the therapeutic efficacy of practolol, a β-adrenoceptor antagonist with poor brain penetration (Bonn, Turner and Hicks 1972).

The observation that oxprenolol positively affected emotional experience not related to experimentally-induced anxiety in man whilst not affecting emotional responses induced by the anxiety-provoking conditions (Lui *et al.* 1978) may find a parallel in the effects of propranolol in animal models of anxiety. Despite an early report of activity of propranolol in a conditioned emotional response, which was apparently independent of β-adrenoceptor antagonism (Bainbridge and Greenwood 1971), little or no anxiolytic-like action has been observed in food-motivated conflicts (Sepinwall, Grodsky, Sullivan and Cook 1973; Wise *et al.* 1973) punished exploratory activity (Aron *et al.* 1971) or licking conflict (Petersen and Lassen 1981; Gardner and Piper 1982). Thus, assuming that propranolol is capable of blocking some central serotonin receptors, the pathways which innervate them are not involved in the responses in these models of anxiety.

Thus, in general, serotonin antagonists show inconsistent anxiolytic-like effects in animal models. Furthermore, similar to studies of serotonin agonists and depletors, an anti-anxiety effect of serotonin is suggested by the brain-stimulation-evoked fear model, in which serotonin antagonists may increase the fear-like behaviour.

11.5.3   Clinical activity of serotonin antagonists

There is very little clinical evidence to test the hypothesis that serotonin antagonists would either reduce or induce anxiety. It is not surprising that the serotonin irritation syndrome, which includes anxiety symptoms, is reduced by methysergide (Giannini 1979) but this may not have similarities with primary anxiety symptoms. There has been a report of anxiolytic effects of cyproheptadine and pizotifen (Banki 1977) although sedative effects were also reported and could be responsible for the apparent anxiolytic effect. A recent preliminary clinical trial of a selective 5-HT$_2$ antagonist seretiazine showed it to be effective against anxiety and depression in patients with mixed symptoms (Janssen 1983). Further studies of specific serotonin antagonists are required to establish the effects of this group of compounds in anxiety.

11.5.4   Interpretation problems with serotonin antagonists

The presence of serotonin receptors at different locations (Fig. 11.1) leaves it possible that the overall effect of these antagonists may not be marked reduction of central serotonergic systems. Blockade of receptors on nerve terminals (presynaptic) or serotonin cell bodies (autoreceptors) may lead to enhanced synaptic release of serotonin which may lessen the post-synaptic antagonism particularly with competitive antagonists. The net effect will depend on the tonic nature of receptor activation and the relative potency of the antagonists at the presynaptic receptors. The antagonists tested in models of anxiety may have different potencies in different brain areas. Metergoline is potent in the hypothalamus, methysergide and cypro-heptadine are active but cinanserin is not (Cox and Ennis 1982; Martin and Sanders-Bush 1982a). In the frontal cortex cinanserin is effective but metergoline, methysergide and cyproheptadine are without effect (Mounsey et al. 1982). Similarly, the antagonists have differential activities on different postsynaptic receptors (Fuller, Mason and Molloy 1980; Whitaker and Seeman 1978; Peroutka and Snyder 1981; Leysen and Tollenaere 1982; Segal and Weinstock 1983) and do not block all of the actions of ionophoretically applied serotonin (Haigler and Aghajanian 1977).

Cyproheptadine and methysergide inhibit the firing of raphe neurones when applied locally (Haigler and Aghajanian 1974) although contradic-tory results have been reported for methiothepin (Tebecis 1972). Whether this is related to serotonin receptor effects or α-adrenoceptor block is not known. Such an inhibitory action would seem to exclude a reduction of the receptor blockade by the antagonists resulting from non-serotonin-mediated neuronal feedback to the serotonin cells, or from block of serotonin autoreceptors. Another reservation stems from the suggestion that some of the serotonin antagonists possess partial agonist properties at some sites (Colpaert, Niemegeers and Janssen 1982).

In general it can be concluded that the presently available serotonin antagonists are a less certain means of reducing the effects of all central serotonergic pathways, as has been the aim in many of the studies described. It cannot presently be determined whether these agents effec-tively block the serotonergic pathways that are involved in responses in some of the models of anxiety.

11.5.5   *Antidepressants with serotonin antagonist properties*

Despite these reservations it is still tempting to associate the anxiolytic activities of some antidepressants with their serotonin antagonist proper-ties. Amitriptyline reduces anxiety as well as depression in patients with mixed symptoms (Johnstone, Cunningham, Owens, Frith et al. 1980; Davidson, Linnoila, Raft and Turnbull 1981) and has been reported to be effective in recurrent anxiety attacks and 'night terrors' (Logan 1979). Amitriptyline displaces binding of ligands to serotonin receptors (Gay and

Von Voigtlander 1982; Leysen and Tollenaere 1982) and possesses central antiserotonin actions (Maj, Lewandowska and Rawlow 1979a). Little activity was observed with amitryptyline in animal models of anxiety. It was inactive in suppressed eating due to novel food (Poschel 1971), fluid-motivated conflict in squirrel monkeys (Lippa et al. 1979a) and licking conflict in rats (Table 11.1) although it has been reported to be effective in a similar procedure (Van Riezen, Pinder, Nickolson et al. 1981). Punished exploration in mice was increased, but the specificity of this response must be questioned in that stimulants such as amphetamine, cocaine, caffeine and trihexyphenidyl were all significantly effective (Aron et al. 1971). Whether the anxiolytic effects of this drug are a result of a primary anti-depressant effect or due to sedative properties (Ogren, Cott and Hall 1981) remains to be determined.

Mianserin has anxiolytic properties in patients with mixed depression and anxiety (Brogden, Heel, Speight and Avery 1978) and in patients with acute or chronic primary anxiety (Murphy 1978; Conti and Pinder 1979). This drug preferentially binds to $5-HT_2$ receptors (Peroutka and Snyder 1981; Leysen and Tollenaere 1982) and possesses central antiserotonin effects (Maj, Sowinska, Baran, Gancarczyk and Rawlow 1978), although partial agonism has also been suggested (Colpaert et al. 1982). Mianserin was reported to be effective in food-motivated conflict and licking conflict (Van Reizen et al. 1981) although this latter effect has not been confirmed (Table 11.1).

Trazodone possesses anxiolytic properties in patients with mixed depression and anxiety and in patients with predominant or primary anxiety (Brogden, Heel, Speight and Avery 1981; Ayd and Settle 1982). It binds to serotonin receptors (Clements-Jewery, Robson and Chidley 1980; Kendall, Taylor and Enna 1983) and blocks central serotonin-mediated effects (Baran, Maj, Rogoz and Skuza 1979; Maj, Palider and Rawlow 1979b; Clements-Jewery et al. 1980). In animal models trazodone decreases avoidance behaviour (Sansone et al. 1983) but it is not clear whether this effect could represent anxiolytic activity. Trazodone was ineffective in a licking conflict (Table 11.1). However, a serotonin agonist, m-chlorophenylpiperazine, is a major metabolite of trazodone (Melzacka, Rurak and Vetulani 1980; Caccia, Ballabio, Samanium, Zanini and Grattini 1981) and it may reduce the net serotonin antagonism produced by the drug (Sansone, Melzacka, Hano and Vetulani 1983).

It is interesting that these antidepressants show anxiolytic effects in clinical use but are not consistently active in animal models of anxiety. It seems less likely that all the reported anxiolytic effects could result from a reversal of depression with concomitant relief of associated anxiety symptoms. It is possible that sedative properties of these drugs (Ogren et al. 1981) quieten the patients (perhaps due to α-adrenoceptor antagonism) and lead to an apparent anxiolytic effect. The mechanisms behind the observed clinical activity of these drugs remain to be elucidated.

11.5.6   *Buspirone*

A recent investigation has shown that buspirone, which has been reported to possess anxiolytic properties (Goldberg and Finnerty 1979) interacts with serotonin receptors (Glaser and Traber 1983). A correlation between these two effects was proposed, but in view of the previous discussion, such a link is far from certain and requires rigorous confirmation.

11.6   Benzodiazepines and serotonin

11.6.1   *Serotonin turnover and neurone activity*

The theory that serotonin was involved in the mediation of the anxiolytic effects of benzodiazepines was first given impetus by the observation that, on sub-acute dosing, a decrease in noradrenaline turnover induced by oxazepam showed tolerance similar to sedative effects of the drug whilst a decrease in serotonin turnover persisted (Stein *et al.* 1975, 1977*b*). The lack of tolerance to the decrease in serotonin turnover was correlated with a similar lack of tolerance of anxiolytic actions of these drugs (Goldberg, Manian and Efron 1967; Margules and Stein 1968). Benzodiazepines depress serotonin synthesis and turnover, decrease the egress of 5-HIAA, the major metabolite of serotonin, and may increase the steady state levels of serotonin (Koe 1979). Marked increases in mouse brain serotonin and 5-HIAA induced by clonazepam were not, however, observed after sub-acute dosing (Jenner, Chadwick, Reynolds and Marsden 1975). The reduction of serotonin turnover in rats after sub-acute dosing of benzo-diazepines has recently been confirmed using diazepam (Collinge, Pycock and Taberner 1983). It has been suggested that the decrease in serotonin turnover reflects reduced activity of serotonin-containing neurones and thus decreased release of serotonin. This has been directly observed by measuring release of [^{3}H]-serotonin in projection areas of serotonin neurones with a push pull cannula (Soubrié *et al.* 1983). Release was reduced after 10 mg/kg i.p. of chlordiazepoxide in *encephale isolé* but not anaesthetized cats. Furthermore, chlordiazepoxide probably reduces the activity of serotonin neurones by an action within the raphe nuclei as super-fusion of the dorsal raphe with the benzodiazepine antagonist Ro15-1788 prevented the effect of intraperitoneal chlordiazepoxide.

Benzodiazepines do decrease the activity of neurones in the raphe nuclei of *encephale isolé* rats and conscious cats (Trulson, Preussler, Howell and Fredrickson 1982; Laurent, Mangold, Humbel and Haefely 1983) although no such inhibitory effect had been observed in anaesthetized or immobilized rats (Gallager 1978). Effects of benzodiazepines on serotonin neurones have been associated with enhancement of neurotransmission mediated by release of GABA, possibly onto raphe neurones (Gallager 1978; Soubrié *et al.* 1981; Forchetti and Meek 1981; Sainati and Lorens 1983; Collinge *et al.* 1983). However, a recent study of dorsal raphe cell activity in freely moving cats showed no decrease in neuronal activity induced by benzodiazepines unless decreased electromyographic activity and ataxia were also present (Trulson *et al.* 1982). Lower doses of the benzodiazepines had no effect on the raphe cell activity suggesting that the suppression of cell activity at higher doses is more closely related to general motor behaviour than to anxiolytic properties of the drugs. This conclusion is supported by the lack of activity of the triazolopyridazine CL218872 in inhibiting serotonin-mediated hyperthermia (Lippa *et al.* 1979 *b*). Benzo-diazepines inhibit the hyperthermia and this response has been used as a measure of their anti-serotonin properties. CL218872 is a ligand for benzodiazepine binding sites and possesses anxiolytic but not muscle relaxant properties. Its lack of inhibition of serotonin-mediated hyper-thermia may be more related to its lack of central depressant properties, questioning the involvement of serotonin in its anxiolytic effects (Lippa, Critchett, Sano *et al.* (1979 *b*).

It is, however, possible that benzodiazepines may reduce serotonin turn-over with very little effect on the activity of serotonin cell bodies, depending on the relative strengths of feedback systems. They may reduce serotonnn release at nerve terminals. Potassium-evoked [³H]-serotonin release from hippocampal synaptosomes is inhibited by diazepam although spon-taneous serotonin release was enhanced (Balfour 1980). Similarly, Collinge and Pycock (1982) observed reductions in both spontaneous and evoked release of [³H]-serotonin from slices of cerebral cortex, although similar release from raphe slices was increased. However, superfusion of serotonergic nerve terminals of the substantia nigra and caudate nucleus with chlordiazepoxide in *encephale isolé* cats did not alter local release of [³H]-serotonin (Soubrié *et al.* 1983).

11.6.2 *Serotonin inhibition and benzodiazepines*

Injection of benzodiazepines into the dorsal raphe nucleus leads to release of responding for food suppressed by punishment (Thiebot *et al.* 1980; Soubrié *et al.* 1981) leaving it possible that these drugs may affect the raphe nuclei in producing similar responses after oral administration. However, if the anti-punishment effects of benzodiazepines were mediated to a greater extent or entirely via inhibition of serotonin pathways these drugs should be without effect if the functioning of the serotonin pathway has already been inhibited. Lesion of central serotonin pathways with intraventricular 5,6-DHT led to an increase in punished drinking. However, the anti-punishment activity of diazepam was additive with this response (Lippa *et al.* 1979*a*). The release of punished behaviour by chlordiazepoxide observed by Tye *et al.* (1977) was reduced after 5,7-DHT lesions. However, the punishment level had been increased to reverse the increase in punished responding induced by the lesion and this could explain the reduced response to the benzodiazepine. Cook and Sepinwall (1975*a*) observed no anticonflict activity of PCPA free acid but the response to chlordiazepoxide was abolished. The response to chlordiazepoxide recovered with a similar time course to the serotonin levels. In contrast to this, PCPA methyl ester hydrochloride had an anti-punishment effect and the response to chlordiazepoxide was not reduced. The degree of serotonin depletion was similar with both forms of PCPA. Punished lever pressing for food was increased by PCPA but the anti-punishment activity of chlordiazepoxide was not reduced (Shephard *et al.* 1982; S. E. Green *et al.* 1983). Thus with the exception of the one observation of Cook and Sepinwall (1975*a*), depletion of central serotonin does not reduce the activity of benzodiazepines in conflict models of anxiety. It is always possible that, despite the depletion, there is still some function in the serotonin pathways remaining to be further inhibited.

The serotonin agonist 5-MeODMT might be expected to prevent the actions of benzodiazepines on serotonin systems by two mechanisms. As a postsynaptic agonist it would replace the reduced serotonin function and by inactivating serotonin neurones (Aghajanian 1978) it would prevent any further such effect by the benzodiazepines. Although 5-MeODMT reversed the anti-punishment effects induced by PCPA it had no effect on that induced by chlordiazepoxide (Shephard *et al.* 1982). These authors suggested that benzodiazepines act on conflict behaviour at a site distal to serotonergic synapses.

11.6.3   Increased serotonin function and benzodiazpines

Intraventricular injection of 5 μg serotonin reduced the anti-conflict activity of oxazepam in six of eight rats tested (Stein *et al.* 1973). Interpretation of this effect is complicated by the potential dual effect of the exogenous serotonin on serotonin pathways—as a postsynaptic agonist and an inhibitor of serotonin cell activity. Furthermore, both punished and unpunished behaviour were reduced such that the reduction of responding after oxazepam could be a non-specific effect. Both serotonin reuptake inhibitor WY25093 and 5-HTP reduced increases in punished responding after chlordiazepoxide but 5-HTP also substantially reduced unpunished responding (S. E. Green *et al.* 1983). Again this effect could be non-specific behavioural depression. However, the reduction of both punished and unpunished responding evoked by injection of carbachol into the dorsal raphe nucleus was reversed by oxazepam (Stein *et al.* 1973, 1977*b*). These data do little to clarify the role of serotonin in the anticonflict activity of benzodiazepines.

11.6.4 *Serotoninomimetic actions of benzodiazepines*

Some benzodiazepines induce serotonin agonist-like head twitches in mice (Nakamura and Fukushima 1976; Boissier and Dumont 1981) or potentiate head twitches induced by intracerebral injection of serotonin (Nakamura and Fukushima 1977). This effect tends to be at high doses of the benzo-diazepines and is not related to their *in vitro* activity on benzodiazepine binding (Boisser and Dumont 1981). It would follow that this effect is thus not related to the activity of the benzodiazepines in models of anxiety or as anxiolytics in clinical use.

Diazepam has been shown to stimulate the uptake of L-tryptophan by brain slices (Hockel, Muller and Wollert 1979) and such an effect may lead to enhanced release of serotonin, but this would not seem consistent with observed decreases in serotonin turnover (Koe 1979). The central sero-toninomimetic effects of benzodiazepines remain to be explained but they may not occur at doses showing anxiolytic effects and may not therefore be of any consequence in this respect.

11.7   Summary

Serotonin is involved in the responses in some models of anxiety (e.g. conflict procedures, stress-induced ultrasounds and social interaction) and reduction of the function of central serotonin pathways leads to an anxiolytic-like effect. Lesions of serotonin pathways or depeletion of sero-tonin have induced this effect but serotonin receptor antagonists have yielded much less consistent effects. More recently developed models of anxiety based on punished responding have, by design, improved sensi-tivity to anxiolytics (particularly benzodiazepines). However, they have shown fewer positive results after reductions of central serotonin function, whether due to depeletion of transmitter, lesions of neurones or block of postsynaptic receptors (Kilts *et al.* 1981; Petersen and Lassen 1981; Gardner, results presented here). Taking all results together, the statement of Sepinwall (1983) that 'There are a few instances of negative findings, but these are far outweighed by the numerous positive findings' is no longer representative of the present situation. It has not been established that these responses in models of anxiety represent potential anxiolytic activity in clinical use.

Some observations suggest the presence of functionally opposed serotonin pathways with respect to anxiety and other behaviours. A fear
response induced by stimulation of the periaqueductal region of the brain
is reduced by enhancement of central serotonin function and enhanced by
its reduction. There may be different relative contributions of such pathways in different types of anxiety.

Although there is evidence which is consistent with the view that benzodiazepines exert their anxiolytic effects by reduction of central serotonin
function, recent evidence calls this thesis into question and it cannot be
considered proven.

Acknowledgements

I wish to thank my colleagues who collaborated in the research reported
here: D. C. Piper for licking conflict, R. Deacon for food-motivated conflict
and A. P. Guy for social interaction.

References

Aghajanian, G. K. (1978). LSD and other hallucinogenic indoleamines:
    Preferential action upon presynaptic serotonin receptors. In *Bio-
    chemistry of mental disorders* (eds. E. Usdin and A. J. Mandell) p. 101.
    Marcel Dekker, New York.
Albert, D. J. and Walsh, M. L. (1982). The inhibitory modulation of
    agonistic behaviour in the rat brain: A review. *Neurosci. Biobehav. Rev.* **6**,
    125.
Ananth, J., Solyom, L. and Bryntwick, S. (1979). Chlorimipramine therapy
    for obsessive-compulsive neurosis. *Am. J. Psychiat.* **136**, 100.
Anderson, C. D., Pasquier, D. A., Forbes, W. B. and Morgane, P. J. (1977).
    Locus coeruleus-to-dorsal raphe input examined by electrophysio-
    logical and morphological methods. *Brain Res. Bull.* **2**, 209.
Aprison, M. H. and Ferster, C. B. (1961). Neurochemical correlates of
    behaviour. II. Correlation of brain monoamine oxidase activity with
    behavioural changes after iproniazid and 5-hydroxytryptophan. *J.
    Neurochem.* **6**, 350.
Aron, C., Simon, P., Larousse, C. and Boissier, J. R. (1971). Evaluation of a
    rapid technique for detecting minor tranquilisers. *Neuropharmacology*
    10, 459.
Ayd, F. J. and Settle, E. C. (1982). Trazodone: A novel, broad-spectrum
    antidepressant. In *Modern problems in pharmacopsychiatry.* (ed. H. E.
    Lehmann) Vol. 18, p. 49. Karger, Basel.

Bainbridge, J. G. and Greenwood, D. J. (1971). Tranquilising effects of propranolol demonstrated in rats. *Neuropharmacology* **10**, 453.

Balfour, D. J. K. (1980). Effects of GABA and diazepam on ^{3}H-serotonin release from hippocampal synaptosomes. *Eur. J. Pharmac.* **68**, 11.

Banki, C. M. (1977). Correlation of anxiety and related symptoms with cerebrospinal fluid 5-hydroxyindoleacetic acid in depressed women. *J. Neural Transm.* **41**, 135.

Baraban, J. M. and Aghajanian, G. K. (1980*a*). Suppression of serotonergic neuronal firing by α-adrenoreceptor antagonists: Evidence against GABA mediation. *Eur. J. Pharmac.* **66**, 287.

—— —— (1980*b*). Suppression of firing activity of 5-HT neurones in the dorsal raphe by alpha-adrenoreceptor antagonists. *Neuropharmacology* **19**, 355.

—— —— (1981). Noradrenergic innervation of serotonergic neurons in the dorsal raphe: Demonstration by electron microscopic autoradiography. *Brain Res.* **204**, 1.

Baran, L., Maj, J., Rogoz, Z. and Skuza, G. (1979). On the central anti-serotonin action of trazodone. *Pol. J. Pharm. Pharmac.* **31**, 25.

Barofsky, A-L., Grier, H. C. and Pradhan, T. K. (1980). Evidence for regulation of water intake by median raphe serotonergic neurons. *Physiol. Behav.* **24**, 951.

Barrett, J. E. and Brady, L. S. (1983). The benzodiazepine antagonist Ro15-1788 antagonises the behavioural effects of cyproheptadine. *Pharmacologist* **25**, Abst. 613.

Billingsley, M. L. and Kubena, R. K. (1978). The effects of naloxone and picrotoxin on the sedative and anticonflict effects of benzodiazepines. *Life Sci.* **22**, 897.

Blackshear, M. A., Steranka, L. R. and Sanders-Bush, E. (1981). Multiple serotonin receptors: Regional distribution and effect of raphe lesions. *Eur. J. Pharmac.* **76**, 325.

Blakely, T. A. and Parker, L. F. (1973). The effects of parachlorophenyla-lanine on experimentally induced conflict behaviour. *Pharmac. Biochem. Behav.* **1**, 609.

Blier, P. and de Montigny, C. (1983). Effects of quipazine on pre- and postsynaptic serotonin receptors: Single cell studies in the rat CNS. *Neuropharmacology* **22**, 495.

Blundell, J. E. (1977). Is there a role for serotonin (5-hydroxytryptamine) in feeding? *Int. J. Obesity* **1**, 15.

—— and Latham, C. J. (1979*a*). Serotonergic influences on food intake: Effect of 5-hydroxytryptophan on parameters of feeding behaviour in deprived and free-feeding rats. *Pharmac. Biochem. Behav.* **11**, 431.

—— —— (1979*b*). Pharmacology of food and water intake. In *Chemical influences on behaviour* (eds. K. Brown and S. J. Cooper) p. 201.

Academic Press, London.

Boissier, J. R. and Dumont, C. (1981). Quelques aspects comportementaux de la pharmacologie des benzodiazepines. *Actualités Chim. Ther.* **8**, 57.

Bonn, J., Turner, P. and Hicks, D. L. (1972). Beta adrenergic receptor blockade with practolol in the treatment of anxiety. *Lancet* **i**, 814.

Bovier, P. Broekkamp, C. L. and Lloyd, K. G. (1981). Anxiolytic anticonvulsant and other centrally acting drugs on aversive dorsal periaqueductal gray stimulation in rats. *Neurosci. Letts.* Suppl. **7**, 8243.

Bradley, P. B. and Gladman, J. R. F. (1981). Microiontophoretic study of the interaction between β-adrenoceptor antagonists and 5-hydroxytryptamine in the rat brain stem. *Br. J. Pharmac.* **73**, 245P.

Bradshaw, C. M., Stoker, M. J. and Szabadi, E. (1983). Comparison of the neuronal responses to 5-hydroxytryptamine, noradrenaline and phenylephrine in the cerebral cortex: Effects of haloperidol and methysergide. *Neuropharmacology* **22**, 677.

Brantigan, C. O. Brantigan, T. A. and Joseph, N. (1982). Effects of beta blockade and beta stimulation on stage fright. *Am. J. Med.* **72**, 88.

Brogden, R. N., Heel, R. C., Speight, T. M. and Avery, G. S. (1978). Mianserin: A review of its pharmacological properties and therapeutic efficacy in depressive illness. *Drugs* **16**, 273.

—— —— —— —— (1981). Trazodone: A review of its pharmacological properties and therapeutic use in depression and anxiety. *Drugs* **21**, 401.

Caccia, S., Ballabio, M., Samanin, R., Zanini, M. G. and Garattini, S. (1981). *m*-Chlorophenylpiperazine, a central 5-hydroxytryptamine agonist, is a metabolite of trazodone. *J. Pharm. Pharmac.* **33**, 447.

Campbell, A. B., Brown, R. M. and Seiden, L. S. (1971). A selective effect of *p*-chlorphenylalanine on fixed ratio responding. *Physiol. Behav.* **7**, 853.

Carey, R. J. (1976). Effects of selective forebrain depletions of norepinephrine and serotonin on the activity and food intake effects of amphetamine and fenfluramine. *Pharmac. Biochem. Behav.* **5**, 519.

Cazala, P. and Garriques, A. M. (1983). Effects of apomorphine clonidine or 5-methoxy-*N,N*-dimethyltryptamine on approach and escape components of lateral hypothalamic and mesencephalic central gray stimulation in two inbred strains of mice. *Pharmac. Biochem. Behav.* **18**, 87.

Cerrito, F. and Raiteri, M. (1979). Serotonin release is modulated by presynaptic autoreceptors. *Eur. J. Pharmac.* **57**, 427.

Cervantes, M. Ruelas, R. and Beyer, C. (1983). Serotonergic influences on EEG synchronisation induced by milk drinking in the cat. *Pharmac. Biochem. Behav.* **18**, 851.

Charney, D. S., Heninger, G. R. and Redmond, D. E. (1983). Yohimbine induced anxiety and increased noradrenergic function in humans:

Effects of diazepam and clonidine. *Life Sci.* **33,** 19.

Clarke, A. and File, S. E. (1981). Neurotoxin lesions of the lateral septum and changes in social and aggressive behaviours. *Br. J. Pharmac.* **74,** 766.

—— —— (1982). Effects of ACTH, benzodiazepines and 5-HT antagonists on escape from periaqueductal grey stimulation in the rat. *Prog. Neuro-Psychopharmac. Biol. Psychiat.* **6,** 27.

Clements-Jewery, S., Robson, P. A. and Chidley, L. J. (1980). Biochemical investigations into the mode of action of trazodone. *Neuropharmacology* **19,** 1165.

Clineschmidt, B. V. (1979). MK212: A serotonin-like agonist in the CNS. *Gen. Pharmac.* **10,** 287.

Cole, J. O., Hartmann, E. and Brigham, P. (1980). L-Tryptophan: Clinical studies. *McLean Hosp. J.* **1,** 37.

Collinge, J. and Pycock, C. J. (1982). Differential actions of diazepam on the release of [^{3}H]5-hydroxytryptamine from cortical and midbrain raphe slices in the rat. *Eur. J. Pharmac.* **85,** 9.

—— —— and Taberner, P. V. (1983). Studies on the interaction between cerebral 5-hydroxytryptamine and γ-aminobutyric acid in the mode of action of diazepam in the rat. *Br. J. Pharmac.* **79,** 637.

Colpaert, F. C., Niemegeers, C. J. E. and Janssen, P. A. J. (1982). A drug discrimination analysis of lysergic acid diethylamide (LSD): In vivo agonist and antagonist effects of purported 5-hydroxytryptamine antagonists and of pirenperone, a LSD antagonist. *J. Pharmac. Exp. Ther.* **221,** 206.

Commissaris, R. L., Lyness, W. H. and Rech, R. H. (1981). The effects of *d*-lysergic acid diethylamide (LSD), 2,5-dimethoxy-4-methylamphetamine (DOM), pentobarbital and methaqualone on punished responding in control and 5,7-dihydroxytryptamine-treated rats. *Pharmac. Biochem. Behav.* **14,** 617.

—— and Rech, R. H. (1982). Interactions of metergoline with diazepam, quipazine and hallucinogenic drugs on a conflict behaviour in the rat. *Psychopharmacology* **76,** 282.

Conti, L. and Pinder, R. M. (1979). A controlled comparative trial of mianserin and diazepam in the treatment of anxiety states in psychiatric out-patients. *J. Int. Med. Res.* **7,** 285.

Cook, L. and Sepinwall, J. (1975*a*). Psychopharmacological parameters of emotion. In *Emotions—their parameters and measurement* (ed. L. Levi) p. 379. Raven Press, New York.

—— —— (1975*b*). Behavioural analysis of the effects and mechanisms of action of benzodiazepines. In *Mechanisms of actions of benzodiazepines* (eds. E. Costa and P. Greengard) p. 1. Raven Press, New York.

Cooper, S. J. (1983). Effects of chlordiazepoxide on drinking compared in rats challenged with hypertonic saline, isoproterenol or polyethylene glycol. *Life Sci.* **32**, 2453.

Couch, J. R. (1970). Responses of neurones in the raphe nuclei to serotonin, norepinephrine and acetylcholine and their correlation with an excitatory synaptic input. *Brain Res.* **19**, 137.

—— (1976). Further evidence for a possible excitatory serotonergic synapse on raphe neurones of pons and lower midbrain. *Life Sci.* **19**, 761.

Cox, B. and Ennis, C. (1982). Characterisation of 5-hydroxytryptamine autoreceptors in the rat hypothalamus. *J. Pharm. Pharmac.* **34**, 438.

Cremata, V. Y. and Koe, B. K. (1966). Clinical-pharmacological evaluation of p-chlorophenylalanine: A new serotonin depleting agent. *Clin. Pharmac. Ther.* **7**, 768.

Crow, T. J. and Deakin, J. F. W. (1977). Role of tryptaminergic mechanisms in the elements of the behavioural syndrome evoked by tryptophan and a monoamine oxidase inhibitor. *Br. J. Pharmac.* **59**, 461P.

—— —— (1981). Affective change and the mechanisms of reward and punishment: A neurohumoural hypothesis. In *Developments in psychiatry* (eds. C. Perris, G. Struwe and B. Jansson) Vol. 5, p. 536. Elsevier, Amsterdam.

Davidson, J., Linnoila, M., Raft, D. and Turnbull, C. D. (1981). MAO inhibition and control of anxiety following amitriptyline therapy. *Acta Psychiat. Scand.* **63**, 147.

Davis, M. (1980). Neurochemical modulation of sensory-motor reactivity: Acoustic and tactile startle reflexes. *Neurosci. Biobehav. Rev.* **4**, 241.

Deakin, J. F. W. and Dashwood, M. R. (1981). The differential neurochemical bases of the behaviours elicited by serotonergic agents and by the combination of a monoamine oxidase inhibition and L-dopa. *Neuropharmacology* **20**, 123.

—— and Green, A. R. (1978). The effects of putative 5-hydroxytryptamine antagonists on the behaviour produced by administration of tranylcypromine and L-dopa to rats. *Br. J. Pharmac.* **64**, 201.

de Montigny, C. and Aghajanian, G. K. (1977). Preferential action of 5-methoxytryptamine and 5-methoxydimethyltryptamine on presynaptic serotonin receptors: A comparative iontophoretic study with LSD and serotonin. *Neuropharmacology* **16**, 811.

Duka, T., Cumin, R., Haefely, W. and Herz, A. (1981). Naloxone blocks the effect of diazepam and meprobamate on conflict behaviour in rats. *Pharmac. Biochem. Behav.* **15**, 115.

Engelman, K., Lovenberg, W. and Sjoerdsma, A. (1967). Inhibition of serotonin synthesis by p-chlorophenylalanine in carcinoid syndrome. *New Eng. J. Med.* **277**, 1103.

Ennis, C. and Cox, B. (1982). Pharmacological evidence for the existence of two distinct serotonin receptors in rat brain. *Neuropharmacology* **21**, 41.

—— Kemp, J. D. and Cox, B. (1981). Characterisation of inhibitory 5-hydroxytryptamine receptors that modulate dopamine release in the striatum. *J. Neurochem.* **36**, 1515.

Evans, L. and Moore, G. (1981). The treatment of phobic anxiety by zimelidine. *Acta Psychiat. Scand.* **63**, supple. 290, 342.

—— Best, J., Moore, G. and Cox, J. (1980). Zimelidine—A serotonin uptake blocker in the treatment of phobic anxiety. *Prog. Neuro-Psychopharmac.* **4**, 75.

File, S. E. (1980). The use of social interaction as a method for detecting anxiolytic activity of chlordiazepoxide-like drugs. *J. Neurosci. Meth.* **2**, 219.

—— (1981). Behavioural effects of serotonin depletion. In *Metabolic disorders of the nervous system* (ed. F. C. Rose) p. 429. Pitman Medical, London.

—— and Deakin, J. F. W. (1980). Chemical lesions of both dorsal and median raphe nuclei and changes in social and aggressive behaviour in rats. *Pharmac. Biochem. Behav.* **12**, 855.

—— and Hyde, J. R. G. (1977). The effects of parachlorophenylalanine and ethanolamine-*O*-sulphate in an animal test of anxiety. *J. Pharm. Pharmac.* **29**, 735.

—— —— (1978). Can social interaction be used to measure anxiety? *Br. J. Pharmac.* **62**, 19.

—— and Vellucci, S. V. (1978). Studies on the role of ACTH and of 5-HT in anxiety, using an animal model. *J. Pharm. Pharmac.* **30**, 105.

—— Hyde, J. R. G. and MacLeod, N. K. (1979). 5,7-Dihydroxytryptamine lesions of dorsal and median raphe nuclei and performance in the social interaction test of anxiety and in a home cage aggression test. *J. Affect. Disord.* **1**, 115.

—— James, T. A. and MacLeod, N. K. (1981). Depletion in amygdaloid 5-hydroxytryptamine concentration and changes in social and aggressive behaviour. *J. Neural Transm.* **50**, 1.

Forchetti, C. M. and Meek, J. L. (1981). Evidence for a tonic GABA ergic control of serotonin neurones in the median raphe nucleus. *Brain Res.* **26**, 208.

Fujiwara, J. and Otsuki, S. (1974). Subtypes of affective psychoses classified by response to amine precursors. *Fol. Psychiat. Neurol. Jap.* **28**, 93.

Fuller, R. W., Mason, N. R. and Molloy, B. B. (1980). Structural relationships in the inhibition of [^{3}H] serotonin binding to rat brain membranes *in vitro* by 1-phenyl-piperazines. *Biochem. Pharmac.* **29**, 833.

—— Snoddy, H. D., Mason, N. R. and Molloy, B. B. (1978). Effect of 1-(*m*-

trifluoromethylphenyl)-piperazine on ^{3}H-serotonin binding to membranes from rat brain in vitro and on serotonin turnover in rat brain in vivo. *Eur. J. Pharmac.* **52**, 11.

Gallager, D. W. (1978). Benzodiazepines: Potentiation of a GABA inhibitory response in the dorsal raphe nucleus. *Eur. J. Pharmac.* **49**, 133.

Gardner, C. R. (1984). Inhibition of ultrasonic distress volcalisation in rat pups by chlordiazepoxide and diazepam. *Drug Dev. Res.* (In press).

—— and Guy, A. P. (1983). Behavioural effects of RU24969, a 5HT₁ receptor agonist, in the mouse. *Br. J. Pharmac.* **78**, 96P.

—— —— (1984). A social interaction model of anxiety sensitive to acutely administered benzodiazepines. *Drug. Dev. Res.* **4**, 207.

—— and Piper, D. C. (1982). Effects of adrenoreceptor modulation on drinking conflict in rats. *Br. J. Pharmac.* **75**, 50P.

—— —— and Kidd, S. B. (1981). Increased anxiety behaviour after prior experience of punishment during drinking in thirsty rats. In *Quantification of steady state operant behaviour* (eds. C. M. Bradshaw, E. Szabadi and C. F. Lowe) p. 425. Elsevier, North Holland.

Geller, I. and Blum, K. (1970). The effects of 5-HT on para-chlorophenyl-alanine (p-CPA) attenuation of "conflict" behaviour. *Eur. J. Pharmac.* **9**, 319.

—— and Seifter, J. (1960). The effects of meprobamate, barbiturates, d-amphetamine and promazine on experimentally induced conflict in the rat. *Psychopharmacology* **1**, 482.

—— Hartmann, R. J. and Croy, D. J. (1974). Attenuation of conflict behaviour with cinanserine, a serotonin antagonist: Reversal of the effect with 5-hydroxytryptophan and α-methyl tryptamine. *Res. Comm. Chem. Path. Pharmac.* **7**, 165.

Gerson, S. C. and Baldessarini, R. J. (1980). Motor effects of serotonin in the central nervous system. *Life Sci.* **27**, 1435.

Geyer, M. A., Puerto, A., Menkes, D. B., Segal, D. S. and Mandell, A. J. (1976). Behavioural studies following lesions of the mesolimbic and mesostriatal serotonergic pathways. *Brain Res.* **106**, 257.

—— Warbritton, J. D., Menkes, D. B., Zook, J. A. and Mandell, A. J. (1975). Opposite effects of intraventricular serotonin and bufotenin on rat startle responses. *Pharmac. Biochem. Behav.* **3**, 687.

Giannini, A. J. (1979). Serotonin irritation syndrome: A hypothesis. *Int. J. Psychiat. Med.* **9**, 199.

Glaser, T. and Traber, J. (1983). Buspirone: Action on serotonin receptors in calf hippocampus. *Eur. J. Pharmac.* **88**, 137.

Gloger, S., Grunhaus, L., Birmacher, B. and Troudart, T. (1981). Treatment of spontaneous panic attacks with chlomipramine. *Am. J. Psychiat.* **138**, 1215.

Goldberg, H. L. and Finnerty, R. J. (1979). The comparative efficacy of buspirone and diazepam in the treatment of anxiety. *Am. J. Psychiat.* **136**, 1184.

Goldberg, M. E., Manian, A. A. and Efron, D. H. (1967). A comparative study of certain pharmacological responses following acute and chronic administration of chlordiazepoxide. *Life Sci.* **6**, 481.

Gothert, M. and Schlicker, E. (1983). Autoreceptor-mediated inhibition of ^{3}H-5-hydroxytryptamine release from rat brain cortex slices by analogues of 5-hydroxytryptamine. *Life Sci.* **32**, 1183.

Graeff, F. G. (1974). Tryptamine antagonists and punished behaviour. *J. Pharmac. Exp. Ther.* **189**, 344.

—— (1976). Effect of cyproheptadine and combinations of cyproheptadine and amphetamine on intermittently reinforced lever pressing in rats. *Psychopharmacology* **50**, 65.

—— and Rawlins, J. N. P. (1979). Dorsal periaqueductal gray punishment, septal lesions and the mode of action of minor tranquilisers. *Pharmac. Biochem. Behav.* **12**, 41.

—— and Schoenfeld, R. I. (1970). Tryptaminergic mechanisms in punished and non-punished behaviour. *J. Pharmac. Exp. Ther.* **173**, 277.

—— and Silveira Filho, N. G. (1978). Behavioural inhibition induced by electrical stimulation of the median raphe nucleus of the rat. *Physiol. Behav.* **21**, 477.

Gray, D. D. and von Voigtlander, P. F. (1982). Effects of analgesics, endorphins and other drugs in a ^{3}H-serotonin binding assay. *Drug Dev. Res.* **2**, 29.

Gray, J. A. (1976). The behavioural inhibition system: A possible substrate for anxiety. In *Theoretical and experimental basis of the behaviour therapies* (eds. J. S. Feldman and P. L. Broadhurst) p. 3. John Wiley, New York.

Green, A. R. and Grahame-Smith, D. G. (1974). The role of brain dopamine in the hyperactivity syndrome produced by increased 5-hydroxytryptamine synthesis in rats. *Neuropharmacology* **13**, 949.

—— —— (1976). (−)-Propranolol inhibits the behavioural responses of rats to increased 5-hydroxytryptamine in the central nervous system. *Nature* **262**, 594.

—— Guy, A. P. and Gardner, C. R. (1984). The behavioural effects of RU 24969, a suggested 5HT$_1$ receptor agonist in rodents and the effect on the behaviour of treatment with antidepressants. *Neuropharmacology* **23**, 655.

—— Johnson, P. and Nimgaonkar, V. L. (1983a). Interactions of β-adrenoreceptor agonists and antagonists with the 5-hydroxytryptamine$_2$ (5-HT$_2$) receptor. *Neuropharmacology* **22**, 657.

—— Youdim, M. B. H. and Grahame-Smith, D. G. (1976). Quipazine: its effects on rat brain 5-hydroxytryptamine metabolism, monoamine oxidase activity and behaviour. *Neuropharmacology* **15**, 173.

—— O'Shaughnessy, K., Hammond, M., Schachter, M. and Grahame-Smith, D. G. (1983*b*). Inhibition of 5-hydroxytryptamine mediated behaviours by the putative 5HT$_2$ antagonist pirenperone. *Neuropharmacology* **22**, 573.

Green, S. E., Hodges, H. M. and Summerfield, A. (1983). Evidence for the involvement of brain GABA and serotonin systems in the anticonflict effects of chlordiazepoxide in rats. *Br. J. Pharmac.* **79**, 265P.

Gumulka, W., Samanin, R., Valzelli, L. and Consolo, S. (1971). Behavioural and biochemical effects following the stimulation of the nucleus raphe dorsalis in rats. *J. Neurochem.* **18**, 533.

Haigler, H. J. and Aghajanian, G. K. (1974). Peripheral serotonin antagonists: Failure to antagonise serotonin in brain areas receiving a prominent serotonergic input. *J. Neural. Trasm.* **35**, 257.

—— —— (1977). Serotonin receptors in the brain. *Fed. Proc.* **36**, 2159.

Hard, E., Ahlenius, S. and Engel, J. (1983). Effects of neonatal treatment with 5,7-dihydroxytryptamine of 6-hydroxydopamine on the ontogenic development of the audiogenic immobility reaction in the rat. *Psychopharmacology* **80**, 269.

—— Engel, J. and Musi, B. (1982). The ontogeny of defensive reactions in the rat: Influence of the monoamine transmission systems. *Scand. J. Psychol.* Suppl. 1, 90.

Harrison-Read, P. E. (1983). "Wet dog shake" behaviour in rats may reflect functionally opposed indoleaminergic systems involving different 5-HT receptors. *Br. J. Pharmac.* **78**, 92P.

Harvey, J. A. and Lints, C. E. (1971). Lesions in the median forebrain bundle: Relationships between pain sensitivity and telencephalic content of serotonin. *J. Comp. Physiol. Psychol.* **74**, 28.

Herman, Z. S. (1975). Behavioural changes induced in conscious mice by intracerebroventricular injection of catecholamines, acetylcholine and 5-hydroxytryptamine. *Br. J. Pharmac.* **55**, 351.

Hjorth, S., Carlsson, A., Lindberg, P., Sanchez, D., Wikstrom, H., Arvidsson, L. E., Hacksell, U. and Nilsson, J. L. G. (1982). 8-Hydroxy-2-(di-*n*-propylamino) tetralin, 8-OH-DPAT, a potent and selective simplified ergot congener with central 5-HT-receptor stimulating activity. *J. Neural. Transm.* **55**, 169.

Hockel, S. H. J., Muller, W. E. and Wollert, U. (1979). Diazepam increases L-tryptophan uptake into various regions of the rat brain. *Res. Comm. Psychol. Psychiat. Behav.* **4**, 467.

Hoebel, B. G. (1977). The psychopharmacology of feeding. In *Handbook of psychopharmacology* (eds. L. L. Iversen, S. D. Iversen and S. H. Snyder) Vol. 8, p. 55. Plenum Press, New York.

Hole, K., Fuxe, K. and Jonsson, G. (1976). Behavioural effects of 5,7-dihydroxytryptamine lesions of ascending 5-hydroxytryptamine pathways. *Brain Res.* **107**, 385.

Hunt, P. and Oberlander, C. (1981). The interaction of indole derivatives with the serotonin receptor and non-dopaminergic circling behaviours. In *Serotonin—current aspects of neurochemistry and function* (ed. B. Haber) p. 547. Plenum Press, New York.

—— Nedelec, L., Euvrard, C. and Boissier, J. R. (1981). Tetrahydro-pyridinyl indole derivatives as serotonin analogues which may differentiate between two distinct receptor sites. *Proc. 8th Int. Congr. Pharmac.*, Tokyo, p. 659, Abs. 1434.

Iversen, S. D. (1980). Animal models of anxiety and benzodiazepine actions. *Arzneim. Forsch.* **30**, 852.

Jacobs, B. L. (1976). An animal behaviour model for studying central serotonergic synapses. *Life Sci.* **19**, 777.

—— and Klemfuss, H. (1975). Brain stem and spinal cord mediation of a serotonergic behavioural syndrome. *Brain Res.* **100**, 450.

—— Asher, R. and Dement, W. C. (1973). Electrophysiological and behavioural effects of electrical stimulation of the raphe nuclei in cats. *Physiol. Behav.* **11**, 489.

—— Heym, J. and Rasmussen, K. (1983). Raphe neurons: Firing rate correlates with size of drug response. *Eur. J. Pharmac.* **90**, 275.

Jacoby, J. H., Howd, R. A., Levin, M. S. and Wurtman, R. J. (1975). Mechanisms by which quipazine, a putative serotonin receptor agonist, alters brain 5-hydroxyindole metabolism. *Neuropharmacology* **15**, 520.

Janssen, P. A. J. (1983). The psychopharmacological profiles and the therapeutic properties in psychiatric practice of a series of potent and selective 5-HT$_2$-receptor blockers. Abstract, *VII World Congress Psychiat.*, Vienna.

Jenner, P., Chadwick, D., Reynolds, E. H. and Marsden, C. D. (1975). Altered 5-HT metabolism with clonazepam, diazepam and diphenyl-hydantoin. *J. Pharm. Pharmac.* **27**, 707.

Johnstone, E. C., Cunningham Owens, D. G., Frith, C. D., McPherson, K., Dowie, C., Riley, G. and Gold, A. (1980). Neurotic illness and its response to anxiolytic and antidepressant treatment. *Psychol Med.* **10**, 321.

Jones, R. S. G. (1982). Responses of cortical neurones to stimulation of the nucleus raphe medianus: A pharmacological analysis of the role of indoleamines. *Neuropharmacology* **21**, 511.

Kelly, D. D. (1980). Clinical review of beta-blockers in anxiety. *Pharmakopsychiat.* **13**, 259.

—— (1983). The role of endorphins in stress-induced analgesia. *Ann. N.Y. Acad. Sci.* **398**, 260.

Kendall, D. A., Taylor, D. P. and Enna, S. J. (1983). [^{3}H]-Tetrahydro-trazodone binding: Association with serotonin$_1$ binding sites. *Mol. Pharmac.* **23**, 594.

Kilts, C. D., Commissaris, R. L. and Rech, R. H. (1981). Comparison of anti-conflict drug effects in three experimental animal models of anxiety. *Psychopharmacol.* **74**, 290.

Kiser, R. S. and Lebovitz, R. M. (1975). Monoaminergic mechanisms in aversive brain stimulation. *Physiol. Behav.* **15**, 47.

—— Brown, C. A., Sanghera, M. K. and German, D. C. (1980). Dorsal raphe nucleus stimulation reduces centrally elicited fearlike behaviour. *Brain Res.* **191**, 265.

—— German, D. C. and Lebovitz, R. M. (1978). Serotonergic reduction of dorsal central gray area stimulation produced aversion. *Pharmac. Biochem. Behav.* **9**, 27.

Koczkas, S., Holmberg, G. and Wedin, L. (1981). A pilot study of the effect of the 5-HT-uptake inhibitor, zimelidine, on phobic anxiety. *Acta Psychiat. Scand.* **63**, suppl. 290, 328.

Koe, K. (1979). Biochemical effects of antianxiety drugs on brain mono-amines. In *Anxiolytics, industrial pharmacology* (eds. S. Fielding and H. Lal) Vol. 3, p. 173. Futura, New York.

Koella, W. P. and Czicman, J. (1966). Mechanism of the EEG-synchron-ising action of serotonin. *Am. J. Physiol.* **211**, 927.

Kostowski, W. (1971). The effects of some drugs affecting brain 5HT on electrocortical synchronisation following low frequency stimulation of brain. *Brain Res.* **31**, 151.

—— Giacolone, S., Garattini, S. and Valzelli, L. (1969). Electrical stimu-lation of midbrain raphe: Biochemical, behavioural and bioelectric effects. *Eur. J. Pharmac.* **7**, 170.

Lansdown, M. J. R., Nash, H. L., Preston, P. R., Wallis, D. I. and Williams, R. G. (1980). Antagonism of 5-hydroxytryptamine receptors by quipazine. *Br. J. Pharmac.* **68**, 525.

Laurent, J. P., Mangold, M., Humbel, U. and Haefely, W. (1983). Reduction by two benzodiazepines and pentobarbitone of the multiunit activity in substantia nigra, hippocampus, nucleus locus coeruleus and nucleus raphe dorsalis of *encephale isolé* rats. *Neuropharmacology* **22**, 501.

Leander, J. D. (1983). Effects of punishment-attenuating drugs on depri-vation-induced drinking: Implications for conflict procedures. *Drug Dev. Res.* **3**, 185.

Leone, C. M. L., de Aguiar, J. C. and Graeff, F. G. (1983). Role of 5-hydroxytryptamine in amphetamine effects on punished and un-punished behaviour. *Psychopharmacol.* **80**, 78.

Leroux, A. G. and Myers, R. D. (1975). Action of serotonin microinjected into hypothalamic sites at which electrical stimulation produced aversive responses in the rat. *Physiol. Behav.* **14**, 501.

Leysen, J. E. and Tollenaere, J. P. (1982). Biochemical models for serotonin receptors. In *Ann. Rep. Med. Chem.* (ed. J. McDermed) Vol. 17, p. 1. Academic Press, New York.

Lippa, A. S., Nash, P. A. and Greenblatt, E. N. (1979*a*). Pre-clinical neuropsychopharmacological testing procedures for anxiolytic drugs. In *Anxiolytics, industrial pharmacology* (eds. S. Fielding and H. Lal) Vol. 3, p. 41. Futura, New York.

—— Crichett, D., Sano, M. C., Klepner, C. A., Greenblatt, E. N., Coupet, J. and Beer, B. (1979*b*). Benzodiazepine receptors: Cellular and behavioural characteristics. *Pharmac. Biochem. Behav.* **10**, 831.

Llewelyn, M. B., Azami, J. and Roberts, M. H. T. (1983). Effects of 5-hydroxytryptamine applied into nucleus raphe magnus on nociceptive thresholds and neuronal firing rate. *Brain Res.* **258**, 59.

Logan, D. G. (1979). Antidepressant treatment of recurrent anxiety attacks and night terrors. *Ohio State Med. J.* **75**, 653.

Lovick, T. A. and Robinson, J. P. (1983). Ascending and descending projections from nucleus raphe magnus in the rat. *J. Physiol. Lond.* **338**, 13P.

Lui, K. S., Debus, G. and Janke, W. (1978). Studies on the effect of oxprenolol on experimentally induced anxiety. *Arzneim. Forsch.* **28**, 1305.

Maj, J., Lewandowska, A. and Rawlow, A. (1979*a*). Central antiserotonin action of amitriptyline. *Pharmakopsychiat.* **12**, 281.

—— Palider, W. and Rawlow, A. (1979*b*). Trazodone, a central serotonin antagonist and agonist. *J. Neural. Transm.* **44**, 237.

—— Sowinska, H., Baran, L., Gancarczyk, L. and Rawlow, A. (1978). The central antiserotonin action of mianserin. *Psychopharmacology* **59**, 79.

Malick, J. B., Doren, E. and Barnett, A. (1977). Quipazine-induced head twitch in mice. *Pharmac. Biochem. Behav.* **6**, 325.

Margules, D. L. and Stein, L. (1968). Increase in "antianxiety" activity and tolerance of behavioural depression during chronic administration of oxazepam. *Psychopharmacologia* **13**, 74.

Martin, L. L. and Sanders-Bush, E. (1982*a*). Comparison of the pharmacological characteristics of $5HT_1$ and $5HT_2$ binding sites with those of serotonin autoreceptors which modulate serotonin release. *Naunyn-Schmiedeberg's Arch. Pharmac.* **321**, 165.

—— —— (1982*b*). The serotonin autoreceptor: Antagonism by quipazine. *Neuropharmacology* **21**, 445.

Massari, V. J., Tizabi, Y. and Jacobowitz, D. M. (1979). Potential noradrenergic regulation of serotonergic neurones in the median raphe nucleus. *Exp. Brain Res.* **34**, 177.

Melzacka, M., Rurak, A. and Vetulani, J. (1980). Preliminary study of the

320   Colin R. Gardner

biotransformation of two new drugs, trazodone and etoperidone. *Pol. J. Pharmac.* **32**, 551.

Middlemiss, D. N. and Fozard, J. R. (1983). 8-Hydroxy-2-(di-*n*-propylamino) tetralin discriminates between subtypes of 5-HT₁ recognition site. *Eur. J. Pharmac.* **90**, 151.

—— Blakeborough, L. and Leather, S. R. (1977). Direct evidence for an interaction of adrenergic blockers with the 5-HT receptor. *Nature (Lond.)* **267**, 289.

Morato de Carvalho, S., de Aguiar, J. C. and Graeff, F. G. (1981). Effect of minor tranquilisers, tryptamine antagonists and amphetamine on behaviour punished by brain stimulation. *Pharmac. Biochem. Behav.* **15**, 351.

Moriyama, M., Ichimaru, Y. and Gomita, Y. (1982). Effects of serotonergic drugs on operant behaviour with intracranial aversive stimulation in rats. *Jap. J. Pharmac.* **32**, 254P.

Morley, J. E., Levine, A. S. and Rowland, N. E. (1983). Stress induced eating. *Life Sci.* **32**, 2169.

Mosko, S. S., Haubrich, D. and Jacobs, B. L. (1977). Serotonergic afferents to the dorsal raphe nucleus: Evidence from HRP and synaptosomal uptake studies. *Brain Res.* **119**, 269.

Mounsey, I., Brady, K. A., Carroll, J., Fisher, R. and Middlemiss, D. N. (1982). K⁺-Evoked [³H]-5-HT release from rat frontal cortex slices: The effect of 5-HT agonists and antagonists. *Biochem. Pharmac.* **31**, 49.

Murphy, J. E. (1978). Mianserin in the treatment of depressive illness and anxiety states in general practise. *Br. J. Clin. Pharmac.* **5**, 81S.

Nagayama, H., Hingtgen, J. N. and Aprison, M. H. (1981). Postsynaptic action by four antidepressive drugs in an animal model of depression. *Pharmac. Biochem. Behav.* **15**, 125.

Nahorski, S. R. and Willcocks, A. L. (1983). Interactions of β-adrenoceptor antagonists with 5-hydroxytryptamine subtypes in rat cerebral cortex. *Br. J. Pharmac.* **78**, 107P.

Nakamura, M. and Fukushima, H. (1976). Head twitches induced by benzodiazepines and the role of biogenic amines. *Psychopharmacology* **49**, 259.

—— —— (1977). Effect of benzodiazepines on central serotonergic neuron systems. *Psychopharmacology* **53**, 121.

Nelson, D. L. (1982). Central serotonergic receptors: Evidence for heterogeneity and characterization by ligand-binding. *Neurosci. Biobehav. Rev.* **6**, 499.

Oberlander, C. (1983). Effects of a potent 5-HT agonist, RU 24969, on the mesocorticolimbic and nigrostriatal dopamine systems. *Br. J. Pharmac.* **80**, 675P.

—— and Boissier, J. R. (1981). Haloperidol blocks hyperlocomotion but not the circling behaviour induced by serotonin agonist RU 24969. *Proc. 8th Int. Cong. Pharmac.* Tokyo, Abst. 839.

Ogren, S. O., Cott, J. M. and Hall, H. (1981). Sedative/anxiolytic effects of antidepressants in animals. *Acta Psychiat. Scand.* **63**, suppl. 290, 271.

Parati, E. A., Zanardi, P., Cocchi, D., Caraceni, T. and Muller, E. E. (1980). Neuroendocrine effects of quipazine in man in healthy state or with neurological disorders. *J. Neural. Transm.* **47**, 273.

Patel, J. B. and Malick, J. B. (1980). Effects of isoproterenol and chlordiazepoxide on drinking and conflict behaviours in rats. *Pharmac. Biochem. Behav.* **12**, 819.

Patkina, N. A. and Lapin, J. P. (1976). Effect of serotonergic drugs on positive and negative reinforcing systems in cats. *Pharmac. Biochem. Behav.* **5**, 241.

Peroutka, S. J. and Snyder, S. H. (1981). [^{3}H] Mianserin: Differential labelling of serotonin$_2$ and histamine$_1$ receptors in rat brain. *J. Pharmac. Exp. Ther.* **216**, 142.

Petersen, E. N. and Lassen, J. B. (1981). A water lick conflict paradigm using drug experienced rats. *Psychopharmacology* **75**, 236.

Pilcher, C. W. T., Stolerman, I. P. and D'Mello, G. D. (1978). Aversive effects of narcotic antagonists in rats. In *Characteristics and function of opioids* (eds. J. M. Van Ree and L. Terenius) p. 437. Elsevier, North Holland.

Pinder, R. M., Brogden, R. N., Sawyer, P. R., Speight, T. M. and Avery, G. S. (1975). Fenfluramine: A review of its pharmacological properties and therapeutic efficacy in obesity. *Drugs* **10**, 241.

Poschel, B. P. H. (1971). A simple and specific screen for benzodiazepine-like drugs. *Psychopharmacologia* **19**, 193.

Przewlocka, B., Stala, L. and Scheel-Kruger, J. (1979). Evidence that GABA in the nucleus dorsalis raphe induces stimulation of locomotor activity and eating behaviour. *Life Sci.* **25**, 937.

Puhringer, W., Wirz-Justice, A. and Lancranjan, I. (1976). Mood elevation and pituitary stimulation after i.v. L-5HTP in normal subjects: Evidence for a common serotonin mechanism. *Neurosci. Letts.* **2**, 349.

Pycock, C. J., Horton, R. W. and Carter, C. J. (1978). Interactions of 5-hydroxytryptamine and γ-aminobutyric acid with dopamine. *Adv. Biochem. Psychopharmac.* **19**, 323.

Quik, M. and Azmitia, E. (1983). Selective destruction of the serotonergic fibres of the fornix-fimbria and cingulum bundle increases 5-HT$_1$ but not 5-HT$_2$ receptors in rat midbrain. *Eur. J. Pharmac.* **90**, 377.

Robichaud, R. C. and Sledge, K. L. (1969). The effects of *p*-chlorophenylalanine on experimentally induced conflict in the rat. *Life Sci.* **8**, 965.

Sainati, S. M. and Lorens, S. A. (1982). Intra-raphe muscimol induced hyperactivity depends on ascending serotonin projections. *Pharmac. Biochem. Behav.* **17**, 973.

Samanin, R., Mennini, T., Ferraris, A., Bendotti, C., Borsini, F. and Garattini, S. (1979). *m*-Chlorophenylpiperazine: A central serotonin agonist causing powerful anorexia in rats. *Naunyn-Schmiedeberg's Arch. Pharmac.* **308**, 159.

Sansone, M., Melzacka, M., Hano, J. and Vetulani, J. (1983). Reversal of depressant action of trazodone on avoidance behaviour by its metabolite *m*-chlorophenylpiperazine. *J. Pharm. Pharmac.* **35**, 189.

Schenberg, L. C. and Graeff, F. G. (1978). Role of the periaqueductal gray substance in the antianxiety action of benzodiazepines. *Pharmac. Biochem. Behav.* **9**, 287.

Segal, D. S. (1976). Differential effects of para-chlorophenylalanine on amphetamine-induced locomotion and stereotypy. *Brain Res.* **116**, 267.

Segal, M. and Weinstock, M. (1983). Differential effects of 5-hydroxy-tryptamine antagonists on behaviour resulting from activation of different pathways arising from the raphe nuclei. *Psychopharmacology* **79**, 72.

Sepinwall, J. (1983). Behavioural studies related to the neurochemical mechanisms of action of anxiolytics. In *Anxiolytics: Neurochemical, behavioural and clinical perspectives* (eds. J. B. Malick, S. J. Enna and H. I. Yamamura) p. 147. Raven Press, New York.

—— and Cook, L. (1980). Relationship of γ-aminobutyric acid (GABA) to antianxiety effects of benzodiazepines. *Brain Res. Bull.* **5**, 839.

—— Grodsky, F. S., Sullivan, J. W. and Cook, L. (1973). Effects of propranolol and chlordiazepoxide on conflict behaviour in rats. *Psychopharmacologia* **31**, 375.

Shephard, R. A., Buxton, D. A. and Broadhurst, P. L. (1982). Drug interactions do not support reduction of serotonin turnover as the mechanism of action of benzodiazepines. *Neuropharmacology* **21**, 1027.

Siegel, J. and Brownstein, R. A. (1975). Stimulation of n. raphe dorsalis central gray and hypothalamus: Inhibitory and aversive effects. *Physiol. Behav.* **14**, 431.

Sinden, J. D. and Atrens, D. M. (1978). 5-Methoxy-*N,N*-dimethyltryptamine: Differential modulation of the rewarding and aversive components of lateral hypothalamic self stimulation. *J. Pharm. Pharmac.* **30**, 268.

Squires, R. F., Casida, J. E., Richardson, M. and Saederup, E. (1983). [^{35}S]-*t*-Butyl bicyclophosporothionate binds with high affinity to brain-specific sites coupled to γ-aminobutyric acid-A and ion recognition sites. *Mol. Pharmac.* **23**, 326.

Soubrié, P., Thiebot, M. H., Jobert, A. and Hamon, M. (1981). Serotoninergic control of punished behaviour: Effects of intra-raphe microinjections of chlordiazepoxide, GABA and 5-HT on behavioural suppression in rats. *J. Physiol. (Paris)* **77**, 449.

—— Blas, C., Ferron, A. and Glowinski, J. (1983). Chlordiazepoxide reduces *in vivo* serotonin release in the basal ganglia of encephale isole but not anaesthetised cats: Evidence for a dorsal raphe site of action. *J. Pharmac. Exp. Ther.* **226**, 526.

Soulairac, A. and Soulairac, M. L. (1970). Effects of amphetamine-like substances and L-DOPA on thirst, water intake and diuresis. In *Amphetamines and related compounds* (eds. E. Costa and S. Garattini) p. 819. Raven Press, New York.

Stein, L., Belluzzi, J. D. and Wise, C. D. (1977a). Benzodiazepines: Behavioural and neurochemical mechanisms. *Am. J. Psychiat.* **134**, 665.

—— Wise, C. D. and Belluzzi, J. D. (1975). Effects of benzodiazepines on central serotonergic mechanisms. In *Mechanisms of action of benzodiazepines* (eds. E. Costa and P. Greengard) p. 29. Raven Press, New York.

—— —— —— (1977b). Neuropharmacology of reward and punishment. In *Handbook of psychopharmacology* (eds. L. L. Iversen, S. D. Iversen and S. H. Snyder) Vol. 8, p. 25. Plenum Press, New York.

—— —— and Berger, B. D. (1973). Anti-anxiety action of benzodiazepines: Decrease in activity of serotonin neurones in the punishment system. In *The benzodiazepines* (eds. S. Garattini, E. Mussini, and L. O. Randall) p. 299. Raven Press, New York.

Taeber, K., Appel, E., Badian, M., Palm, D., Rupp, W., Schofer, J. and Sittig, W. (1980). DAF induced stress, betablockers and benzodiazepines. *Proc. World. Conf. Clin. Pharmac. Ther. London* Abst. 897.

Tebecis, A. K. (1972). Antagonism of 5-hydroxytryptamine by methiothepin shown in microiontophoretic studies of neurones in the lateral geniculate nucleus. *Nature (Lond.)* **238**, 63.

Tenen, S. S. (1967). The effects of *p*-chlorophenylalanine, a serotonin depletor, on avoidance acquisition, pain sensitivity and related behaviour in the rat. *Psychopharmacology* **10**, 204.

Terman, G. W., Lewis, J. W. and Liebeskind, J. C. (1983). Opioid and non-opioid mechanisms of stress analgesia: Lack of cross-tolerance between stressors. *Brain Res.* **260**, 147.

Thiebot, M. H., Hamon, M. and Soubrié, P. (1982). Attentuation of induced anxiety in rats by chlordiazepoxide: Role of the raphe dorsalis benzodiazepine binding sites and serotonergic neurones. *Neuroscience* **7**, 2287.

—— Jobert, A. and Soubrié, P. (1980). Chlordiazepoxide and GABA injected into raphe dorsalis release the conditioned behavioural suppression induced in rats by a conflict procedure without nociceptive component. *Neuropharmacology* **19**, 633.

Trimble, M., Chadwick, D., Reynolds, E. H. and Marsden, C. D. (1975). L-

324   Colin R. Gardner

5-Hydroxytryptophan and mood. *Lancet* **i**, 583.

Trulson, M. E., Preussler, D. W., Howell, G. A. and Fredrickson, C. J. (1982). Raphe unit activity in freely moving cats: Effects of benzodiazepines. *Neuropharmacology* **21**, 1045.

Tye, N. C., Everitt, B. J. and Iversen, S. D. (1977). 5-Hydroxytryptamine and punishment. *Nature (Lond.)* **268**, 741.

—— Iversen, S. D. and Green, A. R. (1979). The effects of benzodiazepine and serotonergic manipulations on punished responding. *Neuropharmacology* **18**, 689.

Van Riezen, H., Pinder, R. M., Nickolson, V. J., Hobbelen, P., Zayed, I. and Van der Veen, F. (1981). Mianserin Hydrochloride. In *Pharmacological and biochemical properties of drug substances* (ed. M. E. Goldberg) Vol. 3, p. 1. American Pharmaceutical Association, Washington.

Vetulani, J., Sansone, M., Bednarczyk, B. and Hano, J. (1982). Different effects of 3-chlorophenylpiperazine on locomotor activity and aquisition of conditioned avoidance response in different strains of mice. *Naunyn-Schmiedeberg's Arch. Pharmac.* **319**, 271.

Wada, J. A. and Matsuda, M. (1971). Learned escape behaviour induced by brain electrical stimulation and various neuroactive agents. *Exp. Neurol.* **32**, 357.

Walletschek, H. and Raab, A. (1982). Spontaneous activity of dorsal raphe neurones during defensive and offensive encounters in the tree shrew. *Physiol. Behav.* **28**, 697.

Wallis, D. (1981). Neuronal 5-hydroxytryptamine receptors outside the central nervous system. *Life Sci.* **29**, 2345.

Wang, R. Y. and Aghajanian, G. (1978). Collateral inhibition of serotonergic neurones in the rat dorsal raphe nucleus: Pharmacological evidence. *Neuropharmacology* **17**, 819.

Weinstock, M., Weiss, C. and Gitter, S. (1977). Blockade of 5-hydroxytryptamine receptors in the central nervous system by β-adrenoceptor antagonists. *Neuropharmacology* **16**, 273.

Whitaker, P. M. and Seeman, P. (1978). High affinity ^{3}H-serotonin binding to caudate: Inhibition by hallucinogens and serotoninergic drugs. *Psychopharmacology* **59**, 1.

Wilbur, R. and Kulik, F. (1981). Gray's cybernetic theory of anxiety. *Lancet* **ii**, 803.

Winter, J. C. (1972). Comparison of chlordiazepoxide, methysergide and cinanserin as modifiers of punished behaviour and as antagonists of *N,N*-dimethyltryptamine. *Arch. Int. Pharmacodyn.* **197**, 147.

Wirtshafter, D. and Asin, K. E. (1981). Shock-induced suppression of drinking following electrolytic median raphe lesions. *Physiol. Psychol.* **9**, 263.

Wise, C. D., Berger, B. D. and Stein, L. (1973). Evidence of α-nor-

325 Serotonin in animal models of anxiety

adrenergic reward receptors and serotonergic punishment receptors in the rat brain. *Biol. Psychiat.* **6,** 3.

Yaryura-Tobias, J. A. and Bhagavan, H. N. (1977). L-Tryptophan in obsessive compulsive disorder. *Am. J. Psychiat.* **134,** 1298.

Young, G. A. (1980). Naloxone enhancement of punishment in the rat. *Life Sci.* **26,** 1787.

12

The effects of drugs on serotonin-mediated behavioural models

A. RICHARD GREEN AND DAVID J. HEAL

12.1 Introduction

The value of any simple behavioural change which occurs as a result of increasing the function of a central neurotransmitter is that it enables an investigator to examine not only the possible function of the neurotransmitter, but also (arguably) more importantly, to examine the effect of various drugs on the function of the transmitter.

While ligand-receptor binding studies are capable of demonstrating whether a drug interacts with a neurotransmitter receptor, they do not show whether the compound is acting as an agonist or an antagonist at this site. Furthermore, if the number of neurotransmitter receptors changes following longer term administration of a drug, it is also impossible to determine from this finding alone whether the observed change (up- or down-regulation) is likely to have functional significance; that is, do the animals show a greater or lesser response to the neurotransmitter? Clearly, animal behavioural models do not provide comprehensive answers to such problems and results obtained still have to be treated with caution. Nevertheless, when used in conjunction with other biochemical and pharmacological observations, they can provide a useful indication of the way in which drugs alter neurotransmitter function. In certain respects behavioural models are analogous to gut-bath preparations, except that in the case of the former we are using a whole animal as a 'bioassay' rather than an isolated tissue.

With respect to serotonin we are fortunate in that there are behavioural changes occurring which can be shown convincingly to result from the increased activity of this neurotransmitter within the brain. In contrast, however, there are no satisfactory behavioural models for assessing central β-adrenoceptor function, nor indeed are there any widely used models for the amino acid neurotransmitters. Equally, it should be remembered that serotonergic models are only 'bioassay' systems and are not models of any psychiatric state. Consequently, the effects of, for example, an antidepressant drug on a chosen behavioural model simply suggests possible interactions with central serotonergic function and not its therapeutic mode of action in alleviating depression (although, of course, such a conclusion may sometimes be inferred when taken with other data).

A major problem with the animal models of serotonin function is that the behaviour does not result simply from the stimulation of serotonergic receptors. Other neurotransmitters are involved in the response and, therefore, a change in the function of a neurotransmitter other than serotonin could influence the response being observed. This consideration must always be kept in mind when interpreting results and has been a strong reason for investigating the influence of other neurotransmitters on the model being used. These neurotransmitter interactions will be discussed in this chapter.

A further complication is the recent evidence of different types of serotonin receptor (see particularly Chapter 4) and data which are now accumulating from animal models indicates that these receptor sub-types are functionally distinct. This is also discussed in this chapter.

Finally, an examination is made of the effects of various treatments (particularly antidepressants) on the behavioural changes associated with stimulation of the main serotonin receptor sub-types to determine whether there was an association with the changes in receptor number which have been reported to occur on the basis of ligand-receptor binding studies. The presence of any such association is important since it inevitably stengthens any proposal that the receptor changes may have functional significance.

12.2 The hyperactivity model in rats

12.2.1 *Drugs producing behavioural change*

The behavioural changes which follow administration of a monoamine oxidase inhibitor and L-tryptophan were first described by Hess and Doepfner (1961) and subsequently by Grahame-Smith (1971a). The behaviours observed in the animals are listed in Table 12.1. The behavioural changes noted in this table by asterisks have been scored by observers in more recent studies (see, for example, Section 12.2.3) whilst other components of the syndrome, such as locomotor activity, circling and forepaw treading, have been recorded using Animex or Automex activity meters which rely on capacitance changes. Using Animex meters, it has been shown that the recorded activity is proportional to the rate of increase of serotonin concentration in the brain (Grahame-Smith 1971a). If one is relying on activity meters it is, of course, vital also to observe the animals.

328 A. Richard Green and David J. Heal

Table 12.1 Behavioural changes seen in rats following administration of tranycypromine plus L-tryptophan or serotonin agonists

Disperse	Body tremor
Head weaving*	Athotosis
Piloerection	Ataxia
Proptosis	Forepaw treading*
Salivation	Trunk weaving
Hindlimb abduction*	Unresponsive to visual stimuli
Straub tail*	Opisthotonus
Compulsive movement	(Fits) High dose L-Trp only
Random circling	(Death) High dose L-Trp only

The behaviours marked * plus reactivity were rated in some studies (see text).

Providing that the change observed in the experimental group (for example, after drug treatment) is merely one of enhancing or inhibiting the whole response, then automated counting is feasible. However, if only some of the behavioural changes induced by the increased serotonin function are altered, then automated counting can produce quite misleading results (see Section 12.1.3 on antagonists). Such problems have become apparent over the years as the model has been examined in greater detail.

The behavioural changes result from increased brain serotonin concentrations and can be produced by L-tryptophan plus a monoamine oxidase inhibitor (Grahame-Smith 1971a) or alternatively, by 5-HTP administered either at high dose alone or at lower doses when preceded by a monoamine oxidase inhibitor or peripheral decaboxylase inhibitor (Modigh 1972). L-Tryptophan alone does not produce the syndrome, presumably because the serotonin being formed is being metabolized intraneuronally by MAO. However, administration of 5-HTP bypasses the rate-limiting enzyme (tryptophan hydroxylase; see Section 1.2) and therefore at high dose, presumably forms serotonin sufficiently rapidly to 'spill over' into the synaptic cleft and stimulate the receptors, even though intraneuronal MAO has not been inhibited.

A study with selective MAO inhibitors suggested that normally both type A and type B MAO must be inhibited before L-tryptophan produced the behavioural changes (Green and Youdim 1975). It was suggested that if type A alone was inhibited, the form normally thought to metabolize serotonin (Section 1.2), then type B MAO had the capability *in vivo* of continuing to metabolize the amine. This view has received subsequent biochemical support (Fowler and Tipton 1982; see also Section 1.2). In addition, it was observed that MAO had to be inhibited by more than 80 per cent for L-tryptophan to produce the behavioural syndrome.

The behavioural changes induced by increased rat brain serotonin concentrations following precursor loading are potentiated by serotonin uptake inhibitors such as clomipramine (Green and Grahame-Smith 1975; Modigh 1973) and fluoxetine (Hwang and Van Woert 1980).

Other compounds which are structurally related to serotonin will also produce the behavioural syndrome. These include 5-MeODMT (Grahame-Smith 1971*b*), 5-methoxytryptamine (Green, Hughes and Tordoff 1975), tryptamine (Foldes and Costa 1975; Marsden and Curzon 1978), and LSD when given at high dose (Trulson, Ross and Jacobs 1976*a*). The behaviour has also been shown to occur following administration of various piperazine derivatives which lack an indole structure, namely, quipazine (Green, Youdim and Grahame-Smith 1976*b*) and MK-212 (Clineschmidt 1979). Possible structural similarities between the piperazine and indole compounds accounting for the observations have been discussed elsewhere (Green, Hall and Rees 1981).

Finally, it has been observed that behavioural changes can be induced by two drugs which release serotonin, fenfluramine and *p*-chloroamphetamine (Trulson and Jacobs 1976; Green and Kelly 1976).

12.2.2   *Neurotransmitters involved in the syndrome and the anatomical site of initiation of the behaviour*

The fact that serotonin initiates the syndrome is evident from the previous section in so far as it is also produced by a variety of serotonin-mimetic drugs. Furthermore, prior inhibition of serotonin synthesis with PCPA prevents the production of the syndrome initiated by administration of either tranylcypromine/L-tryptophan or *p*-chloroamphetamine (Grahame-Smith 1971*a*; Jacobs 1974; Trulson and Jacobs 1976). The administration of a MAO inhibitor and L-tryptophan also markedly increases tryptamine concentrations in both the periphery and brain and because MAO has been inhibited, the peripherally-formed tryptamine can be transported into brain and produce a central effect. Marsden and Curzon (1978) have shown that the role of tryptamine is probably that of directly stimulating serotonin receptors or releasing serotonin, or both. Dourish and Greenshaw (1983) have also postulated a joint role for serotonin and tryptamine. They observed that when both amines were given together intracerebroventricularly, the syndrome was produced but not when either compound was given separately. However, since dose-response curves were not obtained, it could have been that the effects of the two compounds were merely additive. Certainly the appearance of the whole syndrome in cases where tryptamine is not formed (5-HTP administration, serotonin agonists, or releasing drugs), suggests that it does not have to be present.

With regard to the involvement of noradrenaline, there is, at present, no evidence for a direct role. Depletion of brain noradrenaline content with disulfiram (Green and Grahame-Smith 1974) or lesions of the median forebrain bundles and locus coeruleus with 6-hydroxydopamine (Green and Deakin 1980), both failed to alter the behavioural responses to serotonin or serotonin agonists.

Nevertheless, there is substantial evidence that both β-adrenoceptor antagonists and agonists alter the serotonin-mediated behavioural responses.

To deal with β-antagonists first. In 1976, Green and Grahame-Smith reported that the non-selective β-adrenoceptor antagonist propranolol, at high dose, antagonized the behaviour elicited by tranylcypromine/L-tryptophan or 5-MeODMT. The antagonist activity resided in the (−)-propranolol isomer, that is, the one with β-adrenoceptor antagonist action. The dose could be reduced if the rapid metabolism of the drug in the liver was inhibited. In a follow-up study, Costain and Green (1978) found that all non-selective (β_1 plus β_2) β-adrenoceptor antagonists inhibited the serotonin-mediated behaviour in rats. In contrast, selective inhibitors (β_1- or β_2-selective) had no effect. Most of these selective compounds (atenolol, practolol) have poor brain penetration, but metoprolol (β_1-selective) and butoxamine (β_2-selective) do enter the brain efficiently and these had no effect on the serotonin-mediated behaviour. (Subsequently, the same lack of effect has been found with the new β_2-selective compound ICI 118,551; Green, unpublished observations).

In view of these data and evidence by other workers of propranolol antagonizing serotonin-mediated behaviour (see the review of Weinstock 1980), there are now substantial arguments in favour of the hypothesis that the inhibition of serotonin-mediated behaviour by the β-adrenoceptor antagonists is a direct effect and not mediated by an inhibition of central noradrenergic function. Furthermore, Schechter and Weinstock (1974) had previously suggested that propranolol was an antagonist of serotonin receptors from studies using the rat stomach and uterus preparations. This suggestion was strengthened by the report of Middlemiss, Blakeborough and Leather (1977) who observed that non-selective (but not selective) β-adrenoceptor antagonists compete with [3H]-serotonin for binding to its receptor sites. With the recent discovery that there are serotonin receptor sub-types, the effects of β-adrenoceptor antagonists have now been re-examined. Propranolol showed a stereoselective inhibition of [3H]-spiperone binding in the frontal cortex while selective (β_1 or β_2) antagonists did not inhibit this 5-HT$_2$-receptor binding (Green, Johnson and Nimgaonkar 1983a). Nahorski and Willcocks (1983) obtained similar data but in addition, found that β-adrenoceptor antagonists were quite potent inhibitors at the 5-HT$_1$ receptor site.

Overall, therefore, it seems probable that the inhibition of serotonin-mediated behaviour by β-adrenoceptor antagonists results from a direct inhibition of the serotonin receptor and does not result from their effects on central noradrenergic systems.

Interestingly, after the β-adrenoceptor antagonists had been shown to inhibit serotonin-mediated behaviour in rats, it was demonstrated that acute administration of $β_2$-adrenoceptor agonists, such as salbutamol, or the more liposoluble clenbuterol, markedly enhanced the behavioural responses of the rats to administration of tranylcypromine/L-tryptophan (Fig. 12.1) or quipazine (Cowen, Grahame-Smith, Green and Heal 1982; Nimgaonkar, Green, Cowen et al. 1983). This suggests that, for reasons which are at present unclear, the administration of a β-adrenoceptor agonist enhances the postsynaptic response to a serotonin agonist. Neither a noradrenergic lesion nor depletion of serotonin with PCPA prevented this enhancement occurring (Nimgaonkar et al. 1983).

An investigation of the effects of $β_2$-adrenoceptor agonists on 5-HT_2-receptor binding did not reveal any effect of the drugs on binding either in vitro (Green et al. 1983a) or in the frontal cortex taken from rats pre-treated with clenbuterol in vivo (Green, Heal, Johnson, Lawrence and Nimgaonkar 1983c). Enhancement of the 5-HTP-induced head twitch in mice has also been observed after clenbuterol (Section 12.3.2). At present, therefore, the indications are that the $β_2$-adrenoceptor agonists enhance serotonin-mediated behaviour by acting at postsynaptic β-adrenoceptors which modulate the serotonin-mediated behaviour. It is, however, not certain that the β-adrenoceptors involved are actually of the $β_2$-sub-type since at the dose of agonist administered selectivity is likely to have been lost.

It has been suggested that GABA might also be involved in serotonergic behavioural responses since increasing the concentration of this inhibitory transmitter with amino-oxyacetic acid (AOAA) inhibited the behavioural response (Green, Tordoff and Bloomfield 1976a). However, a subsequent study did not confirm this (Sloviter, Drust and Connor 1978) and suggested that the inhibition seen was of locomotor activity and when individual behaviours were scored, no clear inhibition of the responses was observed.

A similar controversy has surrounded the involvement of dopamine in the syndrome (Jacobs 1976; Sloviter et al. 1978), which again partly stems from the fact that early observations were made using automated activity counts alone, whereas later studies have concentrated on scoring the individual behavioural changes induced by serotonin. In a comprehensive and careful study which examined many of the problems associated with the controversy, Deakin and Dashwood (1981) showed that α-methyl-p-tyrosine did indeed slow the appearance of the syndrome, confirming an

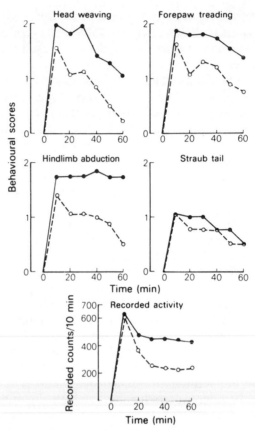

Fig. 12.1 Effect of clenbuterol (5 mg/kg) pre-treatment on the behavioural responses and recorded activity of pairs of rats following administration of quipazine (15 mg/kg). Rats were pre-treated with saline (O) or clenbuterol (●) (5 mg/kg i.p.). Quipazine (25 mg/kg i.p.) was injected 15 min later (time zero) and the behavioural responses measured at 10 min intervals thereafter. Results shown as mean behavioural scores (8 animals in each group) against time after quipazine injection. Head weaving was enhanced ($p < 0.01$) during the period 20–60 min. Forepaw treading ($p < 0.02$ at 20 min; $p < 0.05$ during period 30–60 min), hindlimb abduction ($p < 0.05$ at 20 min; $p < 0.01$ during period 30–60 min) and recorded activity ($p < 0.05$ at 30, 50 and 60 min; $p < 0.02$ at 40 min) were also enhanced by clenbuterol pre-treatment. Total recorded activity also increased during the 60 min following quipazine ($p < 0.025$). (Reproduced from Cowen *et al.* (1982) with permission from Macmillan Journals)

earlier study, but was almost certainly doing so by inhibiting the rate of serotonin synthesis (most probably by competing with L-tryptophan for entry into the brain). α-Methyl-*p*-tyrosine did not inhibit the majority of the behavioural changes induced by serotonin agonists. These authors suggested that head weaving, forepaw treading and hindlimb abduction were 'purely' serotonergic whereas the hyperactivity and hyper-reactivity involved a dopaminergic link postsynaptic to the serotonin neurones initiating the syndrome. This hypothesis explains why dopamine antagonists apparently inhibited the syndrome when measured using automated activity meters; although higher doses of neuroleptics probably do inhibit all behaviours by also having some serotonin antagonist actions (Peroutka and Snyder 1980).

The division of the behavioural changes into those which are serotonergic only and those which are serotonergic with a postsynaptic dopaminergic involvement has received further support in studies on serotonin antagonists, such as methysergide or pirenperone (Green *et al.* 1983*a*; Section 12.2.3) and the putative 5-HT$_1$ agonist RU 24969 (Green, Guy and Gardner 1984; Section 12.5.1), and the possible inter-relationships are discussed further in Section 12.5.1 and shown in Fig. 12.6.

Two studies have investigated the neuroanatomical site of initiation of the behaviour. The first, by Jacobs and Klemfuss (1975), examined the effects of sectioning the brain and showed that most of the behavioural changes still occurred in animals sectioned at the level of the pons-medulla. The authors concluded, therefore, that many of the behaviours (including head weaving, forepaw treading and hindlimb abduction) were hindbrain or spinally mediated. An identical conclusion was also reached by Deakin and Green (1978) who employed the somewhat different approach of lesioning serotonergic pathways in the hindbrain and cord with 5,7-DHT. After 10 days, when the postsynaptic receptors were showing denervation super-sensitivity, they examined which behavioural components were enhanced following administration of a serotonin agonist. It was found that it was these behaviours, which Jacobs and Klemfuss (1975) had suggested were hindbrain or spinally mediated, which were in fact enhanced.

The observation that the behaviours are initiated by hindbrain mechanisms is a continuing problem in so far as most ligand-receptor binding studies are performed on cortical or other forebrain tissue. Trying to equate receptor and functional changes is therefore difficult. However, available evidence suggests that the site of initiation of the behaviour in other commonly used models ('wet dog shakes' in rats, head twitch in mice) is also in the hindbrain regions (see Section 12.3.1).

12.2.3 *Antagonists*

The lack of availability of suitable centrally acting serotonin antagonists has been a problem for several years and not only for studies on serotonin-mediated behaviour. Compounds known to be potent antagonists in peripheral tissues have been shown in iontophoretic investigations to have

little effect in blocking inhibitory responses in areas with a dense serotonergic input (Haigler and Aghajanian 1977).

We therefore investigated (Green *et al.* 1981) the effects of a range of
serotonin antagonists on the behaviour induced by serotonin (i.e. following
tranylcypromine/L-tryptophan), the indole-structured agonist 5-MeODMT
(following tranylcypromine pre-treatment) and the phenylpiperazine,
quipazine. The antagonists known to be potent in peripheral tissues,
namely methysergide and metergoline, both antagonized the head weaving,
forepaw treading and hindlimb adduction. However, the locomotor
and reactivity components were markedly enhanced with the animals
displaying sudden bursts of locomotor activity, and this resulted in an
increase in the response measured by the activity meters (Fig. 12.2).
Furthermore, methysergide and metergoline inhibited the behaviours
listed above irrespective of whether they were elicited by injection of
tranylcypromine/L-tryptophan, 5-MeODMT or quipazine.

In contrast, propranolol and methiothepin not only inhibited the head
weaving, forepaw treading and hindlimb adduction but also reduced the
locomotor activity component (Fig. 12.2). Although in the case of methiothepin it was possible to attribute the latter effect to its dopamine
antagonist properties (Lloyd and Bartholini 1974), this was not so for
propranolol. This, therefore, suggested that head weaving, forepaw treading and hindlimb adduction were mediated by one serotonin receptor
sub-type while locomotor activity was mediated by another and, furthermore, that propranolol was a non-selective antagonist for both. The
probability that propranolol is a serotonin antagonist has already been
discussed (Section 12.2.2) and this hypothesis is examined in more detail
later (Section 12.5.1).

In contrast to the antagonists discussed above, weaker serotonin
antagonists, such as mianserin, cinanserin and cyproheptadine, would only
antagonize the behavioural responses when initiated by the weak agonist
quipazine, but not when the behaviours were initiated by the potent
agonists serotonin or 5-MeODMT.

Of course, this division was not absolute and increasing these dose of
cyproheptadine led to an inhibition of 5-MeODMT-mediated behaviour.
Nevertheless, the study did suggest that one had to be cautious of ascribing
either a role (or lack of one) to serotonin in a behavioural or other physiological model unless several antagonists had been examined.

More recently, drugs with suggested specific 5-HT$_2$ receptor antagonism
have been developed. These drugs include ketanserin, pirenperone and
ritanserin (R 55667). The problem here has been that whilst they have
been termed 'selective', this has referred merely to their selectivity for
5-HT$_2$ rather than for 5-HT$_1$ receptors and has not necessarily implied a
low affinity for other neurotransmitter receptor sites. Ketanserin is a potent

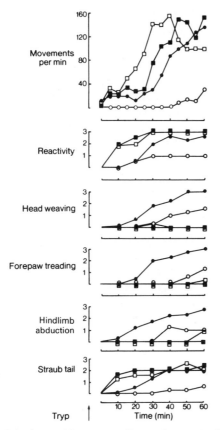

Fig. 12.2 Effects of methysergide, metergoline and propranolol on the behaviour elicited by tranylcypromine and L-tryptophan in rats: (●) saline pre-treated; (□) methysergide (10 mg/kg); (■) methiothepin (5 mg/kg); (○) (−)-propranolol (20 mg/kg). Differences between drug and control group are significant (p <0.05 or better) at 3 or more time points for all behaviours except Straub tail. (Reproduced from Deakin and Breen (1978) with permission from Macmillan Journals)

α_1-adrenoceptor antagonist and we have found that pirenperone is a potent dopamine antagonist (Green, O'Shaughnessy, Hammond, Schächter and Grahame-Smith 1983*d*). Preliminary work with ritanserin, however, suggests that this drug has a rather greater selectivity for the 5-HT$_2$ receptors than those of other neurotransmitters and this drug may become a valuable tool in future studies.

In 1981, on the basis of their studies on the inhibition of the head twitch behaviour in mice (Section 12.3) by a range of compounds, Peroutka, Lebovitz and Snyder (1981) suggested that this particular behaviour was

5-HT$_2$-receptor-mediated. In our recent study with pirenperone (Green *et al.* 1983*a*), we observed that this compound, a putative 5-HT$_2$-receptor antagonist, indeed inhibited head twitch behaviour with an IC$_{50}$ of 76 μg/ kg. Using these results, we gave a dose of pirenperone of 100 μg/kg and examined the effects on both tranylcypromine/L-tryptophan and quipazine-evoked behaviour. The head weaving, forepaw treading and hindlimb abduction were markedly attenuated by pirenperone pre-treatment. There was a slight enhancement of the locomotor response but this was not marked (in contrast to the effects of methysergide pre-treatment). This seemed a little surprising in view of the observations that methysergide and pirenperone are almost equipotent as inhibitors of 5-HT$_2$-receptor binding *in vitro* (Green *et al.* 1983*a*). However, pirenperone also proved to be a potent dopamine antagonist (the IC$_{50}$ for inhibiting methylamphetamine locomotion was around 100 μg/kg (Green *et al.* 1983*d*) and this, therefore, probably explains this apparent anomaly. Our recent experience with ritanserin has indicated that this compound has little dopamine antagonist activity and inhibits both 5-HTP-induced head twitch in mice and quipazine-induced head weaving forepaw treading and hindlimb abduction in rats. Again, however, there was no enhancement of reactivity.

Overall, therefore, these data suggest that head weaving, forepaw treading and hindlimb abduction are probably 5-HT$_2$-receptor-mediated whilst the hyperactivity and hyper-reactivity are not. In addition, the observations lend support to the hypothesis advanced by Deakin and Dashwood (1981) that all the behavioural changes elicited in the rat were serotonergic but that the hyperlocomotion and hyper-reactivity resulted from stimulation of serotonin receptors which required a postsynaptic dopamine link for the behavioural expression. This view has been further strengthened by some of our most recent work on a putative 5-HT$_1$ agonist RU 24969 (Green *et al.* 1984; Section 12.5).

There have been several studies on the effect of longer-term antagonist administration and its ability to produce behavioural super-sensitivity. Lesioning with 5,7-DHT has shown that super-sensitive behavioural responses can be produced (Trulson, Eubanks and Jacobs 1976*b*; Deakin and Green 1978). Enhancement of the behaviour also occurs after PCPA administration (Green *et al.* 1981). Longer-term administration of propranolol (Stolz and Marsden 1982), chlorpromazine (Green 1977) or metergoline (Fuxe, Agnati and Everitt 1975; Samanin, Mennini, Ferraris, Bendotti and Borsini 1980; Stolz, Marsden and Middlemiss 1983) results in increased behavioural responses, but in contrast mianserin administration produces a fairly long-lasting inhibition of behaviour (Blackshear and Sanders-Bush 1982; Section 6.4). Various tricyclic antidepressants have also been shown to produce enhanced responses when withdrawn after long-term administration and this is discussed further in Section 12.7.3.

12.3   Head twitch response in mice and 'wet dog shake' behaviour in rats

12.3.1   *Drugs initiating the behaviour*

In 1963, Corne, Pickering and Warner proposed a novel model for assessing central serotonergic function based on the observation that injection of high doses of 5-HTP to mice produced a characteristic head twitch behaviour. The responses were quantifiable and closely correlated with brainstem, but not whole brain, serotonin concentrations. Although this behaviour appeared similar to the pinna-reflex response of mice, its susceptibility to the effects of serotonin antagonists and other drugs, such as neuroleptic and antidepressant drugs, indicated that the behaviours were in fact quite distinct (Corne *et al.* 1963).

Head twitch responses in mice have also been shown to occur following the administration of the putative serotonin agonist LSD (Corne and Pickering 1967), quipazine (Malick, Dosen and Barnett 1977) and 5-MeODMT (Friedman and Dallob 1979). Interestingly, Malick *et al.* (1977) have shown that quipazine induces head twitch responses in mice by two distinct actions, the first being a direct stimulation of the serotonin receptor and the second being a release of the neurotransmitter itself.

Bédard and Pycock (1977) extended this model to include the rat by demonstrating that these animals would display a behaviour similar to the serotonin-induced head twitch response following injection of 5-HTP and peripheral decarboxylase inhibitor, α-methyldopahydrazine (carbidopa). This behaviour was aptly named the 'wet dog shake' and Bédard and Pycock (1977) described this as consisting of 'a paroxystic shudder of head, neck and trunk reminiscent of the purposeful movements seen in dogs'. Again, visually similar behavioural effects can be induced by tactile stimulation of the pinna (Askew, Liebrecht and Ratner 1969) and pharmacologically, by withdrawal from morphine (Martin, Wikler, Eades and Pescor 1963) or by systemic or central injection of thyrotropin-releasing hormone (TRH) (Prange, Breese, Cott *et al.* 1974; Wei, Sigel, Loh and Way 1985). Wet dog shaking behaviour is also elicited by injection of several serotonin agonists such as LSD and quipazine (Bédard and Pycock 1977; Vetulani, Bednarczyk, Reichenberg and Rokosz 1980; Matthews and Smith 1980) or 5-MeODMT (Bédard and Pycock 1977; Matthews and Smith 1980) and again appears to be mediated via serotonergic systems in the hindbrain (Bédard and Pycock 1977).

The effects of various serotonin antagonists on both the head twitch response in mice and the wet dog shake behaviour in rats have been extensively studied and the results are summarized in Table 12.2. There is general agreement that both behavioural syndromes, whether elicited by serotonin or its agonists, are inhibited by all the classical serotonin antagonists. In the last few years, however, considerable interest has been aroused by the hypothesis that serotonin receptors in the brain could be subclassified into either 5-HT$_1$ or 5-HT$_2$, based on the affinities of [^{3}H]-serotonin and [^{3}H]-spiperone for these two receptor sub-types respectively (Peroutka and Snyder 1979; Blackshear, Steranka and Sanders-Bush 1981; Chapter 4).

Table 12.2 Various serotonin antagonists which have been reported to attenuate the head twitch responses in mice and 'wet dog shake' behaviour in rats

Antagonist	Head twitch response	'Wet dog shakes'
Methysergide	5-HTP[a]	5-HTP[d]
Methysergide	5-HTP/MAOI[c]	5-HTP/carbidopa[e, f]
Methysergide	Quipazine/MAOI[c]	5-HTP/MAOI[j]
Methysergide	LSD[b]	—
Metergoline	5-HTP[i]	5-HTP[d]
Metergoline	—	5-HTP/carbidopa[e]
Methiothipin	5-HTP/MAOI[c]	5-HTP[d]
Methiothipin	Quipazine/MAOI[c]	—
Cinanserin	5-HTP[i]	5-HTP[d]
Cinanserin	5-HTP/MAOI[c]	—
Cinanserin	Quipazine/MAOI[c]	—
Cyproheptadine	5-HTP[i]	5-HTP[d]
Cyproheptadine	LSD[b]	5-HTP/carbidopa[e]
Cyproheptadine	—	Quipazine[g]
Cyproheptadine	—	LSD[g]
2-BromoLSD	5-HTP[a]	5-HTP[d]
2-BromoLSD	LSD[b]	—
Pizotifen	—	5-HTP[d]
Pirenperone	5-HTP/carbidopa[h]	5-HTP[d]
Pirenperone	—	5-HTP/carbidopa[f]
Ketanserin	—	5-HTP/carbidopa[f]

Data from: [a]Corne *et al.* (1963), [b]Corne and Pickerine (1967), [c]Malick *et al.* (1977), [d]Colpaert and Janssen (1983), [e]Mathews and Smith (1980), [f]Yap and Taylor (1983), [g]Vetulani *et al.* (1980), [h]Green *et al.* (1983*d*), [i]Peroutka *et al.* (1981), [j]Drust and Connor (1983).

Selective antagonists for these receptor sub-types have been developed and two drugs which are reported to have a very high specificity for 5-HT$_2$ receptors are ketanserin (Leysen, Niemegeers, Van Neuten and Laduron 1982; Laduron, Janssen and Leysen 1982) and pirenperone (Green *et al.* 1983*d*). Both compounds are extremely potent inhibitors of serotonin-induced wet dog shake in rats (ED$_{50}$ pirenperone 3 µg/kg, ketanserin 25 µg/kg; Yap and Taylor 1983; ED$_{50}$ pirenperone 85 µg/kg; Colpaert and Janssen 1983). Furthermore, pirenperone is also a potent inhibitor of the head twitch response of mice (ED_{50} pirenperone 76 μg/kg; Green *et al.* 1983*d*). Together these results provide good evidence to indicate that both of these behavioural syndromes are mediated by $5\text{-}HT_2$ receptors in the brain.

12.3.2   *The influence of various neurotransmitters on the serotonin-induced headshake responses in mice and rats*

There is convincing biochemical and anatomical evidence to suggest that central serotonin function is influenced by several other neurotransmitters, and furthermore, that a special link exists between this neurotransmitter and noradrenaline. Anatomically serotonergic neurones in the midbrain raphe nuclei are innervated by noradrenergic nerve terminals (Fuxe 1965; Swanson and Hartman 1975) and its has been suggested from electro-physiological studies that the tonic activity of the noradrenergic neurones innervating the dorsal raphe nuclei maintain the firing of these serotonergic neurones (Svensson, Bunney and Aghajanian 1975; Baraban and Aghajanian 1975; Baraban and Aghajanian 1980). Lesioning the locus coeruleus or inhibiting dopamine β-hydroxylase using α-methyl-*p*-tyrosine both increase rat brain 5-HIAA concentration (Johnson, Kim and Boukma 1972; Kostowski, Samanin, Bareggi, More, Garattini and Valzelli 1974). Conversely, *in vitro* release of [³H]-serotonin from brain slices is inhibited by stimulation of α_2-adrenoceptors (Frankhuyzen and Mulder 1980; Göthert and Huth 1980; Maura, Gemignani and Raiteri 1982) while *in vivo*, clonidine administration has been shown to decrease cortical 5-HIAA concentrations (Reinhard and Roth 1982). In contrast, however, stimulation of central β-adrenoceptors using salbutamol or clenbuterol has been shown to increase serotonin turnover (Waldmeier 1981; Nimgaonkar *et al.* 1983). The evidence, however, is that this latter change is mediated by an increase in tryptophan availability from the periphery to the brain (Nimgaonkar *et al.* 1983).

There is also considerable behavioural data to suggest that central noradrenergic function influences serotonin-mediated behavioural changes. Administration of either of the α_2-adrenoceptor agonists, clonidine and guanabenz, has been reported to inhibit the serotonin-induced head twitch response in mice, while the antagonists yohimbine and piperoxan elicited the opposite effect (Handley and Brown 1982). Similarly, Bednarczyk and Vetulani (1978) demonstrated that the wet dog shakes produced by 5-HTP or 5-methoxytryptamine were decreased by clonidine, although Bédard and Pycock (1977) observed no effect. Furthermore, Bednarczyk and Vetulani (1978) not only showed that the effect was probably mediated by postsynaptic α_2-adrenoceptors, since the inhibition was not alleviated by 6-hydroxydopamine lesioning, but also that the locus for this attenuation was almost certainly 'down-stream' of the serotonin receptors since it was also unaffected by disruption of serotonergic neurones using 5,6-DHT.

The influence of central α_1-adrenoceptor-mediated noradrenergic function on the head twitch syndrome is, however, much more controversial, for although Handley and Brown (1982) have suggested that α_1-adrenoceptors are facilitatory on the serotonin-induced head twitch response from studies using both agonists (phenylephrine, methoxamine) and antagonists (prazosin, thymoxamine), studies using the drug WB 4101, showed no effect in either mice (Ortmann, Martin, Radeke and Delini-Stula 1981) or rats (Matthews and Smith 1980). However, this discrepancy may be explained by recent studies which have suggested that, in contrast to prazosin, WB 4101 is not a totally specific α_1-adrenoceptor antagonist (Massingham, Dubocovich, Shepperson and Langer 1981; Palacios 1984). While inhibition of central β-adrenoceptors using a variety of selective and non-selective β_1- and β_2-antagonists had no effect either on the serotonin-mediated head twitch response of mice (Ortmann et al. 1981; Handley and Singh 1984) or wet dog shakes in rats (Bédard and Pycock 1977), stimulation of β_2- and also probably β_1-adrenoceptors produced a marked enhancement of the head twitch responses in mice (Ortmann et al. 1981; Nimgaonkar et al. 1983; Handley and Singh 1984). Since in all these studies the head twitch syndrome was elicited by 5-HTP in combination with either carbidopa or MAO inhibitors, it is at present unclear whether this is a presynaptic effect on serotonin release or if this is an effect 'downstream' of the serotonin receptor. In the case of enhancement of the serotonin-mediated behaviour by clenbuterol in rats, this effect has been shown to be postsynaptic since it has been shown to occur when the behaviour is initiated by 5-MeODMT or quipazine (Cowen et al. 1982; Nimgaonkar et al. 1983; Section 12.2.2). Furthermore, it is not known whether stimulation of β-adrenoceptors enhances the wet dog shakes in rats although it has been shown to enhance other serotonin-mediated behaviours (Ortmann et al. 1981; Cowen et al. 1982; Nimgaonkar et al. 1983; Section 12.2.2).

Dopamine appears to have an inhibitory effect on the serotonin-mediated head shake behaviour in rodents since Corne and Pickering (1963) demonstrated that the syndrome in mice was attenuated by methylamphetamine or amphetamine while Bédard and Pycock (1977) have reported that apomorphine and amphetamine produce similar effects in rats.

In contrast, manipulation of cholinergic systems using either antagonists or agonists had no effect upon the wet dog shaking behaviour in rats (Bédard and Pycock 1977; Matthews and Smith 1980).

12.3.3 *Effect of destruction of central serotonergic neurones on serotonin-mediated wet dog shakes in rats*

Although there is no data currently available for the head twitch response in mice, three studies have shown that in rats destruction of central serotonergic neurones using either 5,7-dihydroxytryptamine or the less selective 5,6-dihydroxytryptamine leads to enhanced wet dog shakes to serotonin (Drust and Connor 1983), its agonist 5-methoxytryptamine (Bednarczyk and Vetulani 1978) or 5-HTP (Barbeau and Bédard 1981).

12.4   Myoclonus in guinea pigs

Administration of L-5-HTP to guinea pigs induces myoclonus (Klawans, Goetz and Weiner 1973; Chadwick, Hallett, Jenner and Marsden 1978; Volkman, Lorens, Kindel and Ginos 1978). This behaviour is also elicited by L-tryptophan after monoamine oxidase inhibition and 5-MeODMT.

In 1981 Luscombe, Jenner and Marsden published a comprehensive study on guinea pig myoclonus and this will now be discussed in some detail.

The minimum dose of 5-HTP required to clearly elicit myoclonus was found to be 20 mg/kg (in animals pre-treated with the peripheral decar-boxylase inhibitor carbidopa, 25 mg/kg) and the dose necessary for continuous myoclonus was 75 mg/kg (Fig. 12.3). Tryptamine (with MAO

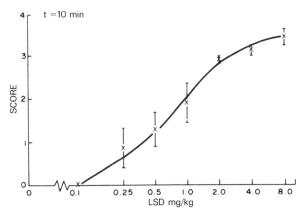

Fig. 12.3   Dose-dependent induction of myoclonus by lysergic acid diethylamide tartrate (LSD). Animals were scored 10 min after LSD administration, the time of peak response of the majority of tested doses. The values represent the means of scores obtained (±1 S.E.M.) for at least 6 animals for each dose. (Reproduced from Luscombe *et al.* (1981) with permission from Pergamon Press)

inhibitor pre-treatment), *N,N*-dimethyltryptamine, 5-MeODMT and LSD
were all found to elicit dose-dependent myoclonus.

In contrast, the piperazine-derivative agonists; quipazine MK-212 and
1-(*m*-trifluoromethylphenyl)piperazine only produced intermittent and
weak myoclonus at very high toxic doses.

Inhibition of the 5-HTP-induced myoclonus was produced by pre-
treatment with metergoline and cyproheptadine and this inhibition was
dose-dependent. In contrast, mianserin only inhibited myoclonus at high
doses (20 mg/kg). Methysergide also had very little effect, as did BW
501C67 except at high dose.

The myoclonus response was potentiated by pre-treatment with clomipramine, paroxetine and ORG 6582 whilst temoxetine, fluoxetine and
desmethylimipramine produced little or no enhancement. While desmethylimipramine is a relatively weak serotonin uptake inhibitor,
fluoxetine and femoxetine are both potent in inhibiting uptake of this
neurotransmitter. Whether the apparent anomaly in the action of these
uptake inhibitors is due to regional variations in their effects as suggested
by the authors requires substantiation. The results obtained have been
summarized in Table 12.3.

Table 12.3 Summary of the ability of serotonin agonists or precursors to induce
myoclonus, serotonin antagonists to antagonize 5-HTP-induced myoclonus and
serotonin re-uptake blockers to potentiate 5-HTP-induced myoclonus in the guinea
pig

Serotonin agonists or precursors	
Did produce myoclonus	*Did not produce myoclonus*
5-Methoxy-*N,N*-dimethyltryptamine	Quipazine
N,N-dimethyltryptamine	MK-212
d-LSD	1-(*m*-trifluoromethylphenyl)-piperazine
5-HTP (+ carbidopa)	
Tryptophan (+ pargyline)	Tryptophan
Tryptamine (+ pargyline)	Tryptamine
Serotonin antagonists	
Potently antagonized myoclonus	*Weak or no effect on myoclonus*
Methergoline	Mianserin
Cyproheptadine	Methysergide
	BW 501C67
Serotonin re-uptake blockers	
Strongly potentiated myoclonus	*Little or no potentiation of myoclonus*
Chlorimipramine	Femoxetine
Paroxetine	Fluoxetine
ORG 6582	Desmethylimipramine

Reproduced from Luscombe *et al.* (1981) with permission from Pergamon Press.

Finally, it should be noted that neither dopamine nor noradrenaline antagonists inhibited the behavioural response (Luscombe *et al.* 1981).

In a recent study these same workers (Luscombe, Jenner and Marsden 1984) examined the distribution of both 5-HT$_1$ and 5-HT$_2$ receptors in the brainstem, the area of the brain responsible for the initation of the myoclonus response in guinea pigs (Chadwick *et al.* 1978). Luscombe and colleagues (1984) found that this area possessed specific binding sites for [^{3}H]-serotonin (the 5-HT$_1$ ligand) but that specific [^{3}H]-spiperone (the 5-HT$_2$ ligand) binding was both weak and inconsistent. Furthermore, the [^{3}H]-serotonin binding was displaced potently by indole-containing serotonin agonists but not by the piperazine-containing agonists. They therefore concluded that the behaviour is 5-HT$_1$-receptor-mediated.

The value of this particular model may be in investigating the action of drugs on a functional 5-HT$_1$-receptor-mediated event. It is of limited interest in terms of examining the effects of drugs on myoclonus *per se* since some types of myoclonus are in fact treated with 5-HTP (Chadwick 1982).

12.5   Studies using the putative 5-HT$_1$ receptor agonist RU 24969

12.5.1   *Studies in rats*

The first step towards 'dissecting out' the complex behavioural syndrome which occurs with serotonergic stimulation in rats came with the availability of the selective 5-HT$_2$ antagonists (Section 12.2.3). The next step occurred with the behavioural studies on the suggested serotonin agonist RU 24969 (5-methoxy-3(1,2,3,6,tetrahydropyridin-4-yl)1H indole (Fig. 10.5). This drug is a potent displacing agent at the 5-HT$_1$ binding site (Hunt and Oberlander 1981) having an IC$_{50}$ of 6.8 nM.

When this compound was given to rats it produced very marked locomotor activity which occurred in bursts, and the animals also showed hyper-reactivity (Green *et al.* 1984). The response seen was dose-dependent (Fig. 12.4). Similar behaviour was also observed when rats were given non-selective serotonin agonists (5-MeODMT or quipazine) after pre-treatment with reasonably selective 5-HT$_2$ receptor antagonists metergoline or pirenperone) (Section 12.2.3).

The behavioural activity to RU 24969 was not attenuated by PCPA pre-treatment and indeed was slightly enhanced by methysergide or metergoline administration (Fig. 12.5). Haloperidol inhibited the response somewhat, suggesting a dopamine link postsynaptic to the serotonin receptors initiating the behaviour. Propranolol produced only a slight non-significant increase in the mean response (Fig. 12.5). These results have been described in detail elsewhere (Green *et al.* 1984). Oberlander and Boissier (1981) also observed inhibition of the RU 24969-induced locomotor response by haloperidol and recently Oberlander (1983) and

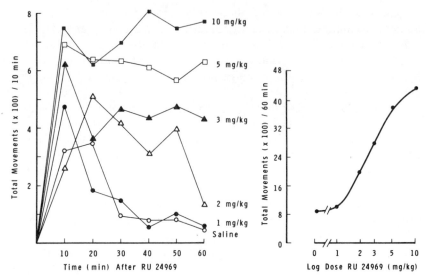

Fig. 12.4 Dose-response curves of the hyperlocomotor response of rats to RU 24969. Each point is the mean ($n=3$) recorded response in 10 min, following various doses of drug. The log dose response curve is also shown.

Tricklebank (1984*b*) have further evidence for the involvement of dopamine in the response and Oberlander (1983) suggested that the mesolimbic pathways are the site of action.

The enhancement of the RU 24969 response by two potent 5-HT_2 antagonists (methysergide and metergoline) suggested to us that this was probably by inhibition of a 5-HT_2-receptor-mediated system which, itself, exerted a tonic inhibitory control over dopamine-mediated behaviour. These drugs have previously been shown to enhance dopamine-mediated behaviour (Green *et al.* 1981).

Overall, therefore, these results led us to suggest the relationship in the behaviour which is shown in Fig. 12.6, since this figure incorporates both the current data and that of the pirenperone experiments (Section 12.2.3). It is a further refinement of the hypothesis of Deakin and Dashwood (1981).

One problem in these studies, however, was the observation that propranolol did not inhibit RU 24969-induced locomotor activity. This compound is a reasonable antagonist at both 5-HT_2 (Green *et al.* 1983*a*) and 5-HT_1 (Nahorski and Willcocks 1983) receptors. At the outset, therefore, it seemed reasonable to propose that propranolol, which inhibits not only the 5-HT_2-mediated behaviour (forepaw treading, head weaving and hindlimb abduction) but also the hyperlocomotion and hyper-reactivity would be doing so by antagonizing both 5-HT_2 and 5-HT_1 receptors. Since

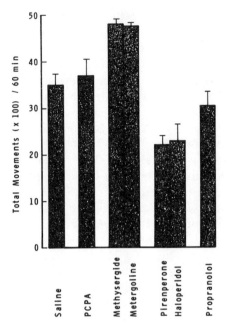

Fig. 12.5   Total recorded activity in 60 min following RU 24969 (5 mg/kg) in rats and effect of pre-treatment with PCPA (200 mg/kg, 48 h earlier), methysergide (10 mg/kg, 30 min), metergoline (2.5 mg/kg, 30 min), pirenperone (100 µg/kg, 30 min), haloperidol (100 µg/kg, 30 min) or (—)-propranolol (20 mg/kg, 30 min). Methysergide and metergoline enhanced the response (p <0.05) whilst pirenperone and haloperidol inhibited it (p <0.05).

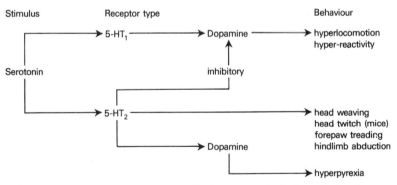

Fig. 12.6   Postulated receptor types mediating serotonin-induced behaviour.

propranolol did not antagonize the behavioural effects of the 5-HT$_1$ agonist, RU 24969, it is at present unclear why this drug inhibited the hyperlocomotion and hyper-reactivity.

On the basis that the behavioural changes produced by RU 24969 are identical to some of the components in the syndromes induced by other serotonin agonists, that the effects are not inhibited by prior serotonin depletion and that the drug decreases the rate of serotonin synthesis (Euvrard and Boissier 1980; Green *et al.* 1984), it seems reasonable to suggest that this compound is a serotonin agonist. The binding data indicates that the changes produced are likely to be mediated via a 5-HT_1-like recognition site. However, there do appear to be sub-populations of 5-HT_1 binding sites (Nelson, Pedigo and Yamamura 1981; Pedigo, Yamamura and Nelson 1981; Tricklebank 1984a) and the fact that RU 24969 had, in binding studies, a Hill slope of considerably less than one, suggested that there may be receptor heterogeneity. The serotonin auto-receptors have been suggested to be of the 5-HT_1 sub-type (Martin and Sanders-Bush 1982; Göthert and Schlicker 1983; Section 2.5). However, propranolol did not prevent RU 24969 decreasing the rate of serotonin synthesis (Green *et al.* 1984) and it therefore appears that RU 24969 may be acting at a sub-population of 5-HT_1 receptors at which propranolol is not an antagonist and very recent data has led to the suggestion that propranolol is a 5-HT_{1A} antagonist (Middlemiss 1984) while RU 24969 may be acting at 5-HT_{1B} binding sites (Cortés, Palacios and Pazos 1984). This is discussed more fully in Section 12.6.

12.5.2   *Studies in mice*

The main behavioural effect of RU 24969 in mice was again that of markedly increasing locomotor activity (Gardner and Guy 1983). This activity, however, occurred more continuously in this species and rather less in bursts as was seen in rats. Again, the effect was long-lasting reaching a peak level two hours after a dose of either 5 or 10 mg/kg. At no time was any head twitch response observed. The behaviour was antagonized by metergoline but the ED_{50} was approximately three times that required to inhibit the MK-212-induced head twitch responses. Finally, propranolol was seen to be a weak antagonist of the hyperlocomotion (Gardner and Guy 1983; Green *et al.* 1984).

Overall, therefore, these data lend further support to the suggestion made by several different research groups (Peroutka *et al.* 1981; Ortmann, Bischoff, Radeke, Buech and Delini-Stula 1982; Green *et al.* 1983d) that the head twitch behaviour in mice is 5-HT_2-receptor-mediated.

12.5.3   *Studies in guinea pigs*

As described in Section 12.4, administration of the serotonin precursor 5-HTP produces myoclonic jerking. This effect is also produced by indole-containing but not by piperazine-containing agonists (except at toxic doses). RU 24969 contains both an indole-ring and a piperazine moiety (Fig. 10.5). Furthermore, as detailed in Section 12.4, it has recently been suggested that the myoclonus response is a 5-HT_1-mediated event (Luscombe *et al.* 1984). Therefore, it is perhaps not surprising that RU 24969 also induces myoclonus in guinea pigs, albeit at a fairly high dose. Nevertheless, the response has a long duration (Luscombe *et al.* 1984).

12.6   The behavioural changes associated with the putative 5-HT_1 agonist 8-OH-DPAT

The problems associated with linking serotonergic-mediated behavioural changes with a specific serotonin receptor sub-type have recently been considerably increased. Hjörth, Carlsson, Lindberg *et al.* (1982) described the biochemical and behavioural effects of the ergot compound 8-hydroxy-2-(di-*N*-propylamino)-tetralin (8-OH-DPAT) (Fig. 10.5). This compound, which is neither an indole nor a piperazine, shows many of the properties of a potent serotonin agonist. It decreases serotonin synthesis, even in serotonin-depleted rat brain, and produces some of the behavioural changes which are associated with the effects of serotonin agonists in this species (hindlimb abduction, forepaw treading, and flat body posture). The paradox is, however, that this compound is suggested to be a selective ligand for the serotonin presynaptic receptor, at least in the striatum (Gozlan, El Mestikawy, Pichat, Glowinski and Hamon 1983). Furthermore, Middlemiss and Fozard (1983) suggested, on the basis of ligand-binding studies, that 8-OH-DPAT showed marked selectivity for the 5-HT_1 type of receptor.

If the receptor sub-types hypothesized in Fig. 12.6 are correct, then one would not expect a 5-HT_1 agonist to produce the behaviour reported to occur after administration 8-OH-DPAT. At present, there is not enough data available to clearly analyse the results and to draw solid conclusions. Tricklebank (1984a) has now confirmed that 8-OH-DPAT induces the same behavioural changes as previously reported by Hjörth *et al.* (1982). However, there are some discrepancies between the two studies. Hjörth *et al.* (1982) did not observe any antagonism of the behavioural responses after pre-treatment with reserpine, phenoxybenzamine, propranolol or haloperidol. In contrast, Tricklebank (1984a) reported a dose-related inhibition of ambulation and head weaving, but not of forepaw treading, after reserpine (18 hour previously) and also a reduction of most components of the syndrome by prazosin, haloperidol and sulpiride. When he gave prazosin or haloperidol to rats pre-treated with reserpine 18 hours previously, the remaining 8-OH-DPAT-induced behaviours (forepaw treading and flat body posture) were unaltered. He suggested, therefore, that these responses had both catecholaminergic and non-catecholaminergic components.

The failure of ketanserin to antagonize the behaviours suggested that they were not, at least in this model, 5-HT$_2$ receptor-mediated.

The work of Gozlan *et al.* (1983) demonstrated that while 8-OH-DPAT bound preferentially to presynaptic serotonin autoreceptors in the striatum, it bound to postsynaptic sites in the hippocampus. Interestingly, it also appeared to bind to postsynaptic sites in the brainstem. This is of particular significance since this is the site of initiation for many of the serotonin-mediated behaviours in rats (Serotonin 12.2.2). Further investigation is obviously necessary in this region to determine whether forepaw treading can be initiated by both 5-HT$_1$ and 5-HT$_2$ receptors or whether the classification of receptors and the drugs acting on them (8-OH-DPAT and RU 24969) in this region are different from the classification based on binding data obtained in some other brain regions, but most often the frontal cortex.

Current data has led to the suggestion that 8-OH-DPAT is a 5-HT$_{1A}$ agonist and RU 24969 a 5-HT$_{1B}$ agonist (Tricklebank 1984c; Cortés *et al.* 1984) and Middlemiss (1984) has demonstrated that (−)-propranolol is a very potent 5-HT$_{1A}$ antagonist. Seretiazine the 5-HT$_2$ antagonist does not inhibit the behaviour induced by 8-OH-DPAT, suggesting that the behaviour induced by this drug is not via 5-HT$_2$ receptors.

Finally, there is evidence from lesioning studies that the 5-HT$_1$ receptors can be postsynaptic in so far as the behaviour induced by 8-OH-DPAT and RU 24969 is not inhibited by serotonergic lesions with 5,7-DHT (Blackburn, Bowery, Cox, Hudson, Martin and Price 1984). However, none of this most recent data unequivocally explains our most recent findings (Goodwin and Green 1985). When the behavioural syndrome is produced in rats by injection of 8-OH-DPAT, it is blocked by (−)-propranolol but not by ritanserin. In contrast, when it is produced by injection of quipazine, it is blocked by both (−)-propranolol and ritanserin.

12.7   The effect of antidepressant treatments on serotonin-mediated behaviour and 5-HT$_2$ receptor binding

12.7.1   *Introduction*

Much of the work described in the previous sections has indicated that changes in serotonin function can be detected by changes in the behavioural responses of the animals to serotonin agonists. However, what one really hopes for is that changes in serotonin receptor number might be reflected in a changed behavioural response. Such an association is of value for two reasons. First, because behavioural models provide a rapid and inexpensive way of 'screening' for possible changes in receptor number, before the complexities of the biochemical investigations, and secondly, because a changed behavioural response suggests that the altered receptor number (as measured by ligand-receptor binding) is actually of functional importance.

Over the last few years there has been a marked resurgence of interest in the effects of antidepressant treatments on serotonin function (Chapters 6 and 7) and we have recently been conducting an extensive investigation into the effects of various antidepressant treatments on both the responses of rodents to serotonin agonists and 5-HT_2-receptor binding.

12.7.2   Effects of electroconvulsive shock in rats and mice

In 1976 we first showed that repeated electroconvulsive shock (ECS) enhanced the behavioural responses of rats to tranylcypromine/L-tryptophan or 5-MeODMT (Evans, Grahame-Smith, Green and Tordoff 1976). Subsequently, it was shown that repeated ECS also enhanced the responses of rats to quipazine (Green, Heal and Grahame-Smith 1977) and that when ECS was given in ways which closely mimicked the administration of electroconvulsive therapy (ECT), enhancement still occurred (Costain, Green and Grahame-Smith 1979). Lebrecht and Nowak (1980) reported that repeated ECS enhanced the head twitch response in mice and this has been confirmed by studies in our laboratory (Green *et al.* 1983*c*). There is, therefore, convincing evidence that repeated ECS, but not a single shock, enhances serotonin-mediated behaviour (for review, see Green 1984).

In 1981 two groups showed independently that repeated ECS (once daily for 10 or more days) increased 5-HT_2 receptor number in frontal cortex (Vetulani, Lebrecht and Pilc 1981; Kellar, Cascio, Butler and Kurtzke 1981). However, intermittent ECS (5 ECS spread over 10 days) is as effective in enhancing serotonin-mediated behaviour (Costain *et al.* 1979) as a course of 10 daily ECS. Furthermore it has been found that administration of α-methyl-p-tyrosine, PCPA or pentylenetetrazol during the period of ECS administration prevents the enhancement of the serotonin-mediated behavioural changes (Green, Costain and Deakin 1980; Green, Sant, Bowdler and Cowen 1982). We have, therefore, examined whether the ECS conditions which not only enhanced but also failed to change serotonin-mediated behaviour correlated with changes in [³H]-spiperone binding (the 5-HT_2 ligand) when assessed 24 hours after the last treatment (Green, Johnson and Nimgaonkar 1983*b*).

In Table 12.4 it can be seen that the maximum number of [³H]-spiperone binding sites (B_{max}) was increased in rats when ECS was given either daily

Table 12.4   Effect of various ECS treatment regimes on 5-HT$_2$ receptor binding characteristics in rat frontal cortex

Treatment	Serotonin receptor binding characteristics	
	K_D (nM)	B_{max} (pmol/mg protein)
Control (handled)	1.23±0.09 (3)	350±15 (3)
ECS×1	1.32±0.17 (5)	379±38 (5)
ECS×10	1.63±0.14 (6)	472±32 (6)†
Control (anaesthetized×5)	1.32±0.01 (3)	356± 7 (3)
ECS×5	1.65±0.04 (4)*	494±40 (4)†
Control (PTZ)	1.04±0.07 (4)	275±24 (4)
ECS×5+PTZ	1.02±0.02 (4)	307± 7 (4)

All measurements were made 24 h after the last treatment. ECS×1 is a single ECS, ECS×10 is ECS once daily for 10 days, ECS×5 is 5 ECS spread over 10 days. Pentylenetetrazol (PTZ; 30 mg/kg) was given 3 min before the ECS. ECS×5 treated rats were given halothane anaesthesia. Results are expressed as mean ±S.E.M. values with number of separate experiments given in parentheses. Different from appropriate control: †p <0.05; *p <0.01. Correlation coefficient (r) 0.9 or better on every analysis. (Reproduced from Green *et al.* (1983*b*) with permission from Macmillan Journals)

for 10 days or 5 times over 10 days, but not after ECS had been given once. Furthermore, injection of pentylenetetrazol just before each seizure, which prevented the enhancement of serotonin-mediated behavioural responses, also prevented the increase in 5-HT$_2$ receptor number. Similarly, pre-treatment with either PCPA or α-methyl-*p*-tyrosine before each ECS also prevented the increase in both behaviour and receptor binding.

Lesioning brain noradrenergic pathways with 6-hydroxydopamine prevents the ECS-induced increase in serotonin-mediated behaviour (Green and Deakin 1980). This condition has yet to be examined with binding studies.

Overall, therefore, there appears to be a good correlation between the enhancement of serotonin-mediated behaviour after ECS and the increase in 5-HT$_2$ receptor number.

An obvious problem in making the assumption that these two changes can be associated is that the binding was performed on frontal cortex tissue, an area where the 5-HT$_2$ receptors have been most clearly characterized, whilst the behaviour is probably mediated by the brainstem and spinal cord (Section 12.2.2). However, 5-HT$_2$ receptors are widely distributed in the brain and this also includes the two areas of particular interest, the brain-stem and spinal cord (Peroutka and Snyder 1981).

The 5-HT$_2$ receptor-mediated components of the serotonin-induced behavioural changes in the rat (head weaving, forepaw treading and hindlimb abduction) are not easy to quantify because of the limitations of the scoring scale. However, the head twitch response of mice does not suffer from this particular disadvantage and was therefore used to examine how repeated ECS alters the behavioural dose-response curve (Green *et al.* 1983*c*).

Increasing the dose of 5-HTP give to carbidopa-pre-treated mice produced a linear log dose/log response curve (Fig. 12.7). It was not possible to see if the log dose/response curve reached a plateau as the animals convulsed at the higher doses (greater than 300 mg/kg). Repeated ECS enhanced the response at every dose examined leading to a parallel shift in the log dose/log response curve (Fig. 12.7).

Although we have recently found that $5\text{-}HT_2$ receptor number is increased in mouse frontal cortex (Goodwin, Green and Johnson 1984), the parallel shift in the dose response curve does not appear to be a reflection of a change in receptor number since administration of clenbuterol, a β_2-adrenoceptor agonist (Section 12.2.2) which did not increase $5\text{-}HT_2$

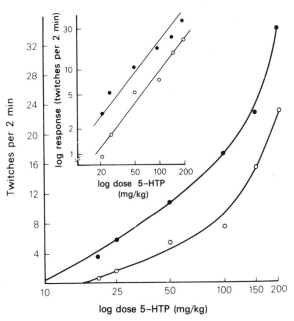

Fig. 12.7 Effect of repeated ECS on the behavioural response to carbidopa and 5-hydroxytryptophan. Mice were given either 5 ECS spread over 10 days (●) or 5 anaesthetic exposures only (○). Twenty-four hours after the last treatment both groups were given carbidopa (25 mg/kg) followed by various doses of 5-hydroxy-tryptophan. Main graph shows the log dose of 5-HTP v. total number of head twitches in 2 min, 30 min after the 5-HTP. Small graph shows the log dose/log response curve. Experimental group significantly different from control group (p <0.05 or better) at every dose. Eight animals used at each dose point. (Reproduced from Green *et al.* 1983*c*, with permission from Macmillan Journals)

receptor number produced an identical parallel shift in the log 5-HTP dose/log head twitch response curve (Green *et al.* 1983*c*).

12.7.3   *Effect of antidepressant drugs in rats and mice*

One of the major advantages of examining the effect of repeated ECS over the other antidepressant treatments is that it does not have the complication of either dosing schedules or the presence of drug in the tissue; it is merely necessary to induce a seizure. Furthermore, there is also no residual drug present to confuse subsequent behavioural experiments. With antidepressant drugs, different dose schedules and rates of metabolism may influence the result obtained. Furthermore, there is also the problem of deciding when to examine the serotonin-mediated behaviour: at a time when the drug has 'cleared' the brain or shortly after the last dose. Moreover, without tissue measurements, it is difficult to know whether there is sufficient drug left in the brain to alter the response seen. Such complications, of course, are also present when trying to interpret ligand-receptor binding studies.

With these problems in mind we recently examined the effects of repeated antidepressant administration on the 5-HTP-induced responses, not only while mice were receiving drugs but also 48 hours after cessation of treatment.

Repeated administration of desmethylimipramine, zimeldine, mianserin or tranylcypromine (given orally dissolved in the drinking water) all decreased the head twitch response of mice whilst the animals were still receiving the drug (Fig. 12.8). After 48 hours withdrawal the mice treated with tranylcypromine, zimeldine and mianserin still showed inhibited head twitch responses while, in contrast, those mice treated with desmethylimipramine displayed an enhancement of this behaviour (Fig. 12.9).

Earlier studies have also observed enhanced serotonin-mediated behaviour in rats after withdrawal from several tricyclic antidepressants (Friedman and Dallob 1979; Friedmand, Cooper and Dallob 1983; Stolz and Marsden 1982; Stolz *et al.* 1983). Microiontophoretic studies have also shown enhanced responses to serotonin after tricyclics and ECS (de Montigny and Aghajanian 1978; Gallagher and Bunney 1979; de Montigny 1981; and see review of Green and Nutt 1983 and Chapter 7).

Can these results be reconciled with binding studies? Several groups have found a decrease in 5-HT$_2$ binding in frontal cortex following longer-term antidepressant drug treatment (Peroutka and Snyder 1980; Kellar *et al.* 1981). In our current study we also observed that 5-HT$_2$ receptor number was decreased in mouse brain, both while the animals were receiving tranylcypromine, zimeldine and desipramine and after withdrawal. The receptor number was also decreased when the animals were

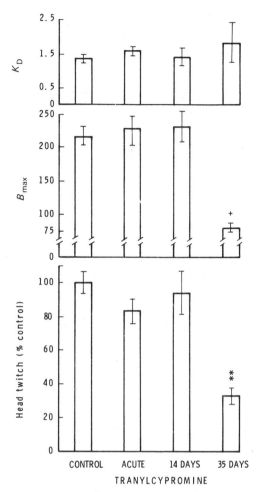

Fig. 12.8 The parameters of [³H]-spiperone binding (K_D:nM; B_{max}:fmol/mg protein), N,N-dimethyltryptamine (15 or 20 mg/kg i.p.), expressed as per cent of untreated controls run simultaneously. Tranylcypromine given as loading dose (6 mg/kg i.p.), then orally (5.6 mg/kg per day). Binding and behaviour determined under control (untreated) condition, 24 h, 14 days and 35 days after starting drug treatment. Results shown as mean ±S.E.M. of at least 8 observations (behaviour) and 3 observations (binding). Different from control groups †p <0.025; **p <0.01.

receiving mianserin administration, but normalized after 48 hours withdrawal (Goodwin, Green and Johnson 1984). There was, therefore, a reasonable, but by no means absolute agreement between 5-HT₂ binding values and behaviour.

The discrepancy after withdrawal in the case of desipramine and mianserin might be explicable in terms of other changes which occur in

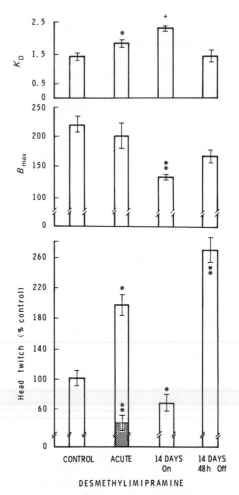

Fig. 12.9   The parameters of [^{3}H]-spiperone binding (K_{D}:nM; B_{max}:fmol/mg protein), together with the mean head twitch response evoked by 5-hydroxytryptophan (100 mg/kg i.p.) or 5-MeODMT (20 mg/kg i.p.; shaded histogram) expressed as per cent or mean of untreated controls run simultaneously. Desipramine given as loading dose (30 mg/kg i.p.), then orally (27 mg/kg per day). Behaviour and binding determined under control (untreated) condition, 24 h and 14 days after starting drug treatment and 48 h after discontinuing chronic (14 days) treatment. Results shown as mean ±S.E.M. of at least 8 observations (behaviour) and 3 observations (binding). Different from control groups *p<0.05; **p <0.01.

noradrenaline function after treatment with these drugs (see, for example, Heal, Lister, Smith, Davies, Molyneux and Green 1983) but we have no direct evidence for this at present.

Since many tricyclics are weak to moderate inhibitors of 5-HT$_{2}$ binding *in vitro* (Peroutka and Snyder 1980; Section 6.2.6), one might question why receptor number seems to decrease after longer-term treatment since it has been generally felt that after withdrawal from an antagonist monoamine receptor site number is increased. In a recent study Blackshear, Friedman and Bush-Sanders (1983) found that following two weeks treatment with various serotonin antagonists 5-HT$_2$ receptor number in frontal cortex was *decreased.* This apparent paradoxical effect did not seem to be due to residual drug in the tissue. However, such findings still do not reconcile with behavioural evidence that repeated administration of metergoline or propranolol leads to enhanced serotonin-mediated behaviour in the rat (Stolz and Marsden 1982). Clearly, we have some way go before binding and behavioural data can be shown to exhibit a good correlation. As we have discussed earlier in the chapter, one very important factor in this equation may be that the behaviour is not initiated in the areas of the rodent brain on which the binding studies are normally performed.

12.8   Conclusions

What behavioural models should be used and what receptor sub-types mediate the behaviour?

Overall, the evidence that the head twitch behaviour in mice is 5-HT$_2$-mediated is good if not conclusive. The behaviour is not produced by the 5-HT$_1$ agonists 8-OH-DPAT or RU 24969, nor is it inhibited by the 5-HT$_1$ antagonist (−)-propranolol (Goodwin and Green 1985). Putative 5-HT$_2$ antagonists are extremely potent in inhibiting head twitch behaviour and there is a reasonable correlation between 5-HT$_2$ receptor number and behaviour (e.g. Goodwin *et al.* 1984).

The wet dog shake behaviour may also be 5-HT$_2$-mediated but is less convenient model to use.

Guinea pig myoclonus may be 5-HT$_1$ receptor mediated (Section 12.4) but which receptor sub-type, that is, 5-HT$_{1A}$ or 5-HT$_{1B}$, is uncertain.

The complex behavioural changes induced in rats probably results from stimulation of several receptor types. Much more information is needed on the drugs being used, particularly the new agonists such as 8-OH-DPAT and RU 24969 and on the interactions between neurotransmitters in the production of the syndrome.

The serotonin-mediated behavioural models have proved of value in clarifying the action of a range of drugs. The fact that there are now questions being raised as to the receptors mediating the behaviour and the interpretation of results obtained does not invalidate the earlier studies; rather it provides evidence for the continuing interest in the models and attempts to maintain the usefulness of the models in investigating serotonin neuropharmacology.

356   A. Richard Green and David J. Heal

References

Askew, H. R., Liebrecht, B. C. and Ratner, S. C. (1969). Effects of stimulus duration and repeated sessions on habituation of the head-shake response in the rat. *J. Comp. Physiol. Psych.* **67**, 497.

Baraban, J. M. and Aghajanian, G. K. (1980). Suppression of firing activity of 5-HT neurons in the dorsal raphé by alpha adrenoceptor antagonists. *Neuropharmacology* **19**, 355.

Barbeau, H. and Bédard, P. (1981). Denervation supersensitivity to 5-hydroxytryptophan in rats following spinal transection and 5,7-dihydroxytryptamine injection. *Neuropharmacology* **20**, 611.

Bédard, P. and Pycock, C. J. (1977). "Wet dog" shake behaviour in the rat: a possible quantitative model of central 5-hydroxytryptamine activity. *Neuropharmacology* **16**, 663.

Bednarczyk, B., and Vetulani, J. (1978). Antagonism of clonidine to shaking behaviour in morphine abstinence syndrome and to head twitches produced by serotonergic agents in the rat. *Pol. J. Pharm. Pharmac.* **30**, 307.

Blackburn, T. P., Bowery, N. G., Cox, B., Hudson, A. L., Martin, D. and Price, G. W. (1984). Lesion of the dorsal raphe nucleus increases the nigral concentration of 5-HT$_1$ receptors. *Br. J. Pharmac.* **82**, 203P.

Blackshear, M. A. and Sanders-Bush, E. (1982). Serotonin receptor sensitivity after acute and chronic treatment with mianserin. *J. Pharmac. exp. Ther.* **221**, 303.

—— Friedman, R. L. and Sanders-Bush, E. (1983). Acute and chronic effects of serotonin antagonists on serotonin binding sites. *Naunyn-Schmiedeberg's Arch. Pharmac.* **324**, 125.

—— Steranka, L. R. and Sanders-Bush, E. (1981). Multiple serotonin receptors: regional distribution and effect of raphé lesions. *Eur. J. Pharmac.* **76**, 325.

Chadwick, D. (1982). Monoamines and epilepsy. In *Psychopharmacology of anticonvulsants* (ed. M. Sandler) p. 79. Oxford University Press, Oxford.

—— Hallett, M., Jenner, P. and Marsden, C. D. (1978). 5-hydroxytryptophan-induced myoclonus in guinea pigs. *J. Neurol. Sci.* **35**, 157.

Clineschmidt, B. V. (1979). MK-212: A serotonin-like agonist in the CNS. *Gen. Pharmac.* **10**, 287.

Colpaert, F. C. and Janssen, P. A. (1983). The head twitch response to intraperitoneal injection of 5-hydroxytryptophan in the rat: antagonist effects of purported 5-hydroxytryptamine antagonists and of pirenperone, an LSD antagonist. *Neuropharmacology* **22**, 993.

Corne, S. J. and Pickering, R. W. (1967). A possible correlation between drug-induced hallucinations in man and a behavioural response in mice.

Psychopharmacologia **11**, 65.

—— —— and Warner, B. T. (1963). A method for assessing the effects of drugs on the central actions of 5-hydroxytryptamine. *Br. J. Pharmac.* **20**, 106.

Cortés, R., Palacios, J. M. and Pazos, A. (1984). Visualisation of multiple serotonin receptors in the rat brain by autoradiography. *Br. J. Pharmac.* **82**, 202P.

Costain, D. W. and Green, A. R. (1978). β-adrenoceptor antagonists inhibit the behavioural responses of rats to increased brain 5-hydroxytryptamine. *Br. J. Pharmac.* **64**, 193.

—— —— and Grahame-Smith, D. G. (1979). Enhanced 5-hydroxytryptamine-mediated behavioural responses in rats following repeated electroconvulsive shock: relevance to the mechanism of the antidepressive effect of electroconvulsive therapy. *Psychopharmacology* **61**, 167.

Cowen, P. J., Grahame-Smith, D. G., Green, A. R. and Heal, D. J. (1982). β-adrenoceptor agonists enhance 5-hydroxytryptamine-mediated behavioural responses. *Br. J. Pharmac.* **76**, 265.

Deakin, J. F. W. and Dashwood, M. R. (1981). The differential neurochemical bases of the behaviours elicited by serotonergic agents and by the combination of a monoamine oxidase inhibitor and L-dopa. *Neuropharmacology* **20**, 123.

—— and Green, A. R. (1978). The effects of putative 5-hydroxytryptamine antagonists on the behaviour produced by administration of tranylcypromine and l-dopa to rats. *Br. J. Pharmac.* **64**, 201.

de Montigny, C. (1981). Enhancement of the 5-HT neurotransmission by antidepressant treatments. *J. Physiol. (Paris)* **77**, 455.

—— and Aghajanian, G. K. (1978). Tricyclic antidepressants: long term treatment increases responsibility of rat forebrain neurones to serotonin. *Science* **202**, 1304.

Dourish, C. T. and Greenshaw, A. J. (1983). Effects of intraventricular tryptamine and 5-hydroxytryptamine on spontaneous motor activity in the rat. *Res. Commun. Psychol. Psychiat. Behav.* **8**, 1.

Drust, E. G. and Connor, J. D. (1983). Pharmacological analysis of shaking behaviour induced by enkephalins, thyrotropin-releasing hormone or serotonin in rats: evidence for different mechanisms. *J. Pharmac. Exp. Ther.* **224**, 148.

Euvrard, C. and Boissier, J. R. (1980). Biochemical assessment of the central 5-HT agonist activity of RU 24969 (a piperidinylindole). *Eur. J. Pharmac.* **63**, 65.

Evans, J. P. M., Grahame-Smith, D. G., Green, A. R. and Tordoff, A. F. C. (1976). Electroconvulsive shock increases the behavioural responses of rats to brain 5-hydroxytryptamine accumulation and central nervous system stimulant drugs. *Br. J. Pharmac.* **56**, 193.

358   A. Richard Green and David J. Heal

Foldes, A. and Costa, E. (1975). Relationship of brain monoamines and locomotor activity in rats. *Biochem. Pharmac.* **24**, 1617.

Fowler, C. J. and Tipton, K. F. (1982). Deaminiation of 5-hydroxytryptamine by both forms of monoamine oxidase in the rat brain. *J. Neurochem.* **38**, 733.

Frankhuyzen, A. L. and Mulder, A. H. (1980). Noradrenaline inhibits depolarization-induced [^{3}H]-serotonin release from slices of rat hippocampus. *Eur. J. Pharmac.* **63**, 179.

Friedman, E. and Dallob, A. (1979). Enhanced serotonin receptor activity after chronic treatment with imipramine or amitriptyline. *Commun. Psychopharmac.* **3**, 89.

—— Cooper, T. B. and Dallob, A. (1983). Effects of chronic antidepressant treatment on serotonin receptor activity in mice. *Eur. J. Pharmac.* **89**, 69.

Fuxe, K. (1965). Evidence for the existence of monoamine neurons in the central nervous system. IV. Distribution of monoamine nerve terminals in the central nervous system. *Acta. Physiol. Scand.* **64**, Suppl. 247, 37.

—— Agnati, L. and Everitt, B. (1975). Effect of metergoline on central monoamine neurones. Evidence for a blockade of 5-HT receptors. *Neurosci. Lett.* **1**, 283.

Gallagher, D. W. and Bunney, W. E. (1979). Failure of chronic lithium treatment to block tricyclic antidepressant induced 5-HT supersensitivity. *Naunyn-Schmiedeberg's Arch. Pharmac.* **307**, 129.

Gardner, C. R. and Guy, A. R. (1983). Behavioural effects of RU 24969, a 5-HT$_1$ receptor agonist, in the mouse. *Br. J. Pharmac.* **78**, 96P.

Goodwin, G. M. and Green, A. R. (1985). A behavioural and biochemical study in mice and rats of putative selective agonists and antagonists for 5-HT$_1$ and 5-HT$_2$ receptors. *Br. J. Pharmac.* (In press).

—— —— and Johnson, P. (1984). 5-HT$_2$ receptor characteristics in frontal cortex and 5-HT$_2$ receptor-mediated head-twitch behaviour following antidepressant treatment to mice. *B. J. Pharmac.* **83**, 235.

Göthert, M. and Huth, H. (1980). Alpha-adrenoceptor-mediated modulation of 5-hydroxytryptamine release from rat brain cortex slices. *Naunyn-Schmiedeberg's Arch. Pharmac.* **313**, 21.

—— and Schlicker, E. (1983). Autoreceptor-mediated inhibition of [^{3}H]-5-hydroxytryptamine release from rat brain cortex slices by analogues of 5-hydroxytryptamine. *Life Sci.* **32**, 1183.

Gozlan, H., El Mestikawy, S., Pichat, L., Glowinski, J. and Hamon, M. (1983). Identification of presynaptic autoreceptors using a new ligand ^{3}H-PAT. *Nature* **305**, 140.

Grahame-Smith, D. G. (1971*a*). Studies *in vivo* on the relationship between brain tryptophan, brain 5-HT synthesis and hyperactivity in rats treated with a monoamine oxidase inhibitor and L-tryptophan. *J. Neurochem.*

18, 1053.

—— (1971*b*). Inhibitory effect of chlorpromazine on the syndrome of hyperactivity produced by L-tryptophan or 5-methoxy-*N,N*-dimethyltryptamine in rats treated with a monoamine oxidase inhibitor. *Br. J. Pharmac.* **43,** 856.

Green, A. R. (1977). Repeated chlorpromazine administration increases a behavioural response of rats to 5-hydroxytryptamine receptor stimulation. *Br. J. Pharmac.* **59,** 367.

—— (1984). Alterations in monoamine-mediated behaviours and biochemical changes after repeated ECS: studies in their possible association. In *ECT: Basic mechanisms* (eds. R. D. Weiner, B. Lerer and R. H. Belmaker) p. 5. John Libbey, London.

—— and Deakin, J. F. W. (1980). Brain noradrenaline depletion prevents ECS-induced enhancement of serotonin- and dopamine-mediated behaviour. *Nature* **285,** 232.

—— and Grahame-Smith, D. G. (1974). The role of brain dopamine in the hyperactivity syndrome produced by increased 5-HT synthesis in rats. *Neuropharmacology* **13,** 949.

—— —— (1975). The effect of diphenylhydantoin on brain 5-hydroxytryptamine metabolism and function. *Neuropharmacology* **14,** 107.

—— —— (1976). (−)-propranolol inhibits the behavioural responses of rats to increased 5-hydroxytryptamine in the central nervous system. *Nature* **262,** 594.

—— and Kelly, P. H. (1976). Evidence concerning the involvement of 5-hydroxytryptamine in locomotor activity produced by amphetamine or tranylcypromine plus L-dopa. *Br. J. Pharmac.* **57,** 141.

—— and Nutt, D. J. (1983). Antidepressants. In *Psychopharmacology* (eds. D. G. Grahame-Smith and P. J. Cowen) p. 1. Excerpta Medica, Amsterdam.

—— and Youdim, M. B. H. (1975). Effects of monoamine oxidase inhibition by clorgyline, deprenil or tranylcypromine on 5-hydroxytryptamine concentrations in rat brain and hyperactivity following subsequent tryptophan administration. *Br. J. Pharmac.* **55,** 415.

—— Costain, D. W. and Deakin, J. F. W. (1980). Enhanced 5-hydroxytryptamine and dopamine-mediated behavioural responses following convulsions. III. The effects of monoamine antagonists on the ability of electroconvulsive shock to enhance responses. *Neuropharmacology* **19,** 907.

—— Guy, A. P. and Gardner, C. R. (1984). The behavioural effects of RU 24969, a suggested 5-HT$_1$ receptor agonist in rodents and the effect on the behaviour of treatment with antidepressants. *Neuropharmacology* **23,** 655.

—— Hall, J. E. and Rees, A. R. (1981). A behavioural and biochemical study in rats of 5-hydroxytryptamine agonists and antagonists with observations on structure-activity requirements for agonists. *Br. J. Pharmac.* **73**, 703.

—— Heal, D. J. and Grahame-Smith, D. G. (1977). Further observations on the effects of repeated electroconvulsive shock on the behavioural responses of rats produced by increases in the functional activity of brain 5-hydroxytryptamine and dopamine. *Psychopharmacology* **52**, 195.

—— Hughes, J. P. and Tordoff, A. F. C. (1975). The concentration of 5-methoxytryptamine in rat brain and its effects on behaviour following its peripheral injection. *Neuropharmacology* **14**, 601.

—— Johnson, P. and Nimgaonkar, V. L. (1983*a*). Interactions of β-adrenoceptor agonists and antagonists with the 5-hydroxytryptamine (5-HT$_2$) receptor. *Neuropharmacology* **22**, 657.

—— —— —— (1983*b*). Increased 5-HT$_2$ receptor number in brain as a probable explanation for the enhanced 5-hydroxytryptamine-mediated behaviour following repeated electroconvulsive shock administration to rats. *Br. J. Pharmac.* **80**, 173.

—— Tordoff, A. F. C. and Bloomfield, M. R. (1976*a*). Elevation of brain GABA concentrations with amino-oxyacetic acid; effect on the hyperactivity syndrome produced by increased 5-hydroxytryptamine synthesis in rats. *J. Neural Transm.* **39**, 103.

—— Youdim, M. B. H. and Grahame-Smith, D. G. (1976*b*). Quipazine: its effects on rat brain 5-hydroxytryptamine metabolism, monoamine oxidase activity and behaviour. *Neuropharmacology* **15**, 173.

—— Heal, D. J., Johnson, P., Lawrence, B. E. and Nimgaonkar, V. L. (1983*c*). Antidepressant treatments: effects in rodents on dose-response curves of 5-hydroxytryptamine- and dopamine-mediated behaviour and 5-HT$_2$ receptor number in frontal cortex. *Br. J. Pharmac.* **80**, 377.

—— Sant, K., Bowdler, J. M. and Cowen, P. J. (1982). Further evidence for a relationship between changes in GABA concentrations in rat brain enhanced monoamine-mediated behaviours following repeated electroconvulsive shock. *Neuropharmacology* **21**, 981.

—— O'Shaughnessy, K., Hammond, M., Schächter, M. and Grahame-Smith, D. G. (1983*b*). Inhibition of 5-hydroxytryptamine-mediated behaviours in the putative 5-HT$_2$ antagonist pirenperone. *Neuropharmacology* **22**, 573.

Haigler, H. T. and Aghajanian, G. K. (1977). Serotonin receptors in the brain. *Fedn. Proc.* **36**, 2159.

Handley, S. L. and Brown, J. (1982). Effects on the 5-hydroxytryptamine-induced head twitch of drugs with selective actions on alpha$_1$- and alpha$_2$-adrenoceptors. *Neuropharmacology* **21**, 507.

—— and Singh, L. (1984). The effect of β-adrenoceptor agonists and antagonists on head-twitch in male mice. *Br. J. Pharmac.* **81**, 127P.

Heal, D. J., Lister, S., Smith, S. L., Davies, C. L., Molyneux, S. G. and Green, A. R. (1983). The effects of acute and repeated administration of various antidepressant drugs on clonidine-induced hypoactivity in mice and rats. *Neuropharmacology* **22**, 983–992.

Hess, S. M. and Doepfner, W. (1961). Behavioural effects and brain amine content in rats. *Archs. Int. Pharmacodyn. Ther.* **134**, 89.

Hjörth, S., Carlsson, A., Lindberg, P., Sanchez, D., Wikström, H., Arvidsson, L-E., Hacksell, V. and Nilsson, J. L. G. (1982). 8-Hydroxy-2-(di-*n*-propylamino)tetralin, 8-OH-DPAT, a potent and selective simplified ergot congener with central 5-HT-receptor stimulating activity. *J. Neural Transm.* **55**, 169.

Hunt, P. J. and Oberlander, C. (1981). The interaction of indole derivatives with the serotonin receptor and non-dopaminergic circling behaviour. In *Serotonin—current aspects of neurochemistry* (ed. B. Haber) p. 547. Plenum Press, New York.

Hwang, E. C. and Van Woert, M. H. (1980). Acute versus chronic effects of serotonin uptake blockers on potentiation of the 'serotonin syndrome'. *Commun. Psychopharmac.* **4**, 161.

Jacobs, B. L. (1974). Evidence for the functional interaction of two central neurotransmitters. *Psychopharmacologia* Berlin, **39**, 81.

—— (1976). An animal behaviour model for studying central serotonergic synapses. *Life Sci.* **19**, 777.

—— and Klemfuss, H. (1975). Brainstem and spinal cord mediation of serotoninergic behavioural syndrome. *Brain Res.* **100**, 450.

Johnson, G. A., Kim, E. G. and Boukma, S. J. (1972). 5-Hydroxy-indole levels in rat brain after inhibition of dopamine-β-hydroxylase. *J. Pharmac. Exp. Ther.* **180**, 539.

Kellar, K. J., Cascio, C. S., Butler, J. A. and Kurtzke, R. N. (1981). Differential effects of electroconvulsive shock and antidepressant drugs on serotonin-2 receptors in rat brain. *Eur. J. Pharmac.* **69**, 515.

Klawans, H. L., Goetz, C. and Weiner, W. J. (1973). 5-Hydroxytryptophan-induced myoclonus in guinea pigs and the possible role of serotonin in infantile myoclonus. *Neurology* (Minneap) **23**, 1234.

Kostowski, W., Samanin, R., Bareggi, S. R., More, V., Garattini, S. and Valzelli, L. (1974). Biochemical aspects of the interaction between midbrain raphe and locus coeruleus in the rat. *Brain Res.* **82**, 178.

Laduron, P. M., Janssen, P. F. M. and Leysen, J. E. (1982). In vivo binding of [^{3}H]-ketanserin on serotonin S_2-receptors in rat brain. *Eur. J. Pharmac.* **81**, 43.

Lebrecht, U. and Nowak, J. Z. (1980). Effect of single and repeated electroconvulsive shock on serotonergic system in rat brain. II. Behavioural studies. *Neuropharmacology* **19**, 1055.

Leysen, J. E., Niemegeers, C. J. E., Van Neuten, J. M. and Laduron, P. M.

362   A. Richard Green and David J. Heal

(1982). [³H]-Ketanserin (R 41 468), a selective ³H-ligand for serotonin receptor binding sites. *Mol. Pharmac.* **21**, 301.

Lloyd, K. G. and Bartholini, G. (1974). The effect of methiothepin on cerebral monoamine neurons. *Adv. Biochem. Psychopharmac.* **10**, 305.

Luscombe, G., Jenner, P. and Marsden, C. D. (1981). Pharmacological analysis of the myoclonus induced by 5-hydroxytryptophan in the guinea pig suggests the presence of multiple 5-hydroxytryptamine receptors in the brain. *Neuropharmacology* **20**, 819.

—— —— —— (1984). Correlation of [³H]-5-hydroxytryptamine (5-HT) binding to rat brain stem preparations and the production and prevention of myoclonus in guinea pig by 5-HT agonists and antagonists. *Eur. J. Pharmac.* **104**, 235.

Malick, J. B., Dosen, E. and Barnett, A. (1977). Quipazine-induced head-twitch in mice. *Pharmac. Biochem. Behav.* **6**, 325.

Marsden, C. A. and Curzon, G. (1978). The contribution of trypatmine to behavioural effects of L-tryptophan in tranylcypromine-treated rats. *Psychopharmacology* **57**, 71.

Martin, L. L. and Sanders-Bush, E. (1982). Comparison of the pharmacological characteristics of 5-HT_1 and 5-HT_2 binding sites with those of serotonin autoreceptors which modulate serotonin release. *Naunyn-Schmiedeberg's Arch. Pharmac.* **321**, 135.

Martin, W. R., Wikler, A., Eades, C. G. and Pescor, F. T. (1963). Tolerance to and physical dependence on morphine in rats. *Psychopharmacologia* **4**, 247.

Massingham, R., Dubocovich, M. L., Shepperson, N. B. and Langer, S. Z. (1981). In vivo selectivity of prazosin but not WB 4101 for postsynaptic alpha-1-adrenoceptors. *J. Pharm. Exp. Ther.* **217**, 467.

Mathews, W. D. and Smith, C. D. (1980). Pharmacological profile of a model for central serotonin receptor activation. *Life Sci.* **26**, 1397.

Maura, G., Gemignani, A. and Raiteri, M. (1982). Noradrenaline inhibits central serotonin release through alpha$_2$-adrenoceptors located on serotonergic nerve terminals. *Naunyn-Schmiedeberg's Arch. Pharmac.* **320**, 272.

Middlemiss, D. N. (1984). Stereoselective blockade at [³H]-5-HT binding sites and at the 5-HT autoreceptor by propranolol. *Eur. J. Pharmac.* **101**, 289.

—— and Fozard, J. R. (1983). 8-Hydroxy-2-(di-*n*-propylamino)-tetralin discriminates between subtypes of the 5-HT_1 recognition site. *Eur. J. Pharmac.* **90**, 151.

—— Blackeborough, L. and Leather, S. R. (1977). Direct evidence for an interaction of adrenergic blockers with the 5-HT receptor. *Nature* **267**, 289.

Modigh, K. (1972). Central and peripheral effects of 5-hydroxytryptophan on motor activity in mice. *Psychopharmacologia* **23**, 48.

—— (1973). Effects of chlorimipramine and protriptyline on the hyperactivity induced by 5-hydroxytryptophan after peripheral decarboxylase inhibition in mice. *J. Neural Transm.* **34**, 101.

Nahorski, S. R. and Willcocks, A. L. (1983). Interactions of β-adrenoceptor antagonists with 5-hydroxytryptamine receptor subtypes in rat cerebral cortex. *Br. J. Pharmac.* **78**, 107P.

Nelson, D. L., Pedigo, N. W. and Yamamura, H. I. (1981). Multiple [³H]-5-hydroxytryptamine binding sites in rat brain. *J. Physiol. (Paris)* **77**, 369.

Nimgaonkar, V. L., Green, A. R., Cowen, P. J., Heal, D. J., Grahame-Smith, D. G. and Deakin, J. F. W. (1983). Studies in the mechanism by which clenbuterol, a β-adrenoceptor agonist, enhances 5-HT-mediated behaviours and increases brain 5-HT metabolism in the rat. *Neuropharmacology* **22**, 739.

Oberlander, L. (1983). Effects of a potent 5-HT agonist, RU 24969, on mesocorticolimbic and nigrostriatal dopamine systems. *Br. J. Pharmac.* **80**, 675P.

—— and Boisser, J. K. (1981). Haloperidol blocks hyperlocomotion but not the circling behaviours induced by the serotonin agonist RU 24969. *Proc. Int. Congr. Pharmac. (Tokyo),* Abstract 839.

Ortmann, R., Martin, S., Radeke, E. and Delini-Stula, A. (1981). Interaction of β-adrenoceptor agonists with the serotonergic system in rat brain. A behavioural study using the L-5-HTP syndrome. *Naunyn-Schmiedeberg's Arch. Pharmac.* **316**, 225.

—— Bischoff, S., Radeke, E., Buech, O. and Delini-Stula, A. (1982). Correlations between different measures of antiserotonin activity of drugs. Study with neuroleptics and serotonin receptor blockers. *Naunyn-Schmiedeberg's Arch. Pharmac.* **321**, 265.

Palacios, J. M. (1984). Light microscopic autoradiographic localization of catecholamine receptor binding sites in brain. *Prog. neuro-Psychopharmac. Biol. Psych.* (In press).

Pedigo, N. W., Yamamura, H. I. and Nelson, D. L. (1981). Discrimination of multiple [³H]-5-hydroxytryptamine binding sites by the neuroleptic spiperone in rat brain. *J. Neurochem.* **36**, 220.

Peroutka, S. J. and Snyder, S. H. (1979). Multiple serotonin receptors: differential binding of [³H]-5-hydroxytryptamine, [³H]-lysergic acid diethylamide and [³H]-spiroperidol. *Molec. Pharmac.* **16**, 687.

—— —— (1980). Long-term antidepressant treatment decreases spiroperidol-labelled serotonin receptor binding. *Science* **210**, 88.

—— —— (1981). Two distinct serotonin receptors: Regional variations in receptor binding in mammalian brain. *Brain Res.* **208**, 339.

—— Lebovitz, R. M. and Snyder, S. H. (1981). Two distinct central serotonin receptors with different physiological functions. *Science* **212**,

827.

Prange, A. J., Breese, G. R., Cott, J. M., Martin, B. R., Cooper, B. R., Wilson, I. C. and Plotnikoff, N. P. (1974). Thyrotropin releasing hormone: antagonism of pentobarbital in rodents. *Life Sci.* **14**, 447.

Reinhard, J. F. and Roth, R. H. (1982). Noradrenergic modulation of serotonin synthesis and metabolism. I. Inhibition by clonidine in vivo. *J. Pharmac. Exp. Ther.* **221**, 541.

Samanin, R., Mennini, T., Ferraris, A., Bendotti, L. and Borsini, F. (1980). Repeated treatment with d-fenfluramine or metergoline alters cortex binding and serotonergic sensitivity in rats. *Eur. J. Pharmac.* **61**, 203.

Schechter, Y. and Weinstock, M. (1974). β-Adrenoceptor blocking agents and responses to adrenaline and 5-hydroxytryptamine in isolated rat stomach and uterus. *Br. J. Pharmac.* **52**, 283.

Sloviter, R. S., Drust, E. G. and Connor, J. D. (1978). Specificity of a rat behavioural model for serotonin receptor activation. *J. Pharmac. Exp. Ther.* **206**, 339.

Stolz, J. F. and Marsden, C. A. (1982). Withdrawal from chronic treatment with metergoline, *dl*-propranolol and amitriptyline enhances serotonin receptor mediated behaviour in the rat. *Eur. J. Pharmac.* **79**, 17.

—— —— and Middlemiss, D. N. (1983). Effect of chronic antidepressant treatment and subsequent withdrawal on ^{3}H-5-hydroxytryptamine and ^{3}H-spiperone binding in rat frontal cortex and serotonin receptor mediated behaviour. *Psychopharmacology* **80**, 150.

Svensson, T. H., Bunney, B. S. and Aghajanian, G. K. (1975). Inhibition of both noradrenergic and serotonergic neurons in brain by the α-adrenoceptor agonist clonidine. *Brain Res.* **92**, 291.

Swanson, L. W. and Hartman, B. K. (1975). The central adrenergic system. An immunofluorescence study of the location of cell bodies and the efferent connections in the rat utilizing dopamine-β-hydroxylase as a marker. *J. Comp. Neurol.* **163**, 467.

Tricklebank, M. D. (1984*a*). Behavioural effects of 8-hydroxy-2-(di-*n*-propylamine) tetralin, a putative 5-HT$_{1A}$ receptor agonist. *Br. J. Pharmac.* **81**, 26P.

—— (1984*b*). Is hyperlocomotion induced by the 5-HT$_1$ agonist, 5-methoxy-3(1,2,3,6-tetrahydropyridin-4-yl)1H indole (RU 24969)? *Br. J. Pharmac.* **81**, 140P.

—— (1984*c*). Central 5-HT receptor subtypes and the behavioural response to 5-methoxy-*N*,*N*-dimethyltryptamine. *Br. J. Pharmac.* **82**, 204P.

Trulson, M. E. and Jacobs, B. L. (1976). Behavioural evidence for the rapid release of CNS serotonin by PCA and fenfluramine. *Eur. J. Pharmac.* **36**, 149.

—— Eubanks, E. E. and Jacobs, B. L. (1976*b*). Behavioural evidence for supersensitivity following destruction of central serotoninergic nerve terminals by 5,7-dihydroxytryptamine. *J. Pharmac. Exp. Ther.* **198,** 23.

—— Ross, C. A. and Jacobs, B. L. (1976*a*). Behavioural evidence for the stimulation of CNS serotonin reception by high doses of LSD. *Psychopharmac. Commun.* **2,** 149.

Vetulani, S., Bednarczyk, B., Reichenberg, K. and Rokosz, A. (1980). Head-twitches induced by LSD and quipazine: similarities and differences. *Neuropharmacology* **19,** 155.

—— Lebrecht, U. and Pilc, A. (1981). Enhancement of responsiveness of the central serotonergic system and serotonin-2 receptor density in rat frontal cortex by electroconvulsive treatment. *Eur. J. Pharmac.* **76,** 81.

Volkman, P. H., Lorens, S. A., Kindel, G. H. and Ginos, J. Z. (1978). L-5-Hydroxytryptophan induced myoclonus in guinea-pigs; a model for the study of central serotonin dopamine interactions. *Neuropharmacology* **17,** 947.

Waldmeier, P. C. (1981). Stimulation of central serotonin turnover by β-adrenoceptor agonists. *Naunyn-Schmiedeberg's Arch. Pharmac.* **317,** 115.

Wei, E., Sigel, S., Loh, H. and Way, E. L. (1975). Thyrotropin-releasing hormone and shaking behaviour in the rat. *Nature* **253,** 739.

Weinstock, M. (1980). Behavioural effects of β-adrenoceptor antagonists associated with blockade of central serotonergic systems. In *Enzymes and neurotransmitters in mental disease* (eds. E. Usdin, J. L. Sourkes and M. B. H. Youdim) p. 431. John Wiley, Chichester.

Yap, C. Y. and Taylor, D. A. (1983). Involvement of 5-HT$_2$ receptors in the wet-dog shake behaviour induced by 5-hydroxytryptophan in the rat. *Neuropharmacology* **22,** 801.

13

The pharmacology of serotonin receptors in invertebrates

ROBERT J. WALKER

13.1 Introduction

The first systematic survey of the occurrence of serotonin in invertebrates was performed by Welsh and Moorhead in 1960. From their study it was apparent that serotonin was present in a wide range of animals and it is likely that, in common with other putative transmitters, it occurs ubiquitously in the animal kingdom (Walker 1982). Unlike the situation with the catecholamines (Walker and Kerkut 1978), the only other indoleamine to occur in any significant amount in animal tissues is tryptamine but a specific physiological role for this amine has not been established. In addition to a possible transmitter function, serotonin also occurs in venoms, e.g. in the venom of the wasp, *Vespa vulgaris* (Jaques and Schachter 1954), in the salivary glands of the octopus (Welsh and Moorhead 1960) and in plants, e.g. in the stinging apparatus of the nettle, *Urtica dioica* (Brittain and Collier 1957).

Serotonin has potent actions on many invertebrate tissues and in particular its action on molluscan hearts has been the subject of considerable study (Hill and Welsh 1966). This was probably the first invertebrate tissue to be investigated in detail with respect to the site of action of serotonin. In most cases the major action is one of excitation but there are clear examples where serotonin is basically inhibitory, e.g. on the heart of *Modiolus* (Wilkens and Greenberg 1973). Molluscan hearts have been employed as bioassays for serotonin since many are particularly sensitive to this amine, including *Mercenaria* (Venus), *Anodonta* and *Helix* (Fange 1955; Gaddum and Paasonen 1955; Kerkut and Laverack 1960; Twarog and Page 1953; Zetler and Schlosser 1954).

There have been a number of excellent reviews recently published on the occurrence and role of serotonin in invertebrates, in particular in the leech nervous system (Lent 1977, 1982; McAdoo 1978) and the gastropod molluscs (Osborne 1978; Pentreath, Berry and Osborne 1982). It is not the intention of the present review to repeat the data contained in these extensive reviews and the reader is referred to them for background literature and to the reviews of Leake and Walker (1980) and Gardner and Walker (1982). The aim of the present review is to examine the pharmacology of invertebrate serotonin receptors and to compare them where appropriate with the situation in the vertebrates. In addition to serotonin receptors on neurones, reference will also be made to peripheral serotonin receptors where the pharmacology has been investigated in detail, e.g. on molluscan hearts and on other tissues to indicate the range of invertebrate systems sensitive to serotonin.

13.2   Gastropod central nervous system

13.2.1   *Serotonin as a transmitter*

There is very good evidence for serotonin as a transmitter in the gastropod molluscs, such as *Aplysia* and *Helix,* particularly associated with the synaptic connections from the giant serotonin-containing neurones in the cerebral ganglia. It has been demonstrated that serotonin is present in these neurones (Sedden, Walker and Kerkut 1968; Cottrell and Osborne 1970) at a level of 4–6 pmol (Weinreich, M. W. McCaman, R. E. McCaman and Vaughan 1973) and can be synthesized by them from tryptophan and 5-hydroxytryptophan (Eisenstadt, Goldman, Kandel, Koike, Koester and Schwartz 1973). The electrophysiological evidence suggests that serotonin is the transmitter from these neurones onto neurones in the buccal ganglia of *Helix* and *Aplysia* (Cottrell and Macon 1974; Paupardin-Tritsch and Gerschenfeld 1973). Following activation of the serotonin-containing neurones, excitatory postsynaptic potentials (epsps) can be obtained from the follower buccal cells (Fig. 13.1). Ionophoretic application of serotonin onto the buccal neurones produced a depolarisation of the cell. It was found that both events were sodium dependent and could be antagonized with morphine. When the animals were pre-treated with reserpine, transmission between the cerebral cells and the buccal cells was impaired.

In addition to morphine, a number of other compounds were tested by Cottrell and Macon (1974) as potential serotonin antagonists. They found that lysergic acid diethylamide (LSD) at a concentration of 1.5×10^{-4}M, had a direct excitatory effect on the buccal neurones but also reduced the response to serotonin. Methysergide and 4-BromoLSD (BOL 148), 2.5×10^{-4}M, produced less direct excitation of the cell and also antagonized the serotonin excitation. For their experiments they selected morphine because at the concentration used, 2×10^{-5}M, it was devoid of direct excitatory effects on the cell but was effective in blocking the serotonin

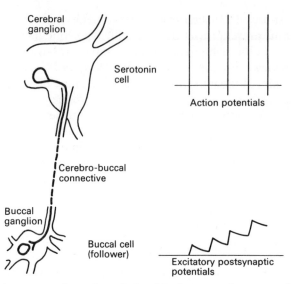

Fig. 13.1 Diagram to show the relationship between the serotonin-containing neurone in the cerebral ganglion of gastropods and a follower neurone in the buccal ganglion. Action potentials in the cerebral neurone induce excitatory postsynaptic potentials (epsps) in the follower neurone. (After Cottrell and Macon 1974)

response. Tryptamine (10^{-6}M) and tubocurarine (1.5×10^{-4}M) also reversibly blocked the action of serotonin without any direct effect on the cell

One of the problems in establishing a compound as a central transmitter is to demonstrate its release. This release has been demonstrated in the case of serotonin from the cerebro-buccal ganglionic ring of *Aplysia* (Gerschenfeld, Hamon and Paupardin-Tritsch 1978). The rate of spontaneous release of serotonin varied between 0.4 and 1.2 pmol per hour. When the serotonin-containing cerebral neurones were activated there was an 80 to 100 per cent increase in the release of serotonin but this was apparent only when an uptake blocker, e.g. chlorimipramine, 1–10 µM, was present in the incubation fluid. Raised external potassium also enhanced the release of serotonin provided an uptake blocker was present. It was also noted that in the presence of chlorimipramine or desmethylimipramine that the duration and amplitude of epsps and ipsps recorded from buccal neurones following activation of the serotonin cerebral cells were enhanced. These experiments provide further data for serotonin as a transmitter in the gastropod central nervous system.

There have been few specific studies on the characteristics of the uptake mechanism into invertebrate serotonin-containing neurones. However evidence suggests that non-hydroxylated tryptamines are not taken up

(Schroder, Neuhoff, Priggemeir and Osborne 1979). Evidence from vertebrate studies suggests that 5-6-dihydroxytryptamine, 5-7-dihydroxy-tryptamine and N-methylated derivatives are transported (Bjorklund, Horn, Baumgarten, Nobin and Schlossberger 1975).

Possible physiological roles for the cerebral serotonin-containing neurones have been investigated. For example, a detailed study of the axon terminals of these neurones has been undertaken by Schwartz and Shkolnik (1981) and extended for glial cells by Goldstein, Weiss and Schwartz (1982). The serotonin cells were injected with a labelled amino sugar precursor of membrane glycoproteins and then studied using electron microscope autoradiography. These glycoproteins are trans-ported rapidly to nerve terminals. It was found that the nerves terminated onto a wide range of tissues including glial cells which form the lining of intraganglionic haemal sinuses, soma of buccal neurones, axon hillock region of buccal neurones, buccal muscle, gut muscle and salivary gland acinar cells. Interestingly, none of the glial cells in the cerebral ganglia were labelled. It is probable that these cells play a role in feeding behaviour. Activation of them can influence buccal ganglion motoneurones which in turn can excite the buccal musculature (Weiss, Cohen and Kupfermann 1978). These authors recorded from the serotonin-containing cells, buccal motoneurones and from the accessory radula closer muscle of the buccal mass of *Aplysia*. They found, when recording extracellularly from the serotonin cells that they were silent while the *Aplysia* were non-feeding and quiescent. If food was introduced, this elicited behavioural signs of arousal in the animal and the serotonin cells began to fire at 1–10 Hz. During biting movements, the serotonin cells decreased their firing rate during each protractor-retractor sequence of the radula. If the cerebral ganglia were bathed in high magnesium Ringer solution, to reduce or block synaptic transmission, then the accessory radula closer muscle contraction was still potentiated when the serotonin cells were activated. This would suggest that in addition to an action via the buccal motoneurones, serotonin can exert a direct effect on the muscle. Weiss *et al.* (1978) therefore suggested that the serotonin cells play a modulatory role in feeding behaviour but do not initiate the behaviour. It is likely that there are two sites of action, one where serotonin increases the amplitude of the excitatory junctional potential (ejp) of the muscle due to activation of the buccal motoneurones and one where serotonin has a direct effect on excitation-contraction coupling which is quite distinct from the enhancement of the ejps.

In another preparation, the brain-salivary gland preparation of *Limax maximum,* serotonin-containing metacerebral ganglion cells influence the firing pattern of buccal neurones which synapse onto salivary gland acinar cells (Copeland and Gelperin 1983). Activation of the serotonin cells excites some buccal neurones while inhibiting others in *Limax* (Gelperin

1975, 1981), a situation similar to that already described for *Helix* and *Aplysia*. This is likely to be the general position in the gastropod molluscs.

13.2.2   *The serotonin receptor*

Serotonin has been shown to have potent actions on central neurones in gastropod molluscs (Gerschenfeld and Tauc 1961; Kerkut and Walker 1961; 1962; Gerschenfeld and Stefani 1966). In particular central neurones from *Helix* and *Aplysia* have been used in these investigations. A structure-activity study on the excitatory receptor for serotonin in *Helix* was performed by Walker and Woodruff (1972). They found that *N*-methyl and α-methyl analogues of serotonin produce a response which was very similar to serotonin but both compounds were slightly less potent. This observation is in agreement with that found by Greenberg (1960*b*) for *Mercenaria* heart where the α-methyl analogue of serotonin was very similar in action to serotonin. However on *Helix* neurones tryptamine itself was devoid of an excitatory effect. This finding is in contrast to that of Greenberg who found that it was about ten times less active than serotonin on the heart of *Mercenaria*. Bufotenin and 5-methoxytryptamine both had serotonin-like activity on *Helix* neurones but were about 50 to 100 times less potent than serotonin. The presence of an hydroxyl group on position 5 of the indole nucleus is further shown to be important for excitatory serotonin-like activity on *Helix* neurones since *N*-methyl-tryptamine has no excitatory effect. Moving the hydroxyl group to position 4 or 6 of the indole nucleus also either greatly reduced potency or converted the response to one of inhibition since psilocin was more than 100 times less active than serotonin while 6-hydroxytryptamine was inhibitory. This data is summarized in Table 13.1. From their study Walker and Woodruff

Table 13.1   Potency ratios for a range of serotonin analogues on *Helix* central neurones excited by serotonin

Compound	Equipotent molar ratio
Serotonin	1
N-Methylserotonin	4.5
α-Methylserotonin	6.7
Bufotenin	45.6
5-Methoxytryptamine	120.0
Tryptamine	inhibition
6-Hydroxytryptamine	inhibition
N-Methyltryptamine	inactive
N'N-Dimethyltryptamine	>10 000

All compounds tested showed less potency than serotonin (from Walker and Woodruff 1972).

(1972) concluded that for potent serotonin-like excitation on *Helix* neurones a compound had to fulfil the following requirements: the presence of an indole nucleus with a hydroxyl group on position 5 and a two carbon side chain with no greater substitution that a methyl group on either the terminal nitrogen or the alpha carbon.

The interesting observation by Walker and Woodruff (1972) that tryptamine and 6-hydroxytryptamine were not only devoid of excitatory serotonin-like activity but were in fact inhibitory on these neurones has been further investigated by Wright and Walker (1984) who investigated the actions of serotonin, 6-HT, and dopamine on cells which were excited by serotonin and by cells which were inhibited by serotonin. The cells excited by serotonin were inhibited by dopamine while the cells inhibited by serotonin were likewise inhibited by dopamine. Wright and Walker (1984) found that low doses of 6-HT and tryptamine did produce threshold excitation but higher doses produced inhibition on cells which were only excited by serotonin. Both compounds were approximately equipotent in terms of inhibition and between 10 and 100 times less potent than dopamine. Tubocurarine, 20 μM, reversibly antagonized the excitatory action of serotonin and converted the 6-HT- and tryptamine-induced excitation into inhibition. In the presence of the dopamine antagonist ergometrine (200 nM) the dopamine inhibitory response was almost abolished while the inhibitory responses to 6-HT and tryptamine were converted to weak excitations. On cells which were inhibited by serotonin, the other three compounds were also inhibitory with the relative potency ratio: dopamine ≫ tryptamine/6-HT > serotonin. Tubocurarine had no antagonist effect on any of these responses while ergometrine reduced or blocked all four responses. In potassium-free Ringer solution the inhibitory responses of all four compounds were enhanced. From these experiments it was concluded that low concentrations of 6-HT and tryptamine could weakly excite serotonin excitatory receptors while higher concentrations act via a dopamine receptor.

There is now evidence for at least six different types of responses to the ionophoretic application of serotonin onto both *Helix* and *Aplysia* neurones (Gerschenfeld and Paupardin-Tritsch 1974*a*). Three of these responses are excitatory while the remaining three are inhibitory. Four of the responses are associated with an increase in membrane conductance, i.e. an increase in permeability while the other two are associated with a decrease in conductance. The properties of these responses are summarized in Table 13.2. Gerschenfeld and Paupardin-Tritsch (1974*a*) examined the pharmacological profiles of these different responses and the results are summarized in Table 13.3. From this table they concluded that each response is associated with the activation of a distinct serotonin receptor since the pharmacological profile of each response is different. No

Table 13.2   Summary of different types of serotonin response of buccal ganglion neurones of *Aplysia* and *Helix*

Response	Effect	Reversal potential (mV)	Ionic mechanism
A	Fast deplorization	0	Increase in sodium conductance
A′	Slow depolarization	0	Increase in sodium conductance
B	Slow hyperpolarization	−75	Increase in potassium conductance
C	Fast hyperpolarization	−56	Increase in chloride conductance
α	Depolarization	−75	Decrease in potassium conductance
β	Hyperpolarization	−30	Decrease in sodium and potassium conductance

The reversal potential values for A and A′ were estimated by extrapolation. (After Gerschenfeld and Paupardin-Tritsch 1974*a*).

Table 13.3   Pharmacological profile of four serotonin responses of *Helix* and *Aplysia* buccal neurones

Antagonist	Receptor Type			
	A	A′	B	C
Tubocurarine	block	none	none	block
LSD	block	none	block	block
Tryptamine	block	none	block	block
7-Methyltryptamine	block	none	none	none
Bufotenin	block	block	block	none
5-Methoxygramine	none	none	block	none
Neostigmine	none	none	none	block

In most cases the block was obtained with 10^{-5} M antagonist. (After Gerschenfeld and Paupardin-Tritsch 1974*a*).

compound was found which would block all four responses though several could block three out of the four responses. Three of the compounds, 7-methyltryptamine, 5-methoxygramine and neostigmine only block one response i.e. A, B and C responses respectively. They were unable to find any specific antagonist for the serotonin responses associated with conductance decreases.

In their second paper Gerschenfeld and Paupardin-Tritsch (1974*b*) investigated the possible physiological role for the receptors already described. The follower cells chosen in this study were in the buccal ganglia of *Aplysia*. They found evidence for a physiological role for the A and A′ excitatory receptors, the B receptors and also possibly for the β receptors. Thus serotonin released from a single neurone can activate at least four different responses onto buccal neurones in *Aplysia*.

Currently an attempt is being made to analyse particularly the excitatory actions of serotonin on *Helix* neurones in terms of the mammalian classification into 5-HT_1 and 5-HT_2 receptors (Bokisch, Gardner and Walker 1983) following the proposals of Peroutka and Snyder (1979, 1982) and Leysen (Chapter 4). The study on *Helix* neurones has been undertaken using both agonists and antagonists. MK-212 has been proposed as an example of a 5-HT_2 agonist (Gardner and Guy 1983). This compound has agonist activity on *Helix* neurones excited by serotonin but is 60–70 times less potent than serotonin. Although less potent, the form of the depolarization resembled that of serotonin. In contrast RU 24969, a potent 5-HT_1 agonist (Hunt, Nedelec, Euvrard and Boissier 1981; Green, Guy and Gardner 1984), had a slow weak excitatory effect which could be an indirect action. Quipazine also has a fast depolarizing action on these neurones with a potency ratio about 200 times less potent than serotonin. This observation that on *Helix* neurones MK-212 is more potent than quipazine is in agreement with findings on mammalian central neurones and contrasts with the position on mammlian uterus, where quipazine is more potent than MK-212 (Clineschmidt 1979). MK-212, RU 24969 and quipazine have also been tested on the *Helix* heart (Boyd and Walker, unpublished data). On this tissue MK-212 is around 50 times less potent than serotonin while RU 24969 is more than 500 times less potent. Quipazine has very weak agonist activity, but antagonizes the action of serotonin. It would be of interest to compare the potencies of MK-212 and quipazine on a range of invertebrate peripheral tissues, MK-212 has already been shown to be inactive on earthworm body wall muscle (Gardner 1981*a*). Both MK-212 and RU 24969 were also examined as potential antagonists against serotonin, acetylcholine and dopamine responses. MK-212 had a pA_2 value of 5.7 against serotonin excitatory responses. pA_2 values of 4.2, 4.6 and 4.4 were obtained against acetylcholine excitation, inhibition and dopamine inhibition respectively. Thus MK-212 exhibited selectivity as a serotonin antagonist in addition to acting as an agonist on *Helix* neurones excited by serotonin. RU 24969 was also examined for its antagonist action against serotonin, acetylcholine and dopamine and pA_2 values determined. The pA_2 values against serotonin excitation, acetylcholine excitation and inhibition and dopamine inhibition were 4.5, 4.6, 4.6 and 5.2 respectively. In contrast to the situation with MK-212, RU 24969 showed a specificity towards dopamine rather than serotonin. This observation is supported by some preliminary experiments on *Helix* neurones inhibited by serotonin

Table 13.4 A summary of the potencies of a range of antagonists against ionophoretic application of serotonin on *Helix* neurones excited by serotonin

Compound	Concentration tested (μM	% Inhibition
2-BromoLSD	5	20
	50	75
Cinanserin	5	30
	50	70
Cyproheptadine	6	10
	60	85
Fenfluramine	75	43
	380	75
	750	100
Methysergide	34	56
Phentolamine	5	28
	50	83
Propranolol	70	43
Quipazine	60	25
	300	71
Tryptamine	50	75

Taken from Bokisch (unpublished data).

and dopamine, where MK-212 has weak or no agonist activity at concentrations where RU 24969 has agonist activity and where in the presence of RU 24969, the serotonin response is attenuated. It is possible that both serotonin and RU 24969 are acting at a dopamine receptor on these neurones. Recently ketanserin has been proposed as a selective 5-HT$_2$ antagonist (Janssen 1983) and this compound has been tested on *Helix* excitatory responses. On these cells ketanserin antagonizes the excitatory action of serotonin at concentrations in the 5–50 μM range. A range of potential serotonin antagonists were also examined on the serotonin excitatory response of these neurones and the results summarized in Table 13.4. As can be seen from this table 2-BromoLSD, cianserin, cyproheptadine, methysergide, phentolamine and tryptamine all have a similar potency as antagonists of serotonin excitation on *Helix* neurones. Fenfluramine and quipazine also block but higher concentrations are required to produce the same degree of antagonism. Thus the classical serotonin antagonists block serotonin excitation on *Helix* neurones, the receptor which is analogous to the A receptor of Gerschenfeld and Paupardin-Tritsch (1974*a*).

13.3 Leech central nervous system

The major study involving serotonin and leech central neurones is centred on the Retzius cells, a pair of which are present in each segmental ganglion.

Table 13.5   Equipotent molar ratios (EPMR) of tryptamine-like compounds as compared with serotonin on *Hirudo* Retzius cells

Compound	Potency ratio
Serotonin	1.0
3(2-aminoethyl)5-hydroxyindene	2.7
α-Methylserotonin	3.1
N-Methylserotonin	6.5
N,N-Dimethylserotonin (Bufotenin)	10.0
N-Methyltryptamine	15.3
N,N-Dimethyltryptamine	16.7
5-Fluorotryptamine	26.8
N,N-Dimethyl-5-methoxytryptamine	31.9
5-Chlorotryptamine	54.7
5-Methoxytryptamine	77.1
Tryptamine	188.5
5-Methyltryptamine	205.5
6-Hydroxytryptamine	295.8
4-Hydroxytryptamine	1031.0
α-Methyltryptamine	1289.0

All compounds were less potent than serotonin (after Smith and Walker 1974).

As early as 1903, Poll and Sommer observed that these cells reduced chromium salts. The leech nervous system was subsequently examined using fluorescence microscopy and it was found that these cells fluoresced yellow indicating the presence of serotonin (Kerkut, Sedden and Walker 1967; Ehinger, Falck and Myhrberg 1968). This work was extended by Marsden and Kerkut (1969) and Rude (1969) and a total of seven neurones were found to contain serotonin in each ganglion. The final confirmation using chromatographic and microspectro-fluorimetric analysis of the presence of serotonin in Retzius cells was made by Rude, Coggeshall and van Orden (1969).

Serotonin has a direct inhibitory action when applied to Retzius cells (Kerkut and Walker 1967) and this effect is associated with an increase in conductance to chloride ions (Walker and Smith 1973). A structure-activity study on this response was performed by Smith and Walker (1974) and the results are summarised in Table 13.5. The addition of a methyl group either on the nitrogen or the alpha carbon of the ethylamine side chain only results in a small drop in potency. The addition of two methyl groups to give bufotenin results in a 10-fold decrease in potency in contrast to other preparations where bufotenin is often either equipotent or more potent than serotonin. Removal of the hydroxyl group on position 5 of the indole nucleus results in a considerable loss of potency. But the addition of a methyl group to tryptamine increases the potency and NN-dimethyltryptamine is only slightly less potent than bufotenin. Substitution of the hydroxyl group on position 5 by a chloro-, fluoro- or a methoxy group reduces potency by 55, 27 and 77 times respectively. Moving the hydroxyl group to position 4 or 6 of the indole nucleus renders the compound even less potent than tryptamine. One interesting observation was that an indene nucleus is of the same order of potency as the indole. A number of potential antagonists of serotonin were also examined on Retzius cells (Smith and Walker 1975). Several of these had direct actions on cell activity including dibenamine, 2-BromoLSD, cyproheptadine, tosylate (BW 545 C64) and phenyldiguanide. Out of these only dibenamine possessed clear antagonist properties at a concentration of 10^{-6}M. Methysergide (10^{-4}M) also partially antagonized serotonin without any direct effect on the cell membrane potential. Several other compounds which have been shown to block serotonin on invertebrates were also tested including tryptamine, 7-methyltryptamine, tubocurarine, bufotenin, neostigmine and iproniazid but were devoid of antagonist activity. Both morphine and atropine (10^{-4}M) were reasonably good antagonists of serotonin on Retzius cells and in addition morphine possessed a direct inhibitory effect on cell activity.

There is good evidence for serotonin as a transmitter in the leech nervous system but this will not be reviewed here in detail (see Leak and Walker 1980). The Retzius cells send axons out to the periphery (Smith, Sunderland, Leake and Walker 1975; Mason and Leake 1978) and a number of possible functions for serotonin have been suggested, including control of dermal mucus secretion (Lent 1973), relaxation of longitudinal muscles of body wall (Mason, Sundarland and Leake 1979) and activation of comlex swimming pattern (Willard 1981). Recently leech Retzius cells have been cultured and their ability to synthesize and release serotonin investigated (Henderson 1983). Pressure sensory cells have also been cultured and these respond to serotonin, low doses producing a slow chloride-dependent hyperpolarization while higher doses produce a faster depolarization. Chloride-dependent potentials have been evoked in pressure sensory cells by impulses in Retzius cells in culture (Fuchs, Henderson and Nicholls 1982). It is likely that serotonin acts as a transmitter onto these sensory cells.

Serotonin has a hyperpolarizing action on leech neuropile glial cells and this event is chloride mediated (Walz and Schlue 1982). This hyperpolarization was completely blocked by methysergide (10^{-4}M) whilst atropine at a similar concentration had no antagonist action. 5-Methoxygramine also failed to block the serotonin hyperpolarization. Bufotenin had a direct inhibitory effect on the glial cells and was approximately equipotent with serotonin.

The actions of serotonin and dopamine have been further studied on leech Retzius neurones by Sunderland, Leake and Walker (1980, 1982).

Both compounds hyperpolarized the membrane potential via the same ionic mechanism, an increase in permeability to chloride ions (Sunderland, Leake and Walker 1979) but it was assumed that they exerted this action via different receptors. In an attempt to resolve this point, a series of antagonists were examined for possible specific action either serotonin or dopamine (Sunderland *et al.* 1980). In this study atropine and morphine were found to be the most potent antagonists for serotonin but they were equally effective against dopamine. Fluphenazine, metoclopramide and ergometrine were weak antagonists but they appeared to block both dopamine and serotonin equally. Strychnine was a more potent antagonist but again blocked both amines to an equal extent. As a result of these experiments, together with a structure-activity study of the requirements for the dopamine inhibitory response on Retzius cells (Sunderland *et al.* 1982) which complemented the serotonin structure activity study (Smith and Walker 1974), a model for a joint dopamine-serotonin receptor was proposed (Fig. 13.2). Four binding sites are proposed for the amines, site 1 which is an

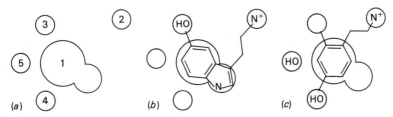

Fig. 13.2 Diagram of the proposed model of a joint serotonin-dopamine receptor on the leech Retzius neurone. (*a*) shows the active sites of the receptor; site 1 is an aromatic binding site; site 2 is a nitrogen binding site; sites 3, 4 and 5 are hydroxyl binding sites. (*b*) shows the possible interaction with serotonin. (*c*) shows the possible interaction with dopamine. (After Sunderland *et al.* 1982).

aromatic binding site, 2 is a terminal nitrogen binding site and 3 and 4 are hydroxyl group binding sites. A fifth binding site may be required in the case with *Haemopis* for dopamine since the presence of two hydroxyl groups on the ring may be more important with this species. In the case of *Hirudo*, dopamine is approximately equipotent with tyramine while noradrenaline and octopamine are also approximately equipotent. As can be seen in Figs. 2*b* and *c* respectively, it is possible for both serotonin and dopamine to fit this model. Serotonin would bind with sites 1, 2 and 3 while dopamine would bind with sites 1, 2, 4 and possibly an additional site for *Haemopis*.

13.4   Lamellibranch and gastropod heart muscle

One of the earliest serotonin structure-activity studies performed on
invertebrate tissue was that of Greenberg (1960a, b), using the heart
muscle of the clam (quahog) *Mercenaria (Venus) mercenaria.* The serotonin
response of *Mercenaria* heart muscle has three components: an increase in
amplitude, an increase in frequency and an increase in the resting tone of
the muscle. The threshold concentration for serotonin excitation is around
10^{-9}M. At concentrations between 10^{-9}M and 10^{-6}M, the serotonin effect is
largely one of increase in amplitude of the beat. As the concentration
increases however so does the increase in tone of the muscle together with
some increase in frequency. Above 10^{-6}M the relative increase in tone
becomes larger. The *Mercenaria* heart muscle appears to exhibit two
distinct responses to serotonin. At low concentrations, the major effect is a
positive inotropic one while at higher concentrations, the major effect is
an increase in tone. The effect of serotonin ($3-10\times10^{-7}$M) can be antagonized by 2-BromoLSD (2.5×10^{-5}M) but the heart can be directly excited by
this compound. If the heart is pre-treated with 10^{-5}M serotonin then it fails
to respond to lower concentrations of serotonin, i.e. it exhibits tachyphylaxis. The heart is also tachyphylactic to serotonin analogues, e.g.
tryptamine, bufotenin and 5-hydroxy-α-methyltryptamine. Although
desensitized hearts were insensitive to up to 10^{-5}M serotonin, they do
respond to higher concentrations but with a decrease in amplitude. Such
responses were almost identical to those normally produced by catecholamines.

Tryptamine was about 10 times less active than serotonin and was slower
in onset. The recovery from tryptamine also took a long time, particularly
at high concentrations of the amine. Table 13.6 summarizes the potencies
for a range of compounds related to serotonin. Tryptamine analogues with
methyl or ethyl groups produced effects which were slow to develop and
were irreversible. At low concentrations these compounds were equi-
active with serotonin but at higher concentrations they often failed to reach
the same maximum response as serotonin. With the addition of a hydroxyl
on position 5, the compounds became reversible and the speed of the
response greatly increased. Bufotenin was more active than serotonin, by
about 35 times. Of all the compounds tested by Greenberg, 5-hydroxy-α-
methyltryptamine produced a response which most resembled serotonin.
Substitution in the two position of the indole ring reduced potency. The
length of the side chain was also critical for activity, e.g. when the side chain
was reduced by one carbon, e.g. gramine or increased by one carbon, e.g.
3-(3-dimethylaminopropyl) indole, there was a drastic reduction in
potency. Gramine was in fact a weak antagonist of serotonin. Compounds
such as tryptophan, 5-hydroxytryptophan, 5-hydroxyindol-3-ylacetic acid,

Table 13.6   A summary of the structure-activity potencies of serotonin analogues obtained from a structure-activity study on the heart of *Mercenaria*

Compound	EPMR
Serotonin	1
Tryptamine	9.9
N-Methyltryptamine	3.7
N-Ethyltryptamine	9.1
N'N'-Dimethyltryptamine	10.7
N'N'-Diethyltryptamine	7.9
α-Methyltryptamine	8.6
Bufotenin (N'N'-Dimethylserotonin)	0.028
5-Hydroxy-α-methyltryptamine	6.0
5-Hydroxy-2-methyltryptamine	31.4
5-Methoxy-2-methyltryptamine	43.8
Gramine	>10 000
3-(3-Dimethylaminopropyl)indole	2000

EPMR, equipotent molar ratio. (After Greenberg 1960*b*).

indol-3-ylacetic acid and indole-3-ylpropionic acid were either inert or more than 10 000 times less active than serotonin.

The agonist effect of LSD was of interest since the threshold concentration for an increase in amplitude of the heart is 10^{-16} to 10^{-15}M. This concentration may produce an almost maximum response for this compound which can take up to four hours for completion. A diagramatic representation comparing the LSD and serotonin responses is shown in Fig. 13.3. The effect of LSD on the heart of *Mercenaria* is of interest in terms of its extremely low threshold for activity. This means that the molecule (Fig. 13.4*b*) must fit very precisely into the receptor and indicates the preferred conformation of serotonin for maximal activity. The spatial arrangement of the lysergic acid molecule has been defined by Cookson (1953). The lysergic acid nucleus is planar up to a line joining C5 and C8 across the upper (D) ring; along this line the D ring is folded so that N6 (which corresponds to the primary amino nitrogen atom of serotonin) is out of the plane by 1 to 1.5 Å; the distance between C12 (corresponding to C5 in serotonin) and N6 is between 5.5 and 6.0 Å. So the only unstrained conformation possible for ring D is a boat, Fig. 13.4*c* (Cookson 1953). Of the many possible conformations of serotonin, one exists which satisfies the spatial requirements defined by lysergic acid but this conformation is not the most probable one. However if the free energy barrier between this conformation and the most probable one is between 4000 and 6500 cal/mol, then for every 13^3 to 10^5 molecules with the probable conformation, one will have the planar conformation. Now LSD requires 10^6 to 10^7 fewer molecules than the tryptamine analogues to produce a threshold response.

380   Robert J. Walker

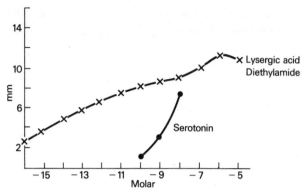

Fig. 13.3 Graph to show partial dose-response curves for lysergic acid and serotonin excitation on the heart of *Mercenaria*. The ordinate represents the increase in amplitude of the heart recorded in mm while the absicca represents the concentration of agonist applied. (After Greenberg 1960*b*)

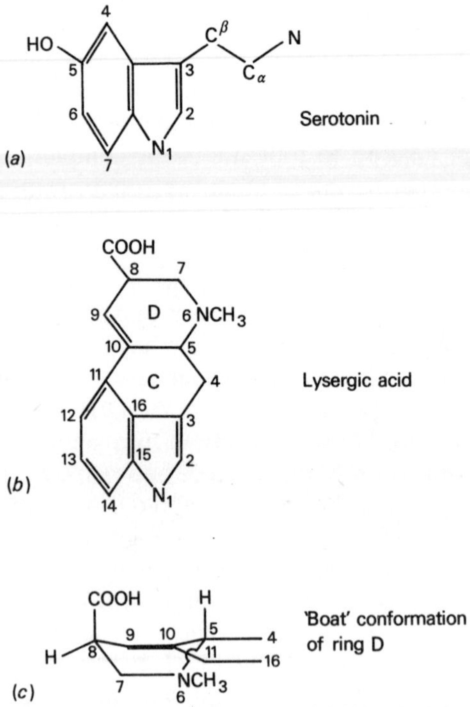

Fig. 13.4 Diagram to compare the structures of serotonin, lysergic acid and the 'boat' conformation of lysergic acid. (After Cookson 1953)

If it is assumed that LSD is 10 to 100 times more potent than these analogues then Greenberg considers that out of 10^3 to 10^5 tryptamine molecules, one would be sufficient to produce the observed effects. Greenberg concludes by stating that the improbability of the proposed conformation can be explained partly by the great difference in the number of molecules required for the LSD response and that to tryptamine analogues. Greenberg states that the serotonin receptor must have two regions; one flat, 11×9 Å to accommodate the indole or benzene rings; secondly a contiguous ovoid depression, 6×4 Å and up to 3.5 Å deep should be present to accommodate the group in the lysergic acid D ring and the terminal amino group of the tryptamine side chain which are folded out of plane of the indole ring.

Another serotonin structure-activity study on the heart of a lamellibranch, *Tapes,* has been performed by Chong and Phillis (1965). They found that removal of the hydroxyl group to give tryptamine reduced potency 13-fold while moving the hydroxy group had a greater effect on potency. For example, 4-HT, 6-HT and 7-HT were 100, 35 and 150 times less potent than serotonin. Substitution of a methoxy group at position 5 of the indole nucleus reduced potency 7-fold while adding a methoxy group to tryptamine at position 6 reduced potency to zero. On this heart bufotenin was only equipotent with serotonin in contrast to the *Mercenaria* heart where it was about 35 times more potent than serotonin.

The actions of a number of derivatives of lysergic acid have been investigated in detail on *Mercenaria* heart by Wright, Moorhead and Welsh (1962). They found that derivatives without substitution in position 1 or 2 of the lysergic acid molecule tended to increase the amplitude and frequency of the heart in a similar manner to serotonin. LSD was the most potent compound in this group which confirms the finding of Greenberg (1960b). Lysergic acid derivatives with substitution in position 1 or 2 had a weak excitatory action or were devoid of agonist activity but possessed specific serotonin antagonist properties. Of the compounds in this group methysergide showed the best antagonist properties, blocking serotonin in a molar ratio of about one to one. Methysergide was also a specific antagonist of LSD with 2-BromoLSD also being a good serotonin antagonist.

Wright *et al.* (1962) have also commented on the very high potency of LSD as an excitant on the *Mercenaria* heart, 10 ml of a 10^{-16}M solution exciting the heart for several hours. They calculated that this solution contained about 600 000 molecules of LSD. The heart of *Mercenaria* is composed of approximately 100 000 muscle fibres (Greenberg 1958) which suggests that 6 molecules of LSD per fibre are sufficient to activate the heart. It is likely that considerably less than this number are required. In addition to the three negative binding sites proposed for serotonin on the *Mercenaria* heart by Greenberg (1960b), Wright *et al.* (1962) suggested that the amino-alcohol of lysergic acid may provide a further binding group, possibly to an adjacent serotonin receptor on the heart muscle. Such a situation might in part account for the high potency of lysergic acid derivatives as serotonin agonists or antagonists.

Evidence for a role for serotonin as an excitatory transmitter onto the heart of *Mercenaria* came from the study of Loveland (1963). He found that methysergide antagonized both the response to applied serotonin and to stimulation of the acceleratory nerve to the heart. Following reserpine administration which decreased endogenous levles of serotonin, stimulation of the acceleratory nerve from such animals failed to activate the heart. Further evidence in support of an excitatory role for serotonin in gastropod hearts came from the experiments of S.-Rozsa and Perenyi (1966). They set up a two heart 'Loewi'-type experiment and found that following stimulation of one heart, the perfusate from this heart contained material which activated the second heart. This activation could be antagonized by 2-BromoLSD. Using spectrophotometry and chromatography, they detected serotonin in the perfusate from the activated heart.

Structure-activity studies on gastropod hearts indicate that tryptamine has a serotonin-like action but is less potent. For example, on the heart of the limpet Patella, tryptamine has a threshold of 10^{-8}M, while the threshold for serotonin is 10^{-9}M (Leake, Evans and Walker 1971). On this preparation 5-methyltryptamine had the same order of potency as tryptamine. Methysergide and 2-BromoLSD both antagonized the action of serotonin and in addition methysergide had a direct inhibitory effect on the heart. There have been two studies on the heart of *Helix*. Bufotenin and 5-methoxytryptamine were found to be more potent than serotonin while the order of potency for other tryptamine analogues was serotonin $>\alpha$-methylserotonin $>$ 4-HT $>$ psilocybin $>$ tryptamine $>$ 6-HT $>N,N$-dimethyltryptamine (Bertaccini and Zamboni 1961). α-Methyltryptamine was inactive. Osborne (1982) found the following order of potency of tryptamine analogues on *Helix* heart: serotonin $\gg$ 5-methoxytryptamine $>$ tryptamine $>$ 5,6-dihydroxytryptamine $>$ 6-HT $>$ 5,7-dihydroxytryptamine.

The effect of serotonin on heart muscle membrane potentials of *Helix* has been investigated by Kiss and S.-Rozsa (1978). They found that serotonin could both depolarize, 33 per cent, and hyperpolarize, 67 per cent, the muscle fibres. They examined the effects of a range of antagonists on these responses. Most of the compounds were tested at a concentration of 100 μM. They found that LSD, nicotine and bufotenin acted as partial competitive antagonists; methysergide, 7-methyltryptamine, ergobasine, tubocurarine and atropine as mixed antagonists; morphine as a noncompetitive antagonist. The hyperpolarizing responses had an equilibrium potential of -40 to -45 mV and appeared to be associated with chloride and potassium. This response was antagonized by LSD, methysergide, tubocurarine, ergobasine, 7-methyltryptamine and atropine. The depolarizing response had an apparent reversal potential of around 0 mV and was probably associated with conductance increases to sodium and calcium ions. It was antagonized by LSD, 7-methyltryptamine, nicotine, morphine and methysergide. The authors concluded that it was possible there are different serotonin receptors mediating the two events.

The central control of the heart and aortae in *Aplysia* by specific motoneurones in the abdominal ganglion has been investigated by Liebeswar, Goldman, Koester and Mayeri (1975). One of these cells, RB-HE, is serotonergic and when activated accelerates the myogenic heart beat, as does serotonin (10^{-9}M) when applied to the heart. LSD (10^{-8}–10^{-6}M), 2-BromoLSD (10^{-6}–10^{-5}M) and methysergide (10^{-6}–2×10^{-5}M) were examined for their ability to antagonize serotonin and RB-HE actions. However these compounds had direct effects on the heart and their degree of block was highly variable. The best antagonism was obtained with cinanserin (2–4×10^{-5}M) which reversibly antagonized both the effect on the heart of activating cell RB-HE and the application of serotonin.

13.5   Other peripheral tissues

In addition to its action on molluscan heart muscle serotonin also acts on a wide range of invertebrate peripheral systems but it is not within the remit of this review to discuss these in detail. However a few examples will be quoted particularly where pharmacological studies have been included in the study. Serotonin, 10^{-8}M threshold, had a potent excitatory action on cilia of lamellibranch gills, e.g. *Modiolus* and *Mytilus* (Gosselin 1961). 2-BromoLSD reduced both the excitatory effect of exogenous serotonin and the excitation following nerve stimulation (Aiello and Guideri 1966). Calcium was required to maintain cilia beating and both serotonin and dopamine mobilized calcium from depots within the gill epithelium (Paparo and Murphy 1975). The cilio-inhibitory effect of excess calcium in the presence of serotonin was enhanced when 2-BromoLSD was added. Low frequency electrical stimulation of the cerebro-visceral connective or application of serotonin to the visceral ganglion increased activity of cilia of *Mytilus* gill (Catapane, Stefano and Aiello 1978). These effects can be antagonized with both 2-BromoLSD and methysergide at concentrations of about 10^{-5}M. Serotonin stimulated the endogenous respiration of gills of the lamellibranch, *Modiolus* and this action was antagonized by 2-BromoLSD. LSD mimicked the action of serotonin but was about 10 times less potent (Moore, Milton and Gosselin 1961).

Serotonin excited the rectum of the lamellibranch *Tapes* (Phillis 1966) and this effect was antagonized by methysergide. Tryptamine was about 40 times less active than serotonin. Moving the hydroxyl group from carbon 5 of the indole nucleus also reduced potency, e.g. 4-HT, 6-HT and 7-HT were 220, 75 and 95 times less active than serotonin. Substitution of a methoxy group for the hydroxyl group on position 5 reduced potency 12-fold while removing the hydroxyl group and adding a methoxy group on position 6, reduced potency some 2000-fold. Bufotenin was equipotent with serotonin on *Tapes* rectum. Serotonin at concentrations in excess of 3×10^{-8}M has been found to excite the rectum of the clam *Mercenarea* (Greenberg and Jegla 1963) an action mimicked by tryptamine and LSD and antagonized by methysergide and 2-BromoLSD. Tryptamine was about 100 times less potent than serotonin, a finding which agreed reasonably well with the potency obtained for *Tapes* mentioned above.

Serotonin and tryptamine excited the rhythmic activity of the glochidia larvae of *Anodonta* (Labos, Salanki and S.-Roza 1964). LSD had a direct excitatory action on the larval activity but it also reduced the excitatory action of tryptamine.

Another molluscan muscle which was excited by serotonin was the penis retractor muscle of *Strophocheilos* (Jaeger 1963). Tryptamine was about 100 times less potent than serotonin while bufotenin was about equipotent with serotonin. LSD activated the muscle while 2-BromoLSD relaxed it and antagonized the action of serotonin. In contrast serotonin relaxed the penis retractor muscle of *Helix* and may be an inhibitory transmitter at this site (Wabnitz and von Wachtendonk 1976).

The action of serotonin on the heart of the cockroach has been investigated by Collins and Miller (1977). They found that serotonin (10^{-6}M) increased the heart rate. Tryptamine was about 50 times less potent than serotonin. The antagonist BW 501c (Wellcome) antagonized the effect of serotonin in an apparently competitive manner.

The anterior byssal retractor muscle (ABRM) of the mussel, *Mytilus,* contracts in the presence of acetylcholine. This contraction can be reversed to a state of relaxation by serotonin, 10^{-7}M (Twarog 1960). It is likely that serotonin acts as a relaxing agent onto the ABRM, being released following stimulation of the nerve innervating this muscle (Satchell and Twarog 1978) and this relaxation may involve a decrease in free intracellular calcium levels (Twarog 1966). LSD is also capable of relaxing the ABRM (Twarog 1959). The structural requirements for this relaxing effect of serotonin have been investigated; bufotenin had a similar potency to serotonin as a relaxing agent while 6-HT and tryptamine were approximately 100 and 1000 times respectively less potent (Twarog and Cole 1972). In addition to serotonin, dopamine also relaxed the ABRM but was about 5 times less potent. The action of 2-BromoLSD, methysergide and an organic mercurial, mersalyl, have been studied as potential antagonists of serotonin and dopamine action on the ABRM (Twarog, Muneoka and

Ledgere 1977). They found that mersalyl was more effective in blocking the relaxing action of serotonin that 2-BromoLSD or methysergide. In contrast, 2-BromoLSD and methysergide were more potent than mersalyl as dopamine antagonists. It is suggested that mersalyl may block serotonin by combining with a sulphydryl group at or near the site on the receptor to which the indole nitrogen attaches.

Serotonin can induce secretion from the salivary glands of a number of insects, including the blowfly, *Calliphora,* and is probably the neurohormone responsible for causing salivary secretion in this species (Berridge 1972; Berridge and Prince 1974; Berridge and Heslop 1982; House 1980). Serotonin is present in the nervous system Berridge and Patel 1968) and is the most potent stimulant of secretion, for example at a concentration of 5×10^{-9}M it increased the rate of fluid secretion from a basic rate of 0.5–1.0 nl min to 40 nl min (Berridge 1970). This action of serotonin can be mimicked by cAMP (10^{-2}M). The structural requirements for this action of serotonin have been investigated by Berridge (1972). Of the ethylamine analogues tested only amylamine showed a strong activation of salivary secretion but was 5×10^{7} times less potent than serotonin. The indole nucleus without the ethylamine side chain had little or no activity and removal of the hydroxyl group from position 5 to give tryptamine resulted in a fall of potency of about 1000. Bufotenin and dimethyltryptamine were about 10 and 10 000 times less active than serotonin. Moving the hydroxyl group to position 4 or 6 also reduced potency by 1000 to 10 000 fold. Replacing the hydroxyl group at position 5 by a chloro, methoxy or methyl group reduced potency by factors of 10, 100 and 1000 respectively. Increasing or decreasing the length of the side chain to give 3-aminopropyl indole and gramine respectively reduced potency in the former case while gramine was inactive. Both compounds were antagonists, gramine being the more potent. From this very thorough study Berridge proposed a serotonin-receptor interaction model. He concluded that for potent serotonin-like activity the receptor required a molecule with a precise interaction between the indole nucleus and a hydrophobic site, hydrogen bonding involving the hydroxyl group at position 5, an electrostatic bond between the positively charged nitrogen on the ethylamine side chain and an anionic site on the receptor.

The action of serotonin on *Calliphora* salivary gland secretion can be mimicked by LSD at a comparable concentration to that of serotonin but the response lasts much longer (Berridge and Prince 1974). The long duration of the response to LSD could be terminated by treating the preparation with serotonin, tryptamine or gramine.

It has been suggested by Berridge (1979) that calcium ions and cAMP act as secondary messengers in the serotonin activation of *Calliphora* salivary glands. Serotonin may act on the basal membrane to open calcium channels and increase the synthesis of cAMP. The former leads to an influx of calcium while the latter leads to the liberation of stored calcium. Both events result in an elevation of intracellular calcium levels which act on the apical membrane of the gland cells to cause release of enzyme, activation of potassium transport and to open chloride channels. Raised intracellular calcium levels also open chloride channels in the basal membrane.

However serotonin is not the most likely candidate associated with the activation of fluid secretion in all insect salivary glands. For a review on this subject the reader is referred to House (1980). In, e.g. the cockroach *Nauphoeta cinerea,* dopamine is probably the transmitter (House and Smith 1980; Ginsborg and House 1980).

Serotonin (10^{-8}M) has been shown to stimulate fluid secretion by the Malpighian tubules of insects, e.g. *Rhodnius prolixus* and *Carausius morsosus* (Maddrell, Pilcher and Gardiner 1971). The structural requirements for this response were very precise since tryptamine, 6-HT, 4-HT, 5-methyltryptamine, 5-methoxytryptamine and 5-chlorotryptamine are inactive. The addition of one or two methyl groups to the nitrogen group of the ethylamine side chain to give *N*-methylserotonin and bufotenin respectively, reduced potency by 10- and 20-fold respectively. Interestingly in this study, it was found that *N*-acetylserotonin did activate the tubules of *Rhodnius* but not *Carausius,* indicating species variation in the structural requirements for activation of this serotonin receptor. Cyclic AMP also stimulated fluid secretion by the Malpighian tubules of both *Rhodnius* and *Carausius.* A number of tryptamine analogues were found to inhibit the stimulation of fluid secretion due to both serotonin and to diuretic hormone including tryptamine, 5-methyltryptamine, 5-methoxytryptamine, 5-chlorotryptamine and 2-BromoLSD as well as tranylcypromine, iproniazid and tyramine. Lysergic acid possessed neither stimulatory nor inhibitory actions at this site.

Serotonin also stimulates secretion from isolated Malpighian tubules of the butterfly, *Papilio demodocus* (Nicolson and Miller 1983), but is less potent than cAMP at this site. Cyproheptadine failed to significantly antagonize the effect of serotonin. In contrast serotonin does not stimulate fluid secretion from locust Malpighian tubules (Rafaeli and Mordue 1982).

Serotonin can influence motility and metabolism in parasitic worms. For example, serotonin stimulated carbohydrate metabolism in the cestode *Hymenolepis* in terms of increased glucose uptake and glycogen utilization but did not appear to alter the levels of glucose and glycolytic intermediates (Rahman, Mettrick and Podesta, 1983). Serotonin enhanced rhythmic contractions and motility in this species and this effect was antagonized by methysergide (Mettrick and Cho 1982). Serotonin also enhanced motility in the liver fluke, *Fasciola,* and 2-BromoLSD antagonized this effect (Mansour 1959). Cell-free particles from *Fasciola* contain a serotonin-sensitive adenylate cyclase (Northup and Mansour 1978*a, b*; Mansour 1979). Serotonin stimulated this enzyme 25–30 fold compared with basal activity, the concentration of serotonin required to produce half maximal stimulation being 2.1 μM. Any substitution on the basic serotonin molecule reduced activity, e.g. removal of the hydroxyl group to give tryptamine reduced potency and addition of *N*-methyl groups reduced potency further. *N',N*-Dimethyltryptamine had almost no agonist activity. *N*-Methylserotonin was slightly less potent than serotonin but the introduction of a second methyl group on the terminal nitrogen to give bufotenin, reduced potency further. Lysergic acid derivatives were poor agonists compared with serotonin but possessed much higher affinity than serotonin. LSD was the most potent, activating maximally about 25 per cent of the maximum serotonin activity. However, half maximal activation by LSD occurred with 46 nM and this action of LSD was stereospecific since the L isomer was devoid of activity. 2-BromoLSD acted as a high affinity antagonist of serotonin but failed to activate the cyclase and in fact lowered the basal level of the enzyme. 2-BromoLSD had a K_i value of 28 nM for serotonin antagonism. Activation of *Fasciola* adenylate cyclase by serotonin or LSD is dependent on the presence of guanine nucleotides (Northup and Mansour 1978*b*). There are high levels of serotonin in the schistosome, *Schistosoma,* but it is likely that this is derived from the host via a high affinity uptake system rather than synthesis by *Schistosoma* tissues (Bennett and Bueding 1973). Tryptamine can also be taken up by this mechanism. Serotonin uptake can be inhibited by 5 μM 8-β [(carbobenzoxyamino)methyl]-1,6-dimethyl-10-α-ergoline but not by 2-BromoLSD, 10 μM. Serotonin has a stimulatory effect on *Schistosoma* motility and this action can be antagonized by methysergide (1 mM) and dihydroergotamine (0.1 mM) (Hillman, Olsen and Senft 1974). Hycanthone increased the level of serotonin in *Schistosoma* apparently via stimulation of a low affinity uptake system for serotonin (Chou, Bennett, Pert and Bueding 1973). This compound appears to reduce the ability of *Schistosoma* to store serotonin in neuronal structures and under these conditions serotonin is localized only in extraneuronal tissues. The action of serotonin on motor activity of *Schistosoma* has been investigated by Tomosky, Bennett and Bueding (1974) who found that serotonin and tryptamine were equipotent as excitatory agents on this preparation. Interestingly moving the hydroxyl group from position 5 of the indole nucleus to give 4-HT, 6-HT or 7-HT all caused a fall in potency compared with serotonin and tryptamine. *N*-Methylserotonin, bufotenin and 5-methoxytryptamine were all intermediate in potency between 4-HT and serotonin. 2-BromoLSD was a potent antagonist of serotonin at this site while LSD was devoid of antagonist properties.

Serotonin facilities field-stimulated responses of earthworm body wall and induces rhythmic, spontaneous contractions, 5-Methoxytryptamine has a similar action but is 9 times less potent (Gardner 1981 *a*, *b*). 7-Methyltryptamine, tryptamine and bufotenin were all equi-active but less potent than serotonin whilst quipazine and TMPP were weak excitants and MK-212 inactive. Fenfluramine and *p*-chloroamphetamine which both release serotonin also excited the preparation but were respectively 42 and 12 times less active than serotonin. Likewise the serotonin uptake blockers fluoxetine and chlorimipramine excited the preparation but were respectively 71 and 39 times less potent than serotonin. In the presence of the monoamine oxidase inhibitor, nialamide, 5-hydroxytryptophan was 87 times less potent than serotonin as an excitant. The antidepressants mepiprazole and trazodone also potentiated field-stimulated responses and induced spontaneous contractions with potency ratios respectively of 38 and 47 compared to serotonin. The author suggested that this preparation might make a useful model for the study of compounds acting on serotonergic transmission.

Serotonin acts at a number of peripheral sites in crustacea. For example it has been shown to increase the rate and strength of the heart beat, facilitated the release of transmitter from nerve endings, and act directly on muscle fibres to produce contractures (Kravitz, Glusman, Harris-Warrick, Livingstone, Schwarz and Goy 1980). Serotonin may be involved in posture control in these animals. Serotonin enhanced the amplitude of excitatory junctional potentials of *Astacus* muscle and this effect was associated with an increase in amplitude of the synaptic current (Fischer and Florey 1983).

13.6   Cyclic nucleotide and binding studies

The distribution of receptors for serotonin and dopamine in *Aplysia* and *Helix* tissues have been analysed using [³H]-LSD binding and adenylate cyclase stimulation techniques (Drummond, Bucher and Levitan 1978; 1980*a*, *b*). Using [³H]-LSD binding Drummond *et al.* (1980*a*) have identified dopamine and serotonin receptors in a particulate fraction derived from the central nervous system of *Helix*. In this system dopamine and serotonin sensitive [³H]-LSD binding can be studied independently and the affinities of a range of agonists and antagonists for these two binding sites determined. Certain dopamine and serotonin sensitive agonists were found to discriminate well between the two populations of binding sites. For example, serotonin, *N*-methylserotonin, 5-methoxytryptamine and bufotenin were more potent against the serotonin than the dopamine site. Interestingly removal of the hydroxyl group from position 5 or moving it to position 6,. removed the ability of the compound to discriminate between the two sites. Apart from apomorphine and piribedil, dopamine derivatives were also able to distinguish between the two sites. Ergometrine showed a preference for the dopamine site but cyprohepta-dine, methysergide and propranolol were unable to show a preference for the serotonin site. In general the neuroleptics examined were unable to show a preference for one site compared to the other. Thus in contrast to the agonists, antagonists were poor at discriminating between the two receptor sites.

Drummond *et al.* (1980*a*) also studied the effect of a range of compounds against serotonin adenylate cyclase activity and for any intrinsic activity of the compounds themselves. The results are summarized in Table 13.7. Out of the compounds examined only *N*-methylserotonin and bufotenin produced a similar degree of cyclase activation. Tryptamine, 5-methoxytryptamine and 6-HT were partial agonists producing respec-tively 36, 63 and 37 per cent of the maximum response to serotonin. Thus either removal or changing the position of the ring hydroxyl group produces a 10- to 20-fold increase in the EC_{50} concentration required for cyclase activation. *d*-LSD was a competitive antagonist with a K_i of 0.01 μM but in contrast to the findings with *Aplysia*, it was not a partial agonist for *Helix* nerve tissue. Quipazine also failed to have an intrinsic activity. There was a good correlation between the ability of a compound to act as an antagonist or agonist in the serotonin sensitive adenylate cyclase assay and their ability to displace serotonin-sensitive [^{3}H]-LSD binding. A number of neuroleptics were inhibitors in a stereoselective manner in both assays. The authors concluded that in gastropod tissues, serotonin sensitive [^{3}H]-LSD

Table 13.7   Effect of compounds on serotonin sensitive adenylate cyclase activity from *Helix* nervous tissue

Compound	Serotonin-sensitive adenylate cyclase activity EC_{50} or K_i nM
Serotonin	1650
N-Methylserotonin	1550
Butotenin	1250
Tryptamine	18 300
5-Methoxytryptamine	6200
6-Hydroxytryptamine	26 000
Dopamine	100 000
Apomorphine	12 000
d-LSD	10
l-LSD	10 000
d-Butaclamol	21
cis-Flupenthixol	100
Quipazine	4000

Only the first 6 compounds possessed any intrinsic activity. (From Drummond *et al.* 1980*a*).

binding is related to the serotonin receptor which is coupled to adenylate cyclase.

In their second paper, Drummond *et al.* (1980*b*) found there was a high level of specific [³H]-LSD binding in all the ganglia and nerves examined. The ability of serotonin and dopamine to inhibit [³H]-LSD binding varied depending on the tissue examined, in muscle most of the binding was sensitive to serotonin, while in the nervous system a number of ganglia contained up to 50 per cent dopamine sensitive binding. The concentration of compounds required to inhibit 50 per cent of the binding [IC_{50}] were obtained for serotonin sensitive [³H]-LSD binding and expressed as K_i values. The results are summarized in Table 13.8, where data from *Aplysia* central nervous system and gill muscle and *Helix* central nervous system have been compared with data from rat central nervous system studies (Bennett and Snyder 1976). Overall, the values are remarkably similar between the gastropod and rat tissues. Drummond *et al.* (1980*b*) found that serotonin stimulated adenylate cyclase approximately three-fold with K_a values ranging from 0.8 to 3 μM. There was a good correlation between the amount of serotonin sensitive [³H]-LSD binding and the amount of serotonin sensitive adenylate cyclase activity in *Aplysia* tissues, which suggests that the two methods are measuring the same serotonin receptor population. *d*-LSD stimulated adenylate cyclase activity in membranes from pedal ganglia while partially inhibiting serotonin stimulation of the enzyme, that is *d*-LSD was exhibiting a partial agonist effect.

There are many examples where cAMP has been implicated in the response associated with serotonin onto invertebrate nerve and muscle cells. Serotonin will stimulate the formation of cAMP in ganglia, connectives and specific cell bodies in the nervous system of *Aplysia* (Cedar

Table 13.8   Serotonin sensitive [³H]-LSD binding in gastropod and rat tissue

Compound	K_i values (μM)			
	Aplysia CNS	Gill muscle	Helix CNS	Rat CNS
Serotonin	2.6	1.5	1.3	2.3
Bufotenin	1.0	1.5	2.1	0.4
5-Methoxytryptamine	2.0	1.5	2.0	2.0
5-Hydroxyindole acetic acid	>100.0	>100.0	>100.0	>100.0
Tryptamine	5.0	17.0	13.0	4.0
6-HT	100.0	240.0	75.0	20.0
d-LSD	0.001	0.00025	0.0012	0.009
l-LSD	7.0	20.0	12.0	20.0

From Drummond *et al.* (1980*b*).

and Schwartz 1972), half maximum activation was obtained with 6 μM serotonin. Both LSD (500 μM) and methysergide (2×10^{-5}M) had little or no effect in terms of blocking this activation of cAMP. Tubocurarine (14 μM) and neostigmine (30 μM) were likewise devoid of blocking action though higher concentrations of both compounds, 140 μM and 300 μM respectively, did appear to reduce serotonin activation although these effects may be non-specific.

Serotonin has been implicated in heterosynaptic facilitation in the central nervous system of *Aplysia* (Shimahara and Tauc 1977). That is, serotonin would be released onto presynaptic endings and enhance the release of transmitter from the test cell onto the follower cell. It is suggested that serotonin activates adenyl cyclases in the test cell nerve terminals, this leading to the synthesis of cAMP. The accumulation of cAMP enhances active permeability to calcium ions, which causes increased calcium inflow during the action potential and the enhanced calcium entry increases the size of the synaptic event recorded in the follower cell.

Serotonin has been shown to hyperpolarize R15 neurone in *Aplysia* (Drummond, Benson and Levitan 1980c). This response is associated with an increase in potassium conductance, with a reversal potential of around −75 to −80 mV, the value of the potassium equilibrium potential. The potassium channel modulated by serotonin is an anomalous or inward rectifier (Benson and Levitan 1983) which is different from the fast transient potassium currents. These effects of serotonin can be mimicked by cAMP applied either extra- or intracellularly. Serotonin can also cause changes in the phosphorylation of specific proteins in cell R15 (Lemos, Novak-Hofer and Levitan 1982). The effect of serotonin on the membrane potential and on adenylate cyclase can be mimicked by bufotenin and N-methylserotonin which were approximately equiactive with serotonin and by 5-methoxytryptamine, tryptamine and 6-HT which were less potent. D-LSD was very potent in stimulating the adenyl cyclase but less potent in inducing hyperpolarisation of the membrane potential, whilst L-LSD is less active than the D-isomer in both systems. *d*-Butaclamol was considerably more potent than its 1-isomer as a serotonin antagonist on both systems. It is of interest pharmacologically that tryptamine and 6-HT were approximately equiactive on both systems and more than 10 times less potent than serotonin.

In a number of *Helix* neurones, including F1, F6, E1, E2, E13 and D2, serotonin can induce an inward current which is associated with a decrease in potassium conductance (an increase in membrane resistance associated with depolarization of the membrane potential) (Deterre, Paupardin-Tritsch, Bockaert and Gerschonfeld 1982). The reversal potential for this current is around −50 mV. This response can also be induced by dopamine and cAMP. When the response in terms of inward current is maximal for one compound, the response to the other two is blocked. Both serotonin and dopamine stimulate adenylate cyclase in these cells and their effects are additive which suggests that the two amines activate the enzyme via two distinct receptors although in this study it was not possible to find specific antagonists for serotonin and dopamine.

Serotonin can also induce a voltage sensitive calcium current in identified cell clusters of LB and RB neurones of *Aplysia* (Pellmar and Carpenter 1980). The maximum amplitude of this current is around −5 to 0 mV and is absent at membrane potentials more negative than −40 mV. In the same neurones intracellular injection of cAMP appears to mimic the action of serotonin (Pellmar and Carpenter 1981). Pellmar (1981) has investigated in detail the possibility that serotonin might activate an adenylate cyclase system in these cells, however these studies cast doubt concerning the link between serotonin and cAMP, e.g. intracellular injection of guanylyl imidodiphosphate which activates cAMP neither mimicked nor enhanced the serotonin inward current. Various phospho-diesterase inhibitors all antagonized the serotonin response while the adenylate cyclase antagonist, dithiobisnitrobenzoic acid had no effect on the serotonin current. Neither tubocurarine nor neostigmine reduced this inward current while methysergide elicited a direct inward current. Thus this investigation failed to provide evidence for a second messenger role for cAMP at this site. Clearly one must interpret with caution the observation that cAMP can mimic the action of a putative transmitter at a particular site and one cannot automatically assume a physiological role.

The action of serotonin has also been studied on transmission from mechanoreceptor sensory neurones of the gill withdrawal reflex of *Aplysia* onto motoneurones and interneurones (Siegelbaum, Camardo and Kandel 1982). Both serotonin and cAMP have a facilitatory effect on this transmission, both compounds close single potassium channels in the cell body of the sensory neurone. These channels are active at resting membrane potential. This closure can be prolonged and complete and can account for the increase in duration of the action potential, calcium influx and transmitter release. Cinanserin (3×10^{-4}M) but not LSD, methysergide or tubocurarine can antagonize this facilitatory effect of serotonin on the gill reflex (Brunelli, Castellucci and Kandel 1976).

It is also probable that the effect of serotonin released from the metacerebral cells in *Aplysia* onto the buccal musculature is via the activation of adenylate cyclase (Weiss, Mandelbaum, Schonberg and Kupferman 1979). If cell-free homogenates are made of the muscle than the addition of serotonin produces a dose related increase in the accumulation of cAMP. If the serotonin-containing cells are stimulated then there is also an increase in cAMP synthesis. Bursts of activity in the buccal ganglion motoneurones produce contractions which are enhanced by analogues of cAMP. In the presence of phosphodiesterase inhibitors, e.g. RO 20-1724, the effect of stimulation of the metacerebral cells on the contractions of the buccal muscles is enhanced.

Weiss and Drummond (1981) examined the stimulation of adenylate cyclase by dopamine and serotonin in a particulate fraction from *Aplysia* gill homogenates. Dopamine augmented activity 3- to 5-fold with an EC_{50} of 10 μM while serotonin increased activity by 15- to 20-fold with an EC_{50} of 1 μM. A structure activity study was made for both amines and that obtained for serotonin is shown in Table 13.9. From this table it can be seen that methoxylation of position 5 caused only a slight loss of potency. Removal of the hydroxyl group resulted in a greater loss in potency and efficacy. Interestingly the presence of two hydroxyl groups destroyed activity. It would appear that for serotonin-like activity positions 5, 6 and 7 of the indole ring and the ethylamine side chain are essential. In addition to dopamine and serotonin both *d*-LSD and ergotamine activated the cyclase. Chlorpromazine, metergoline, 2-BromoLSD, haloperidol and cyproheptadine all inhibited the ability of serotonin to activate the adenylate cyclase with K_i values in the range 0.33 to 2.63 μM.

In their paper Weiss and Drummond (1981) discuss the sites of action of dopamine and serotonin. They confirm the experiments of Kebabian and colleagues (1979) that at sub-maximal concentrations the responses to both amines are additive, however at saturated concentrations of serotonin, dopamine fails to produce further stimulation. Also from their observations that the dose-response curve for ergotamine activation of cyclase in the presence of both metergoline and chlorpromazine appears to contain two components, they suggest there is a population of receptors, possibly serotonin-like, making up 70–75 per cent of total enzyme stimulation and a second population, possibly dopamine-like, making up 25–30

Table 13.9   Effect of compounds on serotonin sensitive adenylate cyclase activity from *Aplysia* gill homogenates

Compound	Serotonin sensitive adenylate activity	
	EC_{50} μM	Relative efficacy
Serotonin	1	1
5-Methoxytryptamine	3.2	1
5-Methoxy *N,N*-dimethyltryptamine	3.2	0.93
Tryptamine	12.6	0.77
N,N-Dimethyltryptamine	7.5	0.8
6-Methoxytryptamine	23.8	0.36
5,7-Dihydroxytryptamine	inactive	—
5-Methoxytryptophol	50	0.21

The relative efficacy is the fraction of the maximum serotonin activation (from Weiss and Drummond 1981).

per cent. They conclude that in *Aplysia* gill there are two separate receptors, one for dopamine and one for serotonin, and that they have overlapping specificity so that serotonin can activate both receptor classes.

Serotonin also activates adenylate cyclase derived from insect nervous tissue, e.g. from cockroach thoracic ganglia (Nathanson and Greengard 1974). However the magnitude of this activation is less than that of dopamine and octopamine although evidence suggests there are separate receptors for each amine. This activation of adenylate cyclase by serotonin can be blocked by several antagonists including phentolamine, propranolol, haloperidol, 2-BromoLSD, LSD and cyproheptadine. It is interesting that at this site LSD is a potent antagonist of serotonin activation with a K_i value of 5 nM. 2-BromoLSD has a similar K_i value and so this system does not appear to distinguish between the two compounds. Serotonin has no effect on the adenylate cyclase from the light organ of the firefly *Photuris* (Nathanson 1979). It has also been reported that *Lumbricus* nervous tissue contains an adenylate cyclase which is activated by serotonin and LSD (Robertson and Osborne 1979), although in the presence of LSD, the maximum activation by serotonin is reduced.

Serotonin would also appear to have a role in the regulation of neuromuscular transmission in the radular muscles of the marine gastropod, *Rapana thomasiana* (Kobayashi and Muneoka 1980; Muneoka and Kobayashi 1980; Fujiwara and Kobayashi 1983). On the radular protractor muscle, octopamine enhances twitch contractions and contractions induced by acetylcholine and L-gluatamate. In contrast, on the radular retractor muscle, octopamine hyperpolarizes the fibre membrane potential and depresses twitch contractions and contractions induced by L-glutamate and acetylcholine. It is suggested that in the protractor muscle, serotonin may be a modulatory neurotransmitter while in the retractor muscle, it is likely to act as an inhibitory transmitter.

Methysergide was found to block the serotonin induced enhancement of the electrical and mechanical responses to stimulation in the protractor muscle. Likewise in the retractor muscle, methysergide blocked the serotonin induced depression of twitch responses. However methysergide did not antagonize the hyperpolarizing action of serotonin on the muscle. Cyclic nucleotides appear to mimic most of the actions of serotonin on the twitch contractions. Fujiwara and Kobayashi (1983) suggest that serotonin acts on the twitch contraction via the activation of an adenylate cyclase and cAMP then activates an electrogenic sodium pump. This receptor can be blocked by methysergide. In addition serotonin can directly hyperpolarize the muscle fibres via an increase in chloride conductance (Kobayashi and Hasimoto 1982).

Serotonin enhances buccal muscle E1 contraction but inhibits buccal muscle E2 contractions in *Aplysia* (Ram, Gole, Shukla and Greenberg

1983). It is probable that cAMP mediates the former, but not the latter, effect of serotonin.

13.7   Discussion

From the studies reviewed in this chapter a number of interesting points emerge. Firstly it is probable that the basic requirements for a molecule to activate the serotonin receptor are universal and constitute an indole or indene nucleus, an ethylamine side chain with minimum substitution on the terminal nitrogen, and a hydroxyl group on position 5 of the indole nucleus. In all cases so far examined, moving the hydroxyl group from position 5, to position 4, 6 or 7 of the indole nucleus has a very considerable effect on potency, again stressing the importance of this site in the molecule for inter-action with the receptor. A possible deviation from this pattern involves the situation where tryptamine and serotonin are equipotent. Here it is probable that the receptor lacks a site requiring activation by the hydroxyl group and so can be considered as a tryptamine rather than a serotonin receptor, e.g. activation of motor activity in *Schistosoma* (Tomosky *et al.* 1974). However, such examples are surprisingly rare and so the number of sites where tryptamine may be a transmitter are likely to be few compared with serotonin. Ionophoretic studies on mammalian cerebral cortex neurones show little correlation between serotonin and tryptamine suggesting separate sites of action (Jones 1982). Tryptamine was found to enhance inhibitory actions of serotonin but to antagonize excitatory actions on mammalian cortical neurones (Jones and Boulton 1980; Jones 1983). This latter observation supports findings from invertebrate neurone studies. Thus further studies are required regarding a physiological role for tryptamine. The potency between serotonin and tryptamine range from tryptamine being 10 times less potent to it being either more than 1000 times less potent or inactive. This latter is the situation on several serotonin sensitive neuronal receptors in both invertebrates and vertebrates, e.g. gastropod neurones and neuronal receptors associated with mouse duo-denum (Drakontides and Gershon 1968). On this latter preparation, tryptamine mimics the action of serotonin on the duodenal muscle as it does on invertebrate muscles, e.g. *Mercenaria* heart muscle and methyser-gide anagonizes the action of serotonin on these muscle cells. It is also likely that tryptamine can act on at least some dopamine receptors, a situa-tion that could also be occurring in the mammalian central nervous system. It is interesting that while tryptamine is over 100 times less potent than serotonin on leech Retzius cells, *N*-methylation increases potency such that *N*-methyltryptamine is 6–7 times more potent than tryptamine. This increase in potency following *N*-methylation also occurs on *Mercenaria* heart tissue where *N*-methyltryptamine is about 3 times more potent than tryptamine. Similarly $N'N$-dimethylation of serotonin can enhance potency, e.g. bufotenin is 35 times more potent than serotonin on Mercenaria heart muscle but is considerably less potent than serotonin on gastropod neurones. In most preparations N-methylserotonin and α-methylserotonin are very similar in potency and mode of action to serotonin.

In terms of binding studies using mammalian tissues it has been proposed that there are two forms of the serotonin receptor, 5-HT$_1$ which has a higher affinity for agonists than antagonists and 5-HT$_2$ receptors which bind preferentially to antagonists (Peroutka and Snyder 1979, 1982 and Leysen, Chapter 4). [^{3}H]-Serotonin labels the 5-HT$_1$ population while [^{3}H]-spiperone and [^{3}H]-mianserin both label 5-HT$_2$ receptors. In addition it has been postulated that there are two populations of 5-HT$_1$ receptors, 5-HT$_{1A}$ and 5-HT$_{1B}$ based on displacement experiments of [^{3}H]-5-HT using spiperone. Spiperone has a high affinity for 5-HT$_{1A}$ receptors and a low affinity for 5-HT$_{1B}$ receptors (Pedigo, Yamamura and Nelson 1981). There have been few binding studies involving serotonin receptors in invertebrate tissues. The study of Dummond *et al.* (1980a) used [^{3}H]-LSD which from mammalian studies binds to both 5-HT$_1$ and 5-HT$_2$ receptors and so does not help in deciding whether these invertebrate serotonin receptors are 5-HT$_1$ or 5-HT$_2$ type. Though a comparison of their studies with similar studies on rat central nervous tissue shows good structure-activity correlation which indicates similar receptor requirements. From the experiments on *Helix* neurones it would appear that the serotonin excitatory type 'A' receptor may be considered as a 5-HT$_2$ receptor. This would appear to be the case in terms of both agonist and antagonist studies since MK-212 is more potent than RU 24969 and serotonin is antagonized by the classical serotonin antagonists. It is also postulated that the excitatory actions of serotonin on mammalian neurones may be of the 5-HT$_2$ type while the inhibitory action may be type 5-HT$_1$ (Peroutka and Snyder 1982). These 5-HT$_1$ receptors are possibly linked to adenylate cyclase and so it would be of interest to see whether any of the invertebrate adenylate cyclase activated systems can be associated with activation of 5-HT$_1$ receptors. However, it should perhaps be pointed out that in rat brain RU 24969 has been reported not to stimulate serotonin sensitive adenylate cyclase (Euvrard and Boissier 1980).

As has been shown in mammalian studies, e.g. on rat raphe neurones where LSD is equipotent with serotonin as an inhibitory agonist (Haigler and Aghajanian 1974), lysergic acid derivatives have clear interactions with invertebrate serotonin receptors. LSD would appear to be a serotonin agonist at a number of sites and activates preparations at very low concentrations, e.g. on the heart muscle of *Mercenaria* although it fails to produce the same maximum response as serotonin. In addition the responses last much longer than do these of serotonin and may take longer in onset. But there are exceptions, e.g. lysergic acid is devoid of activity at insect Malpighian tubules. Both 2-BromoLSD and methysergide have antagonist actions on many invertebrate serotonin receptors including neuronal receptors though not all mammalian neuronal serotonin receptors are blocked by methysergide, e.g. neuronal receptors associated with mouse duodenum (Drakontides and Gerson 1968) and mammalian central inhibitory serotonin 5-HT$_1$ responses. Methysergide can also have direct agonist actions on invertebrate tissues which may obscure any antagonist activity.

The invertebrates, particularly the gastropod molluscs and more recently the leech, provide excellent models for the study of serotonergic transmission. These electrophysiological studies could be extended to include serotonin ligands in current use in mammalian studies. In addition further binding and biochemical studies would yield valuable information on the comparison between mammalian and invertebrate serotonin receptor properties. The present review indicates the wide occurrence and extensive role for serotonin in many aspects of invertebrate physiology including control of heart and other organs, motor control in feeding and body posture, and as a central transmitter involving neuronal control of complex behaviours. It is likely that mechanisms such as inactivation are similar in the invertebrates and vertebrates and such uptake systems in invertebrates warrant further study. Finally cyclic nucleotides have been implicated in the mechanism of action of serotonin at many sites in invertebrates and clearly this is an important mechanism of action for this amine. Fortunately the structural requirements for adenylate cyclase activation are simlar to those for the activation of muscles and neurones indicating similar receptor profiles and a good correlation between the diverse techniques employed.

References

Aiello, E. and Guideri, G. (1966). Relationship between 5-hydroxytryptamine and nerve stimulation of ciliary activity. *J. Pharmac. Exp. Ther.* **154,** 517.

Bennett, J. L. and Bueding, E. (1973). Uptake of 5-hydroxytryptamine by *Schistosoma mansoni. Mol Pharmac.* **9,** 311.

—— and Snyder, S. H. (1976). Serotonin and lysergic acid diethylamide binding in rat brain membranes: relationship to postsynaptic serotonin receptors. *Mol. Pharmac.* **12,** 373.

Benson, J. A. and Levitan, I. B. (1983). Serotonin increases an anomalously rectifying K$^+$ current in the *Aplysia* neuron R15. *Proc. Natl. Acad. Sci. USA* **80,** 3522.

Berridge, M. J. (1970). The role of 5-hydroxytryptamine and cyclic AMP in the control of fluid secretion by isolated salivary glands. *J. Exp. Biol.* **53**, 171.

—— (1972). The mode of action of 5-hydroxytryptamine. *J. Exp. Biol.* **56**, 311.

—— (1979). Relationship between calcium and the cyclic nucleotides in ion secretion. In *Mechanisms of Intestinal Secretion* (ed. H. J. Binder) p. 65. Alan R. Liss, New York.

—— and Heslop, J. P. (1982). Receptor mechanisms mediating the action of 5-hydroxytryptamine. In *Neuropharmacology of insects* (eds. D. Evered, M. O'Connor and J. Whelan) p. 260. Pitman, London.

—— and Patel, N. G. (1968). Insect salivary glands: stimulation of fluid secretion by 5-hydroxytryptamine and adenosine-3′,5′-monophosphate. *Science* **162**, 462.

—— and Prince, W. T. (1974). The nature of the binding between LSD and a 5-hydroxytryptamine receptor: A possible explanation for hallucinogenic activity. *Br. J. Pharmac.* **51**, 269.

Bertaccini, G. and Zamboni, P. (1961). The relative potency of 5-hydroxytryptamine like substances. *Archs. Int. Pharmacodyn. Ther.* **133**, 138.

Bjorklund, A., Horn, A. S., Baumgarten, H. G., Nobin, A. and Schlossberger, H. G. (1975). Neurotoxicity of hydroxylated tryptamines: structure-activity relationships. II. *In vitro* studies on monoamine uptake inhibition and uptake impairment. *Acta physiol. Scand.* suppl. 429, 30.

Bokisch, A. J., Gardner, C. R. and Walker, R. J. (1983). The actions of three mammalian serotonin receptor agonists on *Helix* central neurones. *Br. J. Pharmac.* **80**, 512P.

Brittain, R. T. and Collier, H. O. J. (1957). Antagonism of 5-hydroxytryptamine by dock leaf extract. *J. Physiol.* **135**, 58.

Brunelli, M., Castellucci, V. and Kandel, E. R. (1976). Synaptic facilitation and behavioural sensitization in *Aplysia:* Possible role of serotonin and cyclic AMP. *Science* **194**, 1178.

Catapane, E. J., Stefano, G. B. and Aiello, E. (1978). Pharmacological study of the reciprocal dual innervation of the lateral ciliated gill epithelium by the CNS of *Mytilus edulis* (bivalvia). *J. exp. Biol.* **74**, 101.

Cedar, H. and Schwartz, J. H. (1972). Cyclic adenosine monophosphate in the nervous system of *Aplysia californica*. II. Effect of serotonin and dopamine. *J. gen. Physiol.* **60**, 570.

Chong, G. C. and Phillis, J. W. (1965). Pharmacological studies on the heart of *Tapes watlingi,* a mollusc of the family veneridae. *Br. J. Pharmac.* **25**, 481.

Chou, T-C. T., Bennett, J. L., Pert, C. and Bueding, E. (1973). Effect of Hycanthone and of two of its structural analogues on levels and uptake of 5-hydroxytryptamine in *Schistosoma mansoni*. *J. Pharmac. Exp. Ther.*

186, 408.

Clineschmidt, B. V. (1979). MK-212: a serotonin-like agonist in the CNS. *Gen. Pharmac.* **4,** 287.

Collins, C. and Miller, T. (1977). Studies on the action of biogenic amines on cockroach heart. *J. exp. Biol.* **67,** 1.

Cookson, R. C. (1953). The stereochemistry of alkaloids. *Chem. & Ind.* 337.

Copeland, J. and Gelperin, A. (1983). Feeding and a serotonergic inter-neurone activate an identified autoactive salivary neurone in *Limax maximus. Comp. Biochem. Physiol.* **76A,** 21.

Cottrell, G. A. and Macon, J. B. (1974). Synaptic connections of two symmetrically placed giant serotonin-containing neurones. *J. Physiol.* **236,** 435.

—— and Osborne, N. N. (1970). Subcellular localisation of serotonin in an identified serotonin-containing neurones. *Nature (Lond.)* **225,** 470.

Deterre, P., Paupardin-Tritsch, D., Bockaert, J. and Gerschenfeld, H. M. (1982). CAMP-mediated decrease in K^+ conductance evoked by sero-tonin and dopamine in the same neuron: A biochemical and physiological single-cell study. *Proc. Natl. Acad. Sci. USA* **79,** 7934.

Drakontides, A. B. and Gerson, M. D. (1968). 5-Hydroxytryptamine receptors in the mouse duodenum. *Br. J. Pharmac.* **33,** 480.

Drummond, A. H., Benson, J. A. and Levitan, I. B. (1980c). Serotonin-induced hyperpolarization of an identified *Aplysia* neuron is mediated by cyclic AMP. *Proc. Natl. Acad. Sci. USA* **77,** 5013.

—— Bucher, F. and Levitan, I. B. (1978). LSD labels a novel dopamine receptor in molluscan nervous system. *Nature* **272,** 368.

—— —— —— (1980a). d-[^{3}H]-Lysergic acid diethylamide binding to sero-tonin receptors in the molluscan nervous system. *J. Biol. Chem.* **255,** 6679.

—— —— —— (1980b). Distribution of serotonin and dopamine in *Aplysia* tissues: analysis by [^{3}H]-LSD binding and adenylate cyclase stimulation. *Brain Res.* **184,** 163.

Ehinger, B., Falck, B. and Myhrberg, H. E. (1968). Biogenic amines in *Hirudo medicinalis. Histochemie* **15,** 140.

Eisenstadt, M., Goldman, J. E., Kandel, E. R., Koike, H., Koester, J. and Schwartz, J. H. (1973). Intrasomatic injection of radioactive precursors for studying transmitter synthesis in identified neurones of *Aplysia californica. Proc. Natl. Acad. Sci. USA* **70,** 3371.

Euvrard, L. and Boissier, J. R. (1980). Biochemical assessment of the central agonist activity of RU 24969 (a piperidinyl-indole). *Eur. J. Pharmac.* **63,** 65.

Fange, R. (1955). Use of the isolated heart of a fresh-water mussel (*Anodonta cygnea* L.) for biological estimation of 5-hydroxytryptamine.

Experientia **11**, 156.

Fischer, L. and Florey, E. (1983). Modulation of synaptic transmission and excitation-contraction coupling in the opener muscle of the crayfish, *Astacus leptodactylus*, by 5-hydroxytryptamine and octopamine. *J. exp. Biol.* **102**, 187.

Fuchs, P. A., Henderson, L. P. and Nicholls, J. G. (1982). Chemical transmission between individual Retzius and sensory neurones of the leech in culture. *J. Physiol.* **323**, 195.

Fujiwara, M. and Kobayashi, M. (1983). Modulation of neuromuscular transmission by serotonin in the molluscan radular muscles: involvement of cyclic nucleotides. *Comp. Biochem. Physiol.* **75C**, 239.

Gaddum, J. H. and Paasonen, M. K. (1955). The use of some molluscan hearts for the estimation of 5-hydroxytryptamine. *Br. J. Pharmac.* **10**, 474.

Gardner, C. R. (1981*a*). Effect of some antidepressants and flurazepam on an invertebrate model of 5-HT neurotransmission. *Drug Develop. Res.* **1**, 245.

——— (1981*b*). Effect of neurally active amino acids and monoamines on the neuromuscular transmission of *Lumbricus terrestris. Comp. Biochem. Physiol.* **68C**, 85.

——— and Guy, A. P. (1983). Behavioural effects of RU 24969, a 5-HT$_1$ receptor agonist, in the mouse. *Br. J. Pharmac.* **78**, 96P.

——— and Walker, R. J. (1982). The roles of putative neurotransmitters and neuromodulators in annelids and related invertebrates. *Prog. Neurobiol.* **18**, 81.

Gelperin, A. (1975). An identified serotonergic neurone has recriprocal effects on two electrically coupled motoneurones in the terrestrial slug, *Limax maximus. Biol. Bull.* **149**, 426.

——— (1981). Synaptic modulation by identified serotonin neurones. In *Serotonin neurotransmission and behaviour* (eds. B. Jacobs and A. Gelperin) p. 288. MIT Press, Cambridge, Mass.

Gerschenfeld, H. M. and Paupardin-Tritsch, D. (1974*a*). Ionic mechanisms and receptor properties underlying the responses of molluscan neurones to 5-hydroxytryptamine. *J. Physiol.* **243**, 427.

——— ——— (1974*b*). On the transmitter function of 5-hydroxytryptamine at excitatory and inhibitory monosynaptic junctions. *J. Physiol.* **243**, 457.

——— and Stefani, E. (1966). An electrophysiological study of 5-hydroxytryptamine receptors of neurones in the molluscan nervous system. *J. Physiol.* **185**, 684.

——— and Tauc, L. (1961). Pharmacological specificities of neurones in an elementary nervous system. *Nature,* **189**, 924.

——— Hamon, M. and Paupardin-Tritsch, D. (1978). Release of endogenous serotonin from two identified serotonin-containing neurones and the physiological role of serotonin re-uptake. *J. Physiol.* **274,** 265.

Ginsbord, B. L. and House, C. R. (1980). Electrical responses of cockroach salivary gland acinar cells mediated by receptor activation. In *Receptors for neurotransmitters, hormones and pheromones in insects* (eds. D. B. Sattelle, L. M. Hall and J. G. Hildebrand) p. 185. Elsevier/North Holland, Amsterdam.

Goldstein, R. S., Weiss, K. R. and Schwartz, J. H. (1982). Intraneuronal injection of horseradish peroxidase labels glial cells associated with the axons of the giant metacerebral neuron of *Aplysia. J. Neurosci.* **2,** 1567.

Gosselin, R. E. (1961). The cilioexcitatory activity of serotonin. *J. Cell. Comp. Physiol.* **58,** 17.

Green, A. R., Guy, A. P. and Gardner, C. R. (1984). The behavioural effects of RU 24969, a suggested 5-HT receptor agonist in rodents and the effect on the behaviour of treatment with antidepressants. *Neuropharmacology* **23,** 655.

Greenberg, M. J. (1958). *The action of indoles on the heart of* Venus mercenaria. Ph.D. Thesis. Harvard University, Cambridge, Mass.

—— (1960*a*). The responses of the *Venus* heart to catecholamines and high concentrations of 5-hydroxytryptamine. *Br. J. Pharmac.* **15,** 365.

—— (1960*b*). Structure-activity relationship of tryptamine analogues on the heart of *Venus mercenaria. Br. J. Pharmac.* **15,** 375.

—— and Jegla, T. C. (1963). The action of 5-hydroxytryptamine and acetylcholine on the rectum of the Venus clam, *Mercenaria mercenaria. Comp. Biochem. Physiol.* **9,** 275.

Haigler, H. J. and Aghajanian, G. K. (1974). Lysergic acid diethylamide and serotonin: a comparison of effects on serotonergic neurones and neurones receiving a serotonergic input. *J. Pharmac. Exp. Ther.* **188,** 688.

Henderson, L. P. (1983). The role of 5-hydroxytryptamine as a transmitter between identified leech neurones in culture. *J. Physiol.* **339,** 309.

Hill, R. B. and Welsh, J. H. (1966). Heart, circulation and blood cells. In *Physiology of mollusca,* Vol. II, p. 125 (eds. K. M. Wilbur and C. M. Yonge) Academic Press, New York.

Hillman, G. R., Olsen, N. J. and Senft, A. W. (1974). Effect of methysergide and dihydroergotamine on *Schistosoma mansoni. J. Pharmac. Exp. Ther.* **188,** 529.

House, C. R. (1980). Physiology of invertebrate salivary glands. *Biol. Rev.* **55,** 417.

—— and Smith, R. K. (1980). Receptors mediating fluid secretion from the cockroach salivary gland. In *Receptors for neurotransmitters, hormones and pheromones in insects* (eds. D. B. Sattelle, L. M. Hall and J. G. Hildebrand) p. 175. Elsevier/North Holland, Amsterdam.

Hunt, P., Nedelec, L., Euvrard, C. and Boissier, J. R. (1981). Tetrahydro-pyridinyl indole derivatives as serotonin analogues which may differentiate between two distinct receptor sites. Abst. No. 1434. *Proc. 8th Int. Cong. Pharmac., Tokyo.*

Jaeger, C. P. (1963). Physiology of Mollusca. IV. Action of serotonin on the penis retractor muscle of *Strophocheilos oblongus. Comp. Biochem. Physiol.* **8,** 131.

Janssen, P. A. J. (1983). 5-HT$_2$ receptor blockade to study serotonin-induced pathology. *Trends Pharmac. Sci.* **4,** 198.

Jaques, R. and Schachter, M. (1954). The presence of histamine, 5-hydroxytryptamine and a potent slow-contracting substance in wasp venom. *Br. J. Pharmac.* **9,** 53.

Jones, R. S. G. (1982). A comparison of the responses of cortical neurones to iontophoretically applied tryptamine and 5-hydroxytryptamine in the rat. *Neuropharmacology* **21,** 209.

—— (1983). Trace biogenic amines: a possible functional role in the CNS. *Trends Pharmac. Sci.* **4,** 426.

Jones, R. S. G. and Boulton, A. A. (1980). Tryptamine and 5-hydroxytryptamine: Actions and interactions on cortical neurones in the rat. *Life Sci.* **27,** 1849.

Kebabian, P. R., Kebabian, J. W. and Carpenter, D. O. (1979). Regulation of cyclic AMP in heart and gill of *Aplysia* by the two putative neurotransmitters dopamine and serotonin. *Life Sci.* **24,** 1757.

Kerkut, G. A. and Laverack, M. J. (1960). A cardio-accelerator present in tissue extracts of the snail, *Helix aspersa. Comp. Biochem. Physiol.* **1,** 62–71.

—— and Walker, R. J. (1961). The effect of drugs on the neurones of the snail *Helix aspersa. Comp. Biochem. Physiol.* **3,** 143.

—— —— (1962). The specific chemical sensitivity of *Helix* nerve cells. *Comp. Biochem. Physiol.* **7,** 277.

—— —— (1967). The action of acetylcholine, dopamine and 5-hydroxytryptamine on the spontaneous activity of the cells of Retzius of the leech, *Hirudo medicinalis. Br. J. Pharmac.* **30,** 644.

—— Sedden, C. B. and Walker, R. J. (1967). Cellular localisation of monoamines by fluorescence microscopy in *Hirudo medicinalis* and *Lumbricus terrestris. Comp. Biochem. Physiol.* **21,** 687.

Kiss, T. and S.-Rozsa, K. (1978). Pharmacological properties of 5-HT receptors of the *Helix pomatia* L. (Gastropoda) heart muscle cells. *Comp. Biochem. Physiol.* **61C,** 41.

Kobayashi, M. and Hasimoto, T. (1982). Antagonistic responses of the radular protractor and retractor to the same putative transmitters. *Comp. Biochem. Physiol.* **72C,** 343.

—— and Muneoka, Y. (1980). Modulatory actions of octopamine and serotonin on the contraction of buccal muscles in *Rapana thomasiana*-I.

Enhancement of contraction in radula protractor. *Comp. Biochem. Physiol.* **65C,** 73.

Kravitz, E. A., Glusman, S., Harris-Warrick, R. M., Livingstone, M. S., Schwarz, T. S. and Goy, M. F. (1980). Amines and a peptide as neurohormones in lobsters: Actions on neuromuscular preparations and preliminary behavioural studies. *J. exp. Biol.* **89,** 15.

Labos, E., Salanki, J. and S.-Rozsa, K. (1964). Effect of serotonin and other bioactive agents on the rhythmic activity in the glochidia of freshwater mussel *Anodonta cygnea* L. *Comp. Biochem. Physiol.* **11,** 161.

Leake, L. D., Evans, T. G. and Walker, R. J. (1971). The role of catecholamines and 5-hydroxytryptamine on the heart of *Patella vulgata*. *Comp. gen. Pharmac.* **2,** 151.

—— and Walker, R. J. (1980). *Invertebrate neuropharmacology,* pp. 358. Blackie, Glasgow.

Lemos, J. R., Novak-Hofer, I. and Levitan, I. B. (1982). Serotonin alters the phosphorylation of specific proteins inside a single living nerve cell. *Nature (Lond.)* **298,** 64.

Lent, C. M. (1973). Retzius cells: Neuronal effectors controlling mucus release by the leech. *Science* **179,** 693.

—— (1977). The Retzius cells within the central nervous system of leeches. *Prog. Neurobiol.* **8,** 81.

—— (1982). Serotonin-containing neurones within the segmental nervous system of the leech. In *Biology of serotonergic transmission* (ed. N. N. Osborne) p. 431. John Wiley, Chichester.

Liebeswar, G., Goldman, J. E., Koester, J. and Mayeri, E. (1975). Neural control of circulation in *Aplysia*. III. Neurotransmitters. *J. Neurophysiol.* **38,** 767.

Loveland, R. E. (1963). 5-Hydroxytryptamine, the probable mediator of excitation in the heart of *Mercenaria (Venus) mercenaria. Comp. Biochem. Physiol.* **9,** 95.

McAdoo, D. J. (1978). The Retzius cell of the leech, *Hirudo medicinalis.* In *Biochemistry of characterized neurons* (ed. N. N. Osborne) p. 19. Pergamon Press, Oxford.

Maddrell, S. H. P., Pilcher, D. E. M. and Gardiner, B. O. C. (1971). Pharmacology of the Malpighian tubules of *Rhodnius* and *Carausius:* The structure-activity relationship of tryptamine analogues and the role of cAMP. *J. exp. Biol.* **54,** 779.

Mansour, T. E. (1959). The effect of serotonin and related compounds on the carbohydrate metabolism of the liver fluke, *Fasciola hepatica. J. Pharm. Exp. Ther.* **126,** 212.

—— (1979). Chemotherapy of parasitic worms: New biochemical strategies. *Science* **205,** 462.

Marsden, C. A. and Kerkut, G. A. (1969). Fluorescence microscopy of the

404   Robert J. Walker

5-HT and catecholamine-containing cells in the CNS of the leech, *Hirudo medicinalis. Comp. Biochem. Physiol.* **31**, 851.

Mason, A. and Leake, L. D. (1978). Morphology of leech Retzius cells demonstrated by intracellular injection of horseradish peroxidase. *Comp. Biochem. Physiol.* **61A**, 213.

—— Sunderland, A. J. and Leake, L. D. (1979). Effects of leech Retzius cells on body wall muscles. *Comp. Biochem. Physiol.* **63C**, 359.

Mettrick, D. F. and Cho, C. H. (1982). Changes in tissue and intestinal serotonin (5-HT) levels in the laboratory rat following feeding and the effect of 5-HT inhibition on the migratory response of *Hymenolepis diminuta* (Cestoda). *Can. J. Zool.* **60**, 790.

Moore, K. E., Milton, A. S. and Gosselin, R. E. (1961). Effect of 5-Hydroxytryptamine on the respiration of excised lamellibranch gill. *Br. J. Pharmac.* **17**, 278.

Muneoka, Y. and Kobayashi, M. (1980). Modulatory actions of octopamine and serotonin on the contraction of buccal muscles of *Rapana thomasiana*-II. Inhibition of contraction in radula retractor. *Comp. Biochem. Physiol.* **65C**, 81.

Nathanson, J. A. (1979). Octopamine receptors, adenosine 3′,5′-monophosphate and neural control of firefly flashing. *Science* **203**, 65.

Nathanson, J. A. and Greengard, P. (1974). Serotonin-sensitive adenylate cyclase in neural tissue and its similarity to the serotonin receptors: a possible site of action of lysergic acid diethylamide. *Proc. Nat. Acad. Sci. USA* **71**, 797.

Nicolson, S. W. and Millar, R. P. (1983). Effect of biogenic amines and hormones on butterfly Malpighian tubules: dopamine stimulates fluid secretion. *J. insect Physiol.* **29**, 611.

Northup, J. K. and Mansour, T. E. (1978a). Adenylate cyclases from *Fasciola hepatica*. 1. Ligand specificity of adenylate cyclase-coupled serotonin receptors. *Mol. Pharmac.* **14**, 804.

—— —— (1978b). Adenylate cyclases from *Fasciola hepatica*. 2. Role of guanine nuclotides in coupling adenylate cyclase and serotonin receptors. *Mol. Pharmac.* **17**, 820.

Osborne, N. N. (1978). The neurobiology of a serotonergic neuron. In *Biochemistry of characterized neurons* (ed. N. N. Osborne) p. 47. Pergamon Press, Oxford.

—— (1982). Assay, distribution and functions of serotonin in nervous tissues. In *Biology of serotonergic transmission* (ed. N. N. Osborne) p. 7. John Wiley, Chichester.

Paparo, A. and Murphy, J. A. (1975). The effect of calcium on the beating of lateral cilia in *Mytilus edulis*. I. A response to perfusion with 5-HT, DA, BOL and PBZ. *Comp. Biochem. Physiol.* **50C**, 9.

Paupardin-Tritsch, D. and Gerschenfeld, H. M. (1973). Transmitter role of serotonin in identified synpases in *Aplysia* nervous system. *Brain Res.* **58**, 529.

Pedigo, N. W., Yamamura, H. I. and Nelson, D. L. (1981). Discrimination of multiple [³H]-5-hydroxytryptamine binding sites by the neuroleptic spiperone in rat brain. *J. Neurochem.* **36**, 220.

Pellmar, T. C. (1981). Does cyclic 3′,5′-adenosine monophosphate act as a second messenger in a voltage-dependent response to 5-hydroxytryptamine in *Aplysia? Br. J. Pharmac.* **74**, 747.

—— and Carpenter, D. O. (1980). Serotonin induces a voltage sensitive calcium current in neurons of *Aplysia californica. J. Neurophysiol.* **44**, 423.

—— and Carpenter, D. O. (1981). Cyclic AMP induces a voltage-dependent current in neurones of *Aplysia californica. Neurosci. Letts.* **22**, 151.

Pentreath, V. W., Berry, M. S. and Osborne, N. N. (1982). The serotonergic cerebral cells in gastropods. In *Biology of serotonergic transmission* (ed. N. N. Osborne) p. 457. John Wiley, Chichester.

Peroutka, S. J. and Snyder, S. H. (1979). Multiple serotonin receptors: differential binding of ³H-serotonin, ³H-lysergic acid diethylamide and ³H-spiroperidol. *Mol. Pharmac.* **16**, 687.

—— and Snyder, S. H. (1982). Radioactive ligand binding studies: Identification of multiple serotonin receptors. In *Biology of serotonergic transmission* (ed. N. N. Osborne) p. 279. John Wiley, Chichester.

Phillis, J. W. (1966). Regulation of rectal movements in *Tapes watlingi. Comp. Biochem. Physiol.* **17**, 909.

Poll, H. and Somner, A. (1903). Über phaeochrome Zellen im Centralnervensystem des Blutegels. *Arch. Anat. Physiol.* **10**, 549.

Rafaeli, A. and Mordue, W. (1982). The responses of the Malpighian tubules of *Locusta* to hormones and other stimulants. *Gen. & Comp. Endocrin.* **46**, 130.

Rahman, M. S., Mettrick, D. F. and Podesta, R. B. (1983). Effect of 5-Hydroxytryptamine on carbohydrate metabolism in *Hymenolepis diminuta* (Cestoda). *Can. J. Physiol. Pharmac.* **61**, 137.

Ram, J. L., Gole, D., Shukla, U. and Greenberg, L. (1983). Serotonin-activated adenylate cyclase and the possible role of cyclic AMP in modulation of buccal muscle contraction in *Aplysia. J. Neurobiol.* **14**, 113.

Robertson, H. A. and Osborne, N. N. (1979). Putative neurotransmitters in the annelid central nervous system: presence of 5-hydroxytryptamine and octopamine-stimulated adenylate cyclase. *Comp. Biochem. Physiol.* **64C**, 7.

Rude, S. (1969). Monoamine-containing neurones in the central nervous system and peripheral nerves of the leech, *Hirudo medicinalis. J. comp.*

Neurol. **136,** 349.

—— Coggeshall, R. E. and van Orden, L. S. (1969). Chemical and ultra-structural identification of 5-hydroxytryptamine in an identified neurone. *J. Cell. Biol.* **41,** 832.

Satchell, D. G. and Twarog, B. M. (1978). Identification of 5-hydroxy-tryptamine (serotonin) release from the anterior byssus retractor muscle of *Mytilus californicus* in response to nerve stimulation. *Comp. Biochem. Physiol.* **59C,** 81.

Schroder, H. U., Neuhoff, V., Priggemeir, E. and Osborne, N. N. (1979). The influx of tryptamine into snail *(Helix pomatia)* ganglia: comparison with 5-HT. *Malacologia,* **18,** 517–525.

Schwartz, J. H. and Shkolnick, L. J. (1981). The giant serotonergic neuron of *Aplysia:* A multi-targeted nerve cell. *J. Neurosci.* **1,** 606.

Sedden, C. B., Walker, R. J. and Kerkut, G. A. (1968). The localisation of dopamine and 5-hydroxytryptamine in neurones of *Helix aspersa. Symp. zool. Soc. Lond.* **22,** 19.

Shimahara, A. and Tauc, L. (1977). Cyclic AMP induced by serotonin modulates the activity of an identified synapse in *Aplysia* by facilitating the active permeability to calcium. *Brain Res.* **127,** 168.

Siegelbaum, S. A., Camardo, J. S. and Kandel, E. R. (1982). Serotonin and cAMP close single potassium channels in *Aplysia* sensory neurones. *Nature (Lond.)* **299,** 413.

Smith, P. A., Sunderland, A. J., Leake, L. D. and Walker, R. J. (1975). Cobalt staining and electrophysiological studies of Retzius cells in the leech, *Hirudo medicinalis. Comp. Biochem. Physiol.* **51A,** 655.

—— and Walker, R. J. (1974). The action of 5-hydroxytryptamine and related compounds on the activity of Retzius cells of the leech, *Hirudo medicinalis. Br. J. Pharmac.* **51,** 21.

—— —— (1975). Further studies on the action of various 5-HT agonists and antagonists on the receptors of neurones from the leeches, *Hirudo medicinalis* and *Haemopis sanguisuga. Comp. Biochem. Physiol.* **51C,** 195.

S.-Rozsa, K. and Perenyi, L. (1966). Chemical identification of the excita-tory substance released in Helix heart during stimulation of the extracardial nerve. *Comp. Biochem. Physiol.* **19,** 105.

Sunderland, A. J., Leake, L. D. and Walker, R. J. (1979). The ionic mechanism of the dopamine response in Retzius cells of two leech species *(Hirudo medicinalis* and *Haemopis sanguisuga). Comp. Biochem. Physiol.* **63C,** 129.

—— —— —— (1980). Evidence for an amine receptor on the Retzius cells of the leeches *Hirudo medicinalis* and *Haemopis sanguisuga. Comp. Biochem. Physiol.* **67C,** 159.

—— —— —— (1982). Structure-activity studies of the amine receptor on the Retzius cells of the leeches *Hirudo medicinalis* and *Haemopis sanguisuga. Comp. Biochem. Physiol.* **73C**, 347.

Tomosky, T. K., Bennett, J. L. and Bueding, E. (1974). Tryptaminergic and dopaminergic responses of *Schistosoma mansoni. J. Pharmac. Exp. Ther.* **190**, 260.

Twarog, B. M. (1959). The pharmacology of a molluscan smooth muscle. *Br. J. Pharmac.* **14**, 404.

—— (1960). Effects of acetylcholine and 5-hydroxytryptamine on the contraction of a molluscan smooth muscle. *J. Physiol.* **152**, 236.

—— (1966). Catch and mechanism of action of 5-hydroxytryptamine on molluscan muscle: a speculation. *Life Sci.* **5**, 1201.

—— and Cole, R. A. (1972). Relaxation of catch in a molluscan smooth muscle. III. Effects of serotonin, dopamine and related compounds. *Comp. Biochem. Physiol.* **43**, 331.

—— and Page, I. H. (1953). Serotonin content of some mammalian tissues and urine and a method for its determination. *Am. J. Physiol.* **175**, 157.

—— Muneoka, Y. and Ledgere, M. (1977). Serotonin and dopamine as neurotransmitters in *Mytilus:* block of serotonin receptors by an organic mercurial. *J. Pharm. Exp. Ther.* **201**, 350.

Wabnitz, R. W. and von Wachtendonk, D. (1976). Evidence for serotonin (5-hydroxytryptamine) as transmitter in the penis retractor muscle of *Helix pomatia. Experientia* **32**, 707.

Walker, R. J. (1982). Current trends in invertebrate neuropharmacology. *Verh. Dtsch. Zool. Ges.* **75**, 31.

—— and Kerkut, G. A. (1978). The first family (Adrenaline, noradrenaline, dopamine, octopamine, tyramine, phenylethanolamine and phenylethylamine). *Comp. Biochem. Physiol.* **61C**, 261.

—— and Smith, P. A. (1973). The ionic mechanism for 5-HT inhibition on Retzius cells of the leech, *Hirudo medicinalis. Comp. Biochem. Physiol.* **45A**, 979.

—— and Woodruff, G. N. (1972). The effect of bufotenine, psilocybin and related compounds on the 5-hydroxytryptamine excitatory receptors of *Helix aspersa* neurones. *Comp. gen. Pharmac.* **3**, 27.

Walz, W. and Schlue, W. R. (1982). Ionic mechanism of hyperpolarising 5-Hydroxytryptamine effect on leech neuropile glial cells. *Brain Res.* **250**, 111.

Weinreich, D., McCaman, M. W., McCaman, R. E. and Vaughan, J. (1973). Chemical enzymatic and ultrastructural characterisation of 5-hydroxytryptamine-containing neurones from the ganglia of *Aplysia californica* and *Tritonia diomedia. J. Neurochem.* **20**, 969.

Weiss, K. R., Cohen, J. and Kupferman, I. (1978). Modulatory control of buccal musculature by a serotonergic neurone (metacerebral cell) in *Aplysia. J. Neurophysiol.* **41**, 181.

—— Mandelbaum, D. E., Schonberg, M. and Kupferman, I. (1979). Modulation of buccal muscle contractility by serotonergic metacerebral cells in *Aplysia:* Evidence for a role of cyclic adenosine monophosphate. *J. Neurophysiol.* **42,** 791.

Weiss, S. and Drummnnd, G. I. (1981). Dopamine and serotonin sensitive adenylate cyclase in the gill of *Aplysia californica. Mol. Pharmac.* **20,** 592.

Welsh, J. H. and Moorhead, M. (1960). The quantitative distribution of serotonin in invertebrates, especially in their nervous system. *J. Neurochem.* **6,** 146.

Wilkens, L. A. and Greenberg, M. J. (1973). Effects of acetylcholine and 5-hydroxytryptamine and their ionic mechanisms of action on the electrical and mechanical activity of molluscan heart smooth muscle. *Comp. Biochem. Physiol.* **45A,** 637.

Willard, A. L. (1981). Effects of serotonin on the generation of the motor programme for swimming by the medicinal leech. *J. Neurosci.* **1,** 936.

Wright, A. C., Moorhead, M. and Welsh, J. H. (1962). Actions of derivatives of lysergic acid on the heart of *Venus mercenaria. Br. J. Pharmac.* **18,** 440.

Wright, N. J. D. and Walker, R. J. (1984). The possible site of action of 5-hydroxytryptamine, 6-hydroxytryptamine and dopamine on identified neurones in the central nervous system of the snail, *Helix aspersa. Comp. Biochem. Physiol.* **78C**, 217.

Zetler, G. and Schlosser, L. (1954). Über das Vorkommen von 5-Hydroxytryptamine (Entcramin oder Serotonin) im Gehirn von Säugetieren. *Arch. exptl. Pathol. u. Pharmakol. Naunyn-Schmiedeberg's* **222**, 345.

14

Serotonin neuropharmacology: a
review of some current research and
clinical implications

A. RICHARD GREEN

14.1 Introduction

The preceding 13 chapters in this monograph have discussed provocatively
various areas of neuropharmacological research. In this chapter I will
review some of the points that have particularly interested me in editing this
volume and suggest how some of the findings may have clinical implications.

14.2 Serotonin binding sites

Undoubtedly the major 'event' in serotonin neuropharmacology over the
last five years has been the identification of serotonin binding sites which
may be indicative of receptor sub-types. Unfortunately, only too often in
neuropharmacology these days, binding sites are immediately equated with
receptors, a trap pointed out by Leysen (Section 4.1). Before the advent of
ligand-receptor binding, receptor sub-types were identified by the functional responses of physiological systems to specific drugs (for example
nicotinic and muscarinic acetylcholine receptors, α- and β-adrenoceptor
sub-types and even the 'D' and 'M' serotonin receptors in peripheral
tissues). In contrast receptor sub-types are now often 'identified' on the
basis of binding data; with no reference being made at all to the function of
the sites. When one examines the methodology of ligand binding the pitfalls
of ascribing a role to the identified sites become alarmingly clear. For
example, ligand binding methods often involve some fairly harsh techniques being applied to the tissues such as freezing, thawing and detergent
addition. Receptor sub-types could, therefore, be the product of the
preparative techniques.

Historically the development and use of ligand-receptor binding techniques occurred with a marked surge of interest in the dopamine receptor and the effects of neuroleptics. The confusion engendered by the 'identification' of dopamine receptor sub-types by the sole use of ligand binding techniques is illustrated by Table 14.1 which is taken from the review of Jenner and Marsden (1983). It can be seen that characterization is by *in vitro* binding techniques and not by function. It is to be hoped that serotonin neuropharmacologists will not get themselves into the same mess!

Happily there are indicators that pharmacologists working on serotonin sub-types are trying to obtain functional correlates with binding sub-types

Table 14.1   Current classification of brain dopamine receptors based on ligand binding experiments

Classification	Definition
D1	Adenylate cyclase-linked dopamine receptors
D2	Adenylate cyclase-independent dopamine receptors
Agonist sites	High affinity for dopamine agonist but low affinity for agonists
Antagonist sites	Low affinity for dopamine agonists but high affinity for dopamine antagonists
D1	Adenylate cyclase-linked dopamine receptors
D2	Adenylate cyclase-independent dopamine receptors
α	—guanine nucleotide-regulated
β	—guanine nucleotide-independent
D3	Adenylate cyclase-independent dopamine receptors with high affinity for dopamine agonists but low affinity for antagonists
D1	Adenylate cyclase-linked dopamine receptors
D2	Adenylate cyclase-independent dopamine receptors with high affinity for both agonists and antagonists
D3	Adenylate cyclase-independent dopamine receptors with high affinity for agonists but low affinity for antagonists
D4	Adenylate cyclase-independent dopamine receptors with low affinity for agonists but high affinity for antagonists
Sodium-dependent	Adenylate cyclase-independent dopamine receptors where neuroleptic interaction is critically dependent on presence of sodium ions
Sodium-independent	Adenylate cyclase-independent dopamine receptors where neuroleptic interaction is independent of sodium ions

Reproduced from Jenner and Marsden (1983) with permission from Excerpta-Medica.

identified by radioligand binding studies. However, for such studies to be successful appropriate specific sub-type agonists and antagonists are required. Some are becoming available, for example, the suggested 5-HT$_1$ agonists RU 24969 and 8-OH-DPAT and produce functional effects (see Sections 12.5, 12.6 and below). However, information on the binding characteristics of these drugs is sparse at present and there are indications that the binding may be unconventional (Section 4.3.2). There are no specific 5-HT$_2$ agonists. With regard to antagonists there are selective 5-HT$_2$ antagonists, such as ketanserin, pirenperone and ritanserin although the first two at any rate have quite high affinities for other monoaminergic receptors if not the 5-HT$_1$ site. Suggested selective 5-HT$_1$ antagonists are the (−)-isomers of β-adrenoceptor antagonists (Nahorski and Willcocks 1983; Middlemiss and Fozard 1983; Middlemiss 1984; Tricklebank 1984) which is obviously not totally satisfactory given the affinity of the drugs for β-adrenoceptors. In addition these drugs do show at least some affinity for the 5-HT$_2$ receptor (Green, Johnson and Nimgaonkar 1983).

The conflicting views on the function importance of sites are hinted at in the preceding chapters. Moret (Section 2.5) clearly favours the view that the autoreceptor is of the 5-HT$_1$ sub-type but concedes that the evidence is not strong. Leysen (Section 4.8.2) denies any proper functional role at all for the 5-HT$_1$ binding sites on the, not unreasonable, grounds that a site with nanomolar affinity for a neurotransmitter is unlikely to have physiological relevance. However, I think we have all, in our studies, been sufficiently surprised by the diverse techniques used by physiology for its control mechanisms not to regard her hypothesis as firm!

Peroutka (1984) in reviewing the 5-HT$_1$ binding site has ascribed many physiological functions to the 5-HT$_1$ site, including various of the rat behaviours described in Section 12.2. I, in contrast, feel that the evidence is for both 5-HT$_1$ and 5-HT$_2$ sites being involved in the behavioural changes seen in rodents (Section 12.5).

The drugs which have been reported to bind to the 5-HT$_1$ site do produce behavioural and biochemical changes. We cannot be sure that they are doing so by altering serotonin function but such an explanation is the simplest and therefore the most attractive.

RU 24969 produces marked behavioural change, locomotor activity in both mice and rats (Section 12.5) and may decrease punished responding (Section 11.3.6). 8-OH-DPAT produces rather different behaviour (Section 12.6) and both drugs markedly decrease the rate of serotonin synthesis (Table 14.2 and Goodwin and Green 1985) which contrasts with many other serotonin agonists which have modest effects on synthesis. Marsden (Section 9.4.4) has found that RU 24969 decreases the 5-HIAA

Table 14.2   Rate of serotonin synthesis following administration of RU 24969 (10 mg/kg) to rats

	Brain 5-HIAA conc. Saline	Probenecid	Turnover rate (μg serotonin formed h^{-1} g^{-1})
Saline	0.26±0.02 (5)	0.42±0.01 (5)	0.69
RU 24969	0.20±0.02 (5)	0.27±0.02 (5)*	0.29

Synthesis rate was estimated by the accumulation of 5-HIAA in the 60 min following probenecid (200 mg/kg). *Different from control rats injected with probenecid; p <0.01.

peak using *in vivo* voltammetry, which is again consistent with a decrease in synthesis. Finally drug discrimination studies suggest that RU 24969 and 8-OH-DPAT do not share the same profile as quipazine (Section 10.4).

Classification of the receptor sub-types in functional terms using these new drugs is also being attempted in invertebrates (Section 13.7) and in mammalian peripheral tissues, where the story is getting equally complicated (Humphrey 1984; Fozard 1984).

Some attempts are being made to map the binding site sub-types using autoradiography and this has resulted in a further sub-type being proposed—the 5-HT_{1C} site (Cortés, Palacios and Pazos 1984). Here again there are problems since β-adrenoceptor agonists were found to bind to the 5-HT_{1B} receptor (to which RU 24969 also binds) but β-adrenoceptor antagonists have been suggested to be most potent as antagonists at the 5-HT_{1A} binding site (Nahorski and Willcocks 1983; Middlemiss 1984) and we found (−)-propranolol to be a weak antagonist of the behavioural responses induced by RU 24969 (Section 12.6; Green, Guy and Gardner 1984a).

Nevertheless, autoradiography, together with lesioning studies, may start giving clues to the distribution of binding sub-types. As stated earlier 5-HT_1 sub-types have been suggested to be presynaptic, but lesioning suggests that the sites are also postsynaptic (Blackburn, Bowery, Cox, Hudson, Martin and Price 1984).

With the increase in binding site number it seems appropriate to plead for standardization of terminology. In this monograph I have used 5-HT_1 and 5-HT_2 to denote binding sites and S_1, S_2, S_3 for the electrophysiological sub-types (see Section 7.1.2). S_1 and S_2 have also been used for the binding sub-types, but I personally believe this should be avoided to prevent confusion and any unwarranted implication that the S_1 binding site might be equated in any way with the S_1 electrophysiological site (see Section 4.8.6). No serious attempt has yet been made to see whether the binding sites can be related to the electro-physiologically defined sites.

14.3   The control of presynaptic serotonin function

The work described in the first three chapters can, to some extent, be considered together since all three authors are considering the control of the presynaptic functioning of serotonin neuroterminals.

Ray Fuller (Chapter 1) has provided a valuable overview of the synthetic and degradative mechanisms involved in serotonin metabolism and has discussed methods of measuring turnover. Despite the fact that Costa reviewed the problems of measuring and interpreting turnover data years ago (e.g. Costa and Neff 1970) papers still appear claiming to measure turnover when they are doing nothing of the sort. Fuller (Section 1.3) has again pointed out that serotonin/5-HIAA ratios cannot be used to calculate turnover rates.

Fuller (Section 1.5) also points out that agonists reduce the rate of serotonin synthesis but it is worth commenting here that the new putative 5-HT$_1$ agonists are particularly potent in this respect (Section 14.2; Table 14.2).

Increasing synthesis can be achieved by various means (Section 1.6) but interestingly antagonists in general have very little effect and this is certainly in contrast to the situation with dopamine where neuroleptics enhance turnover.

Clinically, alteration of serotonin function could conceivably be achieved in several ways; altering the rate of synthesis, altering the rate of degradation or altering the function of the autoreceptor. The first two approaches have been tried (tryptophan administration and MAO inhibitors), but we do not, as yet, have the serotonergic equivalent of clonidine, which acts on presynaptic α_2-adrenoreceptors. Characterization of the serotonin autoreceptor, as is being done in the experiments described by Moret (Chapter 2) may help this type of approach. Further help in this area may also come with further clarification of the role of the imipramine binding site (Chapter 3).

The benzodiazepine binding site appears to be a modulatory site and drugs acting at this site alter GABA function (Costa 1979). The identification of drugs acting as 'inverse agonists' have led to the observation that drugs acting at this site can both enhance (benzodiazepines), or decrease (inverse agonists such as β-carboline-3-carboxylate ethylester) GABA function (Nutt, Cowen and Little 1982). In a possibly analogous way, drugs acting at the imipramine binding site modulate serotonin uptake. It may be possible to design drugs to modulate function in both directions, particularly if there is an endogenous compound acting at this site, as has been proposed (Barbaccia, Gandolfi, Chuang and Costa 1983). The importance of such work in our ideas on the action of antidepressant drugs cannot by overemphasized (see Section 3.8 and also Section 14.6).

14.4   Phospholipid breakdown

One of the difficulties of ligand receptor binding studies is that of trying to study a functional response of the receptor so identified. Minchin (Section 5.1) has discussed these problems and his chapter reviews the very new approach of examining the effect of serotonin of enhancing the breakdown of inositol phospholipids in the membrane. Much of the basic identification of the mechanisms involved here come from the studies on the control of fluid secretion in the blowfly salivary gland by serotonin (Section 5.3).

In the central nervous system studies on the effect of serotoin-induced metabolism of inositol of phospholipids suggests that the receptor involved might be of the 5-HT$_2$ sub-type (Section 5.4). Since 5-HT$_1$ receptors have been suggested to be predominantly linked to adenylate cyclase (Peroutka, Lebowitz and Snyder 1981) this suggests that the two receptors may have different 'second messenger' or amplification systems.

Work in this area is continuing rapidly but there are already indications that antidepressant drugs might be affecting the PI breakdown in rat brain, consistent with their effects on 5-HT$_2$ binding sites (Kendall and Nahorski 1984 and Section 14.6).

14.5   Studies on serotonin neuropharmacology *in vivo*

To some extent most of the remaining chapters could be said to be encompassed by the heading above although they obviously contain much besides.

The chapter by de Montigny and Blier (Chapter 7) focuses predominantly on the effects of antidepressant treatments on electrophysiological aspects of serotonin neuropharmacology and this research will be considered later (Section 14.6). However, their chapter does outline the way that electrophysiology can be used as an approach to investigate serotonin neuropharmacology and again emphasizes (Section 7.1.2) that the 'receptors' defined electrophysiologically cannot be simply equated with those defined by ligand binding (Section 14.2).

The ability to examine firing rates *in vivo* of microiontophoretically applied serotonin or 5-MeODMT and the effects of treatments on this perameter is obviously very attractive. Particularly interesting was the recovery of firing rates during longer term administration of zimeldine (Section 7.4). Nevertheless, it seems to me that even with these techniques we are still having to measure various aspects of serotonin function—be they presynaptic or postsynaptic and problems still remain in assessing overall synaptic function and even when we put all the data together;

biochemistry, electrophysiology and function problems in getting an overall picture still remain.

Jacobs (Chapter 8) is also examining the electrophysiology of serotonin and has been developing methods of determining firing rates in freely moving conscious cats. The value of this approach is that behaviour and electrophysiology can be considered together (Section 8.3). In addition one can study the effects of drugs on both perameters (Section 8.4) and the potential of this technique is enormous.

The conclusion of Jacobs with regard to the action of hallucinogenics being through an effect of postsynaptic serotonin function (Sectoon 8.4) is shared by Glennon (Chapter 10) using various techniques including that of 'drug discrimination', another *in vivo* behavioural technique.

The studies of Jacobs (Chapter 8) allow *in vivo* investigation of electro-physiological changes induced by drugs acting on serotonin function. In contrast, the methods described by Marsden (Chapter 9) allow *in vivo* measures of serotonin biochemistry. Both methods allow not only examination of changes induced by drugs acting on serotonin but also whether changes in other neurotransmitter systems influence serotonin function (e.g. Section 9.4.3). Both also allow concomitant measurements of behaviour and serotonin—mediated events in conscious freely moving animals and the value of both techniques is now becoming apparent as the techniques become exploited.

The problem with *in vivo* voltammetry is that of absolute identification of the compound being monitored, and in this regard the dialysis technique has a distinct advantage. However, the dialysis tubing has a fairly short 'functional' life *in vivo*.

The use of behavioural techniques are discussed in Chapters 10, 11 and 12. The models we outlined in Chapter 12 are described as whole animal 'bioassay' techniques and should not be confused with the models examined by Gardner (Chapter 11) which are being used as animal models of anxiety and are thus models of a psychiatric state. The work of Glennon (Chapter 10) includes a description of the 'drug discrimination' technique (Section 10.3) which allows detection of particular types of drug and has been used by Glennon not only as part of his studies on hallucinogenic drugs but also to examine some of the new agonists of serotonin receptor sub-types, providing initial evidence that the animals do not recognise 8-OH-DPAT or RU 24969 to be like quipazine; which is consistent with other biochemical and behavioural evidence (Section 14.2).

Given the good characterization of invertebrate serotonin neurones and their pharmacology (Chapter 13) it is perhaps surprising that pharmacologists investigating serotonin pharmacology in vertebrates do not also examine invertebrate pharmacology when appropriate, since the physiology is clearly simpler.

14.6 Serotonin and antidepressant drugs

The action of antidepressant drugs on serotonin biochemistry and function has become increasingly complex and I fear that this monograph is not going to simplify matters. Several chapters discuss antidepressants in some detail—Briley (Chapter 3), Ögren and Fuxe (Chapter 6), de Montigny and Blier (Chapter 7) and Green and Heal (Chapter 12).

To deal with the imipramine binding site first. It now seems clear that imipramine site is associated with, but not identical to the serotonin uptake site. There is some evidence that the site may modulate uptake and Briley proposes that the site is decreased in both platelets and brain of depressed subjects. However, it should be pointed out that this is not a universal finding and investigations still continue as to whether it is state dependent or independent (Section 3.6.3). Both Costa and colleagues (Barbaccia *et al.* 1983) and Langer and colleagues (Langer, Raisman, Tahroin, Scatton, Niddam, Lee and Clanstre 1984) favour the view that there is an endogenous compound which acts at the imipramine binding site. If this proves to be the case it could open up a new era in the treatment of depressive illness.

Ögren and Fuxe (Chapter 6) comprehensively review the change in serotonin receptors following administration of antidepressant drugs. With regard to the $5-HT_2$ site their own work has produced good evidence for a down-regulation of both binding and the functional response of the head twitch behaviour (Section 6.4) and recently similar data has been obtained in our own laboratory (Section 12.7). The problem with endowing this change with therapeutic importance is not merely that not all antidepressant drugs produce this change but rather that a major treament of depression, namely electroconvulsive therapy produces an opposite effect in both behaviour and binding (Section 12.7).

Nevertheless, de Montigny and Blier (Chapter 7) find similar enhancement of electrophysiological responses following repeated treatment of rats with antidepressant drugs and ECS. They also found no change in postsynaptic response with zimeldine, even though both Ögren and Fuxe and Green and Heal report data on this drug decreasing post-synaptic $5-HT_2$ binding and head twitch behaviour. De Montigny and Blier do, however, find evidence for a decrease in the presynaptic receptor sensitivity following zimeldine (Fig. 7.2) and this is consistent with the decrease in $5-HT_1$ binding following this drug (Section 6.3.1). This is one functional response of the $5-HT_1$ site and another may be behaviour elicited by RU 24969 or 8-OH-DPAT. We have looked at the behavioural response to RU 24969 following antidepressant drug administration in rats (Green *et al.* 1984*a*) and are now extending this. One problem with some of these apparently conflicting results may be that responses (be they behavioural, electrophysiological or biochemical) have been studied at different times after the last dose. Behavioural responses can change markedly on drug withdrawal, changing from (for example, after desipramine) an inhibited response to an enhanced response (Goodwin, Green and Johnson 1984). I feel the changes seen on withdrawal may be of little relevance to the therapeutic mechanism of action although they may be important to the understanding of neurotransmitter interactions.

In time it may be possible to reconcile the behavioural, electrophysiological and biochemical data on serotonin following antidepressant drug treatment but at present any ideas I might propose will be more than speculation and unlikely to be better than those indulged in by anyone else!

Even if a change (increase or decrease) in serotonin function may not be easily equated with an antidepressant effect, the importance of serotonin in the changes induced in adrenergic function by antidepressant treatments is now becoming clear. Janowsky, Okada, Manier, Applegate, Sulser and Steranka (1982), Brunello, Barbaccia, Chuang and Costa (1982) and Dumbrille-Rose and Tang (1983) have all shown that serotonergic lesions abolish the down-regulation of β-adrenoceptors produced by administration of desipramine or imipramine. Recently we have confirmed these reports and found that the down-regulation of β-adrenoceptors induced by repeated ECS or clenbuterol is also abolished by a serotonergic lesion (Green, Nimgaonkar and Goodwin 1984b). Whether the β-adrenoceptor density decrease is important in the therapeutic process is unclear but it is a change produced by a range of antidepressant treatments: tricyclics, atypical antidepressants, ECT, β-adrenoceptor agonists and monoamine oxidase inhibitors.

The increase in 5-HT_2 receptor number is not only produced by repeated ECS but also repeated diazepam administration (Green, Johnson, Mountford and Nimgoankar 1985) perhaps reflecting a change in serotonergic function induced by antianxiety drugs (Chapter 11), but, in the light of the failure of benzodiazepines to treat depression, raising further doubts about a simple equation existing between a change in serotonin function and antidepressant action.

14.7   Where now?

To some extent the anser to this question lies in the clinic. We need to know whether selective serotonin antagonists are anxiolytic (Section 11.3) or indeed, in the light of the binding and behavioural data whether serotonin antagonists might have antidepressant properties.

Fuller, at the ACS (1983) meeting in Seattle, reviewed the possible uses of serotonin agonists and I am indebted to him for forwarding a copy of his paper to me. His suggestions included their use as appetite suppressants (the effect of the serotonin releasing drug fenfluramine being well established), as drugs in the treatment of myoclonus (5-HTP is already used), in the treatment of alcoholism (since fluoxetine decreases alcohol consumption is rats; see Rockman, Amit, Brown, Bourque and Ögren 1982) and in analgesia (see Hynes and Fuller 1982).

These suggestions are worthy of further investigation and antagonists at receptor sub-types may further help these clinical approaches.

Such suggestings do not even touch on the investigations that are continuing on the role of serotonin in the periphery. This area is shortly to be reviewed elsewhere (Bradley 1984).

I have only examined a few of these points raised in the preceding chapters. This concluding chapter is not designed to be comprehensive but merely to draw together some of the problems raised. If the reader feels that there are other areas that I should have covered, I am pleased, because it will mean that this monograph has achieved what it was set out to do: stimulate ideas and controversy in those interested in serotonin neuropharmacology!

References

Barbaccia, M. L., Gandolfi, O., Chuang, D. M. and Costa, E. (1983). Modulation of neuronal serotonin uptake by a putative endogenous ligand of imipramine recognition sites. *Proc. Nat. Acad. Sci (USA)* **80,** 5134.

Blackburn, T. P., Bowery, N. G., Cox, B., Hudson, A. L., Martin, D. and Price, G. W. (1984). Lesion of the dorsal raphe nucleus increases the nigral concentration of 5-HT$_1$ receptors. *Br. J. Pharmac.* **82,** 203P.

Bradley, P. B. (1984) (Ed.). 5-HT, peripheral and central receptors and function. *Neuropharmacology* (In press).

Brunello, N., Barbaccia, M. L., Chuang, D. M. and Costa, E. (1982). Down regulation of β-adrenergic receptors following repeated injections of desmethylimipramine: permissive role of serotonergic axons. *Neuropharmacology* **21,** 1145.

Cortés, R., Palacios, J. M. and Pazos, A. (1984). Visualisation of multiple serotonin receptors in the rat brain by autoradiography. *Br. J. Pharmac.* (In press).

Costa, E. (1979). The role of gamma-aminobutyric acid in the action of 1,4 benzodiazepines. *Trends Pharmac. Sci.* **1,** 41.

—— and Neff, N. H. (1970). Estimation of turnover rates to study the metabolic regulation of the steady-state of neuronal monoamines. In *Handbook of neurochemistry* (ed. A. Lajtha) p. 45. Plenum Press, New York.

Dumbrille-Ross, A. and Tang, S. W. (1983). Noradrenergic and serotonergic input is necessary for imipramine-induced changes in beta but
not S_2 receptor densities. *Psychiat. Res.* **9**, 207.

Fozard, J. R. (1984). Neuronal 5-HT receptors in the periphery. *Neuropharmacology* (In press).

Goodwin, G. M. and Green, A. R. (1985). A behavioural and biochemical
study in mice and rats of putative selective agonists and antagonists
for 5-HT_1 and 5-HT_2 receptors. *Br. J. Pharmac.* (in press).

—— —— and Johnson, P. (1984). 5-HT_2 receptor characteristics in frontal
cortex and 5-HT_2 receptor mediated-head twitch behaviour following
antidepressant treatment to mice. *Br. J. Pharmac.* **83**, 235.

Green, A. R., Guy, A. P. and Gardner, C. R. (1984*a*). The behavioural
effects of RU 24969, a suggested 5-HT_1 receptor agonist in rodents and
the effect on the behaviour of treatment with antidepressants. *Neuropharmacology* **23**, 655.

—— Johnson, P. and Nimgaonkar, V. L. (1983). Interactions of β-
adrenoceptor agonists and antagonists with the 5-hydroxytryptamine
(5-HT_2) receptor. *Neuropharmacology* **22**, 657.

—— Nimgaonkar, V. L. and Goodwin, G. M. (1984*b*). β-adrenoceptor
agonists, ECT and other antidepressants: effects on serotonin biochemistry and function. *IUPHAR 9th Int. Congr. (London)* Vol. 3, Chap.
24. Macmillan, Basingstoke.

—— Johnson, P., Mountford, J. A. and Nimgaonkar, V. L. (1985). Some
anticonvulsant drugs alter monoamine-mediated behaviour in mice in
ways similar to electroconvulsive shock: implications for antidepressant
therapy. *Br. J. Pharmac.* (In press).

Humphrey, P. P. A. (1984). 5-HT receptor characterisation. *Neuropharmacology* (In press).

Hynes, M. D. and Fuller, R. (1982). The effect of fluoxetine on morphine
analgesia, respiratory depression and lethality. *Drug Dev. Res.* **2**, 33.

Janowsky, A., Okada, F., Manier, D. H., Applegate, C. D., Sulser, F. and
Steranka, L. R. (1982). Role of serotonergic input in the regulation of the
β-adrenergic receptor-coupled adenylate cyclase system. *Science
(Washington)* **218**, 900.

Jenner, P. and Marsden, C. D. (1983). Neuroleptics. In *Psychopharmacology* (eds. D. G. Grahame-Smith and P. J. Cowen) p. 180. Excerpta
Medica, Amsterdam.

Kendall, D. A. and Nahorski, S. R. (1984). Supression of 5-HT receptor-
mediated inositol phospholipid breakdown in brain by chronic
antidepressant treatment. *Br. J. Pharmac.* (In press).

Langer, S. Z., Raisman, R., Tahraoin, L., Scatton, B., Niddam, R., Lee, C. R.
and Claustre, Y. (1984). Substituted tetrahydro β-carbolines are possible
candidates as endogenous ligands of the [³H]-imipramine recognition

site. *Eur. J. Pharmac.* **98,** 153.

Middlemiss, D. N. (1984). Stereoselective blockade at [^{3}H]-5-HT binding sites and at the 5-HT autoreceptor by propranolol. *Eur. J. Pharmac.* **101,** 289.

—— and Fozard, J. R. (1983). 8-Hydroxy-2-(di-n-propylamine)-tetralin discriminates between subtypes of the 5-HT$_1$ recognition site. *Eur. J. Pharmac.* **90,** 151.

Nahorski, S. R. and Willcocks, A. L. (1983). Interactions of β-adrenoceptor antagonists with 5-hydroxytryptamine receptor subtypes in rat cerebral cortex. *Br. J. Pharmac.* **78,** 107P.

Nutt, D. J., Cowen, P. J. and Little, H. J. (1982). Unusual interactions of benzodiazepine receptor antagonists. *Nature* **295,** 436.

Peroutka, S. J. (1984). 5-HT$_1$ receptor sites and functional correlates. *Neuropharmacology* (In press).

————— Lebovitz, R. M. & Snyder, S. H. (1981). Two distinct central serotonin receptors with different physiological functions. *Science* **212,** 827.

Rockman, G. E., Amit, Z., Brown, Z. W., Bourque, C. and Ögren, S-O. (1982). An investigation of the mechanisms of action of 5-hydroxy-tryptamine in the supression of ethanol intake. *Neuropharmacology* **21,** 341.

Tricklebank, M. D. (1984). Central 5-HT receptor subtypes and the behavioural response to 5-methoxy-N,N-dimethyltryptamine. *Br. J. Pharmac.* **82,** 204P.

Index

lung, possible site of [^{3}H]-imipramine binding sites, 54
LY51641
    inhibition of type A MAO, 2
    inhibition of serotonin oxidation, 11
LY87079, selective localization in serotonin neurones, 12–3
lysergic acid diethylamide
    assessment of autoreceptor sensitivity, 187
    behavioural syndrome, 329
    discriminative stimuli, 265–9
    effect on gastropod buccal cells, 367, 372
    effect on inositol phospholipid breakdown, 124
    effect on *Mercenaria* heart, 379–82
    effect on serotonergic neuronal firing rate, 197–8
    effect on serotonergic unit activity, 209–10
    induction of myoclonus, 342
    inhibitory effect on serotonin release by stimulation of inhibitory receptors, 26–8
    initiation of head twitch response, 337–8
    initiation of wet dog shake behaviour 337–8
    stimulation of α -adrenoceptors, 27
    stimulation of dopamine receptors, 27
[^{3}H]-lysergic acid diethylamide
    binding studies in invertebrates, 388–90
    ligand for study of antidepressant action on 5-HT$_2$ receptors, 158–9
    ligand for study of 5-HT$_2$ binding sites, 84–5
    ligand for study of 5-HT$_2$ binding sites, 106
    potency of drugs to inhibit binding, 94
    serotonin receptor binding studies, 270
[^{3}H]-lysergic acid diethylamide binding sites, effects of antidepressant drugs, 133, 137–8
[^{125}I]-lysergic acid diethylamide, selectivity for 5-HT$_2$ binding sites, 87–88

maprotiline
    effect on serotonin binding sites, 136
    efficacy in depression, 166
mass fragmentographic assays, 219
median raphe nucleus
    electrical and chemical stimulation, 286–7
    localization of serotonergic neurones, 197
median raphe nucleus lesions, anxiety models, 283
    see also dorsal raphe nucleus; raphe nucleus

membrane micelles, interactions with ligands, 92
membrane potentials, serotonin-induced changes in invertebrates, 382–3, 391–2
Mercenaria, heart muscle, serotonin structure-activity studies, 378–81
Mercenaria rectum, action of serotonin
Mercenaria serotonin receptors, 370–4
mescaline, discriminative stimuli, 265–6, 269
metabolic pathway, serotonin, 1–3
metergoline
    antagonism of serotonin-mediated behaviour 334–6
    anxiety studies, 297, 302
    effect on head twitch and wet dog shakes, 338
    effect on 8-OH-DPAT-induced behaviour, 85
    effect on inositol phospholid breakdown, 124
    effect on presynaptic serotonin autoreceptors, 36–7
    effect on serotonin turnover, 9–10
    enhancement of RU 24969-induced behavioural response, 344, 355
    inhibition of 5-HTP-induced myoclonus, 342
    inhibition of RU 24969-induced hyperlocomotion, 346
    ligand for study of 5-HT$_1$ binding sites, 84–5
    reversal of effects of serotoninomimetics, 290
methiothepin
    antagonism of inhibitory effect of exogenous serotonin, 25
    antagonism of LSD-induced inhibition, 26–7
    antagonism, of serotonin mediated behaviour, 334
    anxiety studies, 302
    discriminative studies, 2654
    effect on head twitch and wet dog shakes, 338
    effect on inositol phospholipid breakdown, 124
    increased serotonin turnover, 9–10
5-methoxy-3(1,2,3,6,tetrahydropyridin-4-yl)1H indole
    behavioural responses, 343–7, 411, 416
    drug discrimination studies, 412
    effect on excitatory responses of invertebrate neurones, 373–4
    effect on 5-H1AA oxidation peak, 243–4
    effect on inositol phospholipid breakdown, 124

432   Index

discriminative studies, 267–8
effect on excitatory response of invertebrate neurones, 373–4
effect on inositol phospholipid breakdown, 124
effect on serotonin turnover, 10
induction of myoclonus, 342
initiation of head twitch and wet dog shakes, 337–8
serotonin receptor binding studies, 272
supression of food reward behaviour, 290
quipazine 2-(piperazinyl)quinoline, reduction of rate of serotonin synthesis, 5–6

R 55667, *see* ritanserin
radioimmunoassays, 219
radioligand binding models, correlation between 5-HT$_1$ and 5.HT$_2$ sites, 91
Rapana thomasiana, serotonin regulation of neuromuscular transmission, 394
raphe nucleus
administration of chemicals, 286–7, 294–6
localization of serotonergic neurones in rat, 197
localization of serotonin autoreceptors, 24, 32–6
relationship between serotonin and density of [³H]-imipramine binding sites, 56
see also dorsal raphe nucleus; median raphe nucleus
rapid eye movement sleep
[³H]-imipramine binding sites, 65
serotonergic neuronal activity, 204–5, 207
receptor affinity studies, mechanism of hallucinogenic agents, 256
relaxation behaviour, anxiety models, 283
REM, *see* rapid eye movement sleep
reserpine
action in *Merceneria* heart muscle, 382
effect on 8-OH-DPAT-induced behavioural changes, 347
retina (bovine), [³H]-imipramine binding sites, 54
retzius cells, serotonin studies, 374–8
rhodnius prolixus, serotonin-induced fluid secretion in Malpighian tubules, 386
ritanserin
effect on 8-OH-DPAT-induced behavioural syndrome, 348
effect on serotonin-mediated behaviour, 334–6
radioligand binding studies, 411
Ro 15-1788, effect on responses to cyproheptadine, 297

RU 24969, *see* 5-methoxy-3(1,2,3,6 tetrahydropyridin-4-yl)1h indole

salbutamol, enhanced serotonin-mediated behaviour, 331
salivary gland (blowfly), serotonin interaction with its receptor, 117, 119–20
schistosoma, serotonin-enhanced motor activity, 387
second messenger, 5-HT$_2$ receptors, 117, 125
selegiline, *see* deprenyl
seretiazine, *see* ritanserin
sensory stimuli, effect on serotonin unit activity, 202–3, 205, 213
serotonergic lesions, effect on 5-HT$_2$ sites, 97
serotonergic transmissions, effect on serotonin autoreceptors and postsynaptic binding sites, 40
serotonin
central administration, 288
continuous CSF sampling, 225–8
effect on firing rates, 414–5
effect on gastropod heart muscle membrane potential, 382–3
inhibition of [³H]-serotonin release, 26
in vivo measurement by intracranial dialysis, 221–5
receptor binding profile, 83
regulation of dopamine release, 103
stimulation of adenylate cyclase in *Aplysia*, 393
suppression of food reward behaviour, 290
serotonin antagonists
action in invertebrate peripheral systems, 383–88
binding affinities in relation to inhibition of serotonin release, 104
discriminative studies, 267–8
effects on serotonin synthesis rate, 5–6
5-HT$_1$ sub-types, 348
serotonin analogues
effect on *Helix* neurones, 370
effect on serotonin autoreceptor, 33–5
effect on [³H]-serotonin binding, 33–5
induction of tachyphylaxis in *Mercenaria* heart, 378
serotonin antagonists
actions in invertebrate peripheral systems, 283–8
actions on leech Retzius cells, 376
anxiety studies, 296–304
clinical studies, 301
discriminative studies, 265–9

[³H]-spiperone binding sites, effects of antidepressant drugs, 138–9

stress analgesia, serotonin mediated, 284

stress-induced ultrasounds, anxiety models, 297

strophocheilos, action of serotonin on penis retractor muscle, 384

subjective feelings, role for 5-HT$_2$ receptors, 110

tachyphylaxis, induction in *Mercenaria* heart by serotonin analogues, 378

tapes heart, structure activity studies of serotonin, 381

tapes rectum, action of serotonin, 383–4

temoxetine, effect on 5-HTP-induced myoclonus, 342

tetracyclic antidepressant drugs, electrophysiological studies, 184–5

tosylate, actions on leech Retzius cells, 376

TR3369, *see* indorate

transmitter interactions, *in vivo* serotonin release studies, 242

transneuronal feedback, role in brain monoamine turnover, 9

tranylcypromine
    effect on 5-HT$_2$ receptor number, 352–3
    effect on serotonin-mediated behaviour, 352–3

trazodone
    anxiolytic studies, 303
    effect on serotonin receptors, 136, 142

tricyclic antidepressant drugs
    blockade of serotonin uptake mechanisms, 131
    electrophysiological studies, 183–4
    rapid effect of lithium addition, 190–1
    see also antidepressant treatments

m-trifluoumethylphenylpiperazine
    action at serotonin autoreceptors, 7
    induction of myoclonus, 342

trimipramine, effect on serotonin binding sites, 136

tryptamine
    actions on leech Retzius cells, 376
    actions on *Mercenaria* heart muscle, 378
    antagonist potency in *Helix* neurones, 374
    behavioural syndrome, 329
    effect on gastropod buccal cells, 367, 372
    effect on *Helix* neurones, 371
    induction of tachyphylaxis in *Mercenaria*, 378
    role in activation or serotonin receptor, 395
    structure-activity studies on gastropod heart, 382

tryptamines analogues
    structure activity studies in *Mercenaria* heart muscle, 378–9
    potency studies in leech Retzius cells, 375

L-tryptophan
    acceleration of serotonin synthesis, 7–8
    anxiogenic effects, 291
    effect on brain serotonin levels, 199
    isotopic measurement of serotonin turnover, 3
    structure-activity studies in *Mercenaria* heart muscle, 378–9

tryptophan analogues, inhibiton of serotonin synthesis, 5

tryptophan hydroxylase, role in serotonin biosynthesis, 4–5

L-tryptophan/monoamine oxidase inhibitors
    behavioural changes, 327–9
    induction of myoclonus, 341

tubocurarine
    actions on leech Retzius cells, 376
    effect on gastropod buccal cells, 367, 372
    effect on *Helix* heart muscle membrane potential, 382–3
    effect on *Helix* neurones, 371

ultrasonic vocalization, anxiety model, 284–5

viloxazine, effect on serotonin binding sites, 136

voltammetry
    in vivo measurement of serotonin metabolism 229–41
    in vivo monitoring circadian changes in serotonin turnover, 243
    study of serotonin autoreceptor, 22

wet dog shake behaviour, 337–41, 355

working electrode, 230, 232–41

WY25093, effect on anticonflict activity of oxazepam, 306

yohimbine, anxiogenic activity, 296

zimeldine
    behavioural responses, 416
    effect in development of serotonin receptor supersensitivity 161–2, 163
    effect on firing rates, 414–5